Sustainable Development and Biodiversity

Volume 12

Series editor

Kishan Gopal Ramawat
Botany Department, M.L. Sukhadia University, Udaipur, Rajasthan, India

This book series provides complete, comprehensive and broad subject based reviews about existing biodiversity of different habitats and conservation strategies in the framework of different technologies, ecosystem diversity, and genetic diversity. The ways by which these resources are used with sustainable management and replenishment are also dealt with. The topics of interest include but are not restricted only to sustainable development of various ecosystems and conservation of hotspots, traditional methods and role of local people, threatened and endangered species, global climate change and effect on biodiversity, invasive species, impact of various activities on biodiversity, biodiversity conservation in sustaining livelihoods and reducing poverty, and technologies available and required. The books in this series will be useful to botanists, environmentalists, marine biologists, policy makers, conservationists, and NGOs working for environment protection.

More information about this series at http://www.springer.com/series/11920

Dinesh K. Maheshwari
Editor

Bacterial Metabolites in Sustainable Agroecosystem

Springer

Editor
Dinesh K. Maheshwari
Department of Botany and Microbiology
Gurukul Kangri University
Haridwar, Uttarakhand
India

ISSN 2352-474X ISSN 2352-4758 (electronic)
Sustainable Development and Biodiversity
ISBN 978-3-319-24652-9 ISBN 978-3-319-24654-3 (eBook)
DOI 10.1007/978-3-319-24654-3

Library of Congress Control Number: 2015950025

Springer Cham Heidelberg New York Dordrecht London

Printed on acid-free paper

Springer International Publishing AG Switzerland is part of Springer Science+Business Media (www.springer.com)

Preface

Various groups of beneficial bacteria synthesize a large number of "biomolecules" that allow plants to survive under adverse environmental/abiotic and biotic conditions. Such bacteria govern phytohormone-mediated immune response, manage to regulate hormones, produce biosurfactants which are involved in several important functions for bacteria themselves as well as for the plants and their ecosystem. Thus, bacterial hormones and biosurfactants are identified as effector molecules in plant–microbe interactions, pathogenesis and phytostimulation which can be beneficial either for the bacteria or for the crops so as to warrant sustainability.

The organization of the book is from practice to theory and from basic to applied aspects of bacterial phytohormones and biosurfactants. Some specific bacterial genera, *Azospirillum* and emergence of *Methylobacterium*, in particular, and their potential to support plant growth and development have been documented. To begin with, techniques for isolation and purification of "classical five" microbial phytohormones, namely, auxins, gibberellins, cytokinins, ethylene and abscisic acid are covered here, stressing the need to join the practical with the theoretical thus make the contents alive. Other than these modulators, importance of jasmonic acid and salicylic acid produced by bacteria or plants have also been emphasized. Microorganisms contain over 30 growth-promoting compounds from the cytokinin group and also produce about 100 GAs and other groups of hormones, which are extremely important for plants from seed germination stage to fruit ripening processes.

A scientific linkage and evidence to show bacterial hormones and their impact to act as biofertilizers is also provided. This book provides in-depth insights into bacterial traits required for rhizosphere competence, root colonization and/or endophytic phytohormone secretion which act as a sink of IAA, thus protecting the plants from different environmental stresses.

Some of the chapters emphasized the concepts related to drought and salt tolerance through Abscissic acid and other microbial hormone regulations that provide valuable insight into evolution of microbial interactions with plants under hostile environments. Such suitable strains (consortia) and their application in promoting the growth of healthy and disease-free crops that are eco-friendly in nature.

Biofilm formation and biosurfactant activity of plant-associated bacteria play essential role in bacterial motility, signaling and biocontrol of disease-causing pathogens, their mechanism at both physiological and genetic level is suitably evidenced with the need of green chemicals to study and application of bacteria-mediated biosurfactants has become imperative. Efforts have been made to stress the bioremediation potential of rhamnolipids to eliminate a wide range of pollutants and to promote a sustainable development of our society. The contents lay stress upon microbial world that synthesizes and secretes phytohormones and biosurfactants and emits many volatiles that lead to sustainable agriculture ecosystem.

This book will be useful not only for students, teachers and researchers but also for those interested in biotechnology, microbiology, physiology of plant growth and development, phytoprotection, agronomy and environmental sciences.

I wish to acknowledge all the subject experts, who were instrumental and cooperative to spare their valuable contributions to make this book a success. Thanks are due to my research team members, especially to Mohit and Shrivardhan, who generously assisted the compilation and completion of this task. The credit also goes to my family members, especially to my wife, Dr. Mrs. Sadhana Maheshwari. I extend my sincere thanks to Dr. Mrs. Valeria and her colleagues for their valuable support to facilitate the completion of this volume.

Uttarakhand, India Dinesh K. Maheshwari

Contents

Contributors

Mohit Agarwal Department of Botany and Microbiology, Gurukul Kangri University, Haridwar, Uttarakhand, India

Hossein Ali Alikhani Department of Soil Sciences, University of Tehran, Tehran, Iran

Abdulaziz A. Alqarawi Department of Plant Production, College of Food and Agricultural Sciences, King Saud University, Riyadh, Saudi Arabia

Ummay Amara Department of Soil Science & SWC, PMAS Arid Agriculture University, Rawalpindi, Pakistan

T.N. Arkhipova Institute of Biology, Ufa Research Center of RAS, Ufa, Russia

Dhamodharan Bakkiyaraj Department of Biotechnology, Alagappa University, Science Campus, Karaikudi, Tamil Nadu, India

Aurenivia Bonifacio Department of Biological Science, Center of Natural Science, Federal University of Piaui, Teresina-PI, Brazil

Ruben Bottini Laboratorio de Bioquímica Vegetal, Instituto de Biología Agrícola de Mendoza, Consejo Nacional de Investigaciones Científicas y Técnicas-Universidad Nacional de Cuyo, Chacras de Coria, Argentina

X. Chen College of Life Science, Zhejiang University, Hangzhou, China

Ana Carmen Cohen Laboratorio de Bioquímica Vegetal, Instituto de Biología Agrícola de Mendoza, Consejo Nacional de Investigaciones Científicas y Técnicas-Universidad Nacional de Cuyo, Chacras de Coria, Argentina

Fabio Fernando de Araujo Faculty of Agrarian Sciences, University of Oeste Paulista (UNOESTE), Presidente Prudente-SP, Brazil

Marcia do Vale Barreto Figueiredo Agronomical Institute of Pernambuco IPA/SEAGRI, Recife-PE, Brazil

Pinakin Dhandhukia Ashok and Rita Patel Institute of Integrated Study and Research in Biotechnology and Allied Sciences, Anand, Gujarat, India

Shrivardhan Dheeman Department of Botany and Microbiology, Gurukul Kangri University, Haridwar, Uttarakhand, India

Dilfuza Egamberdieva Institute for Landscape Biogeochemistry, Leibniz Centre for Agricultural Landscape Research (ZALF), Müncheberg, Germany

Hassan Etesami Department of Soil Sciences, University of Tehran, Tehran, Iran

L. Ge Nanyang Environment & Water Research Institute, Nanyang Technological University, Singapore, Singapore

Walter Giordano Departamento de Biología Molecular, Universidad Nacional de Río Cuarto, Río Cuarto, Córdoba, Argentina

Dweipayan Goswami Department of Biochemistry and Biotechnology, St. Xavier's College, Ahmedabad, Gujarat, India

Abeer Hashem Department of Botany and Microbiology, College of Science, King Saud University, Riyadh, Saudi Arabia

Rifat Hayat Department of Soil Science & SWC, PMAS Arid Agriculture University, Rawalpindi, Pakistan

Chaitanya Kumar Jha Department of Microbiology, Government Science College, Surat, Gujarat, India

Mario Andrade Lira Junior Department of Agronomy, Agricultural University of Pernambuco (UFRPE), Recife-PE, Brazil

Rabia Khalid Department of Soil Science & SWC, PMAS Arid Agriculture University, Rawalpindi, Pakistan

G.R. Kudoyarova Institute of Biology, Ufa Research Center of RAS, Ufa, Russia

Sandeep Kumar Regional Experiment Station, Rubber Research Institute of India, Rubber Board, Jalpaiguri, West Bengal, India

Dinesh Kumar Maheshwari Department of Botany and Microbiology, Gurukul Kangri University, Haridwar, Uttarakhand, India

Oscar Masciarelli Departamento de Ciencias Naturales, Universidad Nacional de Río Cuarto, Río Cuarto, Córdoba, Argentina

A.I. Melent'ev Institute of Biology, Ufa Research Center of RAS, Ufa, Russia

Amina Meliani Department of Biology, University of Mustapha Stambouli, Mascara, Algeria

Hossein Mirseyed Hosseini Department of Soil Sciences, University of Tehran, Tehran, Iran

Alwar Ramanujam Padmavathi Department of Biotechnology, Alagappa University, Science Campus, Karaikudi, Tamil Nadu, India

Shunmugiah Karutha Pandian Department of Biotechnology, Alagappa University, Science Campus, Karaikudi, Tamil Nadu, India

Keyur Patel Department of Biotechnology, P.D. Patel Institute of Applied Sciences, Charotar University of Science and Technology, Anand, Gujarat, India

Patricia Piccoli Laboratorio de Bioquímica Vegetal, Instituto de Biología Agrícola de Mendoza, Consejo Nacional de Investigaciones Científicas y Técnicas-Universidad Nacional de Cuyo, Chacras de Coria, Argentina

Emiliano D. Primo Departamento de Biología Molecular, Universidad Nacional de Río Cuarto, Río Cuarto, Córdoba, Argentina

Artenisa Cerqueira Rodrigues Department of Agricultural Engineering and Soil Science, Federal University of Piaui (UFPI), Center of Agrarian Science, Campus of Socopo, Teresina-PI, Brazil

Francisco Ruiz Departamento de Microbiología e Inmunología, Universidad Nacional de Río Cuarto, Río Cuarto, Córdoba, Argentina

Meenu Saraf Department of Microbiology and Biotechnology, University School of Sciences, Gujarat University, Ahmedabad, Gujarat, India

S.N. Tan Natural Sciences and Science Education Academic Group, Nanyang Technological University, Singapore, Singapore

Janki Thakker Department of Biotechnology, P.D. Patel Institute of Applied Sciences, Charotar University of Science and Technology, Anand, Gujarat, India

W.S. Wong Singapore University of Technology and Design, Singapore, Singapore

J.W.H. Yong Singapore University of Technology and Design, Singapore, Singapore

Techniques to Study Microbial Phytohormones

Keyur Patel, Dweipayan Goswami, Pinakin Dhandhukia and Janki Thakker

Abstract Soil is replete with microscopic life such as bacteria, fungi, actinomycetes, protozoa and algae. Microscopic life tends to reside in the rhizosphere of the soil and interact with plants. A microbial–plant interaction occurs due to the microbial ability to produce phytohormones regarded as the "classical five," which are auxin, gibberellin, cytokinins, ethylene and abscisic acid. In addition to these modulators, jasmonic and salicylic acid are also documented as bacterial hormones contributing to a sustainable agro system. Auxins, gibberellins and cytokinins are known to be produced by *Azospirillum* species. Auxin production in fungus such as *Pistolithus tinctorius* leads to promotion of plant growth and different bacterial species show effect on root length by increasing the surface area and induction of gall and tumor formations. Gibberellins are tetracyclic diterpenoid acids that are involved in a number of developmental, reproduction and floral formation in plants, while plant growth promotion and induction of tumor and gall formation are done by cytokinins. *Pseudomonas solanacearum, Mycobacterium hiemalis* and largely spore forming bacteria have shown to form ethylene in culture. Abscisic acid (ABA) is a stress-related signaling molecule reported in all kingdoms of life such as plant-associated bacteria, plant pathogenic

K. Patel · J. Thakker (✉)
Department of Biotechnology, P.D. Patel Institute of Applied Sciences,
Charotar University of Science and Technology, CHARUSAT Campus,
Changa, 388421 Anand, Gujarat, India
e-mail: jankithakker.bt@charusat.ac.in

D. Goswami
Department of Biochemistry and Biotechnology, St. Xavier's College, Ahmedabad
38009, Gujarat, India

P. Dhandhukia
Ashok and Rita Patel Institute of Integrated Study and Research in Biotechnology
and Allied Sciences, ADIT Campus, New Vidyangar, Anand 388121, Gujarat, India

D.K. Maheshwari (ed.), *Bacterial Metabolites in Sustainable Agroecosystem*,
Sustainable Development and Biodiversity 12, DOI 10.1007/978-3-319-24654-3_1

fungi, certain cyanobacteria, algae, lichens, protozoa and sponges. Salicylic acid is synthesized by the fungus *P. patulum* and it is an effective therapeutic agent for plants. SA plays a role in plant response during biotic and abiotic stress. It also regulates physiological and biochemical processes during the plant lifespan. Jasmonic acid is a signaling molecule involved in plant defense reported to be produced by fungus *Lasiodiplodia theobromae*. However, despite the significant research pursued in this area, there are limited reports suggesting strategies that focus on the production, extraction and detection of microbial phytohormones. Here, the present review focuses on the techniques used for isolation and purification of these phytohormones.

Keywords Abscisic acid · Auxin · Cytokinin · Ethylene · Gibberellin · Microbial–plant interaction · Phytohormones · Rhizosphere

1 Introduction

Along with water and air, soil is a major part of the natural environment and vital to the existence of life on the globe. Soil composition is an important feature of nutrient management. Soil minerals and organic matter hold and store nutrients while soil water makes it available to plants. Air trapped in between soil particles plays an integral role, where many of the microorganisms live and require air to undergo the biological processes that release additional nutrients into the soil. Meanwhile, the basic components of soil are minerals, organic matter, water and air. The basic component ratio found in typical soil is approximately 9:1:5:5 of mineral, organic matter, water and air respectively (Hillel 2003). As a matter of fact, soil is very complex. The composition of soil can fluctuate on a day-to-day basis and depends on numerous factors such as water supply, cultivation practice and soil type (Rowell 2014).

The soil blankets are abundant replete sources for microscopic life including bacteria, fungi, actinomycetes, protozoa and algae (Glick 2012). These are classified as beneficial to plant and pathogenic to plants (Zamioudis et al. 2013) (Fig. 1). Beneficial microbes are subsequently divided into the following groups: (A) General plant growth promoters that stimulate plant growth through a variety of mechanisms. For example, approximately 90 % of land plants live in symbiosis with arbuscular mycorrhizal fungi (AMF). Since exudates of fungal hyphae solubilize more phosphorus (P) than root exudates, AMF can enhance plant establishment and increase water and nutrient uptake. (B) Microbial fertilizers are known for specific nutrients like nitrogen (N), Phosphorus (P) and ferric ion (Fe^{3+}). For example, ability to fix N is widespread in both bacteria and archaea (Dekas et al. 2009). (C) Microbial plant growth regulators secrete hormones or hormone-like substances which stimulate plant growth (Table 1).

Hormones and hormone-like substances regulating developmental processes in plants are known as phytohormones. Phytohormone pathways and cross-talk between them play a key role in process coordination and cellular responses

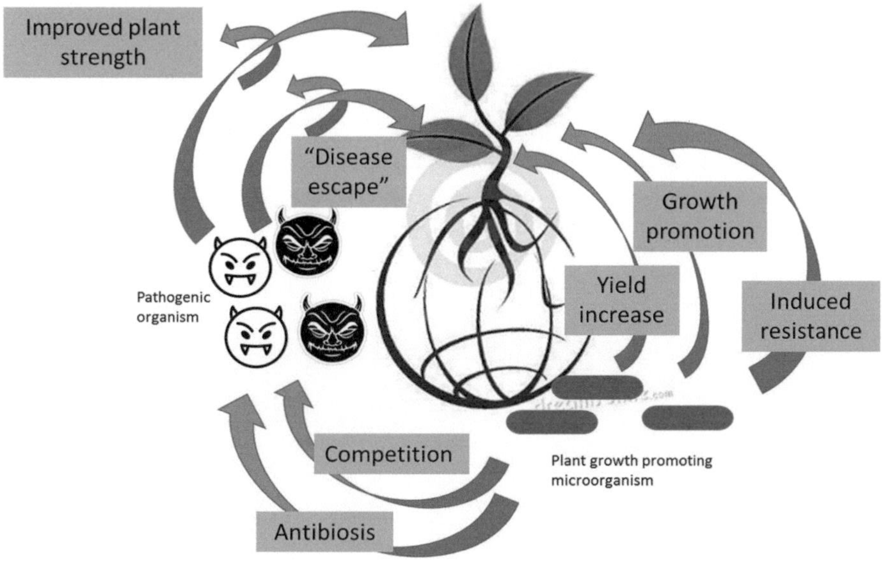

Fig. 1 An overview of mode of action of microorganism and interaction with plants, beneficial organisms and pathogenic organisms (Kilian et al. 2000)

(Moller and Chua 1999; Santner et al. 2009). In vitro many rhizospheric bacteria have the capability to produce hormones such as auxins, gibberellins, cytokinins, ethylene and abscisic acid (Zahir et al. 2003). *Azospirillum* is well known for its ability to produce phytohormones such as gibberellins (Janzen et al. 1992), cytokinins (Tien et al. 1979) and auxins (Reynders and Vlassak 1979; Mascarua-Ezparza et al. 1988; Omay et al. 1993). Ethylene and abscisic acid producing bacteria are known as stress controllers. Jasmonic acid is produced by pathogenic fungus *Lasiodiplodia theobromae* (Dhandhukia and Thakkar 2008; Tsukada et al. 2010) and salicylic acid is synthesized by *P. patulum* (Yalpani et al. 2001).

Plant–microbe interactions in the rhizosphere play a pivotal role in nutrient transformation, mobilization and solubilization from a limited nutrient pool in the world of sustainable agro systems (Hayat et al. 2010). In 2025, the world population is expected to increase by 7–8.3 billion. Therefore, the production of half of the human's calorie intake cereals, especially wheat, rice and maize has to be increased. According to Lu et al. (2015), an unsustainable approach to using chemicals creates a number of worries such as water contamination leading to eutrophication and health risks for humans. Moreover, it results in soil degradation and loss of biodiversity. Currently, plant growth is increased by bacterial molecules that act as hormones and as nutrients to plants in the sustainable agro system (Lugtenberg et al. 2013; Nadeem et al. 2014). In this chapter we aim to focus on the production, extraction and detection strategies of phytohormones along with the mode of action contributed by them in the sustainable agro system.

Table 1 Types of microbes producing different phytohormones and their effect on plants

Phytohormone	Producing microorganisms	Effect on plants	References
Auxins	Fungus: *Pisolithus tinctorius*	Plant growth promotion	Frankenberger and Poth (1987)
	Bacteria: *Azospirillum, Rhizobium, Bradyrhizobium, Gluconacetobacter, Herbaspirillum*	Decrease of root length, increase of root hair development, corn seedlings inoculated showed an increase in free active IAA and IBA	Tien et al. (1979), Atzorn et al. (1988), Badenosch-Jones et al. (1982), Fuentes-Ramírez et al. (1993), Bastián et al. (1998), Fallik et al. (1989)
	Klebsiella	Increase in root branching and root surface	El-Kawas and Adachi (1999)
	Pseudomonas sp. *Kocuria turfanen-sis, Agrobacterium, Erwinia herbicola* pv*gypsophilae*	Induction of gall and tumor formation; shows plant growth promotion in chick pea and green gram	Comai and Kosuge (1980, 1982), Goswami et al. (2013, 2014), Liu et al. (1982), Manulis et al. (1998)
Gibberellin	Fungus: *Gibberella fujikuroi*	"Bakanae" effect in maize, rice and other plants	Rojas et al. (2001), Fernández-Martin et al. (1995)
	Bacteria: *Azospirillum brasilense, Azospirillum lipoferum, Azospirillum brasilense*	Reversion of dwarfism in maize and rice promotion of shoot elongation, growth and root hair density	Fulchieri et al. (1993)
Cytokinins	*Azospirillum*	Plant growth promotion	Tien et al. (1979)
	Pseudomonas syringae pv *savastanoi, Agrobacteriumtumefaciens, Erwinia herbicola*	Induction of gall and tumor formation	Roberto and Kosuge (1987), Lichter et al. (1995)
Jasmonate	Fungus: *Botryosphaeria rhodina, Lasiodiplodia theobromae*	Participate in the metabolism's control, development and protection of plant processes	Weber et al. (1997), Dhandhukia and Thakkar (2007)

2 Bacterial Phytohormones

A secondary metabolite synthesized by bacteria becomes important in biotechnology. For example, one group of secondary metabolites regarded as "classical five" is plant growth promoting hormones being secreted in minute quantity by bacterial species.

Classical five phytohormones produced by bacteria are auxins, gibberellins, cytokinins, ethylene and abscisic acid. Jasmonic acid is produced by fungus *Lasiodiplodia theobromae* and salicylic acid is synthesized by the fungus *P. patulum*. According to Takahashi (1986), other than these, phytohormones produced by microbial origin are helminthosporol and related compound *cis*-sativendiol from *Helminthosporium sativum*; Sclerin, sclerotinin A, B from *Sclerotinia* sp.; malformins A1, A2, B1, B2, C from *Aspergillus* sp.; cotylenol and cotylinin A–F from *Clodosporium* sp.; radiclonic acid from *penicillium* sp., synergist to gibberellins from *Pestalotia cremeraecola* (Fig. 2). Other than higher plants they are not only synthesized by bacteria, but have also been secreted by mosses (Nuray et al. 2002) and fungi (Unyayar et al. 1996; Yürekli et al. 1999). To control plant growth and development, phytohormone—"A signal molecule" acts as a chemical messenger. Instead of their plant response, hormones also regulate expression of the intrinsic genetic potential of plants as a principal agent (Elmerich et al. 2007).

Bacterial phytohormones are necessary to be produced, excreted out and transferred in plant cells during plant–bacterial interaction. To understand the mechanism, one should understand the techniques for its isolation, from which source its nature, exophyte or endophyte, its purification and detection as well as its crosstalk during interaction with the plants will easily be available.

2.1 Auxins

The phytohormone auxin ("Auxeins" in Greek, means "to grow") regulates a plant growth and its developmental process. As reviewed by Moore (1979), the discovery of auxins was the outcome of the phototropism and geotropism experiments done in the nineteenth century. The Dutch botanist F. Went in 1926 discovered auxin and described a bioassay for its detection by "Avena coleoptile curvature

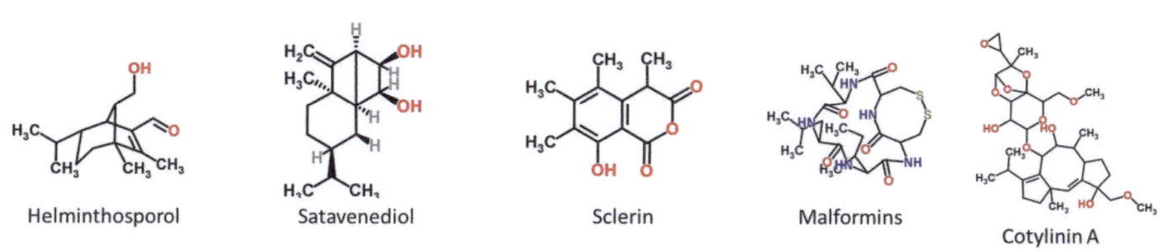

Fig. 2 Structural chemistry of plant growth promoters of bacterial origin (*Source* http://www.chemspider.com/)

test." The biochemists Kögl, Haagen-Smit and Erxleben got an active substance indole-3-acetic acid (IAA) from urine in 1934, which was found to be similar to auxin. Finally in 1935, K.V. Thimann isolated IAA from cultures of the fungus *Rhizopus suinus*.

2.1.1 Indole 3-Acetic Acid (IAA) and Auxins

All auxins have an aromatic skeleton with weak organic acid. IAA, indole-3-butyric acid (IBA) and 4-chloroindole-3-acetic acid (4-CI-IAA) are most of the naturally occurring auxins having an aromatic indole ring (Fig. 3a) (Wightman and Lighty 1982). It has been reported that bacteria synthesize IAA, the most abundant and basic molecule of auxin. Most of the study is concentrated on auxins, especially IAA in bacterial plant hormone production. IAA producing bacteria may be either pathogenic (harmful) or growth promoting (beneficial) to the plant (Dobbelaere et al. 2003). These bacteria can interfere in plant development by disturbing the auxin balance in plant during plant–bacterial interactions. For example, IAA producing *Agrobacterium* spp. (Jameson 2000) and *Pseudomonas savastanoi* pv. *Savastanoi* (Comai and Kosuge 1980) cause tumors and galls in olive and oleander plants. Many studies suggest that IAA, produced by *Azospirillum,* is involved in making morphological and physiological changes in the inoculated plant roots (Tien et al. 1979; Kapulnik et al. 1985; Harari et al. 1988); while the general effect of beneficial bacteria is either direct, i.e. through plant growth promotion, or indirect, i.e. through improving plant nutrition by solubilizing mechanisms and making it available to the plants during plant–bacterial interaction. Deepa et al. (2010) showed that strains of *Enterobacter aerogens* and *Enterobacter cloacae* produced IAA and showed growth promotion in cowpea (*Vigna unguiculata* L.), whereas *Kocuria turfanensis* strain 2M4 was also found to produce IAA and showed growth promoting ability in groundnut (*Arachis hypogaea* L.) in saline soil (Goswami et al. 2014). In some cases, bacterial auxins are found in conjugated form and they are involved in storage, transport and protection from

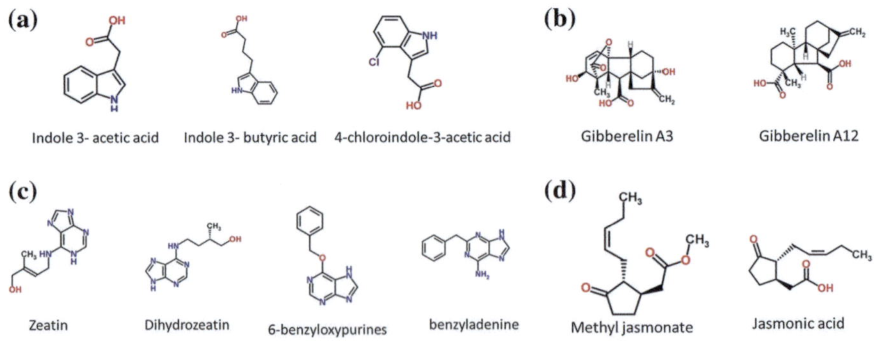

Fig. 3 Structural chemistry of microbial phytohormones. Stoichiometry of auxins (**a**), gibberellins (**b**), cytokinins (**c**) and jasmonates (**d**). (*Source* http://www.chemspider.com)

enzymatic degradation. *Pseudomonas savastanoi* pv. *Savastanoi* has a characterized bacterial indole-3-acetic acid-lysine synthetase gene involved in IAA conjugation. The gene product is involved in the conversion of IAA into IAA-lysine (Glass and Kosuge 1986).

2.1.2 Sources of Microbial IAA

For plant growth promotion, symbiotic association and pathogenesis, widespread soil and plant-associated bacteria are capable to produce IAA. Previously, bacterial auxin production was known for mainly being associated with the pathogenesis. Then it became clear that not only pathogenic bacteria but plant growth promoting bacteria (PGPR) also synthesize IAA (Table 1). Genes responsible for synthesis of IAA are widely distributed and are responsible for metabolic pathways that differ from one bacterium to another. It has been seen that *Pseudomonas syringae* induce tumor and gall formation and *Azospirillum* spp. promote plant growth by production of cytokinin and gibberellin. In the presence of nitrate, *Azospirillum* produces a compound that mimics the effect of IAA in several plant tests (Zimmer et al. 1988).

2.1.3 Production and Detection of IAA and Related Indole Compounds

Indole is generated via indole pyruvic acid by reductive deamination of tryptophan. During the deamination reaction in which tryptophanase catalyzes the amine (–NH_2) group, the tryptophan molecule is removed and the final products of the reaction are indole, pyruvic acid, ammonium (NH_4^+) and energy. Pyridoxal phosphate is required as a coenzyme (Fig. 4).

To detect the production of IAA from the culture, the major requirement is that the medium contains an adequate amount of tryptophan and a pinch of sodium chloride as per Difco (1998) to culture an organism prior to performing indole test (Mac Faddin 1976). For an alternate method of IAA production, casein peptone, sodium chloride and tryptone are also used as medium (Mac Faddin 1976). Simultaneously, determining other characteristics such as motility and the ability to produce hydrogen sulfide as a by-product of metabolism of the bacterium, the sulfide indole motility (SIM) medium is a multi-test agar used to test for indole production (Mac Faddin 1976). Another multi-test agar is motility-indole ornithine

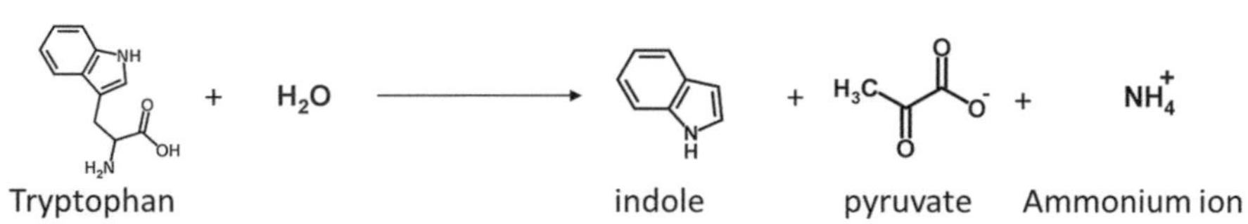

Fig. 4 Indole biosynthesis from precursor tryptophan in microorganisms

(MIO) medium. In addition to testing for indole production, it is used to test for motility and ornithine decarboxylase, causing an increase in pH in the tube. This is indicated by the purple-gray color throughout the tube when a microbe is grown in a medium containing tryptophan and produces an indole, which demonstrates that an organism has the capacity to degrade the tryptophan. Detection of indole, a by-product of tryptophan metabolism, relies on the chemical reaction between indole ring and a reagent used in the test (Hiroya et al. 2004).

2.1.4 Detection of IAA with Biochemical Assay

For the detection of IAA or related indole compounds, two methods are used: Indole spot test and biochemical chromogenic reagent method. For the indole spot test, place several drops of indole spot reagent (DMACA–dimethylaminocinnamaldehyde, HCl and deionized water) on a piece of filter paper. With the help of a wire loop or wooden applicator stick, a portion of an 18–25 h old colony has to be rubbed onto the reagent saturated area of the filter paper; a positive reaction is denoted by the change in color from blue to blue green, or red-violet in the case of *Providencia alcalifaciens*, within 10 s (Miller and Wright 1982). In biochemical chromogenic reagent method, several reagents are used. To test for indole production, 4–5 drops of Kovac's reagent (iso amyl alcohol, DMAB-dimethylaminobenzaldehyde, $HCl_{(conc.)}$ are added to the tube containing incubated culture in tryptophan broth. A positive test indicates cherry red color in the reagent at top of the medium (Isenberg and Sundheim 1958). Eherlich's reagent is alternatively used to Kovac's reagent. It also contains DMAB-dimethyl amino benzaldehyde which reacts to indole and produces a red color. This formulation is more sensitive but it requires flammable solvents or additional toxics.

2.1.5 Detection of IAA by Capillary Electrophoresis

Along with other methods, capillary electrophoresis (CE) is a commonly used analytical technique for separation of solutes and analysis from the biological system. Capillary electrophoresis is a simple technique having high resolving power, low solvent content and has been an active technique in the research area of separation science (Jiang et al. 2006). Neutrally and negatively charged sample matrix is pumped out from the capillary by the reversed potential under reversed electroosmotic flow (EOF). An electrophoretic speed of cationic analytes lower than the magnitude of EOF for a capillary filled with running buffer will migrate to the detector (Kim et al. 2009). Simply, migration flows extend because anions are migrating in the opposite direction to the electroosmotic flow (EOF). Auxins, gibberellins and abscisic acid have carboxylic group which exist as anions at high pH level. Capillary electrophoresis was performed by Chen et al. (2011) on a self-made CE system with a high voltage supplier, a coaxial sheath liquid and HPLC pump to supply constant flow rate of sheath liquid. Another laboratorial syringe pump is also used to condition amino coated capillary with background electrolyte (BGE) and drive out bubbles from the capillary. They are used to rinse capillary with 1 M NaOH solution for 4 h followed by water until the pH value of the outlet maintained 7.0 for cleaning and activate inner surface of capillary for effective attachment of silica skeletons. This step is followed by further flushing with 1 M HCl, water and methanol for 3, 1 and 1 h, respectively and then capillary was dried with nitrogen at 160 °C for 5 h prior to use. Afterwards, 3-aminopropyltriethoxysilane dissolved in dry toluene was filled in a capillary and both the ends of the capillary were sealed with silicone rubber. After filling, the reaction was performed at 105 °C for 24 h, followed by rinsing the capillary with toluene and analytical grade acetonitrile (ACN, containing 20 % water and 5 % formic acid) to remove the residual components. After connecting to CE system, 100 cm of capillary was conditioned with background electrolytes of 30 mM ammonium formate containing 10 % (v/v) acetonitrile with a low voltage of 15 kV for 30 min. ACN-a sheath liquid used with the flow rate of 5 μl min^{-1}. Jiang et al. (2006) separated plant hormones from the biofertilizer by using an uncoated fused silica capillary column with a total length of 43.5 cm with a maintained temperature at 25 °C and applied voltage −25 kV constantly. The sample loaded by pressure injection at 50 mbar for 5 s was detected after extraction at wavelength of 200 nm under UV detection.

2.1.6 Qualitative Detection of IAA by TLC and Paper Chromatography

Thin layer chromatography is a simple technique to detect IAA by using various solvent systems. For the procedure, a sample preparation is done by acidification of supernatant after centrifugation of bacterial culture broth from pH 2.5–3.0 using 1 N HCl and extracted twice by ethyl acetate. After air drying of ethyl acetate, the extract is dissolved in methanol (Mohite 2013; Patil 2011; Swain and Ray 2007; Goswami et al. 2014). These extracts are spotted and allowed to develop using mobile phase isopropanol: ammonia: water (16:3:1 (v/v/v)) (Goswami et al. 2014) for IAA detection for *Kocuria* sp. 2m4, ethyl acetate: chloroform: formic acid (55:35:10 (v/v/v)) or benzene: n-butanol: acetic acid (70:25:5 (v/v/v)) for *Azotobacter* and fluorescent *Pseudomonas*, Propanol: Water (8:2) was used as solvent system for IAA detection from the rhizospheric isolates (Mohite 2013); benzene: n-Butanol: acetic acid (70:25:5 (v/v/v)) used for detection of indole acetic acid by *Azotobacter* sp. (Patil 2011).

2.1.7 Quantitative Determination of Auxins by HPLC

For rapid separation and detection of phytohormones, TLC was the first technique used. Nowadays, HPLC is used with various types of detectors for separation. The types of HPLC depends on the phase of chromatography; phase-dependent chromatography and reverse phase HPLC (RP-HPLC) is also used for IAA quantification. The analyte preparation is performed by taking directly bacterial culture supernatant or bacterial spiked broth filtrate for RP–HPLC analysis. Reversed phase HPLC (RP–HPLC or RPC) has a nonpolar stationary phase that is used in normal phase HPLC and an aqueous, moderately polar mobile phase. It is based on the principle of hydrophobic interaction because of repulsive forces between a polar eluent, the relatively nonpolar analyte and the nonpolar stationary phase (Bansal 2010). This HPLC system is composed of binary pump, fluorimetric detector, auto sampler and a computer. Chromatographic separation of an analyte is performed at ambient temperature in several columns used for analyte separation. These columns or stationary phase are the core of the chromatographic system. These are commercially available in different lengths, bore size and packing materials. The combination of length and packing material of column is correlated with the appropriate mobile phase used to assist in separation of sample compound. A variety of column dimensions are available, including preparative, normal-bore, micro- and mini-bore and capillary columns used for HPLC analysis. The mobile phase eluents ratio, flow rate and total run time is maintained with the injection volume and fluorimetric detector set at excitation and emission wavelengths of 280 and 350 nm, respectively, for indole compounds. Szkop and Bielawski (2013) performed by using bacterial culture filtrate and using C8 column (Symmetry 4.6 × 150 mm, 5 μm, Waters) fitted with a C8 guard column (symmetry 3.9 × 20 mm, 5 μm, waters) using gradient elution at ambient temperature. They used mobile phase with eluent A: eluent B at 80:20 %, changing to 50:50 %, 0:100 % and 80:20 % in 25, 31 and 33 min, respectively. Where eluent A consisted of acetic acid: water (2.5:97.5 % (v/v), pH 3.8 adjusted by addition of 1 mol L^{-1} KOH and eluent B consisted of acetonitrile: water (80:20 % (v/v)). Here, they preferred a gradient elution method (in which two or more solvent systems used for mobile phase differ significantly in polarity and the ratio of solvents is varied in a programmed way either continuously or in a series of steps, after elution is initiated). Instead of isocratic elution method (A separation that having single solvent or solvent mixture of constant ratio), the gradient method ensures a more stable separation, with improved peak shapes and a shorter cycle time.

2.1.8 Detection by HPTLC Method

During spectrophotometric analysis of indolic derivatives from tryptophan, Salkowski reagent is used which reacts with the indolic derivatives and develops color (Glickmann and Dessaux 1995). It is simple but highly inaccurate, because instead of binding with IAA it gives nonspecific color reaction with all the indolic derivatives produced by the bacteria. Although, HPLC is sensitive, it requires high purity of the sample and it makes the process very tedious. Moreover, HPLC is time-consuming for detection and calibration. Thus, Goswami et al. (2015) used the HPTLC technique for detection and quantification of IAA from several PGPRs by using a mobile phase of 50 % isopropanol, 30 % n-butanol, 15 % ammonia, and 5 % water. First, they determined an ability for IAA production of different strains (*Pseudomonas aeruginosa* OG, *K. turfanensis* 2M4, *Kocuria flava* 2M7, *Bacillus subtilis* H6 and *Bacillus licheniformis* A2) using Salkowski reagent based on spectrophotometric method described by Brick et al. (1991). On the other hand, they extracted indolic derivatives from the bacterial strains grown in a nutrient broth supplemented with 1 mg ml^{-1} tryptone. The culture supernatant was acidified by 1 N HCl and extracted thrice using equal volume of ethyl acetate. The fraction was air dried and redissolved it into one-tenth volume of methanol for HPTLC analysis. A quantitative densitometric analysis was performed under deuterium source lamp at 256 nm as indolic derivatives show fluorescence on silica TLC plate $60F_{254}$ and quantification was determined by plotting a peak area under the curve. This method can detect IAA in the range of 100–1000 ng per TLC spot, which suggests that this method is highly sensitive and detects IAA even in low concentrations.

2.1.9 Detection of Microbial Phytohormones by Chromatographic/ Mass Spectroscopy

For the final analysis of IAA, a range of instrumental techniques is used such as HPLC coupled to UV or fluorimetric detection (Crozier et al. 1980) and gas chromatography (GC), which are common chromatographic analytical techniques used for quantitative analysis of phytohormones. According to Ljung et al. (2001) GC combined with mass spectroscopy detects and quantifies smaller range 0.1 mg fresh weight of sample. Plant hormones need to be derived to increase their volatility and improve their thermal stability, most frequently done by methylation (Schneider et al. 1985) or trimethylsilylation (Edlund et al. 1995) prior to gas chromatographic separation. Such derivatization protocol extends considerably analytical protocols. HPLC is more suitable for most plant hormones without derivatization when using a UV detector. The main drawback of using a UV detector is its lower sensitive detection (Nicander et al. 1993). Therefore, HPLC along with mass spectroscopy (MS) is widely used for quantification of IAA and especially when high throughput is required. HPLC-MS mass detection based on an ion trap (Ma et al. 2008; Prinsen et al. 1998) or triple quadruple is highly sensitive and selective. The analyte bound via chemical interaction with the solid phase to the column and remaining interfering substances are removed from the column by washing it with suitable solvent and the analyte is then eluted by strong solvent during solid phase extraction methods which have often replaced the HPLC method. The retention mechanism in solid phase extraction (SPE) is based on van der Waal forces, dipole–dipole forces, hydrogen bonding and ionic interactions during the process. For the eluted sample, peak height or peak area is important for quantification. The peak height or peak area versus concentration of sample is plotted for determination of concentration of a sample. For well-resolved peaks, peak height and area is proportional to the concentration.

2.1.10 Detection by FTIR Analysis

An infrared spectrum observed with absorption peaks of a sample represents the frequency of vibration between the bonds of the atoms; two atoms do not get the exact same infrared spectrum. It is just like the fingerprint of an atom and therefore IR spectroscopy is used for the qualitative analysis of every different kind of material. According to Åmand and Tullin, FTIR instrument having Michelson-type interferometer instead of the monochromator and the slits has a beam of radiation that is further split into two by beam splitter. A path difference between the beams is introduced by allowing them to recombine and can be monitored using an appropriate detector. FTIR spectroscopy is preferred over infrared spectroscopy because of nondestructive, precise enough, rapid, sensitive and greater optical throughput (Griffiths and De Haseth 2007). Goswami et al. (2014) detected IAA from bacterial isolates by dried extract mixed with potassium bromide and analysis recorded at the transmission mode from frequency of 400–4000 cm^{-1} using Thermo Scientific Nicolet FTIR 6700. The spectral analysis generated correlates with the chemical bonds, molecular structure and vibration. Here, FTIR analysis does not provide the quantification but can give the structure analysis. Kamnev et al. (2001) showed that metal ions interact with IAA, which significantly modifies the characteristic FTIR spectra of IAA. Thus, FTIR is used to detect the interaction of other molecules with IAA while the purity of the IAA can also be determined.

2.2 Gibberellins

Gibberellin was recognized by the Japanese scientist, Eiichi Kurosawa (1926), while studying bakanae disease in rice plant. It was first isolated by Teijiro Yabuta and Sumuki form fungal strains and later the name Gibberellin was given by Yabuta in 1935. Gibberellins known as gibberellic acid (GA) has 136 different chemical structures from higher plants, 28 GAs from fungi and only 4 GAs from bacteria characterized as gibberellins (MacMillan 2001). Gibberellins are a type of microbial phytohormone having tetra carbocyclic diterpernes that regulate and influence the growth, developmental process (Zhang et al. 2007), flowering (Cleland and Zeevaart 1970), stem elongation (Suge and Rappaport 1968), seed germination (Ogawa et al. 2003) and enzyme induction (Rogers and Rogers 1999). This class of gibberellin is extremely large and some of these hormones have a similar structure. All these hormones have fluorine compounds with a structure of gibbane (Fig. 3b). The biosynthesis of gibberellins begins with cyclization of an intermediate geranylgeranyl diphosphate (GGPP), which is synthesized via isopentenyl diphosphate (IPP) either coming from mevalonic acid or via the plastid deoxylulose 5-phosphate pathway (Litchtenthaler 1999; Sponsel 2001). This cyclization of GGPP yields copalyl diphosphate (CPP) by enzyme copalyl diphosphate synthase and *ent*-kaurene (*Ent*-K) by *ent*-kaurene synthase (KS) enzyme

(Sun and Kamiya 1994). Finaly, *ent*-K is converted into true gibberellins by several oxidation reactions. GA is a crystalline solid melt at 221–223 °C. GA having pK value of 4.0 and it is a monobasic acid. It is partially soluble in water and ethyl acetate and highly soluble in methanol, ethanol, acetone and alkali reaction with dilute mineral acid at 20 °C converting into allogibberic acid followed by heating with acid at 100 °C forming gibberic acid (West 1960).

2.2.1 Gibberellins Producing Microorganisms

About 20 different types of gibberellins are produced by *Gibberella fujikuori,* of which GA_3 is most abundant. As shown in Table 1, *Azospirillum* spp. is known to produce gibberellins and reverse dwarfism in maize and rice (Cassán et al. 2001) and promote shoot elongation, growth and root hair density (Fulchieri et al. 1993). Srivastava (2003) observed different strains of *F. moniliforme* isolated from different geographical locations showed a range of 0.66–600 mg production of gibberellin per gm dry mycelium weight. *Acinetobacter calcoaceticus* produced gibberellin were able to grow cucumber, Chinese cabbage and crown daisy (Kang et al. 2009). Also, *Acinetobacter, Agrobacterium, Arthrobacter, Azotobacter, Bacillus, Bradyrhizobium japonicum, Clostridium, Flavobacterium, Micrococcus, Pseudomonas, Rhizobium* and *Xanthomonas* are known to produce gibberellins in the rhizosphere (Rademacher 1994; Frankenberger and Arshad 1995; Gutiérrez-Manero et al. 2001; Tsavkelova et al. 2006). *Acinetobacter* sp., *Aerobacter* sp., *Aeromonas* sp., *Agrobacterium* sp., *Arthrobacter* sp., *Bacillus* sp., *Brevibacterim* sp., *Chromobacterium* sp., *Clostridium* sp., *Flavobacterium, Nocardia* sp., *Pseudomonas* sp. and *Streptomyces* sp. are bacteria that produce gibberellin molecule and gibberellin-like substances. Species of *Alternaria, Aspargillus, Cephalosporium, Cladosporium, Cylindrocarpon, Fusarium, Mycelium, Neurospora, Penicillium, Rhizopus, Sphaceloma, Trichoderma* and *Verticillium* are fungi known to produce gibberellins or gibberellin-like substances (Frankenberger and Arshad 1995). *R. japonicum, R. leguminosarum* and *R. Meliloti* are known to produce GA_3 and GA_9 (Katznelson and Cole 1965).

2.2.2 Production and Extraction of Gibberellin

For isolation of gibberellic acid producing microorganisms, it should be cultured in a specific medium from where an organism can increase its biomass that leads to the production of gibberellic acid or gibberellic acid-like substances. For the culture of organisms, cultured a suspension in nutrient agar for bacterial growth and yeast extract mannitol agar for fungal growth for 28 °C at 240 rpm and a total of 73 isolates were cultured. For the maximum growth of GA_3, fungus *G. fujikuroi* was incubated at 72 h, pH 8 and 30 °C in dark on a rotary shaker (Karakoç and Aksöz 2006). For isolation of rhizobacteria a cell suspension was grown in glucose nitrogen-free mineral medium and nutrient agar medium and incubated at

30 °C (Lwin et al. 2012). Pandya and Desai (2014) isolated 59 bacterial isolates on nutrient agar plate supplemented with clotrimazole and incubated at 30 °C for 24 h for production of GA_3.

2.2.3 Spectrophotometric Assay for Detection of Gibberellins

For the gibberellin bioassay, Pandya and Desai (2014) incubated bacterial isolates in a nutrient broth and then centrifuged 48 h old culture at 10,000 rpm for 10–15 min and culture supernatant pH is adjusted to 2.5 using 4 N HCl. Then the culture supernatants were extracted using liquid–liquid (ethyl acetate/$NaHCO_3$) extraction method. According to the method of Uthandi et al. (2010), *F. fujikuroi* culture broth of Czapek-Dox was filtered through pre-weighed filter paper and filtrate was set to pH 2.5 with the help of 10 % HCl. Acidified filtrate is extracted with ethyl acetate (1:3 filtrate to solvent ratio) and collected for estimation of GAs. In these biological assays gibberellic acid is converted into gibberellenic acid which absorbs light at 254 nm wavelength. 25 ml of supernatant and 2 ml of Zn acetate reacted for a while, then 2 ml of potassium ferrocyanide is added and after centrifugation at 10,000 rpm, in 5 ml supernatant equal volume of 30 %, HCl was added and incubated at 20 °C for approximately 1 min. Absorbance was taken at 254 nm in spectrophotometer (Vikram et al. 2007).

2.2.4 Titrimetric Method to Determine Gibberellins

Gibberellic acid from *Fusarium* species was determined by acid–base titration method (Sanchez-Marroquin 1963), in which, gibberellic acid was titrated with 0.1 or 0.25 N NaOH solution using phenolphthalein as an indicator and measured in 10^{-3} of a gram equivalent (milliequivalent) of gibberellic acid.

2.2.5 Qualitative Estimation by TLC and Paper Chromatography

Gibberellin and related substances were extracted and dissolved in 5 ml ethanol and separated by ascending chromatographic technique using isopropanol: 25 % ammonium hydroxide: water (10:1:1 v/v) and spot detection done by spraying 3 % H_2SO_4 in methanol containing 50 mg $FeCl_3$; after heating at 80 °C for 10 min, plates generate greenish spots under UV light (Rangaswamy 2012). According to Bird and Pugh (1958), 19 × 46 cm of Schleicher and Schuell No. 598 or Muktells, Cremer-Tiselius electrophoretic filter paper was used to detect gibberellin A and gibberellic acid. Samples were spotted near one end of a sheet and its opposite end was serrated to permit solvent (10 volume of thiophene-free benzene, 2.5 volume of glacial acetic acid and 5 volume of deionized water on descending chromatogram) to drip off more freely to allow maximum flow.

2.2.6 Qualitative Determination of Gibberellins by HPLC

Polar charges of the gibberellins, i.e. esterification of carboxyl group are done for GC analysis. Here, gibberellins are privatized as 4-bromophenacyl esters and *p*-nitro benzyl esters (Morris and Zaerr 1978; Heftmann et al. 1978) for identification by high performance liquid chromatography (HPLC). Barendse and Van De werken (1980) used ionic separation technique and gradient elution for separation of different gibberellins from plant extracts. For analysis of seven different gibberellins from strains of *Fusarium*, Bhalla et al. (2010) developed the isocratic HPLC system. According to Bhalla et al. (2010) the isocratic system is a better method for detecting small concentrations, because in gradient system the baseline is disturbed during grading of mobile phase. Sample is doubled filtered and dissolved in 80 % MeOH and injected into the reverse phase C18 HPLC column, where gradient of anhydrous MeOH: aqueous acetic acid with the gradient interval of 10 min having flow rate of 1 ml min^{-1} was used. They collected several fractions and aliquots assayed for radioactivity by liquid scintillation counting (Dobert et al. 1992). After centrifugation (21,040 g for 5 min) of fermentation broth, pH of the supernatant was adjusted to 2 with 1 M HCl and supernatant was extracted with ethyl acetate. Washing of organic phase was done with 5 ml of brine (Saline water) and then dried using 1 g of Na_2SO_4 and evaporated. Residue was redissolved in 1 ml of methanol and passed through octadecylsilane column (4.6 mm × 150 mm), where methanol: water (3:1) at flow rate 0.5 ml min^{-1} was fixed. Mass and UV detector were used simultaneously for detection of the component of gibberellins produced by *Fusarium verticilliodes* MTCC 156 (Sharma et al. 2004). While Bhalla et al. (2010) used LiChrospher on RP-18 packed stainless steel column (250 × 4 mm i.d.) and acetonitrile: acidic water (0.01 % H_3PO_4) in the ratio of 60:40 with flow rate of 0.6 ml min^{-1}. Detection of gibberellin produced by *Fusarium* strains was done by photo diode-array detector.

2.3 Cytokinins

Cytokinins, a class of phytohormones play an important role in cell division or cytokinesis in plant roots and shoots and, hence, are named as "Cytokinins." Gottlied Haberlandt discovered in 1913 that a compound found in phloem can stimulate cell division in potato parenchyma. In 1941, Johannes Van Overbeek discovered that the milky endosperm from coconut has a similar mode of action as mentioned earlier by Haberlandt. Later in 1955, Carlos Miller, a student of Folke Skoog's laboratory identified kinetin from herring sperm. During 1961–1963, naturally occurring most active of the cytokinins, zeatin was first isolated from *Zea mays*. Numerous cytokinins have been isolated from tRNA of all organisms having a function of influencing tRNA structure, providing recognition site, affecting the translation efficiency and accuracy and a regulatory role. Distribution of modified at position 37 of tRNA structure was hypermodified, hydrophobic isopentenyl adenosine-"cytokinin" among organisms seems to show inter-kingdom differences (Bartz et al. 1970; Greene 1980; Sprinzl et al. 2005). Cytokinin has both naturally occurring compounds and their synthetic analogues. Kinetin was ascended in original isolates by structural rearrangements, so it is not accepted in naturally occurring cytokinin (Hecht 1980). Cytokinin is mainly divided into two groups (Fig. 3c) based on chemical structure: one is isoprenoid cytokinins which have an isoprenoid side chain at the position of N^6. Examples of isoprenoid cytokinins are zeatin, isopentyl and dihydrozeatin forms. Another type of cytokinin is aromatic cytokinins, which possess a side chain of aromatic (Benzyl or furfuryl) origin. Many natural cytokinin-like compounds that are structurally related to kinetin, such as trans-zeatin, 4-hydroxy-3-methyltrans-2-butenylaminopurine, dihydrozeatin, have been identified in free bases, as glucosides, ribosides, or nucleotides (Entsch et al. 1980). Synthetic substituted purines, such as 4-alkylaminopteridines and 6-benzyloxypurines that are less structurally similar to N^6-substituted adenine possess cytokinin activity. Some of these are reported to be more active than kinetin or benzyladenine (BA). Nowadays, topolin—an aromatic naturally occurring cytokinin is a widely accepted derivative of BA. A few reports mention cytokinin, zeatin and 1, 3-diphenylurea as naturally physiologically active substances found in tissue cultures. Zeatin and 9R-zeatin biosynthesis by *B. japonicum* cultures was reported by Sturtevant and Taller (1989).

2.3.1 Diversity of Microbial Cytokinin Producers

The potential of producing cytokinins is widespread among organisms associated with soil and plants. These bacteria are responsible for plant growth, symbiosis and pathogenesis. Such types of organisms mentioned in Table 1 are known to produce several phytohormones and one of them is cytokinins. Several PGPR, such as *Azospirillum, Azotobacter, Pseudomonas* and *Bacillus,* produce several auxins, gibberellins as well as cytokinins (Dobbelaere et al. 2003). *Actinomycete* and *Streptomyces flavous* isolated from the rhizosphere have been reported in the production of cytokinin (Coppola and Giannattasio 1968). As shown in Table 1, *Erwinia herbicola* is reported by Lichter et al. (1995) as a cytokinin producing bacteria. A variety of bacterial strains such as *A. tumefacians, P. syringeae, P. fluorescens* and *P. putida* were screened as cytokinin producing microorganisms.

2.3.2 Production and Detection of Cytokinin

The production of cytokinin produced by *C. fascians* is demonstrated by Thimann and Sachs (1966). They grew *C. fascians* in a defined purine-free medium. The fungal strain *Rhizopogon ochraceorubens* grown on agar containing modified Hagem's medium and incubated for 2 weeks at room temperature produced cytokinin (Crafts and Miller 1974). *Corynebacterium aurimucosum* was isolated from infected fruit of *Prunus salicina* and grown on the simple nutrient agar medium was reported to produce zeatin (Patel et al. 2012).

2.3.3 Detection of Cytokinin by HPLC

Reverse phase column under acidic condition is used for separation of cytokinin-like compounds because it has relatively hydrophobic sugar conjugates and bases of cytokinin (Tarkowski et al. 2009). The analysis and separation of cytokinin has been done by C_{18} or C_8 columns. Acetic acid or formic acid and their ammonium salts are added to the solvent methanol/acetonitrile for better separation (Ge et al. 2005). As cytokinin exhibits strong UV absorbance between 200 and 300 nm, UV detection is suitable for detection of cytokinin. Narrow bore LC coupled with MS is a relative method for detection of cytokinin used, in which mass spectroscopy coupled with liquid chromatography overcome the low detectability of the cytokinins. Different ionization techniques were used for mass analysis in combination with RP–HPLC including thermospray, electrospray, atmospheric pressure chemical ionization and fast atom bombardment (Novák et al. 2003). Novák et al. (2003), performed LC–MS by dissolving sample in mobile phase and filtered with microfilter and 25 μl of sample was injected on RP column (150 mm × 2.1 mm, 5 m). Solvent A consisted of 15 mM formic acid adjusted to pH 4 by ammonium hydroxide. Solvent B consisted of methanol.

2.3.4 Detection by Immunoaffinity Chromatography

Cytokinin was isolated from the purified sample of *P. polymyxa* cultivated on a medium and the filtrate was treated with methanol. Then cytokinin was isolated from the purified samples by immunoaffinity chromatography using P_{ab} prepared against zeatin riboside and iPR having column temperature of 30–35 °C. The eluate was evaporated in vacuum to less than 300 μl and diluted by PBS and every sample passed through column a second time and new elute was evaporated in vacuum (Timmusk et al. 1999).

2.3.5 Detection of Cytokinin by GC–MS

To detect cytokinin by gas chromatography, glass packed column was used although fused-silica capillary column was used by Tarkowski et al. (2009). Doumas et al. (1989) identified two bacterial cytokinin-like compounds GC–MS using a combination of appropriate fraction from different extractions, dried and redissolved it in 40 mM phosphate buffer. After purification with HPLC grade putative premethylated derivatized cytokinin dissolved in dichloromethane and introduced in GC by on-column injection (30 m 0.32 mm fused silica capillary column coated with DB-5). Temperature has to be increased from 50 to 200 °C to remove the bulk of the solvent and premethylated cytokinin was eluted directly into mass spectrometer by increasing temperature from 200 to 300 °C. Premethylated samples were analyzed by Kamboj et al. (1998) using capillary column coupled directly to the ion source with an inference temperature of 275 °C and helium (He) as carrier gas, inlet pressure at 0.08 MPa. After injecting a sample they maintained GC oven at 60 °C for 1 min with splitter closed and increased to 230 °C.

2.3.6 Cytokinin Detection by Capillary Electrophoresis

Capillary electrophoresis was used for detection of cytokinin-like compounds and it is a suitable method for detection because of high speed, resolving power and fewer requirements of sample and buffer. Different modes of capillary electrophoresis are used for detection such as capillary zone electrophoresis (CZE) and micellar electrokinetic chromatography (MEKC). To obtain good separation, electrolyte composition is extremely important in each mode of CE. Relatively high concentration of nonvolatile buffer results in significant loss of electrospray efficiency and produces ion source contamination, therefore volatile buffer system (ammonium formate) is used for cytokinin detection in CZE-MS (Ge et al. 2006). Small aliquots of sample are a significant challenge in CE, although UV and MS detector have been used for detection of cytokinin (Pacáková et al. 1997).

2.4 Jasmonic Acid

Jasmonates belong to a family of oxylipins, an oxygenated fatty acid possessing one or more oxygen atoms other than those in the carboxyl group. It is widely found in aerobic organisms including plants, animals and fungi. Basically, these types of oxylipins are not stored in the tissue but are produced on demand. Jasmonates arise from the several enzymatic oxygenation reactions of C^{18} and C^{16} tri-unsaturated fatty acids (Wasternack and Kombrink 2009) and can also be produced by fungus like *Lasiodiplodia theobromae* (Demole et al. 1962; Miersch et al. 1991; Dhandhukia and Thakkar 2007). JA production is observed in *Escherichia coli* RC424 and RC-7 strains and also in yeast extract used in culture media for bacterial growth. As shown in (Fig. 3d), jasmonic acid (JA), methyl jasmonate (MeJA) and jasmonoyl-isoleucine (JA-Ile) are best known jasmonates. Jasmonic acids and most of its derivatives have a cyclopentanone ring structure possessing a saturated bond instead of carbon–carbon double bond in the ring structure of cyclopentenone (Acosta and Farmer 2010).

2.4.1 Producers of Microbial Jasmonates

Jasmonic acids are naturally occurring fatty acids and are identified in a wide variety of plant species, although a few fungi also produce jasmonic acid. The first identified jasmonate was from the fungal cell culture of *Lasiodiplodia theobromae* contributing as plant growth regulator (Dhandhukia and Thakkar 2007). Other than this, *B. rhodina* has been found to produce jasmonic acid and its derivatives.

2.4.2 Production and Extraction of Jasmonic Acid

Lasiodiplodia theobromae was grown on potato dextrose agar plate and incubated at 30° C for 48 h. Agar plug was cut and used for inoculation on various medium for media optimization on static condition. After 7 days of growth in static condition, mycelia were separated from the broth by filtration. The filtrate was acidified with 6 mol L^{-1} HCl to pH 3 and 25 ml of broth used for extraction using equal volume of ethyl acetate (Dhandhukia and Thakkar 2008). Buttarello et al. (2014) mentioned bioproduction of *Lasiodiplodia theobromae* in M2 medium. After production, mycelium was removed by vacuum filtration and subjected for fermentation extraction test by adjusting a pH 3.0, then extracted by liquid–liquid partition using ethyl acetate as solvent extractor. The same method was performed by Leite et al. (2014) for production and extraction of jasmonic acid produced by *Botryosphaeria rhodina* from the precursor linolenic acid.

2.4.3 Identification and Detection of Jasmonates by HPTLC

After 7 days of growth in static condition, mycelia of *Lasiodiplodia theobromae* were separated from the broth by filtration. The filtrate was acidified with 6 mol L^{-1} HCl to pH 3 and 25 ml of broth used for extraction using equal volume of ethyl acetate (Dhandhukia and Thakkar 2008). Buttarello et al. (2014) mentioned bioproduction of *Lasiodiplodia theobromae* in M2 medium. After production, mycelium was removed by vacuum filtration and subjected for fermentation extraction test by adjusting a pH 3.0, then extracted by liquid–liquid partition using ethyl acetate as solvent extractor. After the treatment of ethyl acetate, extract was concentrated 100 times for measurement of jasmonates by high performance thin layer chromatography (HPTLC). Concentrated extract was then loaded on silica gel 60 F_{254} aluminum foils using linomate-5 spray on applicator under flow of N_2. Bands were regenerated using solvent system isopropanol: ammonia: water (10:1:1 v/v) (Ueda and Miyamoto 1994).

2.4.4 Identification and Quantification of Jasmonates by Liquid Chromatography

Ultra high pressure liquid chromatography/time-of-flight mass spectrophotometry (UHPLC/TOFMS) was performed by (Glauser et al. 2010) using a micromass LCT premier time-of-flight mass spectrophotometer with an electrospray. Sample was fractionated in C_{18} UPLC column (50 × 1.0 mm i.d. 1.7 μm) and gradient of solvent A (0.1 % formic acid–water) and solvent B (0.1 %

formic acid-acetonitrile) with flow rate of 0.3 ml min^{-1}. To detect jasmonates from sample extracts of *B. rhodina* HPLC was done (Supelcosil C_{18} column (25 cm × 4.6 mm) and solvent system: MeOH: acetic acid (60: 40)) coupled with diode-array detector was used (dos Santos et al. 2014). As preferred by Tsukada et al. (2010), jasmonates were purified and detected by HPLC using preparative TLC, using solvent system n-hexane: ethyl acetate: acetic acid (60:40:1 (v/v/v)); then the resultant residue was purified by HPLC (TSK gel ODS80Ts, TOSOH, 20 mm i.d. × 250 mm) in methanol (80 %) with flow rate of 5.0 ml min^{-1}. Forchetti et al. (2007) used to separate and identify on preparative HPLC system with C18 reverse phase column (μBond pack, 300 × 3.9 mm) coupled with UV-visible spectrometry system with diode arrangement. Elution was performed at 2 ml min^{-1} with gradient of 73 % (v/v) methanol in 1 % (v/v) acetic acid.

3 Conclusion

In the journey of bacterial phytohormones in the era of sustainable agro systems, several basic techniques as well as latest upgraded techniques are being used for detection and identification of phytohormones. From basic techniques such as TLC and spectrophotometry for simple detection to complex techniques such as HPTLC, GC–MS and LC–MS were used for spontaneous qualitative and quantitative detection of phytohormones. HPTLC, first for a while used for qualitative and quantitative detection for IAA and capillary electrophoresis also contributed in the field of bacterial phytohormone analysis. Despite advancements in the development of new techniques to detect microbially produced phytohormones, there is a still a huge scope in developing new techniques that provide better sensitivity, accuracy, robustness and save time. Despite several known protocols to detect several phytohormones, all these protocols are quite less laborious new, simple and quick protocols are needed in order to categorize the rhizobacterial strains as PGPR.

References

Acosta IF, Farmer EE (2010) Jasmonates. The Arabidopsis book/American Society of Plant Biologists 8:e0129. doi:10.1199/tab.0129

Atzorn R, Crozier A, Wheeler CT, Sandberg G (1988) Production of gibberellins and indole-3-acetic acid by *Rhizobium phaseoli* in relation to nodulation of *Phaseolus vulgaris* roots. Planta 175(4):532–538

Badenoch Jones J, Summons RE, Djordjevic MA, Shine J, Letham DS, Rolfe BG (1982) Mass spectrometric quantification of indole-3-acetic acid in *Rhizobium* culture supernatants: relation to root hair curling and nodule initiation. Appl Environ Microb 44(2):275–280

Bartz J, Söll D, Burrows WJ, Skoog F (1970) Identification of the cytokinin-active ribonucleosides in pure *Escherichia coli* tRNA species. Proc Natl Acad Sci 67(3):1448–1453

Bhalla K, Singh SB, Agarwal R (2010) Quantitative determination of gibberellins by high performance liquid chromatography from various gibberellins producing *Fusarium* strains. Environ Monit Assess 167(1–4):515–520

Bric JM, Bostock RM, Silverstone SE (1991) Rapid in situ assay for indole acetic acid production by bacteria immobilized on a nitrocellulose membrane. Appl Environ Microb 57(2):535–538

Cleland CF, Zeevaart JA (1970) Gibberellins in relation to flowering and stem elongation in the long day plant *Silene armeria*. Plant Physiol 46(3):392–400

Comai LUCA, Kosuge TSUNE (1980) Involvement of plasmid deoxyribonucleic acid in indoleacetic acid synthesis in *Pseudomonas savastanoi*. J Bacteriol 143(2):950–957

Comai LUCA, Kosuge TSUNE (1982) Cloning characterization of *iaa*M, a virulence determinant of *Pseudomonas savastanoi*. J Bacteriol 149(1):40–46

Coppola S, Giannattasio M (1968) Cytokinin activity in a rhizospheric actinomyces. J Biol Res 44(22):1913–1915

Crafts CB, Miller CO (1974) Detection and identification of cytokinins produced by mycorrhizal fungi. Plant Physiol 54(4):586–588

Crozier A, Loferski K, Zaerr JB, Morris RO (1980) Analysis of picogram quantities of indole-3-acetic acid by high performance liquid chromatography-fluorescence procedures. Planta 150(5):366–370

Deepa CK, Dastager SG, Pandey A (2010) Isolation and characterization of plant growth promoting bacteria from non-rhizospheric soil and their effect on cowpea (*Vigna unguiculata* (L.) Walp.) Seedling growth. World J Microbiol Biotechnol 26(7):1233–1240

Dekas AD, Poretsky RS, Orphan VJ (2009) Deep-sea archaea fix and share nitrogen in methane-consuming microbial consortia. Science 326:422–426

Demole E, Lederer E, Mercier D (1962) Isolation and determination of methyl jasmonate structure, forming characteristic fragrant jasmine essence. Helv Chim Acta 45(2):675–685

Dhandhukia PC, Thakkar VR (2007) Standardization of growth and fermentation criteria of *Lasiodiplodia theobromae* for production of jasmonic acid. Afr J Biotechnol 6(6):707–712

Dhandhukia PC, Thakkar VR (2008) Response surface methodology to optimize the nutritional parameters for enhanced production of jasmonic acid by *Lasiodiplodia theobromae*. J Appl Microbiol 105(3):636–643

Dobbelaere S, Vanderleyden J, Okon Y (2003) Plant growth-promoting effects of diazotrophs in the rhizosphere. Crit Rev Plant Sci 22(2):107–149

Dobert RC, Rood SB, Blevins DG (1992) Gibberellins and the legume-*Rhizobium* symbiosis I. endogenous gibberellins of lima bean (*Phaseolus lunatus* L.) stems and nodules. Plant Physiol 98(1):221–224

dos Santos AZ, de Mello Innocentini MD, Lourenço MV (2014) Optimization of jasmonates bioproduction, 4th edn. In: BMC Proceedings, vol 8, p 208. doi:10.1186/1753-6561-8-S4-P208

Doumas P, Bonnet-Masimbert M, Zaerr JB (1989) Evidence of cytokinin bases, ribosides and glucosides in roots of Douglas-fir, *Pseudotsuga menziesii*. Tree Physiol 5(1):63–72

Edlund A, Eklof S, Sundberg B, Moritz T, Sandberg G (1995) A microscale technique for gas chromatography-mass spectrometry measurements of picogram amounts of indole-3-acetic acid in plant tissues. Plant Physiol 108(3):1043–1047

El-Khawas H, Adachi K (1999) Identification and quantification of auxins in culture media of *Azospirillum* and *Klebsiella* and their effect on rice roots. Biol Fertil Soils 28(4):377–381

Elad Y (1993) Regulators of ethylene biosynthesis or activity as a tool for reducing susceptibility of host plant tissues to infection by *Botrytis cinerea*. Neth J Plant Pathol 99(3):105–113

Elmerich C, Newton WE (2007) Associative and endophytic nitrogen-fixing bacteria and cyanobacterial associations. Springer

Entsch B, Letham DS, Parker CW, Summons RE, Gollnow BI (1980) Metabolites of cytokinins. In: Skoog F (ed) Plant growth substances 1979. Springer, Berlin Heidelberg, pp 109–118

Fallik E, Okon Y, Epstein E, Goldman A, Fischer M (1989) Identification and quantification of IAA and IBA in *Azospirillum brasilense*-inoculated maize roots. Soil Biol Biochem 21(1):147–153

Fernández-Martín R, Reyes F, Domenech CE, Cabrera E, Bramley PM, Barrero AF, Avalos J, Cerdá-Olmedo E (1995) Gibberellin biosynthesis in *gib* mutants of *Gibberella fujikuroi*. J Biol Chem 270:14970–14974

Forchetti G, Masciarelli O, Alemano S, Alvarez D, Abdala G (2007) Endophytic bacteria in sunflower (*Helianthus annuus* L.): Isolation, characterization and production of jasmonates and abscisic acid in culture medium. Appl Microbiol Biotechnol 76(5):1145–1152

Frankenberger WT, Poth M (1987) Biosynthesis of indole-3-acetic acid by the pine ectomycorrhizal fungus *Pisolithus tinctorius*. Appl Environ Microb 53(12):2908–2913

Frankenberger WT Jr, Arshad M (1995) Phytohormones in soils: microbial production and function. Marcel Dekker Inc., New York, USA

Fulchieri M, Lucangeli C, Bottini R (1993) Inoculation with *Azospirillum lipoferum* affects growth and gibberellin status of corn seedling roots. Plant Cell Physiol 34:1305–1309

Ge L, Yong JWH, Goh NK, Chia LS, Tan SN, Ong ES (2005) Identification of kinetin and kinetin riboside in coconut (*Cocos nucifera* L.) water using a combined approach of liquid chromatography–tandem mass spectrometry, high performance liquid chromatography and capillary electrophoresis. J Chromatogr 829(1):26–34

Ge L, Yong JW H, Tan SN, Yang XH, Ong ES (2006) Analysis of cytokinin nucleotides in coconut (*Cocos nucifera* L.) water using capillary zone electrophoresis-tandem mass spectrometry after solid-phase extraction. J Chrom A 1133(1):322–331

Glass NL, Kosuge T (1986) Cloning of the gene for indoleacetic acid-lysine synthetase from *Pseudomonas syringae* subsp. *savastanoi*. J Bacteriol 166(2):598–603

Glauser G, Boccard J, Rudaz S, Wolfender JL (2010) Mass spectrometry-based metabolomics oriented by correlation analysis for wound-induced molecule discovery: identification of a novel jasmonate glucoside. Phytochem Anal 21(1):95–101

Glick BR (2012) Plant growth-promoting bacteria: mechanisms and applications. Scientifica 2012:963401. doi:10.6064/2012/963401

Glickmann E, Dessaux Y (1995) A critical examination of the specificity of the salkowski reagent for indolic compounds produced by phytopathogenic bacteria. Appl Environ Microbiol 61(2):793–796

Goswami D, Pithwa S, Dhandhukia P, Thakker JN (2014) Delineating *Kocuria turfanensis* 2M4 as a credible PGPR: a novel IAA-producing bacteria isolated from saline desert. J Plant Interact 9(1):566–576

Goswami D, Thakker JN, Dhandhukia PC (2015) Simultaneous detection and quantification of indole-3-acetic acid (IAA) and indole-3-butyric acid (IBA) produced by rhizobacteria from l-tryptophan (Trp) using HPTLC. J Microbiol Meth 110:7–14. doi:10.1016/j.mimet.2015.01.001

Goswami D, Vaghela H, Parmar S, Dhandhukia P, Thakker JN (2013) Plant growth promoting potentials of *Pseudomonas* spp. strain OG isolated from marine water. J Plant Interact 8(4):281–290

Greene EM (1980) Cytokinin production by microorganisms. Bot Rev 46(1):25–74

Griffiths PR, De Haseth JA (2007) Fourier transform infrared spectrometry, vol 171. Wiley, Hoboken

Gutiérrez-Mañero FJ, Ramos-Solano B, Probanza A, Mehouachi J, R Tadeo F, Talon M (2001) The plant-growth-promoting rhizobacteria *Bacillus pumilus* and *Bacillus licheniformis* produce high amounts of physiologically active gibberellins. Physiol Plantarum 111(2):206–211

Harari A, Kigel J, Okon Y (1988) Involvement of IAA in the interaction between *Azospirillum brasilense* and *Panicum miliaceum* roots. Plant Soil 110(2):275–282

Hauserová E, Swaczynová J, Doležal K, Lenobel R, Popa I, Hajdúch M, Strnad M (2005) Batch immunoextraction method for efficient purification of aromatic cytokinins. J Chrom A 1100(1):116–125

Hayat R, Ali S, Amara U, Khalid R, Ahmed I (2010) Soil beneficial bacteria and their role in plant growth promotion: a review. Ann Microbiol 60(4):579–598

Hecht SM (1980) Probing the cytokinin receptor site (s). In: Skoog F (ed) Plant growth substances. Springer, Berlin Heidelberg, pp 144–158

Heftmann E, Saunders GA, Haddon WF (1978) Argentation high-pressure liquid chromatography and mass spectrometry of gibberelline esters. J Chrom A 156(1):71–77
Hillel D (2003) Introduction to environmental soil physics. Elsevier Science, USA
Hiroya K, Itoh S, Sakamoto T (2004) Development of an efficient procedure for indole ring synthesis from 2-ethynylaniline derivatives catalyzed by Cu (II) salts and its application to natural product synthesis. J Org Chem 69(4):1126–1136
Hradecká V, Novák O, Havlíček L, Strnad M (2007) Immunoaffinity chromatography of abscisic acid combined with electrospray liquid chromatography–mass spectrometry. J Chrom B 847(2):162–173
Hyodo H (1991) Stress/wound ethylene. In: Mattoo AK, Suttle JC (eds) The plant hormone ethylene. CRC Press, USA, pp 151–189
Isenberg HD, Sundheim LH (1958) Indole reactions in bacteria. J Bacteriol 75(6):682–690
Jameson PE (2000) Cytokinins and auxins in plant-pathogen interactions–An overview. Plant Growth Regul 32(2–3):369–380. doi:10.1023/A:1010733617543
Janzen RA, Rood SB, Dormaar JF, McGill WB (1992) *Azospirillum brasilense* produces gibberellin in pure culture on chemically-defined medium and in co-culture on straw. Soil Biol Biochem 24:1061–1064
Jiang TF, Lv ZH, Wang YH, Yue ME (2006) Separation of plant hormones from biofertilizer by capillary electrophoresis using a capillary coated dynamically with polycationic polymers. Anal Sci 22(6):811–814
Kamboj JS, Blake PS, Baker DA (1998) Cytokinins in the vascular saps of *Ricinus communis*. Plant Growth Regul 25(2):123–126
Kamnev AA, Shchelochkov AG, Perfiliev YD, Tarantilis PA, Polissiou MG (2001) Spectroscopic investigation of indole-3-acetic acid interaction with iron (III). J Mol Struct 563:565–572
Kang SM, Joo GJ, Hamayun M, Na CI, Shin DH, Kim HY, Lee IJ (2009) Gibberellin production and phosphate solubilization by newly isolated strain of *Acinetobacter calcoaceticus* and its effect on plant growth. Biotechnol Lett 31(2):277–281
Kannangara T, Wieczorek A, Lavender DP (1989) Immunoaffinity columns for isolation of abscisic acid in conifer seedlings. Physiol Plant 75(3):369–373
Kapulnik Y, Okon Y, Henis Y (1985) Changes in root morphology of wheat caused by *Azospirillum* inoculation. Can J Microbiol 31(10):881–887
Karakoç Ş, Aksöz N (2006) Some optimal cultural parameters for gibberellic acid biosynthesis by *Pseudomonas sp*. Turkish J Biol 30(2):81–85
Katznelson H, Cole SE (1965) Production of gibberellin-like substances by bacteria and actinomycetes. Can J Microbiol 11:733–741
Kilian M, Steiner U, Krebs B, Junge H, Schmiedeknecht G, Hain R (2000) FZB24® *Bacillus subtilis*–mode of action of a microbial agent enhancing plant vitality. Pflanzenschutz-Nachr Bayer 1(00):72–93
Kim J, Chun MS, Choi K, Chung DS (2009) Large volume stacking using an EOF pump in NACE-MS. Electrophoresis 30(6):1046–1051
Kurosawa E (1926) Experimental studies on the nature of the substance secreted by the "bakanae" fungus. Nat Hist Soc Formosa 16:213–227
Leite V, Santos A, Innocentini M, Maciel G, Lourenço M (2014) Influence of linolenic acid in the production of jasmonate. BMC Proc 8(4):178
Lichtenthaler HK (1999) The 1-deoxy-D-xylulose-5-phosphate pathway of isoprenoid biosynthesis in plants. Annu Rev Plant Biol 50(1):47–65
Lichter A, Barash I, Valinsky L, Manulis S (1995) The genes involved in cytokinin biosynthesis in *Erwinia herbicola* pv. *gypsophilae*: Characterization and role in gall formation. J Bacteriol 177(15):4457–4465
Liu ST, Perry KL, Schardl CL, Kado CI (1982) *Agrobacterium* Ti plasmid indole acetic acid gene is required for crown gall oncogenesis. Proc Natl Acad Sci U.S.A. 79:2812–2816
Ljung K, Bhalerao RP, Sandberg G (2001) Sites and homeostatic control of auxin biosynthesis in Arabidopsis during vegetative growth. Plant J 28(4):465–474

Lu Y, Song S, Wang R, Liu Z, Meng J, Sweetman AJ, Wang T (2015) Impacts of soil and water pollution on food safety and health risks in China. Environ Int 77:5–15. doi:10.1016/j.envint.2014.12.010

Lugtenberg BJ, Malfanova N, Kamilova F, Berg G (2013) Microbial control of plant root diseases. In: Bruijn (ed) Molecular Microbial Ecology of the Rhizosphere, vol 2. Wiley, pp 575–586

Lwin KM, Myint MM, Tar T, Aung WZM (2012) Isolation of plant hormone (Indole-3-Acetic Acid-IAA) producing rhizobacteria and study on their effects on maize seedling. Eng J 16(5):137–144

Ma Z, Ge L, Lee AS, Yong JWH, Tan SN, Ong ES (2008) Simultaneous analysis of different classes of phytohormones in coconut (*Cocos nucifera* L.) water using high-performance liquid chromatography and liquid chromatography–tandem mass spectrometry after solid-phase extraction. Anal Chim Acta 610(2):274–281

Mac Faddin JF (1976) Biochemical tests for identification of medical bacteria. 88(4):1080–1091. doi:10.1172/JCI115407

MacMillan J (2001) Occurrence of gibberellins in vascular plants, fungi and bacteria. J Plant Growth Regul 20(4):387–442

Manual D (1998) Detroit. MI: Difco Laboratories

Manulis S, Haviv-Chesner A, Brandl MT, Lindow SE, Barash I (1998) Differential involvement of indole-3-acetic acid biosynthetic a pathways in pathogenicity and epiphytic fitness of *Erwinia herbicola* pv. *gypsophilae*. Mol Plant Microbe Interact 11:634–642

Maruyama A, Maeda M, Simidu U (1986) Occurrence of plant hormone (cytokinin)-producing bacteria in the sea. J Appl Bacteriol 61(6):569–574

Mascarua-Esparza MA, Villa-Gonzalez R, Caballero-Mellado J (1988) Acetylene reduction and indoleacetic acid production by *Azospirillum* isolates from Cactaceous plants. Plant Soil 106:91–95

Miersch O, Schneider G, Sembdner G (1991) Hydroxylated jasmonic acid and related compounds from *Botryodiplodia theobromae*. Phytochem 30(12):4049–4051

Miller JM, Wright JW (1982) Spot indole test: evaluation of four reagents. J Clin Microbiol 15(4):589–592

Mohite B (2013) Isolation and characterization of indole acetic acid (IAA) producing bacteria from rhizospheric soil and its effect on plant growth. J Soil Sci Plant Nutr 13(3):638–649

Moller SG, Chua NH (1999) Interactions and intersections of plant signalling pathways. J Mol Biol 293:219–234

Moore TC (1979) Auxins. In: Biochemistry and Physiology of plant hormones, Springer US. 32–89. doi:10.1007/978-1-4684-0079-3_2

Morris RO, Zaerr JB (1978) 4-Bromophenacyl Esters of Gibberellins, Useful Derivatives for High Performance Liquid Chromatography. Anal lett 11(1):73–83

Nadeem SM, Ahmad M, Zahir ZA, Javaid A, Ashraf M (2014) The role of mycorrhizae and plant growth promoting rhizobacteria (PGPR) in improving crop productivity under stressful environments. Biotech Adv 32(2):429–448

Nicander B, Ståhl U, Björkman PO, Tillberg E (1993) Immunoaffinity co-purification of cytokinins and analysis by high-performance liquid chromatography with ultraviolet-spectrum detection. Planta 189(3):312–320

Novák O, Tarkowski P, Tarkowská D, Doležal K, Lenobel R, Strnad M (2003) Quantitative analysis of cytokinins in plants by liquid chromatography–single-quadrupole mass spectrometry. Anal Chim Acta 480(2):207–218

Nuray ERG (2002) Auxin (indole-3-acetic acid), gibberellic acid (GA3), abscisic acid (ABA) and cytokinin (zeatin) production by some species of mosses and lichens. Turk J Bot 26(1):13–18

Ogawa M, Hanada A, Yamauchi Y, Kuwahara A, Kamiya Y, Yamaguchi S (2003) Gibberellin biosynthesis and response during *Arabidopsis* seed germination. Plant Cell 15(7):1591–1604

Omay SH, Schmidt WA, Martin P, Bangerth F (1993) Indoleacetic acid production by the rhizosphere bacterium *Azospirillum brasilense* Cd under in vitro conditions. Can J Microbiol 39:187–192

Pacáková V, Štulík K, Vlasáková V, Březinová A (1997) Capillary electrophoresis of cytokinins and cytokinin ribosides. J Chrom A 764(2):331–335

Pandya ND, Desai PV (2014) Screening and characterization of GA3 producing *Pseudomonas monteilii* and its impact on plant growth promotion. Int J Curr Microbiol App Sci 3(5):110–115

Patel PP, Rakhashiya PM, Chudasama KS, Thaker VS (2012) Isolation, purification and estimation of zeatin from *Corynebacterium aurimucosum*. Eur J Exp Biol 2(1):1–8

Patil V (2011) Production of indole acetic acid by *Azotobacter* sp. Recent Res Sci Technol 3(12)

Playtis AJ, Leonard NJ (1971) The synthesis of ribosyl-cis-zeatin and thin layer chromatographic separation of the cis and trans isomers of ribosylzeatin. Biochem Biophys Res Commun 45(1):1–5

Prinsen E, Van Dongen W, Esmans EL, Van Onckelen HA (1997) HPLC linked electrospray tandem mass spectrometry: a rapid and reliable method to analyse indole-3-acetic acid metabolism in bacteria. J Mass Spectrom 32(1):12–22

Prinsen E, Van Dongen W, Esmans EL, Van Onckelen HA (1998) Micro and capillary liquid chromatography–tandem mass spectrometry: a new dimension in phytohormone research. J Chrom A 826(1):25–37

Rademacher W (1994) Gibberellin formation in microorganisms. Plant Growth Regul 15(3):303–314

Rangaswamy V (2012) Improved production of gibberellic acid by *Fusarium moniliforme*. J Microbiol Res 2(3):51–55

Redig P, Shaul O, Inzé D, Van Montagu M, Van Onckelen H (1996) Levels of endogenous cytokinins, indole-3-acetic acid and abscisic acid during the cell cycle of synchronized tobacco BY-2 cells. FEBS Lett 391(1):175–180

Reynders L, Vlassak K (1979) Conversion of tryptophan to indole acetic acid by *Azospirillum brasilense*. Soil Biol Biochem 11:547–548

Roberto F, Kosuge T (1987) Phytohormone metabolism in *Pseudomonas syringae* subsp. *savastanoi*. In: J E Fox, M Jacob (Eds.), Molecular biology of plant growth control, New York: Alan R Liss, Inc. pp 371–380

Rogers SW, Rogers JC (1999) Cloning and characterization of a gibberellin-induced RNase expressed in barley aleurone cells. Plant Physiol 119(4):1457–1464

Rojas MC, Hedden P, Gaskin P, Tudzynski B (2001) The P450-1 gene of *Gibberella fujikuroi* encodes a multifunctional enzyme in gibberellin biosynthesis. Proc Natl Acad Sci U.S.A. 98:5838–5843

Rowell DL (2014) Soil science: methods & applications. Taylor & Francis, Routledge

Sanchez-Marroquin A (1963) Microbiological production of gibberellic acid in glucose media. J Appl Microbiol 11(6):523–528

Sandberg G, Crozier A (1986) Purification of indole-3-acetic acid in plant extracts by immunoaffinity chromatography. Phytochem 25(2):295–298

Santner A, Calderon-Villalobos LIA, Estelle M (2009) Plant hormones are versatile chemical regulators of plant growth. Nature Chem Biol 5:301–307

Schmitt J, Flemming HC (1998) FTIR-spectroscopy in microbial and material analysis. Int Biodeter Biodegr 41(1):1–11

Schneider EA, Kazakoff CW, Wightman F (1985) Gas chromatography-mass spectrometry evidence for several endogenous auxins in pea seedling organs. Planta 165(2):232–241

Sharma R, Iyer JP, Chakraborti AK, Banerjee UC (2004) Determination of gibberellins in fermentation broth produced by *Fusarium verticilliodes* MTCC 156 by high-performance liquid chromatography tandem mass spectrometry. Biotechnol Appl Biochem 39(1):83–88

Sponsel VM (2001) The deoxyxylulose phosphate pathway for the biosynthesis of plastidic isoprenoids: early days in our understanding of the early stages of gibberellin biosynthesis. J Plant Growth Regul 20(4):332–345

Sprinzl M, Vassilenko KS (2005) Compilation of tRNA sequences and sequences of tRNA genes. Nuc Acids Res 33(1):139–140

Srivastava AC, Ahamad S, Agarwal DK, Sarbhoy AK (2003) Screening of potential gibberellin producing *Fusarium* strains for the hybrid rice production. J Food Agric Environ 1:250–253

Sturtevant D, Taller B (1989) Cytokinin production by a parasponia nodule bacterium. Abstr Annu Meet Am Soc Microbiol 89:300

Suge H, Rappaport L (1968) Role of gibberellins in stem elongation and flowering in radish. Plant Physiol 43(8):1208–1214

Sun TP, Kamiya Y (1994) The Arabidopsis GA_1 locus encodes the cyclase *ent*-kaurene synthetase A of gibberellin biosynthesis. Plant Cell Online 6(10):1509–1518

Swain MR, Naskar SK, Ray RC (2007) Indole-3-acetic acid production and effect on sprouting of yam (*Dioscorea rotundata L.*) minisetts by *Bacillus subtilis* isolated from culturable cowdung microflora. Pol J Microbiol 56(2):103–110

Szkop M, Bielawski W (2013) A simple method for simultaneous RP-HPLC determination of indolic compounds related to bacterial biosynthesis of indole-3-acetic acid. Antonie van Leeuwenhoek 103(3):683–691

Takahashi N (1986) Chemistry of plant hormones. CRC Press, Boca Raton

Tarkowski P, Ge L, Yong JWH, Tan SN (2009) Analytical methods for cytokinins. TrAC Trend Anal Chem 28(3):323–335

Thimann KV, Sachs T (1966) The role of cytokinins in the "fasciation" disease caused by *Corynebacterium fascians*. Am J Bot 53(7):731–739

Tien TM, Gaskins MH, Hubbell DH (1979) Plant growth substances produced by *Azospirillum brasilense* and their effect on the growth of pearl millet (*Pennisetum americanum* L.). Appl Environ Microbiol 37(5):1016–1024

Timmusk S, Nicander B, Granhall U, Tillberg E (1999) Cytokinin production by *Paenibacillus polymyxa*. Soil Biol Biochem 31(13):1847–1852

Tsavkelova EA, Klimova SY, Cherdyntseva TA, Netrusov AI (2006) Microbial producers of plant growth stimulators and their practical use: a review. Appl Biochem Microbiol 42(2):117–126

Tsukada K, Takahashi K, Nabeta K (2010) Biosynthesis of jasmonic acid in a plant pathogenic fungus, *Lasiodiplodia theobromae*. Phytochemistry 71(17):2019–2023

Ueda J, Miyamoto K, Kamisaka S (1994) Separation of a new type of plant growth regulator, jasmonates, by chromatographic procedures. J Chrom A 658(1):129–142

Ulvskov P, Marcussen J, Seiden P, Olsen CE (1992) Immuno affinity purification using monoclonal antibodies for the isolation of indole auxins from elongation zones of epicotyls of red-light-grown Alaska peas. Planta 188(2):182–189

Unyayar S, Topcuoglu SF, Unyayar A (1996) A modified method for extraction and identification of Indole-3-Acetic Acid (IAA), Gibberellic Acid (GA3), Abscisic Acid (ABA) and zeatin produced by *Phanerochaete chrysosporium* ME446. Bulg J Plant Physiol 22(3–4):105–110

Uthandi S, Karthikeyan S, Sabarinathan KG (2010) Gibberellic acid production by *Fusarium fujikuroi* SG2. JSIR 69(3):211–214

Vikram A, Hamzehzarghani H, Alagawadi AR, Krishnaraj PU, Chandrashekar BS (2007) Production of plant growth promoting substances by phosphate solubilizing bacteria isolated from Vertisols. J Plant Sci 2(3):326–333

Wasternack C, Kombrink E (2009) Jasmonates: structural requirements for lipid-derived signals active in plant stress responses and development. 5(1):63–77. doi:10.1021/cb900269u

Weber H, Vick BA, Farmer EE (1997) Dinor-oxo-phytodienoic acid: a new hexadecanoid signal in the jasmonate family. Proc Natl Acad Sci 94(19):10473–10478

Weiler EW (1981) Radioimmunoassay for pmol-quantities of indole-3-acetic acid for use with highly stable [125I]-and [3H] IAA derivatives as radiotracers. Planta 153(4):319–325

West CA (1960) Gibberellins as native plant growth regulators. Annu Rev Plant Phys 11(1):411–436

Wightman F, Lighty DL (1982) Identification of phenylacetic acid as a natural auxin in the shoots of higher plants. Physiol Plant 55(1):17–24

Yalpani N, Altier DJ, Barbour E, Cigan AL, Scelonge CJ (2001) Production of 6-methylsalicylic acid by expression of a fungal polyketide synthase activates disease resistance in tobacco. Plant Cell 13(6):1401–1410

Yürekli F, Yesilada O, Yürekli M, Topcuoglu SF (1999) Plant growth hormone production from olive oil mill and alcohol factory waste waters by white rot fungi. World J Microb Biot 15(4):503–505

Zahir ZA, Arshad M, Frankenberger WT (2003) Plant growth promoting rhizobacteria: applications and perspectives in agriculture. Adv Agron 81:97–168

Zamioudis C, Mastranesti P, Dhonukshe P, Blilou I, Pieterse CM (2013) Unraveling root developmental programs initiated by beneficial *Pseudomonas* spp. bacteria. Plant Physiol 162(1):304–318

Zhang Y, Schwarz S, Saedler H, Huijser P (2007) SPL8, a local regulator in a subset of gibberellin-mediated developmental processes in *Arabidopsis*. Plant Mol Biol 63(3):429–439

Zimmer W, Roeben K, Bothe H (1988) An alternative explanation for plant growth promotion by bacteria of the genus *Azospirillum*. Planta 176(3):333–342

Azospirillum sp. as a Challenge for Agriculture

Artenisa Cerqueira Rodrigues, Aurenivia Bonifacio, Fabio Fernando de Araujo, Mario Andrade Lira Junior and Marcia do Vale Barreto Figueiredo

Abstract Several processes mediated by soil microorganisms play an important role in nutrient cycling. One such process is biological nitrogen fixation (BNF) by representatives of various bacterial phylogenetic groups, which are called diazotrophs. Most studies of the *Azospirillum*-plant association have been conducted on cereals and grasses. Currently, 17 species of *Azospirillum* have been described. However, a great diversity of these bacteria continues to be revealed, and little is known of the potential applications of the many species that have been described. The *Azospirillum*-plant association begins with the adsorption and adherence

A.C. Rodrigues
Department of Agricultural Engineering and Soil Science, Federal University of Piaui (UFPI), Center of Agrarian Science, Campus of Socopo, Teresina-PI, Brazil
e-mail: artenisacerqueira@ufpi.edu.br

A. Bonifacio
Department of Biological Science, Federal University of Piaui, Center of Natural Science, Teresina-PI, Brazil
e-mail: bonifacio.a@live.com

F.F. de Araujo
Faculty of Agrarian Sciences, University of Oeste Paulista (UNOESTE), Presidente Prudente-SP, Brazil
e-mail: fabio@unoeste.br

M.A.L. Junior
Department of Agronomy, Agricultural University of Pernambuco (UFRPE), Recife-PE, Brazil
e-mail: mario.lira@depa.ufrpe.br

M. do Vale Barreto Figueiredo (✉)
Agronomical Institute of Pernambuco IPA/SEAGRI, 1371, Gen. San Martin Avenue, Recife-PE 50761-000, Brazil
e-mail: mbarreto@elogica.com.br

D.K. Maheshwari (ed.), *Bacterial Metabolites in Sustainable Agroecosystem*, Sustainable Development and Biodiversity 12, DOI 10.1007/978-3-319-24654-3_2

process of these bacteria in roots. Involved in these processes is the recognition of bacterial polysaccharides by the host plant, a step that is necessary in successfully forming a positive relationship between roots and *Azospirillum*. The presence of *Azospirillum* in the rhizosphere can minimize the susceptibility to diseases caused by plant pathogens. Furthermore, the ability to produce phytohormones, mainly auxins (indole-3-acetic acid) and other molecules from secondary metabolism has been suggested to underlie the growth response to inoculation by *Azospirillum* species. These positive aspects of *Azospirillum* colonization in the roots are also responsible for the alleviation of plant stress. For all of the above-mentioned reasons, *Azospirillum* are also widely used as commercial inoculants, resulting in a significant economic impact in crop yields in many countries. In fact, solid and liquid formulations containing *Azospirillum* are marketed in various countries, such as Brazil, Argentina, Mexico, Italy, France, Belgium, Africa, Germany, Pakistan, Uruguay, India and the USA. In addition, new formulations containing *Azospirillum*, such as polymeric inoculants (alginate, agar, chitosan and gum), are already used for the improvement of many crops. This chapter summarizes the positive effects of *Azospirillum*-plant interactions and their biological importance for the improvement of agriculture worldwide.

Keywords Azospirillum · Biological nitrogen fixation · Biocontrol · Phytohormones · IAA

1 Introduction

Brazil has a long tradition of research with nitrogen-fixing species, and the attention to nitrogen-fixing *Azospirillum* species increased after their rediscovery by Dobereiner and Dias in the year 1976. *Azospirillum* is one of the best-studied plant growth-promoting rhizobacteria (PGPR) that are normally associated with grasses, rice, wheat and sugarcane (Bashan and De-Bashan 2010; Babalola and Glick 2012; Duca et al. 2014; Glick 2014). Presently, 17 species of *Azospirillum* have been described (in order of discovery): *A. lipoferum*, *A. brasilense*, *A. amazonense*, *A. halopraeferens*, *A. irakense*, *A. largimobile*, *A. doebereinerae*, *A. oryzae*, *A. melinis*, *A. canadense*, *A. zeae*, *A. rugosum*, *A. picis*, *A. thiophilum*, *A. formosense*, *A. fermentarium*, *A. humicireducens* and *A. himalayense* (http://www.bacterio.net/azospirillum.html). Of these, *A. brasilense* and *A. lipoferum* are the most studied and well described.

Azospirillum strains are marketed in various countries, such as Brazil, Argentina, Mexico, Italy, France, Australia, Pakistan, Germany, the USA, Africa, Belgium, India and Uruguay (Hungria et al. 2010; Reis et al. 2011; Mehnaz 2015), mainly as microbial formulations with other microorganisms. The major visual effects of inoculation with *Azospirillum* are changes in root morphology that results in an increase in root elongation, the number of lateral and adventitious roots and the lengthening and branching of root hairs (Bashan and Levanony 1985;

Okon 1985; Baldani and Döbereiner 1986; Okon et al. 1988, 1991; Okon and Labandera-Gonzalez 1994; Cohen et al. 2015). These effects on root morphology permit the roots to take up more water and mineral nutrients, which leads to an increase in plant growth (Fibach-Paldi et al. 2012). Moreover, *Azospirillum* might help plants survive under stressful situations due to the induction of changes in cell wall elasticity, osmotic adjustments and the release of beneficial substances (Richardson et al. 2009; Groppa et al. 2012).

Several PGPR inoculants currently commercialized seem to promote growth through at least one mechanism: suppression of plant disease (bioprotectants), improvement of nutrient acquisition (biofertilizers), or phytohormone production (biostimulants) (Tenuta 2003; Mitter et al. 2013). Understanding the interaction between the consortium of microbial inoculants and plant systems will enable growers to harness more benefits from microbial inoculants for improving plant growth and yield (Raja et al. 2006; Sivasakthivelan and Saranraj 2013). The use of microorganisms with the aim of improving nutrient availability for plants is an important practice and proved necessary for agriculture (Babalola 2010; Figueiredo et al. 2010; Araujo et al. 2011; Rodrigues et al. 2013a, b; Bashan et al. 2014).

2 *Azospirillum*: Involvement of Polysaccharides in Attachment

Azospirillum sp. is a Gram-negative diazotrophic rhizobacteria associated with plant roots (mainly grasses). Species of *Azospirillum* exhibit chemotaxis toward a variety of root exudates like amino acids, organic acids, sugars and aromatic compounds (Okon et al. 1980; Rodriguez-Navarro et al. 2007). Chemotaxis is a widespread function in motile soil bacteria, because it affords cells with the ability to sense and navigate toward the most favorable niches for growth and represents an important attribute for plant-microbe association (Carreño-López et al. 2009; Alexandre 2010). The chemotactic response of *Azospirillum* species toward root exudates represents the first stage in *Azospirillum* colonization of root plants and is followed by attachment (Schelud'ko et al. 2009; Wisniewski-Dyé et al. 2013). Although attachment is already known, the precise mechanism that rules the attachment process remains unexplained due to its great complexity (Jofré et al. 2009; Richardson et al. 2009).

The attachment of *Azospirillum* sp. to plant roots is necessary for the formation of an active association and seems to occur in two distinct and consecutive phases (Belyakov et al. 2012; Fibach-Paldi et al. 2012); (i) adsorption and (ii) anchoring phases. In the adsorption phase, a weak binding occurs between the bacteria and the root cells mediated by the polar flagellum (Rodriguez-Navarro et al. 2007). *Azospirillum* produces one longer polar flagellum and several shorter peritrichous flagella (Fibach-Paldi et al. 2012). The *Azospirillum* polar flagellum is an important component of cell motility and is divided into a basal body, hook and filament (Lerner et al. 2010; Belyakov et al. 2012). The filament is composed of numerous identical flagellin molecules, a protein with the C- and N-terminal conserved and a variable middle region (Belyakov et al. 2012). The variability of middle region of the flagellin is related to their antigenic and adhesive properties and the flagellin glycosylation mediates the symbiosis with eukaryotic organisms (Iwashkiw et al. 2013; Merino and Tomás 2014). In cooperation with the flagellin protein, capsular polysaccharides (CPS) are also involved in the adsorption phase of attachment of the *Azospirillum* to the root surface (Lerner et al. 2010). It is reported that *Azospirillum* secretes CPS, and this polysaccharide seems to mediate the adhesion of bacteria to surfaces (Dutta and Podile 2010). CPS is a type of external polysaccharide bound to the outer membrane by a covalent bond (Wisniewski-Dyé et al. 2013). In the CPS of the *A. brasilense* Sp7, a glycosylated lectin with a molecular mass of 36 kDa was identified with specificity to L-fucose and D-galactose (Sigida et al. 2013). Lectins are sugar-binding proteins that can specifically and reversibly recognize and bind to carbohydrates present on the plant root surface. An outer membrane lectin of 67-kDa and produced by *A. brasilense* Sp7 could be involved in adhesion processes (Mora et al. 2008).

Azospirillum strains produce various lectin types, and are possibly involved in *Azospirillum* cell adhesion to the root surface (Alenkina et al. 2014a). The diversity and complexity of these lectins, probably due to the high pleiotropic capacity of *Azospirillum*, ensures their adaptation to different host plants (Mora et al. 2008). Lectins produced by *A. brasilense* induce several signaling systems in wheat roots as part of the recognition in the initial stages of development of plant-bacteria association following the ligand-receptor interaction principle (Fig. 1). However, specific receptors present in the *Azospirillum* cell surface can bind to

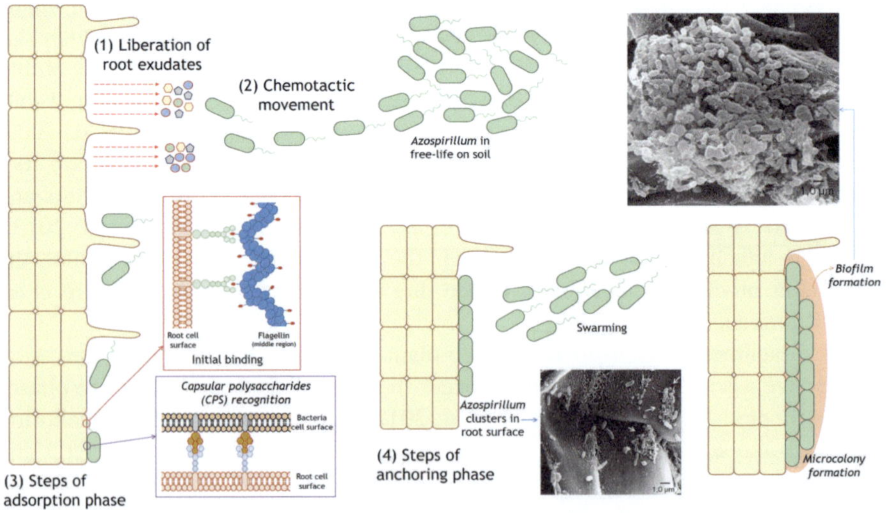

Fig. 1 The steps of attachment of *Azospirillum* sp. to plant roots (the scanning electron micrographs (SEM) are roots colonized by *A. brasilense* and *arrows* indicate zones with granular-like material; SEM from Guerrero-Molina et al. (2012))

the lectins present in the plant root surface (Scheludko et al. 2009). The binding of lectin WGA (wheat germ agglutinin) to cell receptors of *A. brasilense* Sp245 alters bacterial cell metabolism and acts as a signal molecule in the *Azospirillum*-plant association (Aleńkina et al. 2014b).

The second phase of attachment of *Azospirillum* sp. to the plant root surface is mediated by exopolysaccharides (EPS), a polysaccharide weakly associated with the outer cell membrane or totally released into the extracellular medium (Wisniewski-Dyé et al. 2013). In the anchoring phase, the second stage of the attachment process, the *Azospirillum* becomes irreversible and firmly attached to the root surface and other bacteria that are also entrapped, forming clusters at the attachment site (Winik et al. 2009; Guerrero-Molina et al. 2012). The EPS mediates anchoring due to their involvement in the cell-to-cell aggregation phenomenon and its special interaction with bacterial envelope components (Mora et al. 2008). The EPS composition seems to be an important determinant for aggregation ability in *Azospirillum* strains (Fibach-Paldi et al. 2012). This observation is sustained by the fact that in the aggregation phase, the dominant sugar in the EPS structure of *A. brasilense* is L-arabinose (Bahat-Samet et al. 2004; Mora et al. 2008).

Together with CPS and EPS, the lipopolysaccharides (LPS) of *Azospirillum* contribute to the bacteria-plant association due to their responsibility in the immunospecificity of the bacterial cell and thus are involved in direct interactions with plants (Molinaro et al. 2009; Fedonenko et al. 2013; Sigida et al. 2013). LPS is a glycoconjugate present in the cell-surface of *Azospirillum* strains and exhibits a lipid moiety (lipid A), which anchors the molecule in the membrane, and a chain of oligosaccharides covalently linked to lipid A (Sigida et al. 2013; Shelud'ko et al. 2014). The polysaccharide portion includes a central oligosaccharide (core) and an *O*-polysaccharide moiety (OPS) (Fedonenko et al. 2011). The CPS and LPS show marked structural differences, and this is related to the involvement of these structures in different stages of the *Azospirillum*-plant association (Smol'kina et al. 2010; Fedonenko et al. 2011; Shelud'ko et al. 2014).

The OPS structure present in *Azospirillum* LPS is responsible for serological cross-reactions and the basis for the classification into a certain serogroup (Fedonenko et al. 2013). *Azospirillum* strains are divided into three serogroups (Shelud'ko et al. 2014). *Azospirillum* from serogroup I possess a linear homopolymeric D-rhamnan OPS (Boyko et al. 2012), while a heteropolysaccharide OPS occurs in serogroup II (Konnova et al. 2008) that is precipitated with LPS antibodies of *A. brasilense* Sp7 (Sigida et al. 2014). In serogroup III, OPS is composed of a main chain with an oligosaccharide motif formed by three L-rhamnose residues linked with a side chain formed by a D-glucose homopolysaccharide (Fedonenko et al. 2011). The differences in OPS structures are related with host recognition (Fedonenko et al. 2013). In fact, *Azospirillum* strains of serogroup I are usually encountered in association with wheat, while strains of serogroup II and III have an association with other gramineous plant (Sigida et al. 2014).

Azospirillum cells attached to plant roots exhibit a rounded and swollen format, similar to a cyst, and are metabolically active in the rhizosphere (Hou et al. 2014).

After *Azospirillum* cells attachment, the swarming occurs, a process by which a bacterial group rapidly advances on surfaces until specific sites and colonize in a coordinated manner (Verstraeten et al. 2008). In fact, swarming enables the rapid colonization of host tissues and formation of microcolonies which occurs simultaneously with biofilm formation (Bogino et al. 2013). Biofilm is a bacterial aggregate associated with a surface, typically enclosed in an extracellular matrix, and probably composed of microbial EPS (Kadouri et al. 2003; Monds and O'Toole 2009). *Azospirillum brasilense* is reported as a biofilm producer (Guerrero-Molina et al. 2012), and this feature helps *Azospirillum* cells anchor and colonize the root surface (Winik et al. 2009). Overall, swarming and biofilm formation across the roots is important for long-term colonization.

3 *Azospirillum* in the Biological Control of Pathogens

Azospirillum brasilense have been suggested as plant-growth promoting bacteria (Bashan and Holguin 1998). The bacteria of the genus *Azospirillum* inhabit the plant's rhizosphere and sometimes develop an endophytic relationship with the host plant. However, in most cases, their relationship is associative, with a partial supply of nitrogen fixed by the microbial process, known as atmospheric nitrogen biological fixation, and a chemical reaction catalyzed by a dehydrogenase enzyme (Hungria et al. 2010). The success of the relationship also depends on the ability of the bacteria to colonize the host plant rhizosphere, although no consistent results demonstrate specificity in the bacteria-host plant relationship.

Even though the genus *Azospirillum* is not considered directly related to biological control, some reports in the literature show results of moderate biological control in some diseases, such as galls caused by *Agrobacterium tumefaciens* and leaf and vascular diseases caused by bacteria. In addition, research has also shown the genus to inhibit the growth of non-pathogenic microorganisms in the rhizosphere of plants (Somers et al. 2005). The mechanism of action responsible for the benefits found in the growth of plants has also been studied in every bacterial genus of interest, which discusses, among other things, the participation of different microbial molecules involved in such mechanisms (Kloepper et al. 2004). The presence of *A. brasilense* can change plant physiology, especially the production of several phytohormones, such as auxin, gibberellin, cytokinin and ethylene (Dobbelaere et al. 1999).

Some chemical compounds produced by bacteria (when in contact with the plants), such as those of the genus *Azospirillum*, can interfere with plant metabolism and the active frequency of the plant defense. The plant defense system is controlled by different metabolic pathways that are triggered by various chemical factors associated with the presence of chemical and biological inducers (Kuc 1983). The bacterial effect, when present in the plants rhizosphere, can be observed in two specific routes of metabolic defense in plants. These routes include jasmonic and salicylic acid, of which the latter is most often associated with chemical inducer agents or pathogens (Romeiro 2000). With regard to bacteria that promote plant growth, the induced systemic resistance promotion (ISR) is more associated within their mechanism of action, which reinforces the positive effects of these bacteria in promoting growth or biological control of pathogens (Mariano and Kloepper 2000). The ISR, when involved in the action of non-pathogenic microorganisms in the rhizosphere, does not involve the salicylic acid signaling pathway or induction of proteins related to pathogenesis because it is activated in this resistance-signaling pathway of jasmonic acid and ethylene (Hoffland et al. 1995; Pieterse et al. 1998). When plant beneficial bacteria colonize the root system, the constituent molecules of the bacterial cell, or those synthesized by it, elicit a biochemical signal. This signal is translocated to distant sites, which activates genes that code for the synthesis of dynamic resistance components and thus induces the expression of a systemic resistance (Romeiro 2000).

Romero et al. (2003) demonstrated that the activity of phytohormones in two strains of *Azospirillum* sp. promoted growth in tomatoes and modified the plant susceptibility to bacterial diseases. The success of this interaction depends on the plant genotype as well as the pathogen characteristics. In a review of root diseases in cassava, it was determined that the inoculation of *A. brasilense* reduced the incidence rates of disease/infection compared to the non-inoculated controls. This species can restrict the proliferation of other non-pathogenic bacteria in the rhizosphere and bacteria, such as *Pseudomonas syringae* pv. *tomato* (Bashan and De-Bashan 2002), which probably occurs due to competition or the resistance induction phenomenon in the host. Biological control reports provided by *A. brasilense* are related to growth inhibition of *Agrobacterium tumefaciens* and phytopathogenic fungi. This inhibitory activity may be related to compounds detected in the supernatant of *Azospirillum* sp. during growth (Fig. 2); however, the mode of action is not well defined (Somers et al. 2005). According to these authors, the

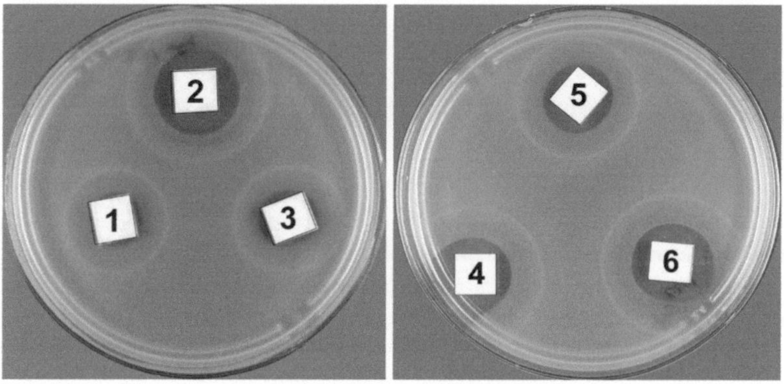

Fig. 2 Determination of the antimicrobial activity of supernatant extracts (10 μL) isolated from *A. brasilense* Sp245 cultures grown in MMAB supplemented with 0.5 mM tyrosine, (*1*) 0.5 mM phenylalanine, (*2*) 0.5 mM tryptophan, (*3*) 0.5 mM phenylpyruvate, (*4*) 0.5 mM prephenate, (*5*) or 0.5 mM chorismate (*6*) by the paper disk method (Somers et al. 2005)

antimicrobial compound produced by *Azospirillum* sp. was identified as phenyl acetic acid of the auxins group. This molecule has been used as a defense mechanism in bacterial competition in its ecological niche. The presence of hormones can alter both plant growth and influence the metabolism and survival of microorganisms in the environment.

Microbial iron chelators known as siderophores are produced by *A. brasilense* and showed antifungal activity in vitro against *Colletotrichum acutatum* M11 strain. Siderophores are molecules that sequester iron from the environment; thus, this essential metal for microbial growth can be crucial in causing deficiency in the growth of certain microbial species. A reduction of anthracnose symptoms in strawberries previously inoculated with *A. brasilense* was also observed. These results suggest the use of bacteria for disease control strategies in the strawberry tree (Tortora et al. 2011). Some chemical compounds (not yet identified) produced by *A. brasilense* sp245 promoted, under controlled conditions, the reduction of mycelial growth of *Rhizoctonia solani* (Russo et al. 2008; Vettori et al. 2010). In another study, chemical inhibitors of fungal growth were also characterized as volatile, and after extracting the growth supernatant of *A. brasilense*, these substances reduced the growth of *Fusarium graminearum* (after addition to the fungal growth medium culture) (Abdulkareem et al. 2014).

With regard to the benefits provided by *Azospirillum* sp. for the biological control of diseases, at this stage, it is still not clear which mechanism of action is critical for the success of this activity. The production of hormones, siderophores and the ability to fix atmospheric nitrogen guarantee benefits and increased competence of *A. brasilense* in colonization and its ability to remain in the rhizosphere of plants. This can promote plant growth and reduce the presence of pathogenic microorganisms in the surrounding soil. Furthermore, the presence of these bacteria in the rhizosphere confers several benefits for plant growth and nutrition, which indirectly increases the resilience capacity.

4 Mechanisms by Which *Azospirillum* Affects Plant Growth: Hormones and Metabolites

Plant-microbe interactions are affected by many different regulatory signals, and the root exudates stand out among them (Spaepen et al. 2009). The root exudates play a key role in the plant-microbe interaction, stimulating the bacterial chemotaxis and mediating the root colonization and the selection of microorganisms driven by the host (Mitter et al. 2013). Therefore, root exudates play an important role in developing microbial communities in the different compartments of plants (Fibach-Paldi et al. 2012). As compensation to root exudates secreted by plants, the microorganism plant association may improve plant growth and health by synthesis of vitamins, antibiotics, enzymes and phytohormones (Cohen et al. 2015). Phytohormones are organic substances that at a very low concentration stimulate a physiological response. Currently, in addition to the five classic plant hormones (auxins, cytokinins, gibberellins, abscisic acid and ethylene), other phytohormones have been identified, such as jasmonate, brassinosteroid, nitric oxide and strigolactone (Shan et al. 2012).

Azospirillum strains have been reported to increase plant growth through the action of carbohydrates, polyamines, amino acids, peptides, lectins and enzymes that are released in the extracellular medium (Cassán et al. 2009a; Richardson et al. 2009; Bashan and De-Bashan 2010). However, many authors report that *Azospirillum* species are able to enhance plant growth due to the self-production of hormone and by inducing synthesis of these compounds in the plant tissues (Chamam et al. 2013; Duca et al. 2014; Cohen et al. 2015). In general, the phytohormones works in complex networks that include responses in cross talk and feedback, and therefore, it is difficult to establish the specific role of a given hormone in the plant response (Glick 2014). *Azospirillum* species are able to produce and secrete phytohormones, mainly auxins, gibberellins (GAs), cytokinins (CK) and nitric oxide, which act as signals and effectors for plant growth promotion (Spaepen et al. 2008; Bashan and De-Bashan 2010; Couillerot et al. 2013; Duca et al. 2014).

Several studies have reported the presence of auxins in the supernatant of *Azospirillum* cultures (Cassán et al. 2014; and references therein). Quantitatively, indole-3-acetic acid (IAA) seems to be the most important auxin produced by *Azospirillum* (Glick 2014; Mehnaz 2015); however, some reports suggest that indole-3-butyric acid (IBA) is also largely produced (Couillerot et al. 2013). In accordance with Duca et al. (2014), IBA probably serves as an important source and reserve of IAA in *Azospirillum* strains. *Azospirillum* sp. produce IAA during all growth stages (Malhotra and Srivastava 2009) and four pathways exist for IAA biosynthesis (Duca et al. 2014): three tryptophan-dependent pathways [indole-3-pyruvic acid (IPA), indole-3-acetamide (IAM) and tryptamine (TAM) pathways] and one tryptophan-independent pathway (Fig. 3). The IPA pathway is of major significance in *Azospirillum* and provides 90 % of the IAA synthesized (Glick 2014).

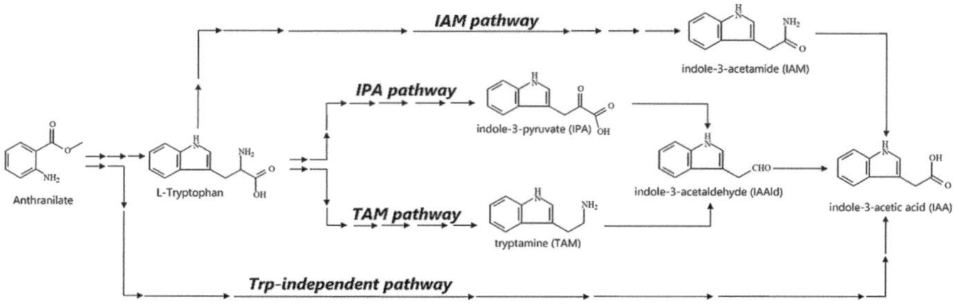

Fig. 3 Indole-3-acetic acid (*IAA*) pathways identified in *Azospirillum* species: the indole-3-acetamide (*IAM*), indole-3-pyruvic acid (*IPA*) and tryptamine (*TAM*) pathway (tryptophan-dependent pathways) and a tryptophan-independent pathway starting with anthranilate (Cassán et al. 2014; Duca et al. 2014)

In the plant-*Azospirillum* interaction, IAA can also be a reciprocal signaling molecule that maintains the symbiotic relationship (Malhotra and Srivastava 2009) and has probably been shaped by co-evolutionary processes between the bacteria and their host plant (Berg 2009; Walker et al. 2011, 2012; Drogue et al. 2012). In fact, some evidence indicates that phytohormone synthesis in *A. brasilense* is strain-specific (Di Salvo et al. 2014). These authors studied the Sp7, Cd, Az39, 40 and 42 strains of *A. brasilense* and reported that the 42M strain displayed higher levels of IAA production than the other strains. Many authors have reported the positive effects of auxin synthesized by *Azospirillum* in plants (Malhotra and Srivastava 2009; Glick 2014 and references therein). IAA regulates the plant cell cycle, tropism, apical dominance and senescence (Mehnaz 2015). In addition, under environmental fluctuations and nutrient limitations, specifically nitrogen, carbon and phosphorus, the IAA levels are increased (Malhotra and Srivastava 2009).

IAA produced by *A. brasilense* alters the root morphology and proliferation in wheat seedlings (Dobbelaere et al. 2001), and is responsible for increases in root and shoot systems. In fact, the inoculation of sugarcane with *Azospirillum* sp. results in significant increases in root dry weight (upper 70 %; Moutia et al. 2010), and wheat inoculated with *A. brasilense* exhibit increases in shoot growth (Spaepen et al. 2008). In addition, IAA affects photosynthesis, the biosynthesis of metabolites and other phytohormones, such as CK and GAs (Ilyas and Bano 2010; Mehnaz 2015). Tien et al. (1979) documented the first report of CK production in *A. brasilense* in 1979; however, little is currently known about the CK produced by *Azospirillum* (Cássan et al. 2014). In plants, CK regulate cell division, and have been associated with shoot and root morphogenesis (Spaepen et al. 2009). Zeatin, the major CK type, has been reported in *A. brasilense* and *A. lipoferum* (Molina-Favero et al. 2007; Esquivel-Cote et al. 2010).

GAs are one class of phytohormones produced and secreted by *Azospirillum* (Mehnaz 2015). GAs are complex compounds composed of terpenes that share a common GA ring (Yamaguchi 2008). Currently, several GA types have been identified; of these, GA_1, GA_3, GA_4 and GA_7 are the types that show functions of phytohormones, and therefore, regulate different aspects of plant growth (Cassán et al. 2014). *Azospirillum* species, specifically *A. brasilense* and *A. lipoferum*, are known to produce GA_1 and GA_3 and GA_3 is the major type of GA identified (Bottini et al. 1989; Jansen et al. 1992; Piccoli and Bottini 1996; Lucangeli and Bottini 1997; Ilyas and Bano 2010). Manivannan and Tholkappian (2013) recorded a production of GA up to 3.3 μg (per 25 mL^{-1} broth) in 20 different *Azospirillum* strains isolated from the tomato rhizosphere.

Although it has been known for a long time that *Azospirillum* synthesizes and metabolizes GAs (Bottini et al. 1989), their mechanism of production is poorly known (Mehnaz 2015). Lucangeli and Bottini (1997) were the first to describe the capacity of *Azospirillum* sp. that produce GAs in plants. GAs promote cell division and the elongation of primary roots, and play an important role in lateral root development (Bottini et al. 2004). In maize, GAs promote shoot elongation and growth and increase root hair abundance (Fulchieri et al. 1993). In rice inoculated with *A. lipoferum*, GA improves nitrogen uptake and increases the dry mass, height and yield (Bottini et al. 2004). In maize plants treated with an inhibitor of GA biosynthesis, results show that GAs produced by *Azospirillum* positively affects plant growth (Lucangeli and Bottini 1997). A study in maize treated with prohexadione (GA biosynthesis inhibitor) by foliar spraying and inoculated with *A. lipoferum* reported an increase in root growth (Cohen et al. 2009).

Data from studies with GA-deficient plants with the dwarf phenotype show that GAs of *A. brasilense* and *A. lipoferum* were responsible for reversal of dwarfism in maize and rice (Cassán et al. 2001). Another important role of GA is the interruption of dormancy during seed germination because its hydrolytic enzymes, α-amylase and protease, are induced in seeds of grasses and cereals, and this facilitates endosperm mobilization (Mehnaz 2015). The seeds of soybean and wheat exhibit an increase in germination when treated with *A. brasilense*, and this response seems related to the high GA production of *A. brasilense* (Bacilio et al. 2003; Cassán et al. 2009b). Furthermore, environmental factors can modify the GA production by *Azospirillum* (Cassán et al. 2014). In this sense, Piccoli et al. (1999) showed that the availability of O_2 and the osmotic potential reduces the GA_3 production in *A. lipoferum* (~50 %), and this response was considered a compensatory mechanism that seems to be activated in water stressed situations.

GAs and abscisic acid (ABA), produced by *Azospirillum* strains, seem to contribute to water stress alleviation in plants (Cohen et al. 2009, 2015) and in plant defense mechanisms (Vacheron et al. 2013). ABA is a phytohormone induced in response to environmental stress, such as water or salt stress (Bauer et al. 2013). The inoculation of maize plants with *A. lipoferum* enhances ABA levels and plant tolerance to drought (Cohen et al. 2009). Likewise, the synthesis of ABA by *Azospirillum* species increases when sodium chloride (commonly used to mimic salt stress) is added to the culture medium (Cohen et al. 2008; Dodd et al. 2010; Cohen et al. 2015). Ilyas and Bano (2010) report that *Azospirillum* strains isolated from water stressed conditions exhibited higher production of ABA. Moreover, *Arabidopsis* seedlings doubled ABA levels when inoculated with *A. brasilense* sp245 (Cohen et al. 2008). These results reinforce idea of a protective role for ABA synthesized by *Azospirillum* (Fig. 4).

Plant tolerance to environmental stress mediated by *Azospirillum* may involve ABA or compatible solutes like proline, polyamines and trehalose (Richardson et al. 2009; Cohen et al. 2015). *Azospirillum* sp. are known to produce polyamines and amino acids in culture media (Cassán et al. 2009a; Bashan and De-Bashan 2010). Spermine, spermidine, putrescine and cadaverine are organic polymers generically named polyamines and are related to root growth and stress mitigation in plants (Gupta et al. 2013). In rice seedlings inoculated with *A. brasilense*, at least in part, cadaverine production induces root growth and mitigates osmotic stress in rice (Cassán et al. 2009a). Maize plants treated with *A. brasilense* modified to overproduce trehalose were more resistant to drought and improved biomass production more than plants treated with a wild type of *A. brasilense* (Rodríguez-Salazar et al. 2009). In addition, in their hosts, *Azospirillum* sp. induce the biosynthesis of phenylacetic acid, bacteriocins and siderophores, which are secondary metabolites with antimicrobial activity (Walker et al. 2011; Vacheron et al. 2013).

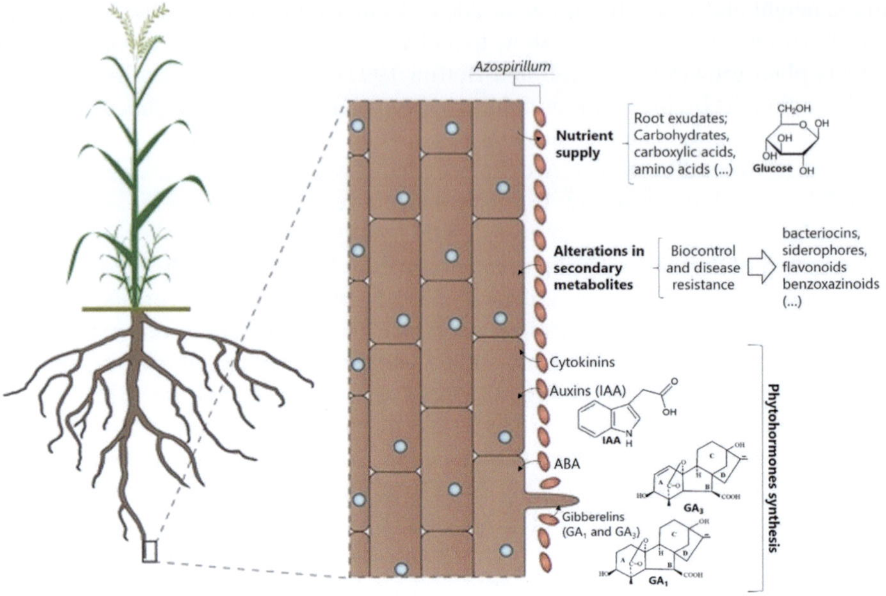

Fig. 4 The positive effects of *Azospirillum* in plant roots. The plants provide nutrients while *Azospirillum* can promote phytohormone synthesis and positive alterations in secondary metabolites. *IAA* Indole-3-acetic acid. *ABA* Abscisic acid

The inoculation of *Azospirillum* modifies the content of secondary metabolites in the host plant, and in accordance with Walker et al. (2011), these alterations are more intense than that in primary metabolism. Alterations in the secondary metabolite profile were reported in rice plants in response to *Azospirillum* inoculation, and the major metabolites affected were flavonoids and hydroxycinnamic derivatives (Chamam et al. 2013). In maize inoculated with *Azospirillum*, variations in benzoxazinoids were registered, which are molecules related to plant resistance against pathogens in both roots and shoots (Walker et al. 2012). These molecules are elicited in the host plants when inoculated with *Azospirillum* and are responsible for disease resistance. Overall, considering the beneficial interactions between plants and *Azospirillum*, their use as an inoculant represents an environmentally friendly strategy for agriculture worldwide.

5 *Azospirillum*: Inoculants and New Products

Plant biostimulants include both substances and microorganisms that enhance plant growth, and the global market is estimated at over 2 million dollars by 2018; however, the definition is still evolving, mostly due to its large scope. Both the American and the European definitions include several potential effects, but specify that these products may not establish nutritional guarantees, although they may increase fertilizer efficiency (Calvo et al. 2014). This broad class of products is especially important for agricultural systems in which external inputs are limited, due to philosophical, ecological, or economic reasons (Bhardwaj et al. 2014).

Bacteria-based inoculants used as plant biostimulants are usually classified as PGPR and include several different mechanisms ranging from BNF to plant protection from pathogens (Maheshwari 2010; Bhattacharyya and Jha 2012; Ahemad and Kibret 2014; Nadeem et al. 2014). One of the major groups of bacteria-based inoculants is collectively known as biofertilizers, and these bacteria improve the nutrient status of the crops through BNF, increasing nutrient availability and/or root growth promotion, and they usually work through a combination of benefits that are hard to experimentally differentiate (Perez-Montano et al. 2014). One of the most widely studied genera of biofertilizers is *Azospirillum* (Laslo et al. 2012; Almaghrabi et al. 2013; Reddy and Saravanan 2013; Nadeem et al. 2014; Perez-Montano et al. 2014).

Inoculants based on *Azospirillum* have already been commercialized in Mexico, France, India and Brazil (Hungria et al. 2010; Sivasakthivelan and Saranraj 2013; Trujillo-Roldán et al. 2013) and have been promoted and studied since at least the 1970s, and thorough review published in 1994 (Okon and Labandera-Gonzalez 1994). The authors indicated a 60–70 % success rate using *Azospirillum* inoculants with a typical gain of 5–30 % for several of the most important cereal crops. However, wide variations in their effects are still routinely found in field experiments (Hungria et al. 2010; Lana et al. 2012; Turan et al. 2012; Castañeda-Saucedo et al. 2013; Romero et al. 2014).

These variations happen due to a wide range of reasons, from inoculant production through storage, field use and the several ecological and environmental effects known to affect the results. A major point that should be considered by both inoculant producers and researchers is that the end-user is mostly interested in the product effect on the crop, not on the bacteria being used to achieve that end, its physiology or ecology. Because these effects largely depend on the strain used and its population on the product when applied, these aspects should probably be the main research focus in this field (Sivasakthivelan and Saranraj 2013). Strain selection should be based on a multitier approach, similar to that conducted with rhizobial legumes or the crop breeding industry (Araújo et al. 2012). Usually, a large (preferably very large) initial population is studied under some kind of environment in which potential gains are maximal and stresses reducing those gains are minimal. This is followed by selection cycles with lower strain numbers, but each time more representative of real agricultural usage and ending with field experiments with several cultivars and environmental conditions.

For *Azospirillum*, the first step might be selection under laboratory controlled pure-culture conditions. That *Azospirillum* species fix nitrogen under these conditions is likely why a large part of the older literature attributes to it, at least in part, the field effects of BNF (Dart 1986; Kennedy et al. 1997), although most field studies indicate that its contribution to crop nitrogen status is relatively small (Lana et al. 2012). The next step is generally conducted under some kind of protected environment such as a greenhouse. There are a large number of papers on this condition, which frequently aim to extend the usage of *Azospirillum* strains to new plant species, to investigate possible mechanisms through which the bacteria affect the plant, or to study a complex mixture of beneficial microorganisms (Viera and Fernandez 2006; Bhattacharjee et al. 2008; Bashan and De-Bashan 2010; Hayat et al. 2010; Jha et al. 2013).

After the protected environment studies, field experiments are most commonly conducted with the major cereals due to their relevance, and their variable results were already mentioned earlier in this section. A large component of this variation may be due to several different mechanisms proposed for *Azospirillum* effects not being adequately adjusted in the experimental design and/or in the results and discussion, leading to an apparent lack of response to the inoculation (Naiman et al. 2009; Pedraza et al. 2009; Hungria et al. 2010; Mostafa and Abo-Baker 2010; Yadegari et al. 2010; Trivedi and Bhatt 2011; Lana et al. 2012; Moghadam et al. 2012; Ferreira et al. 2013; Jha et al. 2013; Perez-Montano et al. 2014). This problem has been mentioned since the 1990s (Okon and Labandera-Gonzalez 1994), and a suggestion given by these authors in their review was a market regulation on inoculant producers.

Under Brazilian law, no commercial inoculant may be used in agricultural fields if not recommended by the Agriculture Ministry (Brasil 2011), which currently recommends *A. brasilense* strains (Table 1) for use in the three major cereal crops of Brazil (wheat, corn and rice). These strains may only be recommended after field trials at several different locations and agricultural years. This legal demand induces a much stronger confidence in the inoculant than may be seen otherwise. One point that should also increase the confidence of the crop grower, with regard to *Azospirillum* inoculant usage, is that most cereal producers also grow soybean and that inoculant usage is widespread and continues to this date (Alves et al. 2003; Phillips 2004).

The very strong adoption of rhizobial inoculants in Brazil for soybean, and to a lesser extent for other legume crops (Zilli et al. 2011), indicates that as long as consistent results can be obtained from *Azospirillum* inoculants under field conditions, we can expect future growth of this biological technique. Another point in which the *Azospirillum* inoculants industry may emulate the rhizobia-based

Table 1 Recommended strains for some cereals designed for use in commercial inoculant production in Brazil (Brasil 2011)

Plant species	Bacterial species	Strain
Triticum spp.	*Azospirillum brasilense*	Ab-V1
Zea mays	*A. brasilense*	Ab-V4
Zea mays and *Triticum* spp.	*A. brasilense*	Ab-V5
Zea mays and *Triticum* spp.	*A. brasilense*	Ab-V6
Zea mays	*A. brasilense*	Ab-V7
Triticum spp.	*A. brasilense*	Ab-V8
Oryza sativa	*A. brasilense*	Ab-V5
O. sativa	*A. brasilense*	Ab-V6

industry is in the constant pursuit of new technologies for the inoculant, including application forms and liquid formulations, which have caused increased inoculant usage in Brazil in the recent years (Albareda et al. 2008; Vieira Neto et al. 2008; Zilli et al. 2010; França et al. 2013). It must be mentioned that there is already the literature on these aspects including shelf life of the product and the use of commercial products (Sivasakthivelan and Saranraj 2013; Trujillo-Roldán et al. 2013).

An additional problem that must be dealt with is the inoculant's adaptation to field conditions and practices, such as genetically modified organisms (GMO), herbicide usage and seed coating with pesticides, which are major causes of concern for the rhizobial inoculant industry (Austin et al. 2006; Bunemann et al. 2006; Gaind et al. 2007; Jacques et al. 2010; Zobiole et al. 2011), but have not been well evaluated up to now.

6 Concluding Remarks

Most studies of the *Azospirillum*-plant association have been conducted on cereals and grasses, while only a few other plant families have been investigated. Recent progress on the understanding of their diversity, colonization ability, action mechanisms, formulation and application of these biological systems should facilitate their development as reliable components in the management of sustainable agricultural. Naturally, the mode of root colonization by *Azospirillum* may vary, depending on the bacterial strain, plant species, environmental conditions and other unidentified factors. Furthermore, the principal mechanism by which *Azospirillum* enhances plant growth is undetermined. However, several possible modes of action have been proposed.

In this regard, efforts have been made by researchers to clearly define and develop commercial inoculants using these organisms with special emphasis on formulations and polymeric carriers. Furthermore, combinations of beneficial bacterial strains that interact synergistically are currently being devised, and numerous recent studies show a promising trend in the field of inoculation technology. The future challenge is to identify management conditions that can contribute to the optimization of several mechanisms of the plant-microorganism interrelationship and that may participate in the association and affect plant growth, including N_2 fixation, hormonal effects, general improvement in root growth and major biocontrol activities.

References

Abdulkareem M, Aboud HM, Saood HM, Shibly MK (2014) Antagonistic activity of some plant growth rhizobacteria to *Fusarium graminearum*. Int J Phytopathol 3(1):49–54

Ahemad M, Kibret M (2014) Mechanisms and applications of plant growth promoting rhizobacteria: current perspective. J King Saud Univ Sci 26(1):1–20

Albareda M, Rodríguez-Navarro DN, Camacho M, Temprano FJ (2008) Alternatives to peat as a carrier for rhizobia inoculants: solid and liquid formulations. Soil Biol Biochem 40(11):2771–2779

Aleńkina SA, Bogatyrev VA, Matora LY, Sokolova MK, Chernyshova MP, Trutneva KA, Nikitina VE (2014a) Signal effects of the lectin from the associative nitrogen-fixing bacterium *Azospirillum brasilense* Sp7 in bacterial-plant root interactions. Plant Soil 381(1–2):337–349

Aleńkina SA, Petrova LP, Sokolova MK, Chernyshova MP, Trutneva KA, Bogatyrev VA, Nikitina VE (2014b) Comparative assessment of inductive effects of *Azospirillum* lectins with different antigenic properties on the signal systems of wheat seedling roots. Microbiology 83(3):262–269

Alexandre G (2010) Coupling metabolism and chemotaxis-dependent behaviours by energy taxis receptors. Microbiology 156(8):2283–2293

Almaghrabi OA, Massoud SI, Abdelmoneim TS (2013) Influence of inoculation with plant growth promoting rhizobacteria (PGPR) on tomato plant growth and nematode reproduction under greenhouse conditions. Saudi J Biol Sci 20(1):57–61

Alves BJR, Boddey RM, Urquiaga S (2003) The success of BNF in soybean in Brazil. Plant Soil 252(1):1–9

Araújo ASF, Leite LFC, Iwata BF, Lira Junior MA, Xavier G, Figueiredo MBV (2012) Microbiological process in agroforestry systems: a review. Agron Sustain Dev 32(1):215–216

Araujo FF, Araújo ASF, Figueiredo MVB (2011) Role of plant growth-promoting bacteria in sustainable agriculture. In: Salazar A, Rios I (eds) Sustainable agriculture: technology, planning and management. Nova Science Publishers, New York

Austin AT, Piñeiro G, Gonzalez-Polo M (2006) More is less: agricultural impacts on the N cycle in Argentina. Biogeochemistry 79(1):45–60

Babalola OO (2010) Beneficial bacteria of agricultural importance. Biotechnol Lett 32:1559–1570

Babalola OO, Glick BR (2012) The use of microbial inoculants in African agriculture: current practice and future prospects. J Food Agri Environ 10(3–4):540–549

Bacilio M, Vazquez P, Bashan Y (2003) Alleviation of noxious effects of cattle ranch composts on wheat seed germination by inoculation with *Azospirillum* spp. Biol Fertil Soils 38:261–266

Bahat-Samet E, Castro-Sowinski S, Okon Y (2004) Arabinose content of extracellular polysaccharide plays a role in cell aggregation of *Azospirillum brasilense*. FEMS Microbiol Lett 237:195–203

Baldani VLD, Döbereiner J (1986) Effect of inoculation of *Azospirillum* spp. on the nitrogen assimilation of field grown wheat. Plant Soil 95:109–121

Bashan Y, De-Bashan LE (2002) Protection of tomato seedlings against infection by *Pseudomonas syringae* pv. *tomato* by using the plant growth-promoting bacterium *Azospirillum brasilense*. Appl Environ Microbiol 68:2637–2643

Bashan Y, De-Bashan LE (2010) How the plant growth-promoting bacterium *Azospirillum* promotes plant growth—a critical assessment. Adv Agron 108:77–136

Bashan Y, Holguin G (1998) Proposal for the division of plant growth-promoting rhizobacteria into two classifications: biocontrol-PGPB (plant growth-promoting bacteria) and PGPB. Soil Biol Biochem 30:1225–1228

Bashan Y, Levanony H (1985) An improved selection technique and medium for the isolation and enumeration of *Azospirillum brasilense*. Can J Microbiol 31:947–952

Bashan Y, De-Bashan LE, Prabhu SR, Hernandez JP (2014) Advances in plant growth-promoting bacterial inoculant technology: formulations and practical perspectives (1998–2013). Plant Soil 378:1–33

Bauer H, Ache P, Lautner S, Fromm J, Hartung W, Al-Rasheid KA, Sonnewald S, Sonnewald U, Kneitz S, Lachmann N, Mendel RR, Bittner F, Hetherington AM, Hedrich R (2013) The stomatal response to reduced relative humidity requires guard cell-autonomous ABA synthesis. Curr Biol 23(1):53–57

Belyakov AY, Burygin GL, Arbatsky NP, Shashkov AS, Selivanov NY, Matora LY, Knirel YA, Shchyogolev SY (2012) Identification of an O-linked repetitive glycan chain of the polar flagellum flagellin of *Azospirillum brasilense* Sp7. Carbohyd Res 361:127–132

Berg G (2009) Plant-microbe interactions promoting plant growth and health: perspectives for controlled use of microorganisms in agriculture. Appl Microbiol Biotechnol 84(1):11–18

Bhardwaj D, Ansari MW, Sahoo RK, Tuteja N (2014) Biofertilizers function as key player in sustainable agriculture by improving soil fertility, plant tolerance and crop productivity. Microb Cell Fact 13:66

Bhattacharjee RB, Singh A, Mukhopadhyay SN (2008) Use of nitrogen-fixing bacteria as biofertiliser for non-legumes: prospects and challenges. Appl Microbiol Biotechnol 80(2):199–209

Bhattacharyya PN, Jha DK (2012) Plant growth-promoting rhizobacteria (PGPR): emergence in agriculture. World J Microb Biotechnol 28(4):1327–1350

Bogino PC, Oliva MDLM, Sorroche FG, Giordano W (2013) The role of bacterial biofilms and surface components in plant-bacterial associations. Int J Mol Sci 14(8):15838–15859

Bottini R, Fulchieri M, Pearce DW, Pharis RP (1989) Identification of Gibberellins A1, A3, and iso-A3 in cultures of *Azospirillum lipoferum*. Plant Physiol 90:45–47

Bottini R, Cassán F, Piccoli P (2004) Gibberellin production by bacteria and its involvement in plant growth promotion and yield increase. Appl Microbiol Biotechnol 65(5):497–503

Boyko AS, Dmitrenok AS, Fedonenko YP, Zdorovenko EL, Konnova SA, Knirel YA, Ignatov VV (2012) Structural analysis of the O-polysaccharide of the lipopolysaccharide from *Azospirillum brasilense* Jm6B2 containing 3-O-methyl-d-rhamnose (d-acofriose). Carbohyd Res 355:92–95

Brasil, SDDA-MDAPE (2011) Normative Ruling no.13 of 2011 March 24. Federal Official Gazette—Section 1. National Press, Brasília

Bunemann EK, Schwenke GD, Van Zwieten L (2006) Impact of agricultural inputs on soil organisms—a review. Aust J Soil Res 44(4):379–406

Calvo P, Nelson L, Kloepper JW (2014) Agricultural uses of plant biostimulants. Plant Soil 383(1–2):3–41

Carreño-López R, Sánchez A, Camargo N, Elmerich C, Baca BE (2009) Characterization of chsA, a new gene controlling the chemotactic response in *Azospirillum brasilense* Sp7. Arch Microbiol 191(6):501–507

Cassán FD, Lucangeli CD, Bottini R, Piccoli PN (2001) *Azospirillum* spp. metabolize [17, 17-2H2] gibberellin A20 to [17, 17-2H2] gibberellin A1 in vivo in dry rice mutant seedlings. Plant Cell Physiol 42(7):763–767

Cassán F, Maiale S, Masciarelli O, Vidal A, Luna V, Ruiz O (2009a) Cadaverine production by *Azospirillum brasilense* and its possible role in plant growth promotion and osmotic stress mitigation. Eur J Soil Biol 45:12–19

Cassán F, Perrig D, Sgroy V, Masciarelli O, Penna C, Luna V (2009b) *Azospirillum brasilense* Az39 and *Bradyrhizobium japonicum* E109, inoculated singly or in combination, promote seed germination and early seedling growth in corn (*Zea mays* L.) and soybean (*Glycine max* L.). Eur J Soil Biol 45:28–35

Cassán F, Vanderleyden J, Spaepen S (2014) Physiological and agronomical aspects of phytohormone production by model plant-growth-promoting rhizobacteria (PGPR) belonging to the genus *Azospirillum*. J Plant Growth Regul 33(2):440–459

Castañeda-Saucedo MC, Gómez-González G, Tapia-Campos E, Maciel ON, Pérez JSB, Silva MLR (2013) Efecto de *Azospirillum brasilense* y fertilización química sobre el crecimiento, desarrollo, rendimiento y calidad de fruto de fresa (*Fragaria x ananassa* Duch). Interciencia 38(10):737–744

Chamam A, Sanguin H, Bellvert F, Meiffren G, Comte G, Wisniewski-Dyé F, Bertrand C, Prigent-Combaret C (2013) Plant secondary metabolite profiling evidences strain-dependent effect in the *Azospirillum-Oryza sativa* association. Phytochem 87:65–77

Cohen AC, Bottini R, Piccoli PN (2008) *Azospirillum brasilense* Sp 245 produces ABA in chemically-defined culture medium and increases ABA content in *Arabidopsis* plants. Plant Growth Regul 54(2):97–103

Cohen AC, Travaglia CN, Bottini R, Piccoli PN (2009) Participation of abscisic acid and gibberellins produced by endophytic *Azospirillum* in the alleviation of drought effects in maize. Botany 87:455–462

Cohen AC, Bottini R, Pontin M, Berli FJ, Moreno D, Boccanlandro H, Travaglia CN, Piccoli PN (2015) *Azospirillum brasilense* ameliorates the response of *Arabidopsis thaliana* to drought mainly via enhancement of ABA levels. Physiol Plant 153:79–90

Couillerot O, Ramírez-Trujillo A, Walker V, von Felten A, Jansa J, Maurhofer M, Défago G, Prigent-Combaret C, Comte G, Caballero-Mellado J, Moënne-Loccoz Y (2013) Comparison of prominent *Azospirillum* strains in *Azospirillum-Pseudomonas-Glomus* consortia for promotion of maize growth. Appl Microbiol Biotechnol 97:4639–4649

Dart PJ (1986) Nitrogen fixation associated with non-legumes in agriculture. Plant Soil 90(1–3):303–334

Di Salvo LP, Silva E, Teixeira KRS, Cote RE, Pereyra MA, García de Salamone IE (2014) Physiological and biochemical characterization of *Azospirillum brasilense* strains commonly used as plant growth-promoting rhizobacteria. J Basic Microbiol 54(12):1310–1321

Dobbelaere S, Croonenborghs A, Thys A, van de Broek A, Vanderleyden J (1999) Phytostimulatory effect of *Azospirillum brasilense* wild type and mutant strains altered in IAA production on wheat. Plant Soil 212:155–164

Dobbelaere S, Croonenborghs A, Thys A, Ptacek D, Vanderleyden J, Dutto P, Labandera-Gonzalez C, Caballero-Mellado J, Aguirre JF, Kapulnik Y, Brener S, Burdman S, Kadouri D, Sarig S, Okon Y (2001) Responses of agronomically important crops to inoculation with *Azospirillum*. Aust J Plant Physiol 28:871–879

Dodd IC, Zinovkina NY, Safronova VI, Belimov AA (2010) Rhizobacterial mediation of plant hormone status. Ann Appl Biol 157(3):361–379

Drogue B, Doré H, Borland S, Wisniewski-Dyé F, Prigent-Combaret C (2012) Which specificity in cooperation between phytostimulating rhizobacteria and plants? Res Microbiol 163(8):500–510

Duca D, Lorv J, Patten CL, Rose D, Glick BR (2014) Indole-3-acetic acid in plant-microbe interactions. Antonie Van Leeuwenhoek 106:85–125

Dutta S, Podile AR (2010) Plant growth promoting rhizobacteria (PGPR): the bugs to debug the root zone. Crit Rev Microbiol 36(3):232–244

Esquivel-Cote R, Ramírez-Gama RM, Tsuzuki-Reyes G, Orozco-Segovia A, Huante P (2010) *Azospirillumlipoferum* strain AZm5 containing 1-aminocyclopropane-1-carboxylic acid deaminase improves early growth of tomato seedlings under nitrogen deficiency. Plant Soil 337(1–2):65–75

Fedonenko YP, Boiko AS, Zdorovenko EL, Konnova SA, Shashkov AS, Ignatov VV, Knirel YA (2011) Structural peculiarities of the O-specific polysaccharides of *Azospirillum* bacteria of serogroup III. Biochemistry (Moscow) 76(7):797–802

Fedonenko YP, Burygin GL, Popova IA, Sigida EN, Surkina AK, Zdorovenko EL, Konnova SA (2013) Immunochemical characterization of the capsular polysaccharide of *Azospirillum irakense* KBC1. Curr Microbiol 67(2):234–239

Ferreira AS, Pires RR, Rabelo PG, Oliveira RC, Luz JMQ, Brito CH (2013) Implications of *Azospirillum brasilense* inoculation and nutrient addition on maize in soils of the Brazilian Cerrado under greenhouse and field conditions. Appl Soil Ecol 72:103–108

Fibach-Paldi S, Burdman S, Okon Y (2012) Key physiological properties contributing to rhizosphere adaptation and plant growth promotion abilities of *Azospirillum brasilense*. FEMS Microbiol Lett 326:99–108

Figueiredo MVB, Seldin L, Araujo FF, Mariano RLR (2010) Plant growth promoting rhizobacteria: fundamentals and applications. In: Maheshwari DK (ed) Plant growth and health promoting bacteria. Springer, Berlin

França CRRS, Lira Junior MA, Figueiredo MVB, Stamford NP, Silva GA (2013) Feasibility of rhizobia conservation by liquid conditioners. Rev Cienc Agron 44(4):661–668

Fulchieri M, Lucangeli C, Bottini R (1993) Inoculation with *Azospirillumlipoferum* affects growth and gibberellin status of corn seedling roots. Plant Cell Physiol 34:1305–1309

Gaind S, Rathi MS, Kaushik BD, Nain L, Verma OP (2007) Survival of bio-inoculants on fungicides-treated seeds of wheat, pea and chickpea and subsequent effect on chickpea yield. J Environ Sci Health, Part B 42(6):663–668

Glick BR (2014) Bacteria with ACC deaminase can promote plant growth and help to feed the world. Microbiol Res 169(1):30–39

Groppa MD, Benavides MP, Zawoznik MS (2012) Root hydraulic conductance, aquaporins and plant growth promoting microorganisms: a revision. Appl Soil Ecol 61:247–254

Guerrero-Molina MF, Winik BC, Pedraza RO (2012) More than rhizosphere colonization of strawberry plants by *Azospirillum brasilense*. Appl Soil Ecol 61:205–212

Gupta K, Dey A, Gupta B (2013) Plant polyamines in abiotic stress responses. Acta Physiol Plant 35(7):2015–2036

Hayat R, Ali S, Amara U, Khalid R, Ahmed I (2010) Soil beneficial bacteria and their role in plant growth promotion: a review. Ann Microbiol 60(4):579–598

Hoffland E, Hakulinen J, van Pelt JA (1995) Comparison of systemic resistance induced by avirulent and nonpathogenic *Pseudomonas* species. Phytopathology 86:757–762

Hou X, McMillan M, Coumans JV, Poljak A, Raftery MJ, Pereg L (2014) Cellular responses during morphological transformation in *Azospirillum brasilense* and Its flcA Knockout Mutant. PLoS ONE 9(12):e114435

Hungria M, Campo RJ, Souza EM, Pedrosa FO (2010) Inoculation with selected strains of *Azospirillum brasilense* and *A. lipoferum* improves yields of maize and wheat in Brazil. Plant Soil 331(1):413–425

Ilyas N, Bano A (2010) *Azospirillum* strains isolated from roots and rhizosphere soil of wheat (*Triticum aestivum* L.) grown under different soil moisture conditions. Biol Fert Soils 46(4):393–406

Iwashkiw JA, Vozza NF, Kinsella RL, Feldman MF (2013) Pour some sugar on it: the expanding world of bacterial protein O-linked glycosylation. Mol Microbiol 89:14–28

Jacques RJS, Procópio SO, Santos JB, Kasuya MCM, Silva AA (2010) Sensitivity of *Bradyrhizobium* strains to glyphosate. Rev Ceres 57(1):28–33

Jansen R, Rood S, Dormar J, McGill W (1992) *Azospirillum brasilense* produces gibberellins in pure culture and chemically-medium and in co-culture on straw. Soil Biol Biochem 24:1061–1064

Jha M, Chourasia S, Sinha S (2013) Microbial consortium for sustainable rice production. Agroecol Sust Food 37(3):340–362

Jofré E, Fischer S, Príncipe A, Castro M, Ferrari W, Lagares A, Mori G (2009) Mutation in ad-alanine-d-alanine ligase of *Azospirillum brasilense* Cd results in an overproduction of exopolysaccharides and a decreased tolerance to saline stress. FEMS Microbiol Lett 290(2):236–246

Kadouri D, Jurkevitch E, Okon Y (2003) Involvement of the reserve material poly-β-hydroxybutyrate in *Azospirillum brasilense* stress endurance and root colonization. Appl Environ Microbiol 69:3244–3250

Kennedy IR, PeregGerk LL, Wood C, Deaker R, Gilchrist K, Katupitiya S (1997) Biological nitrogen fixation in non-leguminous field crops: facilitating the evolution of an effective association between *Azospirillum* and wheat. Plant Soil 194(1–2):65–79

Kloepper JW, Ryu CM, Zhang S (2004) Induced systemic resistance and promotion of plant growth by *Bacillus* spp. Phytopathology 94:1259–1226

Konnova ON, Boiko AS, Burygin GL, Fedonenko YP, Matora LY, Konnova SA, Ignatov VV (2008) Chemical and serological studies of liposaccharides of bacteria of the genus *Azospirillum*. Microbiology 77(3):305–312

Kuc J (1983) Induced systemic resistance in plants to diseases caused by fungi and bacteria. In: Bailey JA, Deveral BJ (eds) The dynamics of host defense. Academic Press, Sydney

Lana MDC, Dartora J, Marini D, Hann JE (2012) Inoculation with *Azospirillum*, associated with nitrogen fertilization in maize. Rev Ceres 59(3):399–405

Laslo É, György É, Mara G, Tamás É, Ábrahám B, Lányi S (2012) Screening of plant growth promoting rhizobacteria as potential microbial inoculants. Crop Prot 40:43–48

Lerner A, Valverde A, Castro-Sowinski S, Lerner H, Okon Y, Burdman S (2010) Phenotypic variation in *Azospirillum brasilense* exposed to starvation. Environ Microbiol Rep 2(4):577–586

Lucangeli C, Bottini R (1997) Effects of *Azospirillum* spp. on endogenous gibberellin content and growth of maize (*Zea mays* L.) treated with uniconazole. Symbiosis 23:63–72

Maheshwari DK (2010) Plant growth and health promoting rhizobacteria. Microbiology Monograph. Springer Heidelberg Dordrecht: London New York, vol 18, p 453

Malhotra M, Srivastava S (2009) Stress-responsive indole-3-acetic acid biosynthesis by *Azospirillum brasilense* SM and its ability to modulate plant growth. Eur J Soil Biol 45(1):73–80

Manivannan M, Tholkappian P (2013) Prevalence of *Azospirillum* isolates in tomato rhizosphere soils of coastal areas of Cuddalore District, Tamil Nadu. Int J Recent Sci Res 4:1610–1613

Mariano RLR, Kloepper JW (2000) Método alternativo de biocontrole: resistência sistêmica induzida por rizobactérias. Rev An Patol Plantas 8:121–137

Mehnaz S (2015) *Azospirillum*: a biofertilizer for every crop. In: Arora NK (ed) Plant microbes symbiosis: applied facets. Springer

Merino S, Tomás JM (2014) Gram-negative flagella glycosylation. Int J Mol Sci 15:2840–2857

Mitter B, Brader G, Afzall M, Compant S, Naveed M, Trognitz F, Sessitsch A (2013) Advances in elucidating beneficial interactions between plants, soil, and bacteria. In: Sparks DL (ed) Advances in Agronomy. Academic Press, Burlington

Moghadam HRT, Salimi H, Kasraie P, Jamshidpour F (2012) Effect of *Azospirillum* and *Azotobacter* in combination with nitrogen chemical fertilizer on wheat [*Triticum aestivum* L]. Res Crop 13(1):29–36

Molina-Favero C, Creus CM, Lanteri ML, Correa-Aragunde N, Lombardo MC, Barassi CA, Lamattina L (2007) Nitric oxide and plant growth promoting rhizobacteria: common features influencing root growth and development. Adv Bot Res 46:1–33

Molinaro A, Newman MA, Lanzetta R, Parrilli M (2009) The structures of lipopolysaccharides from plant-associated Gram-negative bacteria. Eur J Org Chem 34:5887–5896

Monds RD, O'Toole GA (2009) The developmental model of microbial biofilms: ten years of a paradigm up for review. Trends Microbiol 17(2):73–87

Mora P, Rosconi F, Fraguas LF, Castro-Sowinski S (2008) *Azospirillum brasilense* Sp7 produces an outer-membrane lectin that specifically binds to surface-exposed extracellular polysaccharide produced by the bacterium. Arch Microbiol 189:519–524

Mostafa GG, Abo-Baker ABAM (2010) Effect of bio-and chemical fertilization on growth of sunflower (*Helianthus annuus* L.) at South Valley Area. Asian. J Crop Sci 2(3):137–146

Moutia JFY, Saumtally S, Spaepen S, Vanderleyden J (2010) Plant growth promotion by *Azospirillum* sp. in sugarcane is influenced by genotype and drought stress. Plant Soil 337:233–242

Nadeem SM, Ahmad M, Zahir ZA, Javaid A, Ashraf M (2014) The role of mycorrhizae and plant growth promoting rhizobacteria (PGPR) in improving crop productivity under stressful environments. Biotechnol Adv 32(2):429–448

Naiman AD, Latrónico A, García de Salamone IE (2009) Inoculation of wheat with *Azospirillum brasilense* and *Pseudomonas fluorescens*: Impact on the production and culturable rhizosphere microflora. Eur J Soil Biol 45(1):44–51

Okon Y (1985) *Azospirillum* as a potential inoculant for agriculture. Trends Biotech 3:223–228

Okon Y, Labandera-Gonzalez CA (1994) Agronomic applications of *Azospirillum*: an evaluation of 20 years worldwide field inoculation. Soil Biol Biochem 26:1591–1601

Okon Y, Cakmakci L, Nur I, Chet I (1980) Aerotaxis and chemotaxis of *Azospirillum brasilense*: a note. Microb Ecol 6(3):277–280

Okon Y, Kapulnik Y, Sarig S (1988) Field inoculation studies with *Azospirillum* in Israel. In: Subba Rao NS (ed) Biological nitrogen fixation recent developments. Oxford and IBH Publishing Co, New Delhi

Okon Y, Bar T, Tal S, Zaady E (1991) Physiological properties of *Azospirillum brasilense* involved in root growth promotion. In: Polsinelli M, Materassi R, Vincenzini M (eds) Nitrogen fixation: developments in plant and soil. Springer, Netherlands

Pedraza RO, Bellone CH, Carrizo de Bellone S, Boa Sorte PMF, Teixeira KRS (2009) *Azospirillum* inoculation and nitrogen fertilization effect on grain yield and on the diversity of endophytic bacteria in the phyllosphere of rice rainfed crop. Eur J Soil Biol 45(1):36–43

Perez-Montano F, Alias-Villegas C, Bellogin RA, del Cerro P, Espuny MR, Jimenez-Guerrero I, Lopez-Baena FJ, Ollero FJ, Cubo T (2014) Plant growth promotion in cereal and leguminous agricultural important plants: from microorganism capacities to crop production. Microbiol Res 169(5–6):325–336

Phillips PWB (2004) An economic assessment of the global inoculant industry. Crop Management 34:1

Piccoli P, Bottini R (1996) Gibberellin production in *Azospirillumlipoferum* cultures is enhanced by light. Biocell 20(3):185–190

Piccoli P, Masciarelli O, Bottini R (1999) Gibberellin production by *Azospirillumlipoferum* cultured in chemically-defined medium as affected by oxygen availability and water status. Symbiosis 27(2):135–145

Pieterse CMJ, van Pelt JA, Knoester M, Laan R, Gerrits H, Weisbeek PJ, van Loon LCA (1998) Novel signaling pathway controlling induced systemic resistance in *Arabidopsis*. Plant Cell 10:1571–1580

Raja P, Una S, Gopal H, Govindarajan K (2006) Impact of bioinoculants consortium on rice root exudates, biological nitrogen fixation and plant growth. J Biol Sci 6:815–823

Reddy CA, Saravanan RS (2013) Polymicrobial multi-functional approach for enhancement of crop productivity. Adv App Microbiol 82:53–113

Reis VM, Vanderleyden J, Spaepen S (2011) N_2-Fixing endophytes of grasses and cereals. In: Polacco JC, Todd CD (eds) Ecological aspects of nitrogen metabolism in plants. John Wiley & Sons Inc, Hoboken, NJ/USA

Richardson AE, Barea JM, McNeill AM, Prigent-Combaret C (2009) Acquisition of phosphorus and nitrogen in the rhizosphere and plant growth promotion by microorganisms. Plant Soil 321(1–2):305–339

Rodrigues AC, Antunes JEL, Costa AF, Oliveira JP, Figueiredo MVB (2013a) Interrelationship of *Bradyrhizobium* sp. and plant growth-promoting bacteria in cowpea: survival and symbiotic performance. J Microbiol 51:49–55

Rodrigues AC, Silveira JAG, Bonifacio A, Figueiredo MVB (2013b) Metabolism of nitrogen and carbon: optimization of biological nitrogen fixation and cowpea development. Soil Biol Biochem 67:226–234

Rodriguez-Navarro DN, Dardanelli MS, Ruíz-Sainz JE (2007) Attachment of bacteria to the roots of higher plants. FEMS Microbiol Lett 272(2):127–136

Rodríguez-Salazar J, Suárez R, Caballero-Mellado J, Iturriaga G (2009) Trehalose accumulation in *Azospirillum brasilense* improves drought tolerance and biomass in maize plants. FEMS Microbiol Lett 296(1):52–59

Romeiro RS (2000) PGPR e indução de resistência sistêmica em plantas a patógenos. Summa Phytopathologica 26:177–184

Romero AM, Correa OS, Moccia S, Rivas JG (2003) Effect of *Azospirillum*-mediated plant growth promotion on the development of bacterial diseases on fresh-market and cherry tomato. J Appl Microbiol 95:832–838

Romero AM, Vega D, Correa OS (2014) *Azospirillum brasilense* mitigates water stress imposed by a vascular disease by increasing xylem vessel area and stem hydraulic conductivity in tomato. Appl Soil Ecol 82:38–43

Russo A, Vettori L, Felici C, Fiaschi G, Morini S, Toffanin A (2008) Enhanced micropropagation response and biocontrol effect of *Azospirillum brasilense* Sp 245 on *Prunus cerasifera* L. clone Mr.S 2/5 plants. J Biotechnol 134(3–4):312–319

Schelud'ko AV, Makrushin KV, Tugarova AV, Krestinenko VA, Panasenko VI, Antonyuk LP, Katsy EI (2009) Changes in motility of the rhizobacterium *Azospirillum brasilense* in the presence of plant lectins. Microbiol Res 164:149–156

Shan X, Yan J, Xie D (2012) Comparison of phytohormone signaling mechanisms. Curr Opin Plant Biol 15(1):84–91

Shelud'ko AV, Burygin GL, Filip'echeva YA, Belyakov AE, Shirokov AA, Burov AM, Katsy EI, Shchegolev SY, Matora LY (2014) Serological relationships of azospirilla revealed by their motility patterns in the presence of antibodies to lipopolysaccharides. Microbiology 83(1–2):102–109

Sigida EN, Fedonenko YP, Shashkov AS, Zdorovenko EL, Konnova SA, Ignatov VV, Knirel YA (2013) Structural studies of the O-specific polysaccharide(s) from the lipopolysaccharide of *Azospirillum brasilense* type strain Sp7. Carbohyd Res 380:76–80

Sigida EN, Fedonenko YP, Zdorovenko EL, Burygin GL, Konnova SA, Ignatov VV (2014) Characterization of the lipopolysaccharides of serogroup II *Azospirillum* strains. Microbiology 83(4):326–334

Sivasakthivelan P, Saranraj P (2013) *Azospirillum* and its formulations: a review. Int J Microbiol Res 4(3):275–287

Smol'kina ON, Kachala VV, Fedonenko YP, Burygin GL, Zdorovenko EL, Matora LY, Konnova SA, Ignatov VV (2010) Capsular polysaccharide of the bacterium *Azospirillum lipoferum* Sp59b: structure and antigenic specificity. Biochemistry (Moscow) 75(5):606–613

Somers E, Ptacek D, Gysegom P, Srinivasan M, Vanderleyden J (2005) *Azospirillum brasilense* produces the auxin-like phenylacetic acid by using the key enzyme for indole-3-acetic acid biosynthesis. Appl Environ Microb 71(4):1803–1810

Spaepen S, Dobbelaere S, Croonenborghs A, Vanderleyden J (2008) Effects of *Azospirillum brasilense* indole-3-acetic acid production on inoculated wheat plants. Plant Soil 312:1–23

Spaepen S, Vanderleyden J, Okon Y (2009) Plant growth-promoting actions of rhizobacteria. Adv Bot Res 51:283–320

Tenuta M (2003) Plant growth promoting rhizobacteria: prospects for increasing nutrient acquisition and disease control. (en línea). http://www.umanitoba.ca/afs/agronomists_conf/2003/pdf/tenuta_rhizobacteria.pdf

Tien TM, Gaskins MH, Hubbell DH (1979) Plant growth substances produced by *Azospirillum brasilense* and their effect on the growth of pearl millet (*Pennisetumamericanum* L.). Appl Environ Microb 37:1016–1024

Tortora ML, Díaz-Ricci JC, Pedraza RO (2011) *Azospirillum brasilense* siderophores with antifungal activity against *Colletotrichum acutatum*. Arch Microbiol 193:275–286

Trivedi R, Bhatt SA (2011) Impact of biovolume inocula of *Azospirillum* spp. on winter wheat, oat and maize in semi-arid region of North Gujarat. Biotechnol 5(3):186–190

Trujillo-Roldán MA, Valdez-Cruz NA, Gonzalez-Monterrubio CF, Acevedo-Sánchez EV, Martínez-Salinas C, García-Cabrera RI, Gamboa-Suasnavart RA, Marín-Palacio LD, Villegas J, Blancas-Cabrera A (2013) Scale-up from shake flasks to pilot-scale production of the plant growth-promoting bacterium *Azospirillum brasilense* for preparing a liquid inoculant formulation. App Microbiol Biotech 97(22):9665–9674

Turan M, Gulluce M, von Wirén N, Sahin F (2012) Yield promotion and phosphorus solubilization by plant growth-promoting rhizobacteria in extensive wheat production in Turkey. J Plant Nutr Soil Sc 175(6):818–826

Vacheron J, Desbrosses G, Bouffaud ML, Touraine B, Moënne-Loccoz Y, Muller D, Legendre L, Wisniewski-Dyé F, Prigent-Combaret C (2013) Plant growth-promoting rhizobacteria and root system functioning. Front Plant Sci 4:356

Verstraeten N, Braeken K, Debkumari B, Fauvart M, Fransaer J, Vermant J, Michiels J (2008) Living on a surface: swarming and biofilm formation. Trends Microbiol 16(10):496–506

Vettori L, Felici C, Russo A, Morini S, Toffanin A, Cummings S (2010) Biocontrol activity of Azospirillum brasilense Sp 245 against Rhizoctonia solani by in vitro/in vivo tests, DGGE analysis. J Biotechnol 150:503–503

Vieira Neto SA, Pires FR, Menezes CCE, Menezes JFS, Silva AG, Silva GP, Assis RL (2008) Forms of inoculant application and effects on soybean nodulation. Rev Bras Cienc Solo 32(2):861–870

Viera RM, Fernandez F (2006) Practical applications of bacterial biofertilizers and biostimulators. In: Uphoff N, Ball AS, Fernandes E (eds) Biological approaches to sustainable soil systems. CRC Press, Boca Raton

Walker V, Bertrand C, Bellvert F, Moënne-Loccoz Y, Bally R, Comte G (2011) Host plant secondary metabolite profiling shows a complex, strain-dependent response of maize to plant growth-promoting rhizobacteria of the genus Azospirillum. New Phytol 189(2):494–506

Walker V, Couillerot O, von Felten A, Bellvert F, Jansa J, Maurhofer M, Bally R, Moënne-Loccoz Y, Comte G (2012) Variation of secondary metabolite levels in maize seedling roots induced by inoculation with *Azospirillum*, *Pseudomonas* and *Glomus* consortium under field conditions. Plant Soil 356:151–163

Winik BC, Molina MFG, Pedraza RO (2009) Colonization of strawberry (*Fragaria ananassa*) plant tissues by *Azospirillum brasilense*. Acta Microsc 18:675–676

Wisniewski-Dyé F, Drogue B, Borland S, Prigent-Combaret C (2013) *Azospirillum*-plant interaction: from root colonization to plant growth promotion. In: Rodelas González MB, González-López J (eds) Beneficial plant-microbial interactions: ecology and applications. CRC Press, London

Yadegari M, Asadi Rahmani H, Noormohammadi G, Ayneband A (2010) Plant growth promoting rhizobacteria increase growth, yield and nitrogen fixation in *Phaseolusvulgaris*. J Plant Nutr 33(12):1733–1743

Yamaguchi S (2008) Gibberellin metabolism and its regulation. Ann Rev Plant Biol 59:225–251

Zilli JE, Gianluppi V, Campo RJ, Rouws JRC, Hungria M (2010) In-Furrow inoculation with *Bradyrhizobium* alternatively to seed inoculation of soybean. Rev Bras Ciênc Solo 34(6):1875–1881

Zilli JE, Silva Neto ML, França Júnior I, Perin L, Melo AR (2011) Cowpea response to inoculation with *Bradyrhizobium* strains recommended for soybean. Rev Bras Ciênc Solo 35:739–742

Zobiole LHS, Kremer RJ, Oliveira Junior RS, Constantin J (2011) Glyphosate affects chlorophyll, nodulation and nutrient accumulation of second generation glyphosate-resistant soybean (*Glycinemax* L.). Pestic Biochem Phys 99(1):53–60

Emergence of *Methylobacterium* spp. as Potential Organism in Agroecosystems

Chaitanya Kumar Jha, Dinesh Kumar Maheshwari and Meenu Saraf

Abstract *Methylobacterium* spp. includes a group of stringently aerobic, Gram-negative, pink-pigmented, facultatively methylotrophs (PPFM) belonging to α-proteobacteria and are capable of growing on one-carbon compounds, such as formate, formaldehyde, methanol and methylamine or sometimes on multi-carbon compounds like diethyl ether and trimethyl amines. Significance of these bacteria for plant-growth promotion by the possible mechanisms include production of phytohormones, IAA, cytokinins, ACC-deaminase and perform nitrogen metabolism by means of bacterial urease, establish efficient nitrogen (N_2)-fixing symbioses by nodulating legume roots; production of exopolysaccharides (EPS) and Poly-β-hydroxybutyrate (PHB) accumulation and abiotic stress endurance. These organisms induce systemic resistance by production of siderophores and proteins like phenylalanine ammonia lyase, peroxidase, chitinase and β-1,3-glucanase and phenolic compounds. On the other hand, they also promote the biodegradation of polycyclic aromatic hydrocarbon (PAH). In spite of their plant-growth promotional traits, commercialization of the *Methylobacterium* strains as bioinoculant have been hindered constantly.

C.K. Jha
Department of Microbiology, Government Science College, Vankal, Mangrol, Surat 394430, Gujarat, India

D.K. Maheshwari
Department of Botany and Microbiology, Gurukul Kangri University, Haridwar 249404, Uttarakhand, India

M. Saraf (✉)
Department of Microbiology and Biotechnology, University School of Sciences, Gujarat University, Ahmedabad 380009, Gujarat, India
e-mail: sarafmeenu@gmail.com

D.K. Maheshwari (ed.), *Bacterial Metabolites in Sustainable Agroecosystem*, Sustainable Development and Biodiversity 12, DOI 10.1007/978-3-319-24654-3_3

Keywords *Methylobacterium* · Pink-pigmented · Facultatively methylotrophic · Plant-growth promotion · Biocontrol · Quorum sensing · Induced systemic resistance (ISR)

1 Introduction

The omnipresent occurrence of *Methylobacterium* on plant surfaces makes them a model for the study of plant-microbe interaction, and motivating approach for realizing the particular traits that these bacteria having on plant-growth promoting attribute. They utilize the gaseous methanol that is emitted by the plants through the stomata as carbon and energy sources, and promote the growth of their host through release of different metabolites. *Methylobacterium* strains have been localized as endosymbionts within cells in the buds. One species, *Methylobacterium podarium* is thought to be part of the natural human foot microflora. *Methylobacterium* have even been found living inside the human mouth. Actually, the members of PPFM are ubiquitous in nature and found in a variety of habitats including phyllosphere, rhizosphere, root nodules, dust, freshwater, drinking water, lake sediments, etc. (Corpe and Rheem 1989). Their association with more than 70 plant species makes them potential agents for plant-growth promotion and biocontrol against diseases (Holland and Polacco 1994).

The Methylotrophs are defined as those growing on C1 compounds like methanol, formaldehyde, formate and methylamine. Based on their utilization pattern, they are obligate methylotrophs, not able to grow on multi-carbon compounds but if grown on methanol or methylamine but not on methane they are strictly aerobic Gram-negative and classified under two genera, e.g. *Methylophilus* and *Methylobacillus*. In case if they utilize methane then they are called methanotrophs. Methanotrophs are Gram-negative bacteria and classified under five genera: *Methylomonas, Methylobacter, Methylococcus, Methylosinus* and *Methylocystis*. All methanotroph forms extensive intracellular membranes and resting cells, either as cysts or exospores. The intracellular membranes are hypothesizing to be concerned in methane oxidation. On the other hand, *Methylotrophs* growing on C1 compounds and multi-carbon compounds, such as trimethylamine, dimethyl ether, dimethyl carbonate are called facultative methylotroph. A number of Gram-positive and Gram-negative bacterial genera include *Bacillus*, *Acetobacter*, *Mycobacterium*, *Arthrobacter*, *Hyphomicrobium*, *Methylobacterium* and *Nocardia*. Further, most of the *Methylobacterium* species contain property of pigmentation (pink) which is extremely slow and nodulates *Crotolaria podocarpa*. These are capable of growing on one-carbon compounds, such as formate, formaldehyde, methanol and methylamine. Significance of these bacteria as plant-growth promotion by the possible mechanisms include production of phytohormones, such as indole-3-acetic acid (IAA), cytokinins, nitrogen metabolism, nitrogen (N_2)-fixing, contains 1-aminocyclopropane-1-carboxylate (ACC) deaminase, secretes EPS, accumulates PHB and survives in abiotic stress. These beneficial soil bacteria can confer immunity against a wide range of foliar diseases by activating plant defenses, thereby reducing plant susceptibility to pathogen attack (van Loon et al. 1998). For many years, it was considered that beneficial microorganisms could increase plant yield when inoculated in crops; however, it is increasingly appreciated that classic and novel microbial signals may also directly participate in plant morphogenesis. Plant depends on bacteria for the removal of metabolic waste products generated during its growth (Holland 1997). Methanol, a waste product of plants, is a fitting example of this kind of relationship, degraded by PPFMs into simpler compounds, such as ammonium, which eventually return to the plant. Recently, different species of PPFM are reported to be able to benefit plant development using a wide range of mechanisms, including synthesizing compounds to promote plant growth and increasing the uptake of nutrients and acting as biocontrol agents by suppressing plant pathogens in the rhizosphere. Several species of *Methylobacter* namely *Methylobacterium oryzae*, *M. funariae*, *Methylobacterium organophilum*, *Methylobacterium nodulans*, *Methylobacterium populi*, *Methylobacterium extorquens* etc. are reported for plant-growth promotion.

2 Alliance of *Methylobacterium* with Plants

Methylobacterium are root-nodulating symbionts (Jaftha et al. 2002), endophytic (Van Aken et al. 2004) and epiphytic (Omer et al. 2004) on plant surfaces.

It has been considered that plant-*Methylobacterium* association is primeval and permanent (Fedorov et al. 2011), and that plant-associated *Methylobacterium* is a co-evolved phytosymbiont (Kutschera 2007) because of symbiotic interaction. In fact, more than 80 % of viable bacteria isolated from leaf surfaces are members of the genus *Methylobacterium* (Tani et al. 2012).

2.1 Fate of C1 Compounds via Serine Cycle

Numerous studies have established that C1 metabolism plays the key role in the root colonization of *Methylobacterium* (Sy et al. 2005). Enzymes involved in methylotrophy of this microorganism have been identified and characterized in a metaproteomic study (Vorholt 2002). In this pathway, methanol is oxidized by methanol dehydrogenase (MDH) in the periplasmic space of the cell to produce formaldehyde (HCHO), which is then relocated into cytoplasm where part of the formaldehyde is oxidized to carbon dioxide (CO_2) for energy generation, and rest is assimilated via the serine cycle (Fig. 1). The metabolism is characterized into three parts: Part 1 indicates bacterium which oxidizes methanol to formaldehyde is condensed with a tetrahydromethanopterin and further oxidized to formate. Formate reacts with tetrahydropterin and formyltetrahydrofolate is further converted to methylenetetrahydrofolate.

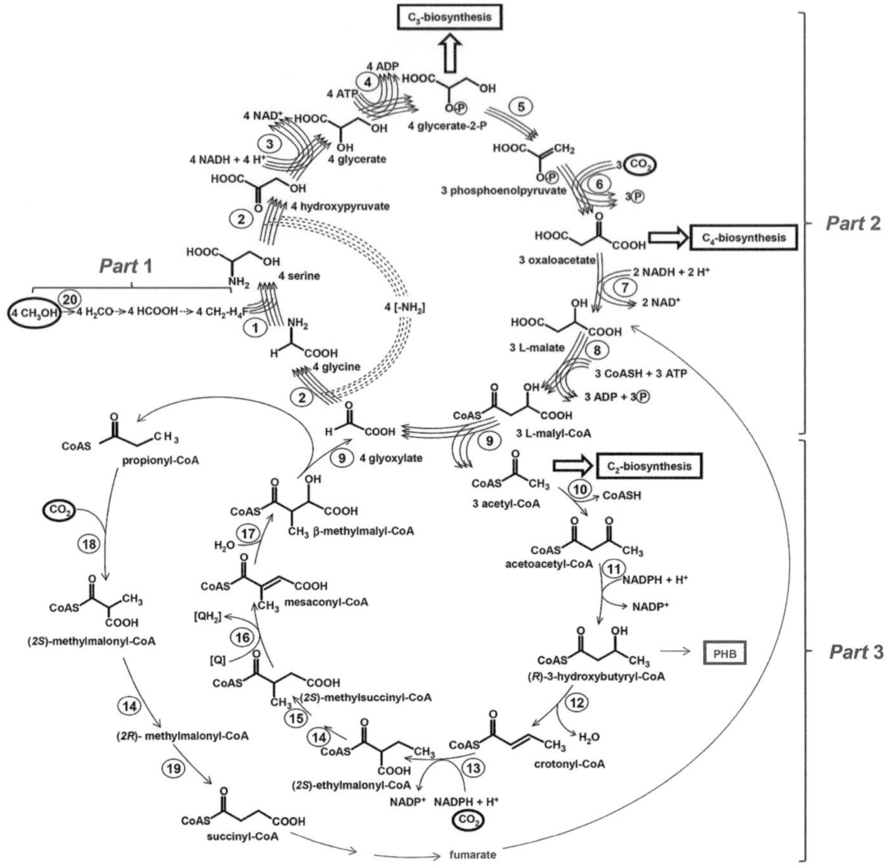

Fig. 1 C1 metabolism of the methylotroph *Methylobacterium extorquens* AM1. Enzymes: *1* serine hydroxymethyl transferase; *2* serine-glyoxylate aminotransferase; *3* hydroxypyruvate reductase; *4* glycerate kinase; *5* enolase; *6* phosphoenolpyruvate carboxylase; *7* malate dehydrogenase; *8* malate-CoA ligase (malate thiokinase); *9*, L-malyl-CoA/b-methylmalyl-CoA lyase; *10* β-ketothiolase; *11* acetoacetyl-CoA reductase; *12* crotonase; *13* crotonyl-CoA carboxylase reductase; *14* ethylmalonyl-CoA/methylmalonyl-CoA epimerase; *15* ethylmalonyl-CoA mutase; *16* methylsuccinyl-CoA dehydrogenase; *17* mesaconyl-CoA hydratase; *18* propionyl-CoA carboxylase; *19* methylmalonyl-CoA mutase; *20* methanol dehydrogenase. *PHB* polyhydroxybutyrate, *Q* quinone (Figure adapted from Smejkalova et al. 2010)

On the other hand, Part 2 involved metabolism during the serine cycle is used for the assimilation of formaldehyde plus bicarbonate and Part 3 contain Acetyl-CoA assimilation and conversion to glyoxylate proceeds via the ethylmalonyl-CoA pathway. The ethylmalonyl-CoA pathway, in connection with the serine cycle, represents an elegant solution of methanol assimilation, where methanol and carbon dioxide contribute nearly equal to cell carbon:

$$\begin{aligned} & 4\,\mathrm{C1}\,(\mathrm{HCHO}) + 5\,\mathrm{CO_2} + 8\,\mathrm{NAD(P)H} + 8\mathrm{H^+} + 8\,\mathrm{ATP} + 1\,\mathrm{Coenzyme} \\ & \rightarrow 1\,\mathrm{C_3}(\mathrm{glycerate{-}2{-}phosphate}) + 1\,\mathrm{C_4}(\mathrm{oxaloacetate}) \\ & + 1\,\mathrm{C_2}(\mathrm{acetyl{-}CoA}) + 8\,\mathrm{NAD(P)^+} + 2\,[\mathrm{H}] + 8\,\mathrm{ADP} + 7\,\mathrm{Pi}. \end{aligned}$$

The assimilation of acetate through the ethylmalonyl-CoA pathway can be expressed by the following equation:

$$\begin{aligned} & \mathbf{3\,acetyl{-}CoA + 2\,CO_2 + 2\,NADPH + 2\,H^+ + ATP} \\ & \mathbf{\rightarrow malate + succinyl - CoA + 2\,CoA + 2[H] + 2\,NADP^+} \\ & \mathbf{+ ADP + Pi}\ \text{(Smejkalova et al. 2010)} \end{aligned}$$

Methylobacterium nodulans is the causal organism of nodulation of the *Crotalaria podocarpa* (Jourand et al. 2004), possesses an *mxa* gene cluster for coding MDH.

3 Mechanism for Plant-Growth Promotion

Methylobacterium spp. are ubiquitous in nature and colonize probably all land plants influencing the growth promotion by fixing atmospheric nitrogen, solubilizing insoluble phosphates, secreting hormones and producing antimicrobial compounds to combat phytopathogens. Induced systemic resistance (ISR), antibiosis, competition for nutrients, parasitism and production of metabolites (hydrogen cyanide, siderophores and enzymes) suppressive to deleterious rhizobacteria are some of the biocontrol mechanism that induce plant growth (Jha et al. 2010).

3.1 Phosphate Solubilization

Long back, Goldstein (2003) proposed direct oxidation of glucose to gluconic acid (GA) as a major mechanism for mineral phosphate solubilization (MPS) in Gram-negative bacteria. As a result of acidification of the surrounding medium, soluble orthophosphate ions ($H_2PO_4^{-1}$ and HPO_4^{-2}) can be readily released. Nowadays, it is widely accepted that a large number of microbes produce a range of low-molecular weight organic acids, such as acetate, lactate, oxalate, tartarate, succinate, citrate, gluconate, ketogluconate, glycolate, etc., which are considered to solubilize insoluble mineral phosphates. It could be assumed that any gene involved in organic acid synthesis might have an effect on this character (Ahemad and Khan 2010). In *Methylobacterium*, by screening genomic libraries of mineral phosphate solubilization (MPS) bacteria for gluconic acid production traps Pyrroloquinoline quinine (PQQ) (Pyrroloquinoline quinine) biosynthesis genes which act as a prosthetic group of bacterial quinoprotein dehydrogenase. PQQ belongs to the family of quinone cofactors that has been recognized as the third class of redox cofactors following pyridine nucleotide and flavin-dependent cofactors (Liu et al. 1992).

It is a prosthetic group required by several bacterial dehydrogenases, including methanol dehydrogenase (MDH) of Gram-negative methylotrophs and some glucose dehydrogenases. PQQ is derived from two amino acids, tyrosine and glutamic acid (Houck et al. 1991) but the pathway for its biosynthesis is unknown. Sequence analysis of this gene (Liu et al.1992) suggested its probable involvement in the synthesis of the enzyme PQQ synthase, which directs the synthesis of PQQ, a cofactor necessary for the formation of the holoenzyme glucose dehydrogenase (GDH)-PQQ. This enzyme catalyzes the formation of gluconic acid from glucose by the direct oxidation pathway (Goldstein 2003).

In *M. extorquens* AM1, the genes for PQQ synthesis are found in two clusters, pqqAB (C/D) E and pqqFG. These gene designations standardize the nomenclature with that of *Klebsiella pneumoniae*. These genes in *Methylobacterium* strains were formerly called pqqDGCBA. In *M. extorquens* AM1, pqqC and pqqD are not separate genes. Instead, they are fused into a single gene, pqqCD.

3.2 Plant Hormone Production

Plant-growth promotion by *Methylobacterium* include synthesis of the major plant hormones IAA and cytokinin, besides breakdown of plant produced ethylene by production of ACC deaminase as stated by (Saraf et al. 2010). In the *Methylobacterium*, genes that encode enzymes related to auxin biosynthesis, such as amine oxidase, aldehyde dehydrogenase, cyanide hydratase, N-acyltransferase, nitrile hydratase, amidase have been reported (Kwak et al. 2014). *Methylobacterium* is able to produce IAA (Ivanova et al. 2001), suggesting that its inoculation can increase IAA accumulation in plants that leads to induce plant growth and development (Madhaiyan et al. 2006a).

Cytokinins can be produced by bacteria by at least two pathways. De novo synthesis involves the direct isopentenylation of AMP catalyzed by dimethyl alkyltransferase (DMAT), which was first characterized in *Agrobacterium tumefaciens* (Golberg et al. 1984) while the second pathway of bacterial cytokinin production involves turnover of modified tRNA which also operate in higher plants. The origin of cytokinins resulting from tRNA degradation involves isopentenylation of adenine by isopentenyl tRNA transferase, the product of the *mia*A gene. In *Methylobacteria,* this modified adenine is subsequently methylated or hydroxylated. It is hypothesized that upon turnover of tRNA the modified adenine residue is released as a free cytokinin. *Methylobacteria* prefer the second pathway for the production of cytokinins. Actually, tRNA is the source of low-level trans-zeatin (active and ubiquitous form of the naturally occurring cytokinins) production. Infact, *M. extorquens* produces the cytokinin trans-zeatin at low levels in pure culture and excrete it into the culture medium (Koenig et al. 2002). Earlier, Ivanova et al. (2000) reported the presence and expression of genes controlling the synthesis and secretion of cytokinins by the PPFM *Methylobacterium mesophilicum* VKM B-2143.

3.3 Nitrogen Fixation

The biological reduction of nitrogen to ammonia (NH_3) can be performed only by some prokaryotes with the presence of the nitrogenase enzyme (Menna et al. 2006). *M. nodulans* was originally isolated from *Crotalaria podocarpa* (Sy et al. 2001) and it was the few nodulating *Methylobacterium* species reported so far (Kwak et al. 2014). *M. nodulans* ORS2060 was reported to contain the *nif*H gene (involved in nitrogen fixation) and to induce N_2-fixing nodules on the 11 leguminous plants (Jourand et al. 2004). Kumar et al. (2009) reported that the ultimate aim of establishing endophytic interaction between diazotrophic bacteria and nonlegumes is to fix N_2 which later transferred the fixed N_2 to the plants. *Azorhizobium caulinodans* and *Methylobacterium* species were capable of N_2-fixing in a free-living condition. It was anticipated that the intercellular colonization of rice might provide a niche for N_2 fixation. They isolated *Methylobacterium* sp. NPFM-SB3 from *Sesbania rostrata* stem nodules possess nitrogenase activity and *nod*A genes.

3.4 1-Aminocyclopropane-1-Carboxylate (ACC) Deaminase Production

Rhizobacterial enzyme, ACC deaminase (EC 4.1.99.4), which regulates ethylene production by metabolizing ACC (an intermediate precursor of ethylene biosynthesis in higher plants) into α-ketobutyrate and NH_3 (Shaharoona et al. 2007). The microbial enzyme ACC deaminase cleaves ACC irreversibly, this being the immediate precursor of ethylene in plants (Saraf et al. 2010). This enzyme facilitates plant growth as a consequence of the fact that it sequesters and cleaves plant produced ACC, thereby lowering the level of ethylene in the plant. In turn, decreased ethylene levels allow the plant to be more resistant to a wide variety of environmental stresses, all of which induce the plant to increase its endogenous level of ethylene; stress ethylene exacerbates the effects of various environmental stresses. ACC deaminase activity is quantified by monitoring the production of either NH_3 or α-ketobutyrate, the products of ACC hydrolysis. However, at present, monitoring the amount of α-ketobutyrate is more widely used by researchers. The presence of ACC deaminase was also verified by Fourier Transform Infrared (FTIR) spectra. FTIR spectra clearly shows the peak at 1683 cm^{-1} which exhibits the presence of ketonic group (–C=O). Whereas, 3452 cm^{-1} peak shows the presence of amino group (–NH_2) (Jha et al. 2012). *Methylobacterium* also carry the *acd*S gene that encodes ACC deaminase enzyme converts ACC into NH_3 and α-ketobutyrate. An analysis of the genomes of *Methylobacterium* species, such as *M. oryzae*, *M. nodulans* and *M. radiotolerans*, contain this ACC deaminase gene (Kwak et al. 2014) and that *M. nodulans* and *M. radiotolerans* are able to use ACC as a nitrogen source by the actions of ACC deaminase, reducing ethylene levels

(Fedorov et al. 2013) and consequently the stress ethylene response in the host plant. More recently, Joe et al. (2014) reported that the ACC deaminase-positive *M. oryzae* CBMB20 with *Azospirillum brasilense* CW903 strain reduced ethylene levels in plants.

3.5 Exoploysaccharides (EPS) Production

EPS play vital roles in a variety of processes among bacteria, such as formation of biofilm (Bhaskar and Bhosle 2005), protection of bacterial cell from desiccation (Pal et al. 1999), maintaining primary cellular functions and antibacterial activity against predators, gelling ability, pollutant degradation kinetics (Fusconi and Godinho 2002), bioremediation activity and plasma substituting capacity (Allison 1998). Breuer and Babel (1999) reported the production of EPS in *M. rhodesiunum* under ammonium limitation conditions. The high amount of PHB accumulation also observed in *Methylobacterium* strains (Alvarez et al. 1996) in some psychrophilic and psychrotrophic crude oil-utilizing marine bacteria, accumulate lipid storage compounds in the cytoplasm under nitrogen limiting conditions when the C:N ratio becomes high. Woo et al. (2012) had compared the growth pattern, floc yield, EPS production and PHB accumulation, resistance to osmotic and acid stress in *Methylobacterium* strains CBMB20, CBMB27, CBMB35 and CBMB110.

4 Biocontrol Potentials

In recent decades, interaction studies have reflected that endophytic microorganisms may enhance plant protection against pathogen attacks. Biocontrol of pathogens can be achieved by several mechanisms viz: ISR, siderophore production, lytic enzyme production, etc. Ardanov et al. (2012) studied the ability of *Methylobacterium* sp. IMBG290 to induce resistance in potato (*Solanum tuberosum* L.) cultivars against *Pectobacterium atrosepticum*, *Phytophthora infestans* and *Pseudomonas syringae* pv. Tomato DC3000, as well as *M. extorquens* DSM13060 in pine (*Pinus sylvestris* L.) against *Gremmeniella abietina*. In earlier studies, Madhaiyan et al. (2006b) observed that seed treatment with *Methylobacterium* sp. induced significant protection against *Aspergillus niger* and *Sclerotium rolfsii* in groundnut. Further, the biocontrol potential of *Methylobacterium* spp. against several fungal pathogens, such as *Fusarium udum*, *Fusarium oxysporum*, *Pythium aphanidermatum* and *Sclerotium rolfsii* was evaluated in vitro by Poorniammal et al. (2009).

4.1 Induced Systemic Resistance (ISR)

The state of enhanced defensive capacity of plant's innate immunity elicited by specific mechanism through *Methylobacterium* against biotic challenges is defined as ISR. Madhaiyan et al. (2004) reported that the treatment with *Methylobacterium* sp. strain PPFM-Os-07 increased activities of various defense-related enzymes like chitinase, Phenylalanine ammonia lyase, β-1,3-glucanase, peroxidase and PR-proteins, which accumulated in paddy with the onset of ISR. This enhanced state of resistance is effective against a broad range of pathogens and parasites (van Loon 2000). The two most clearly defined forms of induced resistance are: (1) SAR and (2) ISR, which can be differentiated on the basis of the nature of the elicitor. SAR is induced either upon infection by an avirulent pathogen or upon restricted infection by a virulent pathogen and depends on the synthesis of salicylic acid (SA) by the host. It is effective against pathogens that are restricted by SA-dependent basal resistance responses. On the other hand, ISR is triggered by selected strains of nonpathogenic rhizobacteria and does not require SA but does depend on the responsiveness of the plant to jasmonic acid (JA) and ethylene. Due to this reason, tolerance to abiotic stresses occurred in plants (Mantelin and Touraine 2004).

4.2 Siderophores Production

The genus *Methylobacterium*, as a member of PPFM, has ubiquitous occurrence in the environment and plays an important role in iron acquisition. Lacava et al. (2008) concluded that *Methylobacterium* spp. have no ability of producing catechol-type siderophores, but are capable to produce hydroxamate-type siderophores. In their study, in vitro growth of *Xylella fastidiosa* subsp. is stimulated by the presence of a supernatant siderophore of endophytic *M. mesophilicum*. Silva Stenico et al. (2005) also reported that a strain of *M. extorquens* isolated from *Citrus sinensis* was able to produce hydroxamate type of siderophore but negative for catechol type. Recently, Vaidehi and Sekar (2012) reported that *Methylobacterium phyllosphaerae* MB-5 and CBMB-27 contained hydroxamate type of siderophore during iron limitation.

4.3 Quorum Sensing

Quorum sensing (QS) systems use N-acyl-homoserine lactones (AHLs) as signaling molecules, commonly found in Gram-negative bacteria that live in association with plants (White and Winans 2007). QS system allows bacteria to function as multicellular organisms, because the extracellular concentration of autoinducer increases with bacteria population growth, after attaining a determinate number. This molecule disseminate back into the bacteria and regulate the transcription of different genes that may be related with the secretion of virulence factors, biofilm formation, sporulation, exchange of DNA and others (Zhu and Sun 2008). Although, several studies demonstrate the importance of the association between *Methylobacterium* plants (Dourado et al. 2012) and that members of the *Methylobacterium* genus produces AHL (Pomini et al. 2009). Recently, Dourado et al. (2013) reported the role of plant exudates and AHL on the expression of bacterial genes that are involved in bacterium plant-interaction. It was observed that AHL induces all analyzed genes *mxa*F, *acd*S, *crt*I and *sss* evade plant-microbe interactions. The gene sodium solute symporter (*sss*) is a transport gene responsible for the symport transport of solute with the sodium (Scier 1998). Genes *crt*I and *acd*S genes are associated with the stress response (Sandmann 2009) and plant metabolism (Hardoim et al. 2008). Phytoene dehydrogenase gene (*crt*I) codifies an enzyme that catalyzes the denaturation reaction resulting in the lycopene synthesis that protects the cell against oxidative damages and the *acd*S gene responsible for the degradation of ACC by ACC deaminase enzyme. Enhanced production of AHL corresponding to biofilms formation cannot be ruled out.

5 Applications

Methylobacterium spp. exhibited a vast range of biotechnological applications in the field of agriculture and industry. Recently, it was established as a potential bioinoculant for the sustainable agriculture.

5.1 Poly-β-Hydroxybutyrate (PHB) Accumulation

Poly-β-hydroxybutyrate (PHB) is an intracellular storage compound, which provides a reserve of carbon and energy in several microorganisms. It has been argued that methanol would appear as an alternative substrate for PHB production because of several advantages including low price and its complete water miscibility. Lopez-Cortes et al. (2008) suggested the presence of bright cytoplasmic inclusions as preliminary step for qualitative PHB determination. *Methylobacterium* sp. shows well-defined brightly refractile cytoplasmatic inclusions under phase contrast microscopy suggesting PHB accumulation. Zahra et al. (2009) reported the production of PHB using methanol by *Methylobacterium extorquens* DSMZ 1340.

5.2 Nitogen-Fixing Biofertilizers

Methylobacterium play a vital role by mediating nutrient transformation from the soil to plants. Rekadwad (2014) reported a thermophilic *M. organophilum*

(N2)-fixing species isolated from hotspring originated mud is able to fix dinitrogen at elevated temperature. During last decade, a first report appeared in the year 2001 by a group of scientists about the symbiotic association of *M. nodulans* to that of *Crotolaria podocarpa*. This root-nodulating bacterium fixes nitrogen in symbiosis with legumes (Sy et al. 2001)

5.3 Seed Germination and Plant-Growth Promotion

The seed resident *Methylobacterium* is a contributing factor to vigor and seed viability. The cytokinins produced by these PPFMs are responsible for their stimulatory effect on germination (Freyermuth et al. 1996). Role of some other compounds to contribute for the enhancement of germination and growth of plants cannot be ruled out. Anitha (2010) reported that increase in maize seeds germination increased by 86 % to those seeds treated with 0.5 mg/l of benzyl adenine and 0.5 mg/l of zeatin (Holland and Polacco 1992). Wei et al. (2014) reported that germination energy and the germination rate decreased with increasing phenanthrene concentrations in wheat. Since date back in 1995, Holland and Polacco granted a patent. They coated seeds with at least one PPFM to improve seed germination, affirming that PPFM can be used to produce cytokinin. Verginer et al. (2010) observed that *M. extorquens* DSM 21961 increase the production of two furanoid compounds, 2,5-dimethyl-4-hydroxy-2H-furanone (DMHF) and 2,5-dimethyl-4-methoxy-2H-furanone in vitro, which are responsible for strawberry flavor.

5.4 Induced Pathogenesis

Methylobacteria induces systemic resistance against diseases due to siderophores and enzymes like phenylalanine ammonia lyase, peroxidase, chitinase and β-1,3-glucanase and phenolic compounds. *Methylobacterium* induce defense programs, such as SAR and ISR, thus reducing phytotoxic microbial communities. Further, bacteria also elicit induced systemic tolerance (IST) to abiotic stress. Proposed signal molecules for PGP by *Methylobacterium* include synthesis of IAA and cytokinin besides breakdown of plant-produced ethylene by bacterial production of ACC deaminase. Although low-molecular weight plant volatiles, such as terpenes, jasmonates and green leaf components have been identified as potential signal molecules for plants and organisms of other trophic levels (Farag and Pare 2002). Additional signals from microbes have been found to play a role in plant morphogenetic processes, including the N-acyl-L-homoserine lactones (AHLs) and volatile organic compounds (VOCs).

5.5 Plant-Growth Promotion

Methylobacteria a significant organism used for plant-growth promotion by producing promising mechanisms which include production of phytohormones, establish efficient (N_2)-fixing symbioses by nodulating legume roots, ACC deaminase-production, exopolysaccharides (EPS) Production, PHB (Poly-β-hydroxybutyrate) accumulation and abiotic stress endurance. In addition, certain isolated *Methylobacterium* strains which produce vitamin B12 suggested stimulating plant development. Methylotrophic bacteria are being associated with plant nitrogen metabolism through bacterial urease production (Holland and Polacco 1994). However, the overall nature of their relationship with plants is as yet poorly understood, and the biological significance of these bacterial species is still under infancy and yet to be fully explored (Abanda-Nkpwatt et al. 2006).

5.6 Bioremediation

Methylobacterium sp. contributes to the bioremediation process via multiple modes of action, because these microorganisms can degrade and mineralize organic xenobiotic compounds allowing them to serve directly as contaminant degraders. The synergistic action of both *Methylobacterium* and the plants lead to increased availability of hydrophobic compounds, affecting their degradation. Ventorino et al. (2014) reported the biodegradation of polycyclic aromatic hydrocarbon (PAH) by *M. populi* VP2, a plant growth promoters. *Methylobacterium* is capable of the metabolism of monochlorinated, dichlorinated and aliphatic substrate. Jing et al. (2008) reported that *Methylobacterium* sp. HJ1 is able to degrade the herbicide 2,2-dichloropropionic acid by removal of the halogen and subsequent metabolism of the product for energy. D,L-2-chloropropionate also supported good growth of the organism, but 3-chloropropionate, monochloroacetate and dichloroacetate were not utilized. Cell-free extracts of the 2,2-dichloropropionate-grown bacteria converted 2,2-dichloropropionate into pyruvate with the release of two chloride ions for each molecule of pyruvate formed.

5.7 A Model Gene Expression System for Recombinant Protein

Research suggests that the *Methylobacterium* is proved as a model organism or an interesting candidate for overexpression of recombinant proteins. Marx and Lidstrom (2001) developed a series of new expression vectors for *M. extorquens* AM1 enabling efficient expression of reporter genes. One of the expression vectors, pCM110, is a 5.8 kb IncP-derived plasmid possessing the strong *M. extorquens* native promoter of methanol dehydrogenase (MDH) (*Pmxa*F). When compared to a similar vector containing only the *lac*Z promoter (*Pla*C), *Pmxa*F led to a 50-fold increase in the expression of the reporter gene *xylE* (Marx and Lidstrom 2001). Belanger et al. (2004) observed the usefulness of two distinct vectors (pRK310 and pCM110) and promoters (*Pla*C and *Pmxa*F) for heterologous expression in a high cell density for fed-batch fermentation process using *M. extorquens* ATCC 55366.

6 Concluding Remarks

Microbes being an integral component of any soil ecosystem provide life to the soil. Methylotrophs are a polyphyletic group of microorganisms capable of utilizing C1 compounds as electron donor and of the most abundant bacteria, which is able to grow on methanol as well as on multi-carbon compounds as sole carbon and energy source. Continued research with colonization and biofilm formation by these bacterial genera also holds potential for developing biofertilizer and biocontrol agents that may be self-perpetuating within the colonizing host plants. Focusing research in these areas may also be aimed to establish *Methylobacterium* sp. as promising plant-growth promoter and a model bioremediator.

References

Abanda-Nkpwatt D, Musch M, Tschiersch J, Boettner M, Schwab W (2006) Molecular interaction between *Methylobacterium extorquens* and seedlings: growth promotion, methanol consumption, and localization of the methanol emission site. J Exp Bot 57(15):4025–4032

Ahemad M, Khan MS (2010) Plant growth promoting activities of phosphate-solubilizing *Enterobacter asburiae* as influenced by fungicide. Eur Asian J Biosci 4:88–95

Allison AG (1998) Exopolysaccharide production in bacterial biofilm. Biofilm J 3(2):1–19

Alvarez MI, Sueldo RJ, Barassi CA (1996) Effect of *Azospirillum* inoculation on coleoptile growth in wheat seedlings under water stress. Cereal Res Commun 24:101–107

Anitha KG (2010) Enhancing seed germination of Mono and Dicotyledons through IAA production of PPFM. Trends Soil Sci Plant Nutr J 1(1):14–18

Ardanov P, Sessitsch A, Haggman H, Kozyrovska N, Pirttila AM (2012) Methylobacterium-induced endophyte community changes correspond with protection of plants against pathogen attack. PLoS ONE 7:10 Article ID e46802

Belanger L, Figueira MM, Bourque D, Morel L, Beland M, Larame L, Groleau D, Miguez DB (2004) Production of heterologous protein by *Methylobacterium extorquens* in high cell density fermentation. FEMS Microbiol Lett 231:197–204

Bhaskar PV, Bhosle NB (2005) Microbial extracellular polymeric substances in marine biogeochemical processes. Curr Sci 88(1):45–53

Breuer U, Babel W (1999) *Methylobacterium rhodesianum* produces poly-3-hydroxybutyrate and after mutagenesis in addition exopolysaccharides. Acta Biotechnol 19:779–786

Corpe WA, Rheem S (1989) Ecology of the methylotrophic bacteria on living leaf surfaces. FEMS Microbiol Ecol 62:243–250

Dourado MN, Andreote FD, Dini-Andreote F, Conti R, Araújo JM, Araújo WL (2012) Analysis of 16S rRNA and *mxa*F genes revealing insights into *Methylobacterium* niche specific plant association. Genet Mol Biol 35:142–148

Dourado MN, Bogas AC, Pomini AM, Andreote FD, Quecine MC, Marsaioli AJ, Araújo WL (2013) Methylobacterium-plant interaction genes regulated by plant exudate and quorum sensing molecules. Brazilian J Microbiol 44(4):1331–1339

Farag MA, Pare PW (2002) C6-green leaf volatiles trigger local and systemic VOC emissions in tomato. Phytochemistry 61:545–554

Fedorov DN, Doronina NV, Trotsenko YA (2011) Phytosymbiosis of aerobic methylobacteria: new facts and views. Microbiology 80(4):443–454

Fedorov DN, Ekimova GA, Doronina NV, Trotsenko YA (2013) 1-aminocyclopropane-1-carboxylate (ACC) deaminases from *Methylobacterium radiotolerans* and *Methylobacterium nodulans* with higher specificity for ACC. FEMS Microbiol Lett 343(1):70–76

Freyermuth SK, Long RL, Mathur S, Holland MA, Holstford TP, Stebbins NE, Morris RO, Polacco JC (1996) Metabolic aspects of plant interaction with commensal methylotrophs. In: Lindstorm M, Tabita R (eds) Microbial growth on C1 compounds. Kluwer Academic Publishers, New York, pp 21–134

Fusconi R, Godinho MJL (2002) Screening for exopolysaccharide producing bacteria from subtropical polluted groundwater. Brazilian J Biol 62(2):363–369

Golberg SB, Flick JS, Rogers SG (1984) Nucleotide sequence of the tmrlocus of *Agrobacterium tumefaciens* pTiT37 T-DNA. Nucl Acids Res 12:4665–4677

Goldstein AH, Lester T, Brown J (2003) Research on the metabolic engineering of the direct oxidation pathway for extraction of phosphate from ore has generated preliminary evidence for PQQ biosynthesis in *Escherichia coli* as well as a possible role for the highly conserved region of quinoprotein dehydrogenases. Biochimica Biophysica Acta 1647:266–271

Hardoim PR, Overbeek LSV, Elsas JDV (2008) Properties of bacterial endophytes and their proposed role in plant growth. Trends Microbiol 16:463–471

Holland MA (1997) Occams razor applied to hormonology. Are cytokinins produced by plants? Plant Physiol 115:865–868

Holland MA, Polacco JC (1992) Urease null and hydrogenase null phenotypes of a phylloplane bacterium reveal altered nickel metabolism in two soybean mutants. Plant Physiol 98:942–948

Holland MA, Polacco JC (1994) PPFMs and other contaminants: is there more to plant physiology than just plant. Annu Rev Plant Physiol Plant Mol Biol 45:197–209

Houck DR, Hanners JL, Unkefer CJ (1991) Biosynthesis of pyrroloquinoline quinone. Biosynthetic assembly from glutamate and tyrosine. J Am Chem Soc 113:3162–3166

Ivanova EG, Doronina NV, Shepeliakovskaia AO, Laman AG, Brovko FA, TrotsenkoIu A (2000) Facultative and obligate aerobic methylobacteria synthesize cytokinins. Mikrobiologia 69(6):764–769

Ivanova EG, Doronina NV, Trotsenko YA (2001) Aerobic methylobacteria are capable of synthesizing auxins. Microbiology 70(4):392–397

Jaftha JB, Strijdom BW, Steyn PL (2002) Characterization of pigmented methylotrophic bacteria which nodulate *Lotononis bainesii*. Syst Appl Microbiol 25(3):440–449

Jha CK, Patel D, Rajendran N, Saraf M (2010) Combinatorial assessment on dominance and informative diversity of PGPR from rhizosphere of *Jatropha curcas* L. J Basic Microbiol 50:211–217

Jha CK, Annapurna K, Saraf M (2012) Isolation of rhizobacteria from *jatropha curcas* and characterization of produced ACC deaminase. J Basic Microbiol 52:285–295

Jing NH, Taha AM, Pakingking RV, Wahabs RAB, Huyop F (2008) Dehalogenase from *Methylobacterium* sp. HJ1 induced by the herbicide 2,2-dichloropropionate (Dalapon). Afr J Microbiol Res 2:32–36

Joe MM, Saravanan VS, Islam MR, Sa T (2014) Development of alginate-based aggregate inoculants of Methylobacterium sp. and *Azospirillum brasilense* tested under in vitro conditions to promote plant growth. J App Microbiol 116(2):408–423

Jourand P, Giraud E, Béna G, Sy A, Willems A, Gillis M, Dreyfus B, De Lajudie P (2004) *Methylobacterium nodulans* sp. nov. for a group of aerobic, facultatively methylotrophic, legume root-nodule-forming and nitrogen-fixing bacteria. Int J Syst Evol Microbiol 54:2269–2273

Koenig RL, Morris RO, Polacco JC (2002) tRNA is the source of low-level trans-Zeatin production in *Methylobacterium* spp. J Bacteriol 184(7):1832–1842

Kutschera U (2007) Plant-associated methylobacteria as co-evolved phytosymbionts. Plant Signal Behav 2:74–78

Kwak MJ, Jeong H, Madhaiyan M, Lee Y, Sa TM, Oh TK, Kim JF (2014) Genome information of *Methylobacterium oryzae*, a plant-probiotic methylotroph in the phyllosphere. PLoS ONE 9(9):106–110

Lacava PT, Silva-Stenico ME, Araújo WL, Simionato AC, Carrilho E, Tsai SM, Azevedo JL (2008) Detection of siderophores in endophytic bacteria *Methylobacterium* spp. associated with *Xylella fastidiosa* subsp. *Pauca*. Pesqagropec Bras Brasília 43(4):521–528

Liu TS, Lee LY, Tai CY, Hung CH, Chang YS, Wolfram JH, Rogers R, Goldstein AH (1992) Cloning of an *Erwinia carotovora* gene necessary for gluconic acid production and enhanced mineral phosphate solubilization in *Escherichia coli* HB101: Nucleotide sequence and probable involvement in biosynthesis of the coenzyme pyrroloquinoline quinine. J Bacteriol 174:5814–5819

Lopez-Cortes A, Lanz-landazuri A, Garcia-Maldonado JQ. (2008) Screening and isolation of PHB producing bacteria in a polluted marine microbial mat. Microb. Ecol. 56:112–120

Madhaiyan M, Poonguzhali S, Senthilkumar M, Sundaram S, Chung H, Yang J, Sundaram S, Sa T (2004) Growth promotion and induction of systemic resistance in rice cultivar Co-47 (*Oryza sativa* L.) by Methylobacterium spp. Bot Bull Acad Sin 45:315–324

Madhaiyan M, Poonguzhali S, Ryu JH, Sa TM (2006a) Regulation of ethylene levels in canola (*Brassica campestris*) by 1-aminocyclopropane-1-carboxylate deaminase-containing *Methylobacterium fujisawaense*. Planta 224(2):268–278

Madhaiyan M, Reddy BVS, Anandham R (2006b) Plant growth-promoting *Methylobacterium* induces defense responses in groundnut (*Arachis hypogaea* L.) compared with rot pathogens. Curr Microbiol 53(4):270–276

Mantelin S, Touraine B (2004) Plant growth-promoting bacteria and nitrate availability: impacts on root development and nitrate uptake. J Exp Bot 55:27–34

Marx CJ, Lidstrom ME (2001) Development of improved versatile broad-host-range vectors for use in methylotrophs and other Gram-negative bacteria. Microbiology 147:2065–2075

Menna P, Hungria M, Barcellos FG, Bangel EV, Hess PN, Martínez-Romero E (2006) Molecular phylogeny based on the 16S rRNA gene of elite rhizobial strains used in Brazilian commercial inoculants. Syst Appl Microbiol 29:315–332

Omer ZS, Tombolini R, Gerhardson B (2004) Plant colonization by pink-pigmented facultative methylotrophic bacteria (PPFMs). FEMS Microbiol Ecol 47(3):319–326

Pal S, Manna A, Paul AK (1999) Production of (poly-hydroxy butyric acid) and exopolysacharide by *Azotobacte rbeijerinckii* WDN—01. World J Microbiol Biotechnol 15:15–21

Pomini AM, Cruz PLR, Gai C, Araújo WL, Marsaioli AJ (2009) Long-chain Acyl-Homoserine lactones from *Methylobacterium mesophilicum*: synthesis and absolute configuration. J Nat Prod 72:2130–2134

Poorniammal R, Sundaram SP, Kumutha K (2009) In vitro biocontrol activity of *Methylobacterium extorquens* against fungal pathogens. Int J Plant Prot 2:59–62

Rekadwad BN (2014) Growth promotion of crop plants by *Methylobacterium organophilum*: efficient bioinoculant and bio-fertilizer isolated from mud. Res Biotechnol 5(5):01–06

Sandmann G (2009) Evolution of carotene desaturation: the complication of a simple pathway. Arch Biochem Biophys 483:169–174

Saraf M, Jha CK, Patel D (2010) The role of ACC deaminase producing PGPR in sustainable agriculture. In: Maheshwari DK (ed) Plant growth and health promoting bacteria. Springer, Berlin Heidelberg, pp 365–385

Scier MH (1998) Proposed independent evolution of different channel and carrier families. In: Poole RK (ed) Advances in microbial physiology. Academic Press, London, pp 81–136

Senthil Kumar M, Madhaiyan M, Sundaram SP, Kannaiyan S (2009) Intercellular colonization and growth promoting effects of *Methylobacterium* sp. with plant-growth regulators on rice (*Oryza sativa* L. Cv CO-43). Microbes Res. 164(1):92–104.

Shaharoona B, Arshad M, Khalid A (2007) Differential response of etiolated pea seedling to 1-aminocyclopropane-1-carboxylate and/or L-methionine utilizing rhizobacteria. J Microbiol 45(1):15–20

Silva Stenico ME, Pacheco FTH, Rodriques JLM, Carrilho E, Tsai SM (2005) Growth and siderophore production of *Xylella fastidiosa* under iron-limited conditions. Microbiol Res 160(1):429–436

Smejkalova H, Erb TJ, Fuchs G (2010) Methanol assimilation in *Methylobacterium extorquens* AM1: demonstration of all enzymes and their regulation. PLoS ONE 5(10):e13001

Sy A, Giraud E, Jourand P, Garcia N, Willems A, De Lajudie P, Prin Y, Neyra M, Gillis M, Boivin-Masson C, Dreyfus B (2001) Methylotrophic *Methylobacterium* bacteria nodulate and fix nitrogen in symbiosis with legumes. J Bacteriol 183:214–220

Sy A, Timmers ACJ, Knief C, Vorholt JA (2005) Methylotrophic metabolism is advantageous for *Methylobacterium extorquens* during colonization of *Medicago truncatula* under competitive conditions. App Environ Microbiol 71(11):7245–7252

Tani A, Sahin N, Matsuyama Y, Enomoto T, Nishimura N, Yokota A, Kimbara K (2012) High-throughput identification and screening of novel *Methylobacterium* species using whole-cell MALDI-TOF/MS analysis. PLoS ONE 7(7):1–13

Vaidehi K, Sekar C (2012) Amino acid conjugated hydroxamate type of siderophore production in *Methylobacterium phyllosphaerae* MB-5. CIBtech J microbiol 1(1):25–30

Van Aken B, Yoon JM, Schnoor JL (2004) Biodegradation of nitro-substituted explosives 2,4,6-trinitrotoluene, hexahydro-1,3,5-trinitro-1,3,5-triazine, an octahydro-1,3,5,7-tetranitro-1,3,5-tetrazocine by a phytosymbiotic *Methylobacterium* sp. associated with poplar tissues (*Populus deltoides* nigra DN34). App. Environ Microbiol 70(1):508–517

Van Loon LC (2000) Systemic induced resistance. In: Slusarenko AJ, Fraser RSS, Van Loon LC (eds) Mechanisms of resistance to plant diseases. Kluwer, Dordrechet, pp 521–574

Van Loon LC, Bakker PAHM, Pieterse CMJ (1998) Systemic resistance induced by rhizosphere 660 bacteria. Annu Rev Phytopathol 36:453–483

Ventorino V, Sannino F, Piccolo A, Cafaro V, Carotenuto R, Pepe O (2014) *Methylobacterium populi* VP2: plant growth-promoting bacterium isolated from a highly polluted environment for polycyclic aromatic hydrocarbon (PAH) biodegradation. Sci World J Article ID 931793, 11 pages

Verginer M, Siegmund B, Cardinale M et al (2010) Monitoring the plant epiphyte *Methylobacterium extorquens* DSM21961 by real-time PCR and its influence on the strawberry flavor. FEMS Microbiol Ecol 74:136–145

Vorholt JA (2002) Cofactor-dependent pathways of formaldehyde oxidation in methylotrophic bacteria. Arch Microbiol 178(4):239–249

Wei H, Song S, Tian H, Liu T (2014) Effects of phenanthrene on seed germination and some physiological activities of wheat seedling. Compets Rendus Biol 337(2):95–100

White EW, Winans SC (2007) Cell-cell communication in the plant pathogen *Agrobacterium tumefaciens*. Phil Trans R Soc B 362:1135–1148

Woo S, Subramanian P, Ramasamy K, Joe MM, Sa T (2012) EPS production, PHB accumulation and abiotic stress endurance of plant growth promoting *Methylobacterium* strains grown in a high carbon concentration. Korean J Soil SciFert 45(4):572–581

Zahra BMH, Ebrahim VF, Abbas SS, Ramin K (2009) Media selection for poly hydroxybutyrate production from methanol by *Methylobacterium extorquens* DSMZ 1340. Iran J Chem Eng 28(3):45–52

Zhu H, Sun SJ (2008) Inhibition of bacterial quorum sensing regulated behaviors by *Tremella fuciformis* extract. Curr Microbiol 57:418–422

Role of Bacterial Phytohormones in Plant Growth Regulation and Their Development

G.R. Kudoyarova, T.N. Arkhipova and A.I. Melenťev

Abstract This review addresses the ability of some soil bacteria to promote plant growth due to production of substances of phytohormonal nature. We discuss possible mechanisms of the action of individual hormones (auxins, cytokinins, abscisic acid, gibberellins, jasmonic acid and salicylic acid) produced either by plants or bacteria on plant growth and development, their supply with mineral nutrients and water and defense responses against phytopathogens.

Keywords Plant growth-promoting bacteria · Rhizosphere · Plant hormones · Auxins · Cytokinins · Abscisic acid · Gibberellins · Jasmonic acid · Salicylic acid

1 Introduction

Promotion of plant growth by a range of bacteria attracts ever increasing attention of researchers. The interest to these bacteria is due, first of all, to possibility of increasing plant yield, which allows recommending their application as fertilizers (Vessey 2003) (Table 1). Beneficial effects on plant growth have been shown for numerous bacterial species and strains isolated from rhizosphere and plant leaf surface (Kishore et al. 2005; Avis et al. 2008). Promotion of plant growth has also been detected under the influence of endophytic bacteria capable of colonizing tissues and residing within plant hosts (Weyens et al. 2009). Representatives of

G.R. Kudoyarova (✉) · T.N. Arkhipova · A.I. Melenťev
Institute of Biology, Ufa Research Center of RAS, Ufa, Russia
e-mail: guzel@anrb.ru

D.K. Maheshwari (ed.), *Bacterial Metabolites in Sustainable Agroecosystem*,
Sustainable Development and Biodiversity 12, DOI 10.1007/978-3-319-24654-3_4

Table 1 Effect of pre-sowing treatments of wheat seeds with *Bacillus subtilis* IB-22 bacterial suspensions (colony forming units per ml) on wheat yield

Parameters	Bacteria concentration, CFU/seed			
	0	10^5	10^6	10^7
Yield (ton ha^{-1})	3.9 ± 0.5	4.5 ± 0.4*	5.6 ± 0.7*	4.8 ± 0.5*
Grain weight (g plant^{-1})	2.9 ± 0.1	3.5 ± 0.1**	4.1 ± 0.2**	3.9 ± 0.3**
Number of spike-bearing tillers	2.4 ± 0.1	3.1 ± 0.1**	3.4 ± 0.2**	3.0 ± 0.2**
Survived plants per bed	185 ± 8	239 ± 11	256 ± 9	231 ± 10

Means ± SE are shown. Yield components measured: grain weight, number of spike-bearing tillers ($n = 50$) and numbers of surviving plants per bed (1.4 m^2 per bed, $n = 10$)
An asterisk (*) indicates significant difference between plants grown from seeds, pretreated and untreated with bacteria (*$P < 0.05$, **$P < 0.01$ and Student's t test)

many species of genera *Azospirillum, Bacillus, Enterobacter, Gluconacetobacter, Paenibacillus* and *Pseudomonas* belong to plant growth-promoting bacteria (PGPB) (Maheshwari 2010). Since PGPB are mostly isolated from natural plant habitats, their application is an attractive alternative to chemical fertilizers enabling reduction of environment contamination (Cakmakci et al. 2007). However, efficacy of the action of plant growth-promoting bacteria depends on plant species, conditions of their growth and other factors (Marulanda et al. 2009), increasing the workload of their testing and developing recommendations for their application. Volume of work increases in geometrical progression due to frequent application of mixtures of several bacterial species and strains (Zaidi and Khan 2005; Pandey and Maheshwari 2007). Clearer understandings of the nature of PGPB action might reduce labor inputs in the development of effective technologies of their application and selection of bacterial combinations of several PGPB species and strains.

Stimulation of plant growth by bacteria is attributed to three of their main properties: (1) production of plant hormones (Dodd et al. 2010); (2) bacteria-related increases in availability of mineral nutrients for plants (Ohkama-Ohtsu and Wasaki 2010) (availability of water may be added to this category, from our point of view); (3) increased resistance to pathogens (Van Loon 2007). These properties may be apparent in different PGPB species or combined in the same species, however, each of them is mostly considered separately (Kannan and Sureendar 2009). Meanwhile, study of interaction of these properties of PGPB may contribute to better understanding the nature of their action on plants. In the present review, we made an attempt to reveal the role of plant hormones produced by microorganisms for the control of not only plant growth, but the availability of mineral nutrients and water as well as plant protection against pathogens. Since, beneficial effect of PGPB is not limited to production of plant hormones, some other mechanisms of plant growth promotion by PGPB will be mentioned in the present review.

2 Production of Plant Hormones by Plant Growth-Promoting Bacteria

Capacity to produce plant hormones is discovered in many plant growth-promoting bacteria. Thus, auxins have been discovered in the culture media of *Azospirillum, Pseudomonas* and others (Spaepen et al. 2007), cytokinins—in that of *Azotobacter vinelandii* (Azcon and Barea 1975), *Pantoea agglomerans* (Omer et al. 2004) and *Bacillus subtilis* (Arkhipova et al. 2005, 2007) (Fig. 1), gibberellins—in *Proteus mirabilis, P. vulgaris, Klebsiella pneumoniae* (Karadeniz et al. 2006), abscisic acid (ABA)—in *Azospirillum brasilense* (Cohen et al. 2009), jasmonic acid—in *Bacillus pumilus* (Forchetti et al. 2007) and *Achromobacter xylosoxidans* (endophytic bacteria of sunflower), salicylic acid—in *Pseudomonas aeruginosa*—(De Meyer et al. 1999) and all those substances belong to the main groups of plant hormones. It is well-known that plant hormones influence plant

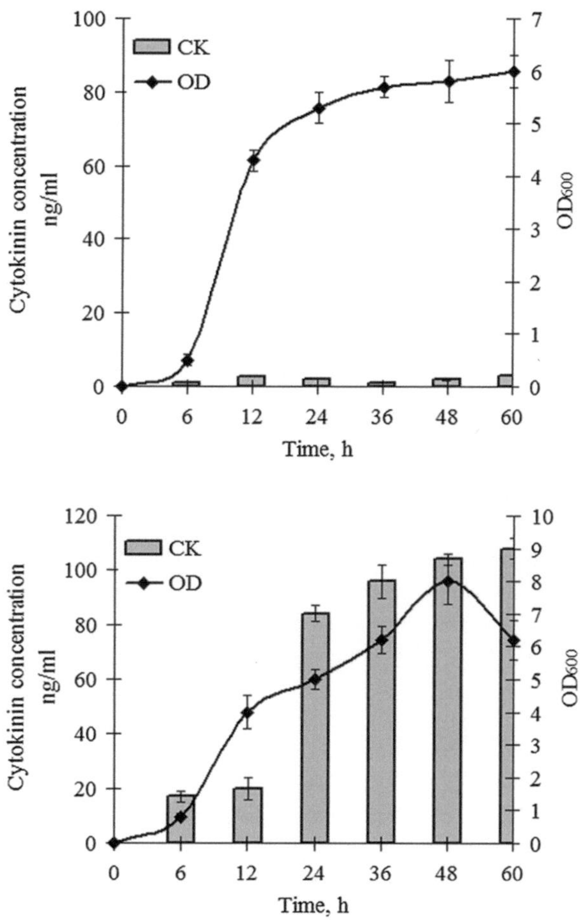

Fig. 1 Growth (expressed as optical density of culture media, OD) and accumulation of cytokinins (zeatin + its riboside and nucleotide) in culture media of cytokinin-nonproducing strain (**a**) and cytokinin-producing strain (**b**). Data are means ±SE of nine replicates

growth and resistance (Bari and Jones 2009; Shakirova et al. 2010). Therefore, it is not surprising that growth-promoting action of bacteria is attributed to their capacity to synthesize plant hormones. However, production of plant hormones in vitro does not mean by itself that PGPB synthesize them under natural environment and that uptake of microbe-mediated hormones by plants may influence hormone level. This explains the importance of reports on the changes in hormone content in plants as influenced by PGPB (Zaidi and Khann 2005). One more approach capable of confirming the importance of plant hormone production by PGPB is the use of mutant organisms (bacteria or plants) that lost sensitivity to hormones (López-Bucio et al. 2007). The results of application of this approach are considered below. Alongside with the production of hormones, capacity to destroy hormones was detected in some PGPB. Thus, there appeared ever growing number of reports on PGPB producing ACC deaminase, which catalyzes decay of 1-aminocyclopropane-1-carboxylate (ACC—precursor of gaseous hormone ethylene) (Shaharoona et al. 2006). Since, ethylene is frequently considered as growth inhibitor, the decline in its production by plants due to decay of its precursor by bacteria, may promote plant growth.

It is of interest that production of plant hormones is characteristic not only of PGPB, but of pathogenic microorganisms, too (Pertry et al. 2009). This paradox may be explained by the difference in the level of plant hormone production between pathogenic and growth-promoting microorganisms. The former are believed to produce more plant hormones than the latter (Persello-Carteaux et al. 2003). It is likely that plant hormones produced by PGPB optimize plant hormonal status, while it is disordered by plant hormones produced by pathogenic microorganisms. Consequences for plants of any changes in hormonal status are discussed below.

3 Effects of Phytohormones on Plant Growth

It is postulated that growth promotion is explained by synthesis of plant hormones of PGPB which is based on an oversimplified assumption that all hormones stimulate plant growth. This assumption is not correct, since hormones are capable of inducing opposite effects depending on the type of hormone and its concentration. Let's consider separately the effects of each group of hormones on plants.

3.1 Auxins

The ability of some bacteria to accumulate auxins in culture media after addition of auxin precursors (tryptophan) was detected by means of Salkowski reagent for indolic substances (Khalid et al. 2004). This comparatively simple approach allowed to discover the capacity to synthesize auxins in many microorganisms

(Persello-Carteaux et al. 2003; Jog et al. 2014). It is necessary to mention that specificity of this reaction was not high (Glickmann and Dessaux 1995), and it is not well recognized for assay of auxins in plants themselves. Still, application of other methods confirmed that inoculation of auxin producing bacteria into rhizosphere increases content of these hormones in plants (Ali et al. 2009). Involvement of auxins in activation of plant growth by PGPB has been confirmed in experiments on plants with genetically disturbed auxin transport. Auxin producing bacteria failed to stimulate the growth of these plants (Choudhary et al. 2009).

Production of auxins by PGPB is usually associated with activation of root growth (Spaepen et al. 2007). Study of cell-type specific developmental markers and employing genetic and pharmacological approaches demonstrated the crucial role of auxin signaling and transport in rhizobacteria-stimulated changes in the root system architecture of *Arabidopsis* (Zamioudis et al. 2013). In this case, it is important to remember that root growth is a complicated process consisting of their elongation and branching, which are differently regulated by hormones (Casson and Lindsey 2003; Wittenmayer et al. 2005). Auxin-induced increase in cell expansion is one of their best known properties (Rayle and Cleland 1980). Apparent paradox is in that exogenous auxins frequently inhibit root elongation (Teale et al. 2005). This property of auxins is well-known to molecular genetics and used for screening auxin-insensitive mutants (e.g. Stepanova et al. 2007). Nevertheless, stimulation of root elongation by auxin producing bacteria (Khalid et al. 2004) is not surprising. For some plant species, low concentrations of auxins have been shown to increase the rate of root elongation (Silva and Davies 2007). Thus, auxin producing bacteria resemble this regularity. PGPB are likely to produce auxins in the optimal range of concentration stimulating root elongation (Vacheron et al. 2013). It is important that high level of auxin production is a characteristic of bacteria that disturb root growth (Persello-Carteaux et al. 2003). Bacteria capable of inactivating auxins partially decreased growth inhibiting action of other microorganisms with excessively high level of auxin production during their co-inoculation (Leveau and Lindow 2005).

Stimulation of root branching by auxins is well-known and not controversial (Casson and Lindsey 2003), making it easy to explain increased branching induced by auxin producing bacteria, although it is more difficult to register this effect of PGPB than changes in root elongation. One more property of auxins is important for explaining plant growth promotion by auxin producing bacteria. It is in auxin-induced stimulation of root hair formation, playing important role in ion uptake (Wittenmayer et al. 2005).

Auxins are capable of influencing growth of not only roots, but shoots too (Rayle and Cleland 1980). This may explain the fact that auxin producing bacteria increased growth of both roots and shoots (Ali et al. 2009). In this case, the effect of bacteria inoculation on shoot growth may also be achieved through increased uptake of mineral nutrients due to faster root growth. This mechanism is discussed below. Here, it is important to emphasize that effect of microbe auxins on shoot growth is infrequently discussed, which is likely to be due to poorer knowledge on the transport of this hormone form roots to shoots compared to that of ABA and cytokinins.

3.2 Cytokinins

Production of cytokinins was discovered in fewer species and strains of PGPB in comparison to that of auxins (Veselov et al. 1998; Dodd et al. 2010). This is likely due to difficulties in cytokinin assay than natural occurrence of cytokinin-producing bacteria. Inoculation of plant rhizosphere with cytokinin-producing bacteria resulted in increased accumulation of biomass of either shoots or roots (Arkhipova et al. 2005, 2006). Elevated cytokinin concentration was detected in plants treated with cytokinin-producing bacteria confirming their effect on plant hormonal status. Simulation of plant growth under the influence of cytokinin-producing bacteria should not appear surprising since cytokinins have been discovered as substances necessary for plant cell division in vitro (Miller et al. 1956). Plant treatment with cytokinin-producing bacteria led to greater increase in biomass of shoots than of roots resulting in reduced ratio of root-to-shoot mass (Arkhipova et al. 2005) (Fig. 2). Such a reaction (allocation to shoot growth) is the characteristic of cytokinins, manifested either in plants treated with exogenous cytokinins or in transgenic plants with increased capacity to synthesize these hormones (Van Loven et al. 1993). According to available literature data, cytokinins inhibit both root elongation and their branching (Werner et al. 2003). Nevertheless, the data on weak development of root system in mutant plants with disturbances in cytokinin signaling (Argyros et al. 2008) serve as an evidence of necessity of cytokinins for normal root growth and development. Still, roots of plants treated with cytokinin-producing bacteria were shorter than in control plants (Arkhipova et al. 2005). The absence of clear inhibitory action of cytokinins produced by these bacteria may be due to their presence in a complex with polysaccharides (Veselov et al. 1998), gradual diffusion of cytokinins from this complex preventing their inhibitory action. As in the case of auxins, the study of hormone producing microorganisms enables better understanding the nature of cytokinin action on root growth.

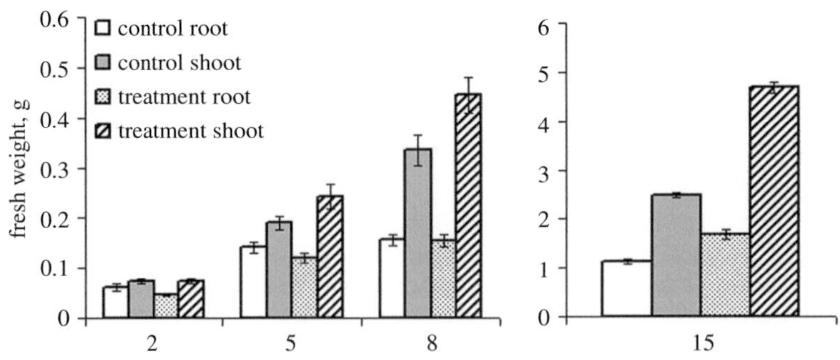

Fig. 2 Time course of fresh weight of roots and shoots, after inoculating the root environment of 12-days-old lettuce plants with *Bacillus subtilis* suspension culture (cytokinin-producing strain)

3.3 Abscisic Acid

Synthesis of this hormone was discovered in some PGPB (Dodd et al. 2010). It is not as easy to couple production of ABA by PGPB with promotion of plant growth as in case of cytokinins or auxins, since most of the researchers still consider this hormone to be a growth inhibitor. Nevertheless, this assumption is being changed. It turned out that mutant plants with decreased capacity to synthesize ABA were small in size, suggesting the necessity of ABA for their normal growth and development (Dodd et al. 2010). Although the mechanism of direct effect of ABA on plant growth is still unclear, there is no doubt that this hormone maintains cell extension due to normalization of water relation in plants (Fricke et al. 2004; Kudoyarova et al. 2011). ABA induces stomatal closure resulting in a decline in water evaporation from leaf surface and enabling plant water economy under drought (Bari and Jones 2009). Alongside with this effect, one more mechanism of ABA action on plant water relations has been revealed recently. It is in the activation of membrane water channels (Maurel et al. 2008). Their increased activity reduces hydraulic resistance and facilitates water flow from roots to shoots. We shall return to this mechanism of ABA action. Since expression of numerous genes implicated in stress response is ABA-dependent (Yoshida et al. 2014), microorganisms producing this hormone are likely to increase plant stress tolerance (Spaepen 2015). Bacterial strains mediating root ABA concentrations and growth by metabolizing ABA were isolated from the rhizosphere of rice (*Oryza sativa*) (Belimov et al. 2014).

3.4 Gibberellins

Discovery of this group of plant hormones is linked to the study of pathogen fungus *Gibberella fujikuroi*. Infection of rice with this pathogen resulted in excessive stem elongation due to the action of gibberellins produced by the fungus (Robert-Seilaniantz et al. 2007). It was shown that plants themselves are capable of synthesizing gibberellins. Modern concept suggests that growth promotion by gibberellins is due to gibberellins-induced degradation of DELLA transcription factor known to repress expression of some genes necessary for maintaining plant growth (Achard et al. 2003). Some PGPB have been shown to synthesize gibberellins (Karadeniz et al. 2006; Kang et al. 2014), and this may also explain how PGPB promote plant growth. The best documented case for the role of gibberellins produced by microbes in plant growth promotion is the reversion of the dwarf phenotype of plants by inoculation with gibberellins-producing *Azospirillum* strains (Spaenen 2015).

3.5 Ethylene

This gaseous hormone plays important role in the control of plant growth. As in the case of ABA, most researchers consider ethylene to be a growth inhibitor (Achard et al. 2003). This is an over simplified assumption, since under certain conditions (flooding and shading) ethylene contributes to maintaining stem elongation (Pierik et al. 2007). However, growth inhibiting action of ethylene appears more frequently. Mutants missing sensitivity to ethylene were characterized by longer roots compared to wild-type plants. Inhibition of root elongation by ethylene has been shown due to stabilization of DELLA-factor. Nevertheless, some reviews emphasize dual action of ethylene and compare it with mythic two-faced Janus (Pierik et al. 2007). Thus, unlike high concentrations of ethylene, its low concentrations activate plant growth (that of root,—first of all). Growth stimulation of the root in seedling stage by the ethylene precursor ACC was dose-dependent and affected by *Herbaspirillum frisingense* inoculation (Straub et al. 2013). Ethylene production has been detected in a range of bacteria (Dodd et al. 2010). However, bacteria, producing ACC deaminase, which catalyzes the decay of ethylene precursor, are more frequently discussed (Belimov et al. 2009). Decay of ACC in soil by this kind of bacteria facilitates its diffusion from roots to rhizosphere decreasing production of ethylene in plants and contributing to root elongation. It is important that unlike synthetic inhibitors, enabling complete block of ethylene action, ACC deaminase producing bacteria do not exclude its action on plants, but decline concentration of ethylene to the level enabling stimulation of root growth. Ethylene is also involved in plant resistance to pathogens (Wang et al. 2002). This aspect of action of ACC deaminase producing bacteria is discussed below. Contrary to the wild-type plants in which *Bacillus megaterium* stimulated growth rates, PGPR caused growth inhibition in ABA-deficient mutant plants (Porcel et al. 2014). Over-accumulation of ethylene detected in ABA-deficient mutant plants indicated that maintenance of normal plant endogenous ABA content may be essential for the growth-promoting action of *B. megaterium* by keeping low levels of ethylene production.

3.6 Jasmonic Acid and Salicylic Acids

According to the modern classification these substances belong to plant hormones (Shakirova et al. 2010). Their main role in plants is in triggering defense responses enabling plant resistance to pathogens. A range of PGPB has been shown to synthesize these compounds (Dodd et al. 2010). Since their effects on plant growth are poorly studied, detailed consideration of these regulators is in the section dedicated to the effects of PGPB on plant defense responses.

In summary, all the above details about the control of plant growth due to microbe-mediated microbe-born hormones indicate the following points. Unlike pathogenic microorganisms or transgenic plants that synthesize excessive amounts of plant hormones and lead to disorder in plant growth and development, PGPBs are characterized by the capacity to optimize plant hormonal system resulting in plant growth promotion. Information concerning exact plant hormones synthesized by certain PGPB may be important for choosing compositions of several strains. Thus, it may be possible to recommend combination of cytokinin-producing bacteria preferentially stimulating shoot growth with those producing auxin or consuming ethylene precursor and in this way activate/facilitate root growth.

4 Role of Plant Hormones Produced by PGPB in the Supply of Plants with Water and Mineral Nutrients

Certain abiotic factor such as water and nutrient deficiency are known to decrease the content of growth stimulating hormones in plants, first of all, cytokinins (Hare et al. 1997). Thus, it may be expected that plant sensitivity to the treatment of PGPB increases under these conditions. Microorganisms adapted to exist under drought and selected under corresponding conditions have been shown to supply plants with hormones (Cohen et al. 2009; Marulanda et al. 2009). Activation of plant growth is beneficial for microorganisms, since it increases root exudation and supply of microorganisms with nutrients (Bais et al. 2006). It is important to find out, if activation of growth of stressed plants is beneficial for plants themselves. Triggering plant protective responses is believed to imply inhibition and not activation of growth (Achard et al. 2003). From this point of view, the decline in the content of growth stimulating hormones and accumulation of growth inhibitors in plants may be considered as an adaptive stress response. Nevertheless, analysis of modern cultivars capable of high crop yields under stress show that they combine high-resistance with comparatively high-growth rate under drought (Collins et al. 2008). Importance of root growth and development under water deficit is obvious and has never been doubted. The matter of debate is the shoot growth under these conditions. It is obvious that inhibition of leaf growth results in formation of smaller leaves contributing to reduced water evaporation from the leaf surface and economic water usage (Bacon 1999). Nevertheless, the decline in leaf area influences detrimentally photosynthesis and plant productivity. To overcome this contradiction, it is important to define clearly the terms. If survival of plants under stressful environment is meant, which is ecological definition of resistance, then inhibition of growth may really contribute to resistance. However, when the matter concerns plant productivity under stressful environment, mechanisms maintaining plant growth are likely to increase stress resistance. It is from this agronomical point of view that the action of PGPB on plants will be discussed.

Since the function of water and ion uptake is fulfilled by plant roots, there is no doubt that activation of their growth by PGPB contributes to increase the availability of water and mineral nutrients for plants, which is particularly important under their deficit. Auxin production detected in a range of bacteria is of interest from this point of view. It is not likely to be just a coincidence that capacity to synthesis auxins were discovered in bacteria capable of activating the uptake of mineral nutrients by plants (Cakmakci et al. 2007). However, the suggestion that effect of these bacteria is limited to activation of root growth and root hair formation is not correct. Low solubility of soil phosphates decreasing their availability for plants is a serious agrochemical problem, which may be solved due to application of PGPB. Solubilization of phosphate-containing inorganic compounds by microorganisms is due to excretion of metabolites that acidify soil solution. An important role in this process is played by the chelating substances: accumulation of citric, gluconic acids, etc. (Whitelaw et al. 1999; Vessey 2003). Similar mechanisms of phosphate solubilization is seen in certain fungus, viz., species of *Fomitopsis* is able to solubilize tri-calcium phosphate, aluminum phosphate and hydroxyapatite. The ability to solubilize under in vitro condition is enhanced in the presence of salinity (Kang et al. 2002).

It is also important to bear in mind that one of the known properties of auxins is in the activation of membrane ATPases (Hager 2003) and manifestation of this property in root epidermis contributes to acidification of soil solution resulting in the increase of solubility of phosphates (Hinsinger et al. 2003). Thus, the role of auxins produced by PGPB may be not only in stimulating root growth, but also in activating the release of hydrogen ions into the soil solution. It was also possible to link stimulation of root growth by PGPB to microbe-borne ACC deaminase, enabling a decline in ethylene syntheses from its precursor, where decay was catalyzed by the enzymes of this class (Belimov et al. 2009). Since, as mentioned above, it is root growth that contributes to adaptation of water deficit, the increase in plant productivity exerted by inoculation of these microorganisms under drought is not surprising.

Although production of abscisic acid (ABA) by PGPB was not frequently detected, the role of this hormone in the control of stress response should attract attention of microbiologists in the search of ABA producing bacteria. To screen for such microorganisms, it may be useful to use a selective media with increased concentration of osmotically active substances stimulating ABA synthesis (Dodd et al. 2010). Inoculation of ABA producing PGPB has been shown to optimize water relations in plants under drought (Cohen et al. 2009), which is likely to contribute in maintaining their productivity. The effect of ABA on water relations is due to either ABA-induced stomatal closure (Bari and Jones 2009), or promotion of water uptake by ABA-activated water channels (Maurel et al. 2008). Moreover, beneficial effect of ABA under conditions of water and ions deficit may be explained by the effect of this hormone on root-to-shoot mass ratio (Chapin 1990). Thus, ABA was shown to induce allocation to root growth, which allows attributing relative activation of root growth under water and ions deficit to accumulation of this hormone.

The role of cytokinins produced by PGPB in the control of root capacity for water uptake may seem less beneficial as compared to that of auxins. Although root mass accumulation increased under inoculation of these bacteria, a reduction of root length was detected in some cases (Arkhipova et al. 2005). Moreover, increased content of cytokinins may influence the uptake of mineral nutrients detrimentally, since cytokinins inhibit activity of nitrate and phosphate transporters (Liu et al. 2009; Maheshwari 2012). Nevertheless, growth promotion by cytokinin-producing bacteria was manifested not only under favorable environment, but under conditions of water and ions deficit (Arkhipova et al. 2007; Arkhipova and Anokhina 2009). It is likely that detrimental consequences of cytokinin production were overcompensated by other properties of PGPB. At any case, in the context of revision of plant resistance concepts (Collins et al. 2008), stimulation of leaf growth by cytokinins (Werner et al. 2003) is likely to contribute to plant yield either under favorable or stressful environment.

One more aspect of cytokinin action may be important for the supply of plants with water and mineral nutrients due to PGPB with special reference to the influence of cytokinins on development of mycorrhiza and nitrogen-fixation. Thus, it is known that cytokinins are necessary for symbiotic relations between nitrogen-fixing nodule bacteria and plants (Lee et al. 2007; Frugier et al. 2008), which is due to stimulation of cell division by cytokinins. Activation of mycorrhiza development and formation of nitrogen-fixing nodules was detected in plants treated with PGPB (Barriuso et al. 2008; Nadeem et al. 2014). Since it is obvious that development of mycorrhiza activates uptake of water and mineral nutrients, while formation of nodules supplies plants with consumable nitrogen, capacity of PGPB to influence plant growth is likely to be due to their beneficial effect on these processes related to some extent to their ability to synthesize hormones.

Summarizing all details produced in this section, we may emphasize that production of plant hormones by PGPB influences root absorbing activity of minerals. This effect is due to either activation of root growth or other processes delinked directly with root growth: acidification of soil solution with root exudates, increase in root hydraulic conductivity and stimulation of formation of mycorrhiza and nitrogen-fixing nodules.

5 Effect of PGPB on Plant Defense Responses to Pathogens

The last but not least mechanism of the effect of PGPB on plants is due to their ability to protect plants against pathogen attacks (Bari and Jones 2009). These effects are partially due to competition of PGPB with plant pathogens leading to secretion of antibiotic substances and hydrolytic enzymes that disturb structure and function of pathogen cells (Melent'ev and Galimzyanova 1999; Melent'ev et al. 2001). Moreover, some PGPB are capable of activating some defense responses of plants themselves (Van Loon 2007). Plants are known to respond to infections by a range of defense reactions. Some of them are aimed on the formation of barriers on the pathway of pathogen, inside the plant. One of the examples of such reactions is hypersensitivity, when penetration of biotrophic pathogens, parasitizing on plant tissue, results in programmed cell death and formation of lignified barrier of dead cells (Van Loon 2007). Another example is the closure of stomature, through which pathogens penetrate inside leaves (Bari and Jones 2009). Production of reactive oxygen species at the site of pathogen penetration and synthesis of protective proteins (proteases and their inhibitors, chitinases and other hydrolases capable of inhibiting development of infection) also belong to defense mechanisms (Aktuganov et al. 2007, 2008). PGPB have been shown to activate these plant defense responses (Bordiec et al. 2011).

The effects of PGPB on plants and not on pathogen itself were most evident in experiments, where inoculation of PGPB into rhizosphere decreased the extent of damage to plants by leaf pathogens (Van Loon 2007). However, it is still unclear how PGPB activate plant defense responses, enabling plant resistance to pathogens. In principle, PGPB produce regulators that are potentially capable to affect the processes enabling plant resistance to pathogens. Thus, implication of salicylic and jasmonic acids in induction of pathogen-related (PR) proteins is well known. The difficulty is in establishing the relation between production of either salicylic or jasmonic acids by PGPB and plant defense responses. Thus, PGPB increased resistance to pathogens in mutant plants that lost sensitivity to these regulators (Dodd et al. 2010). Nevertheless, insight into the transcriptional changes induced by specific rhizobacteria in *Arabidopsis* showed their association with the salicylic acid-dependent resistance response (Van de Mortel et al. 2012). It is necessary to consider the existence of different specific mechanisms of plant defense against the attacks of either biotrophic or necrotrophic pathogens (the latter parasitize on dead plant tissues) that are triggered by different regulators (Bari and Jones 2009). Thus, salicylic acid has been shown to induce defense response aimed against the attacks of biotrophic microorganisms, while jasmonic acid and ethylene—against necrotrophs. Moreover, an antagonism was detected between the cascades of reactions triggered by salicylic and jasmonic acids (Riviere et al. 2008; Bari and Jones 2009). Salicylic acid inhibited defense reaction induced by jasmonic acid and vice versa (Spoel et al. 2007, 2008). These complicated interactions should be considered when attempts are made to decipher protective function of regulators produced by PGPB. Bacteria producing salicylic acid should not be expected to protect against necrotrophic pathogens, while producers of jasmonic acid are not likely to defend plants against biotrophs.

From our point of view, following protective functions of other microbe-borne hormones may be important. Thus, production of ABA may contribute to stomatal closure under penetration of leaf pathogens (Bari and Jones 2009). However, ABA inhibits lignification and other responses induced by salicylic acid. Still some reports provide evidences on involvement of ABA in triggering accumulation of jasmonic acid (Fan et al. 2009). Possibility to protect plants against pathogen infection with the help of PGPB, producing enzymes that destroy ethylene precursor, seems unlikely at first glance, since ethylene is known to induce plant resistance (Pierik et al. 2007). Ethylene has been involved in the induction of programmed cell death during hypersensitive response (Lin et al. 2009). Nevertheless, the role of this hormone is controversial and insensitivity to ethylene increased plant immunity against some infections (Lund et al. 1998). Thus, it is not surprising that bacteria producing ACC deaminase, which destroys ethylene precursor, protected plants against some diseases (Wang et al. 2000).

It is important to emphasize that high auxin concentration may contribute to inhibition of plant defense responses (Spaepen et al. 2007). This explains high level of auxin production by some pathogenic microorganisms. As mentioned above, PGPB also produce auxins, although the level of their synthesis is lower than in pathogenic microorganisms and auxins produced by PGPB are not likely to affect detrimentally plant defense responses. At last, implication of PGPB-produced cytokinins in their protective action on plants is not excluded. For instance, cytokinins have been shown to induce synthesis of salicylic and jasmonic acid (Sano et al. 1996) and to activate lignification (Guo et al. 2005).

Thus, plant hormones produced by PGPB may participate not only in their direct effect on plant growth, but in the increase in availability of mineral nutrients and water as well as protection of plants against pathogen attack. Deeper understanding of the nature of the action of microbe-born plant hormones may contribute to the increase in efficacy of biotechnology of their application focused on the increase in plant resistance and crop yield.

6 Conclusion

Thus, phytohormones produced by microorganisms may exert beneficial effects on plants by influencing their growth, ability for the uptake of water and mineral nutrients, and by increasing their pathogen tolerance and resistance to abiotic stresses. However, further study of the effect of hormone production by microorganisms and their effect on plants is necessary to enable stable and reproducible effect that may be used in agriculture for increasing crop yield.

Acknowledgments This study was supported by the Russian Foundation for Basic Research (grant 14-04-00775; grant 14-04-97049).

References

Achard P, Vriezen WH, Straeten D, Harberd NP (2003) Ethylene regulates *Arabidopsis* development via the modulation of DELLA protein growth repressor function. Plant Cell 15:2816–2825

Aktuganov GE, Galimzyamova NF, Melent'ev AI, Kuz'mina LY (2007) Extracellular hydrolases of strain *Bacillus* sp. 739 and their involvement in the lysis of micromycete cell walls. Microbiology (Mikrobiologiya) 76(4):413–420

Aktuganov G, Melent'ev A, Galimzianova N, Khalikova E, Korpela T, Susi P (2008) Wide-range antifungal antagonism of *Paenibacillus ehimensis* IB-x-b and its dependence on chitinase and β-1,3-glucanase production. Can J Microbiol 54(7):577–587

Ali B, Sabri AN, Ljung K, Hasnain S (2009) Auxin production by plant associated bacteria: impact on endogenous IAA content and growth of *Triticum aestivum* L. Lett Appl Microbiol 48:542–547

Argyros RD, Mathews DE, Chiang YH, Palmer CM, Thibault DM, Etheridge N, Argyros DA, Mason MG, Kieber JJ, Schaller GE (2008) Type B response regulators of *Arabidopsis* play key roles in cytokinin signaling and plant development. Plant Cell 20:2102–2116

Arkhipova TN, Anokhina NL (2009) Effects of wheat plant inoculation with cytokinin-producing microorganisms on plant growth at increasing level of mineral nutrition. Russ J Plant Physiol 56:814–819

Arkhipova TN, Melentiev AI, Martynenko EV, Kudoyarova GR, Veselov SU (2005) Ability of bacterium *Bacillus subtilis* to produce cytokinins and to influence the growth and endogenous hormone content of lettuce plants. Plant Soil 272:201–209

Arkhipova TN, Melent'ev AI, Martynenko EV, Kudoyarova GR, Veselov SY (2006) Comparison of effects of bacterial strains differing in their ability to synthesize cytokinins on growth and cytokinin content in wheat plants. Russ J Plant Physiol 53:507–513

Arkhipova TN, Prinsen E, Veselov SU, Martinenko EV, Melentiev AI, Kudoyarova GR (2007) Cytokinin producing bacteria enhance plant growth in drying soil. Plant Soil 292:305–315

Avis TJ, Gravel V, Antoun H, Tweddell RJ (2008) Multifaceted beneficial effects of rhizosphere microorganisms on plant health and productivity. Soil Biol Biochem 40:1733–1740

Azcon R, Barea J (1975) Sinthesis of auxins, gibberellins and cytokinins by *Azotobacter vinelandii* and *Azotobacter bijernkii* related to effects produced on tomato plants. Plant Soil 43:609–619

Bacon MA (1999) The biochemical control of leaf expansion during drought. Plant Growth Regul 29:101–112

Bais HP, Weir TL, Perry LG, Gilroy S, Vivanco JM (2006) The role of root exudates in rhizosphere interactions with plants and other organisms. Annu Rev Plant Biol 57:233–266

Bari R, Jones JDG (2009) Role of plant hormones in plant defence responses. Plant Mol Biol 69:473–488

Barriuso J, Solano B, Santamaria C, Daza A, Manero FJG (2008) Effect of inoculation with putative plant growth-promoting rhizobacteria isolated from *Pinus* spp. on *Pinus pinea* growth, mycorrhization and rhizosphere microbial communities. J Appl Microbiol 105:1298–1309

Belimov AA, Dodd IC, Hontzeas N, Theobald JC, Safronova VI, Davies WJ (2009) Rhizosphere bacteria containing ACC deaminase increase yield of plants grown in drying soil via both local and systemic hormone signaling. New Phytol 181:413–423

Belimov AA, Dodd IC, Safronova VI, Dumova VA, Shaposhnikov AI, Ladatko AG, Davies WJ (2014) Abscisic acid metabolizing rhizobacteria decrease ABA concentrations in planta and alter plant growth. Plant Physiol Biochem 74:84–91

Bordiec S, Paquis S, Lacroix H, Dhondt S, Barka EA, Kauffmann S, Jeandet P, Mazeyrat-Gourbeyre F, Clement C, Baillieul F, Dorey S (2011) Comparative analysis of defence responses induced by the endophytic plant growth-promoting rhizobacterium *Burkholderia phytofirmans* strain PsJN and the non-host bacterium *Pseudomonas syringae* pv. *pisi* in grapevine cell suspensions. J Exp Bot 62(2):595–603

Cakmakci R, Erat M, Erdogan U, Donmez MF (2007) The influence of plant growth–promoting rhizobacteria on growth and enzyme activities in wheat and spinach plants. J Plant Nutr Soil Sci 170:288–295

Casson SA, Lindsey K (2003) Genes and signaling in root development. New Phytol 158:11–38

Chapin FS (1990) Effects of nutrient deficiency on plant growth: evidence for a centralised stress-response system. In: Davies WI, Jeffcoat B (eds) Plant roots from cells to systems. British Society for Plant Growth Regulation, Bristol, pp 135–148

Choudhary DK, Prakash A, Wray V, Johri BN (2009) Insights of the fluorescent pseudomonads in plant growth regulation. Curr Sci 97(2):170–179

Cohen AC, Travaglia CN, Bottini R, Piccoli PN (2009) Participation of abscisic acid and gibberellins produced by endophytic *Azospirillum* in the alleviation of drought effects in maize. Botany 87:455–462

Collins NC, Tardieu F, Tuberosa R (2008) Quantitative trait loci and crop performance under abiotic stress: where do we stand? Plant Physiol 147:469–486

De Meyer G, Capieau K, Audenaert K, Buchala A, Metraux J-P, Höfte M (1999) Nanogram amounts of salicylic acid produced by the rhizobacterium *Pseudomonas aeruginosa* 7NSK2 activate the systemic acquired resistance pathway in bean. Mol Plant Microbe Interact 12:450–458

Dodd IC, Zinovkina NY, Safronova VI, Belimov AA (2010) Rhizobacterial mediation of plant hormone status. Ann Appl Biol 157:361–379

Fan J, Hill L, Crooks C, Doerner P, Lamb C (2009) Abscisic acid has a key role in modulating diverse plant-pathogen interactions. Plant Physiol 150:1750–1761

Forchetti G, Masciarelli O, Alemano S, Alvarez D, Abdala G (2007) Endophytic bacteria in sunflower (*Helianthus annuus* L.): isolation, characterization, and production of jasmonates and abscisic acid in culture medium. Appl Microbiol Biotechnol 76:1145–1152

Fricke W, Akhiyarova G, Veselov D, Kudoyarova G (2004) Rapid and tissue-specific changes in ABA and in growth rate in response to salinity in barley leaves. J Exp Bot 55:1115–1123

Frugier F, Kosuta S, Murray JD, Crespi M, Szczyglowski K (2008) Cytokinin: secret agent of symbiosis. Trends Plant Sci 13:115–120

Glickman E, Dessaux Y (1995) A critical examination of the specificity of the Salkowski reagent for indolic compounds produced by phytopathogenic bacteria. Appl Environ Microbiol 61:793–796

Guo J, Hu X, Duan R (2005) Interactive effects of cytokinins, light, and sucrose on the phenotypes and the syntheses of anthocyanins and lignins in cytokinin overproducing transgenic *Arabidopsis*. J Plant Growth Regul 24:93–101

Hager A (2003) Role of the plasma membrane H + -ATPase in auxin-induced elongation growth: historical and new aspects. J Plant Res 116:483–505

Hare PD, Cress WA, Van Staden J (1997) The involvement of cytokinins in plant responses to environmental stresses. J Plant Growth Regul 23:79–103

Hinsinger P, Plassard C, Tang C, Jaillard B (2003) Origins of root-mediated pH changes in the rhizosphere and their responses to environmental constraints. Plant Soil 248:43–59

Jog R, Pandya M, Nareshkumar G, Rajkumar S (2014) Mechanism of phosphate solubilization and antifungal activity of *Streptomyces* spp. isolated from wheat roots and rhizosphere and their application in improving plant growth. Microbiology 160:778–788

Kang SC, Chul GH, Lee TG, Maheshwari DK (2002) Solubilization of insoluble inorganic phosphates by a soil inhabiting fungus *Fomitopsis* spp. PS 102. Curr Sci 25:439–442

Kang SM, Waqas M, Khan AL, Lee IJ (2014) Plant-growth-promoting rhizobacteria: potential candidates for gibberellins production and crop growth promotion. In: Use of microbes for the alleviation of soil stresses, vol 1. Springer, New York, pp 1–19

Kannan V, Sureendar R (2009) Synergistic effect of beneficial rhizosphere microflora in biocontrol and plant growth promotion. J Basic Microbiol 49:158–164

Karadeniz A, Topcuoglu SF, Inan S (2006) Auxin, gibberellin, cytokinin and abscisic acid production in some bacteria. World J Microbiol Biotechnol 22:1061–1064

Khalid A, Arshad M, Zahir ZA (2004) Screening plant growth-promoting rhizobacteria for improving growth and yield of wheat. J Appl Microbiol 96:473–480

Kishore GK, Pande S, Podile AR (2005) Phylloplane bacteria increase seedling emergence, growth and yield of field-grown groundnut (*Arachis hypogaea* L.). Lett Appl Microbiol 40:260–268

Kudoyarova G, Veselova S, Hartung W, Farhutdinov R, Veselov D, Sharipova G (2011) Involvement of root ABA and hydraulic conductivity in the control of water relations in wheat plants exposed to increased evaporation demand. Planta 233:87–94

Lee A, Lum MR, Hirsch AM (2007) ENOD40 gene expression and cytokinin responses in the nonnodulating, nonmycorrhizal (Nod-Myc-) mutant, Masym3, of *Melilotus alba* Desr. Plant Signal Behav 2:33–42

Leveau JHJ, Lindow SE (2005) Utilization of the plant hormone indole-3-acetic acid for growth by *Pseudomonas putida* strain 1290. Appl Environ Microbiol 71:2365–2371

Lin Z, Zhong S, Grierson D (2009) Recent advances in ethylene research. J Exp Bot 60(12):3311–3336

Liu T-Y, Chang C-Y, Chiou T-J (2009) The long-distance signaling of mineral macronutrients. Curr Opin Plant Biol 12:312–319

López-Bucio J, Campos-Cuevas JC, Hernández-Calderon E, Velásquez-Becerra C, Farías-Rodríguez R, Macías-Rodríguez LI, Valencia-Cantero E (2007) *Bacillus megaterium* rhizobacteria promote growth and alter root-system architecture through an auxin- and ethylene-independent signaling mechanism in *Arabidopsis thaliana*. Mol Plant Microbe Interact 20(2):207–217

Lund ST, Stall RE, Klee HJ (1998) Ethylene regulates the susceptible response to pathogen infection in tomato. Plant Cell 10:371–382

Maheshwari DK (2010) Plant growth and health promoting bacteria. In: Microbiology monographs, vol 18. Springer, Heidelberg. ISBN:978-3-642-13611-5

Maheshwari DK (2012) Bacteria in Agrobiology: Plant Nutrient Management. Springer, Heidelberg

Marulanda A, Barea J-M, Azcon R (2009) Stimulation of plant growth and drought tolerance by native microorganisms (AM Fungi and Bacteria) from dry environments: mechanisms related to bacterial effectiveness. J Plant Growth Regul 28:115–124

Maurel C, Verdoucq L, Luu DT, Santoni V (2008) Plant aquaporins: membrane channels with multiple integrated functions. Ann Rev Plant Biol 59:595–624

Melent'ev AI, Galimzyanova NF (1999) Effects of metabolites of antagonistic *Bacilli* on spore germination and development of fungi causing common root rot. Appl Biochem Microbiol 35(3):316–320

Melent'ev AI, Aktuganov GE, Galimzyanova NF (2001) The role of chitinase in the antifungal activity of *Bacillus* sp. 739. Mikrobiologiya 70(5):548–552

Miller CO, Skoog F, Okomura FS, Von Saltza MH, Strong FM (1956) Isolation, structure and synthesis of kinetine, a substance promoting cell division. J Am Chem Soc 78:1345–1350

Nadeem SM, Ahmad M, Zahir ZA, Javaid A, Ashraf M (2014) The role of mycorrhizae and plant growth promoting rhizobacteria (PGPR) in improving crop productivity under stressful environments. Biotechnol Adv 32:429–448

Ohkama-Ohtsu N, Wasaki J (2010) Recent progress in plant nutrition research: cross-talk between nutrients, plant physiology and soil microorganisms. Plant Cell Physiol 51(8):1255–1264

Omer ZS, Bjorkman PO, Nicander B, Tillberg E, Gerhardson B (2004) 5-Deoxyisopentenyladenosine and other cytokinins in culture filtrates of the bacterium *Pantoea agglomerans*. Physiol Plant 121:439–447

Pandey P, Maheshwari DK (2007) Bioformulation of *Burkholderia* sp. MSSP with a multi-species consortium for growth promotion of *Cajanus cajan*. Can J Microbiol 53:213–222

Persello-Cartieaux F, Nussaume L, Robaglia C (2003) Tales from the underground: molecular plant–rhizobacteria interactions. Plant Cell Environ 26:189–199

Pertry I, Václavíková K, Depuydt S, Galuszka P, Spíchal L, Temmerman W, Stes E, Schmülling T, Kakimoto T, Van Montagu MCE, Strnad M, Holsters M, Tarkowski P, Vereecke D (2009) Identification of *Rhodococcus fascians* cytokinins and their modus operandi to reshape the plant. Proc Natl Acad Sci USA 106(3):929–934

Pierik R, Sasidharan R, Voesenek LACJ (2007) Growth control by ethylene: adjusting phenotypes to the environment. J Plant Growth Regul 26:188–200

Porcel R, Zamarreño ÁM, García-Mina JM, Aroca R (2014) Involvement of plant endogenous ABA in *Bacillus megaterium* PGPR activity in tomato plants. BMC Plant Biol 14(1):36

Rayle DL, Cleland RE (1980) Evidence that auxin-induced growth of soybean hypocotyls involves proton excretion. Plant Physiol 66:433–437

Riviere M-P, Marais A, Ponchet M, Willats W, Galiana E (2008) Silencing of acidic pathogenesis-related PR-1 genes increases extracellular β-(1/3)-glucanase activity at the onset of tobacco defence reactions. J Exp Bot 59(6):1225–1239

Robert-Seilaniantz A, Navarro L, Bari R, Jones JDG (2007) Pathological hormone imbalances. Curr Opinion Plant Biol 10:372–379

Sano H, Seo S, Koizumi N, Niki T, Iwamura H, Ohashi Y (1996) Regulation by cytokinins of endogenous levels of jasmonic and salicylic acids in mechanically wounded tobacco plants. Plant Cell Physiol 37(6):762–769

Shaharoona B, Arshad M, Zahir ZA (2006) Effect of plant growth promoting rhizobacteria containing ACC-deaminase on maize (*Zea mays* L.) growth under axenic conditions and on nodulation in mung bean (*Vigna radiata* L.). Lett Appl Microbiol 42:155–159

Shakirova FM, Avalbaev AM, Bezrukova MV, Kudoyarova GR (2010) Role of endogenous hormonal system in the realization of the antistress action of plant growth regulators on plants. Plant stress. Global Science Books. doi:10.1007/s00425-010-1286-7

Silva T, Davies PJ (2007) Elongation rates and endogenous indole acetic acid levels in roots of pea mutants differing in internode length. Physiol Plant 129:804–812

Spaepen S (2015) Plant hormones produced by microbes. In: Lugtenberg B (ed) Principles of plant-microbe interactions. Springer International Publishing, pp 247–256

Spaepen S, Vanderleyden J, Remans R (2007) Indole-3-acetic acid in microbial and microorganism-plant signaling. FEMS Microbiol Rev 31:425–448

Spoel SH, Dong X (2008) Making sense of hormone crosstalk during plant immune responses. Cell Host Microbe 3:348–351

Spoel SH, Johnson JS, Dong X (2007) Regulation of tradeoffs between plant defenses against pathogens with different lifestyles. Proc Natl Acad Sci USA 104:18842–18847

Stepanova AN, Yun J, Likhacheva AV, Alonso JM (2007) Multilevel interactions between ethylene and auxin in *Arabidopsis* roots. Plant Cell 19:2169–2185

Straub D, Yang H, Liu Y, Tsap T, Ludewig U (2013) Root ethylene signalling is involved in *Miscanthus sinensis* growth promotion by the bacterial endophyte *Herbaspirillum frisingense* GSF30. J Exp Bot 64:4603–4615

Teale WD, Paponov IA, Ditengou F, Palme K (2005) Auxin and the developing root of *Arabidopsis thaliana*. Physiol Plant 123:130–138

Vacheron J, Desbrosses G, Bouffaud M-L, Touraine B, Touraine B, Touraine B, Moënne-Loccoz Y, Muller D, Legendre L, Wisniewski-Dyé F, Prigent-Combaret C (2013) Plant growth-promoting rhizobacteria and root system functioning. Frontiers Plant Sci. doi:10.3389/fpls.2013.00356

Van De Mortel JE, de Vos RCH, Dekkers E, Pineda A, Guillod L, Bouwmeester K, Van Loon JJA, Dicke M, Raaijmakers JM (2012) Metabolic and transcriptomic changes induced in *Arabidopsis* by the rhizobacterium *Pseudomonas fluorescens* SS101. Plant Physiol 160:2173–2188

Van Loon LC (2007) Plant responses to plant growth-promoting rhizobacteria. Eur J Plant Pathol 119:243–254

Van Loven K, Beinsberger S, Valcke R, Van Onckelen H, Clijsters H (1993) Morphometric analysis of the growth of Phsp70-ipt transgenic tobacco plants. J Exp Bot 44:1671–1678

Veselov SY, Simonyan MV, Ivanova TN, Melent'ev AI (1998) Study of cytokinins produced by microorganisms of the rhizosphere. Appl Biochem Microbiol 34:159–163

Vessey JK (2003) Plant growth promoting rhizobacteria as biofertilizers. Plant Soil 255:571–586

Wang C, Knill E, Glick BR, De'fago G (2000) Effect of transferring 1-aminocyclopropane-1-carboxylic acid (ACC) deaminase genes into *Pseudomonas fluorescens* strain CHA0 and its gacA derivative CHA96 on their growth-promoting and disease-suppressive capacities. Can J Microbiol 46:898–907

Wang KLC, Li H, Ecker JR (2002) Ethylene biosynthesis and signaling networks. Plant Cell (Suppl):131–151

Werner T, Motyka V, Laucou V, Smets R, Van Onckelen H, Schmülling T (2003) Cytokinin-deficient transgenic *Arabidopsis* plants show multiple developmental alterations indicating opposite functions of cytokinins in the regulation of shoot and root meristem activity. Plant Cell 15:2532–2550

Weyens N, Van Der Lelie D, Taqhavi S, Newman L, Vanqronsveld J (2009) Exploiting plant–microbe partnerships to improve biomass production and remediation. Trends Biotechnol 27(10):591–998

Whitelaw MA, Harden TJ, Helyar KR (1999) Phosphate solubilization in solution culture by the soil fungus *Penicillium radicum*. Soil Biol Biochem 31:655–665

Wittenmayer L, Wolfgang Merbach W (2005) Plant responses to drought and phosphorus deficiency: contribution of phytohormones in root-related processes. J Plant Nutr Soil Sci 168:531–540

Yoshida T, Mogami J, Yamaguchi-Shinozaki K (2014) ABA-dependent and ABA-independent signaling in response to osmotic stress in plants. Curr Opin Plant Biol 21:133–139

Zaidi A, Khan MS (2005) Interactive effect of rhizotrophic microorganisms on growth, yield, and nutrient uptake of wheat. J Plant Nutr 28:2079–2092

Zamioudis C, Mastranesti P, Dhonukshe P, Blilou I, Pieterse CMJ (2013) Unraveling root developmental programs initiated by beneficial *Pseudomonas* spp. bacteria. Plant Physiol 162:304–318

Soil Bacteria and Phytohormones for Sustainable Crop Production

Ummay Amara, Rabia Khalid and Rifat Hayat

Abstract Plant growth-promoting rhizobacteria (PGPRs) synthesizes and exports phytohormones which are called plant growth regulators (PGRs). These PGRs may play regulatory role in plant growth and development. PGRs are organic substances that influence physiological processes of plants at extremely low concentrations. Among five classes of well-known PGRs, namely auxins, gibberellins, cytokinins, ethylene and abscisic acid, the most common, best characterized and physiologically active auxin in plants is indole-3-acetic acid (IAA) that stimulate both rapid (e.g. increases in cell elongation) and long-term (e.g. cell division and differentiation) responses in plants. Some bacteria also release indole-3-butyric acid (IBA), Trptophan and tryptophol, or indole-3-ethanol (TOL) that can indirectly contribute to plant growth promotion. On the other hand, cytokinins are usually present in small amounts, but enhance cell division leading to root hair formation and root development. Microorganisms have been found to contain over 30 growth-promoting compounds of the cytokinin group and about 90 % of microorganisms found in the rhizosphere are capable of releasing cytokinins when cultured in vitro. Soil bacteria also produce gibberellins (GAs) and over 100 GAs are known. The most widely recognized gibberellin is GA3 (gibberellic acid), and the most active GA in plants is GA_1, which is primarily responsible for stem elongation. In addition, abscisic acid (ABA) has been detected by radioimmunoassay in supernatants of bacterial cultures held responsible for stomatal closure. Its presence in the rhizosphere could be extremely important for crop survival under a water-stressed soil environment, such as is found in arid and semiarid climates. Ethylene is a potent plant growth regulator that affects many aspects of plant growth, development and senescence. In addition to its recognition

U. Amara · R. Khalid · R. Hayat (✉)
Department of Soil Science & SWC, PMAS Arid Agriculture University,
Rawalpindi, Pakistan
e-mail: hayat@uaar.edu.pk

D.K. Maheshwari (ed.), *Bacterial Metabolites in Sustainable Agroecosystem*,
Sustainable Development and Biodiversity 12, DOI 10.1007/978-3-319-24654-3_5

as a ripening hormone, ethylene promotes formation of adventurous root and root hair, stimulates germination and breaks dormancy of seeds. Soil bacteria promote plant growth especially seed germination by lowering the levels of ethylene in plants/seed rhizosphere. The enzyme 1-aminocyclopropane-1-carboxylate (ACC) deaminase, hydrolyzes ACC, the immediate biosynthesis precursor of ethylene in plants. The products of this hydrolysis, ammonia and α-ketobutyrate, can be used by the bacterium as a source of nitrogen and carbon for growth. Soil bacterium acts as a sink for ACC and thus lowers ethylene level in plants, preventing some of the potentially deleterious consequences of high ethylene concentrations. Soil bacteria along PGPRs also play an important role in production of phosphatases, β-glucanase, dehydrogenase, antibiotic, solubilization of mineral nutrients, stabilization of soil aggregates, improving in soil organic matter and soil structure. PGRs producing soil bacteria help in reduction of/supplementing the need for chemical fertilizers N and P for sustainable crop productivity.

Keywords Soil bacteria · Phytohormones · IAA · Ethylene · ACC deaminase · Cytokinin · Gibberellins · ABA

1 Introduction

The soil supports large and energetic microbial population capable of exerting beneficial effects on plant growth. The importance of microbial population for maintenance of root health, nutrient uptake, tolerance of environmental stress and crop responses has been recognized and well-documented. The rhizosphere bacteria exert on beneficial effects ranging from direct mechanisms to an indirect effects and play an important role in growth of plants are termed plant growth-promoting rhizobacteria (PGPRs). Indirect effects are related to production of metabolites, such as antibiotics, siderophores, or hydrogen cyanide (HCN) that decreases the growth of phytopathogens and other deleterious microorganisms. Direct effects are dependent on production of plant growth regulators, or improvement in plant nutrient uptake (Kloepper 1993; Glick 1995) and synthesis of phytohormones (Glick 1995) like IAA (indole acetic acid), auxin, gibberellins, cytokinins and ethylene (Zhang et al. 1997; Cattelan et al. 1999). The effect of phytohormones is direct, as they stimulate root growth, providing more sites for infection and nodulation (Garcìa et al. 2004). Significant increases in growth and yield of agronomically important crops in response to inoculation with phytohormones yielding PGPRs have been reported (Chen et al. 1994; Amara and Dahdoh 1997; Biswas et al. 2000; Hilali et al. 2001; Asghar et al. 2002). Several species of bacteria like *Agrobacterium, Alcaligenes, Arthrobacter, Azospirillum, Azotobacter, Bacillus, Burkholderia, Caulobacter, Chromobacterium, Enterobacter, Erwinia, Flavobacterium, Klebsiella, Micrococcus Pseudomonas* and *Serratia* have been reported to serve as PGPRs and improve the plant growth (Hayat et al. 2010; Bhattacharyya and Jha 2012; Ahemad and Kibret 2014).

PGPR produces and distributes phytohormones which may play regulatory role in plant growth and development. These regulators are organic substances that stimulate physiological processes of plants at very low concentrations (Dobbelaere et al. 2003). There are five different classes of known phytohormones, namely auxins, abscisic acid, cytokinins, ethylene and gibberellins (Zahir et al. 2004). Auxin is the most dominant phytohormone among all and indole-3-acetic acid (IAA) is physiologically the most active auxin in plant. The hormone is identified to regulate both rapid and long-term responses in crop plants e.g. stimulate cell elongation, cell differentiation and division (Hagen 1990), apical domination, tropistic responses, flowering, fruit ripening and senescence. Regulating these processes by auxin is believed to involve auxin-induced changes in gene expression (Guilfoyle et al. 1998). In this regard, the use of PGPRs has found a potential role in developing sustainable systems in crop production (Shoebitz et al. 2009; Sturz and Nowak 2000). We previously reviewed the beneficial soil bacteria along with the detail mechanisms and their role in sustainable crop production (Hayat et al. 2010, 2012). The particular mechanisms of phytohormones-mediated enhancement of plant growth includes: (i) the ability to produce ACC deaminase to reduce the level of ethylene in the root of the developing plants thereby increasing the root length and growth; (ii) ability to produce hormones like auxin, abscisic acid (ABA), gibberellic acid and cytokinins; (iii) antagonism against phytopathogenic microorganisms by producing siderophores, ß-1-3-glucanase, chitinases, antibiotics, fluorescent pigments and cyanide; (iv) enhanced resistance to drought and oxidative stress and production of water soluble B group vitamins niacin, pantothenic acid, thiamine, riboflavine and biotin. Phytohormones producing PGPRs can play an essential role in helping plants establish and grow in nutrient-deficient conditions. Their application can favor a reduction of agro-chemical use and support eco-friendly crop production. Trials with rhizosphere-associated PGP species indicated yield increases in wheat, rice, maize, sugar cane, sugar beet, legumes, canola, vegetables and conifer species (Hayat et al. 2010, 2012). In this way, PGPRs are becoming attractive alternates for bioinoculants and utilized as an additive to chemical fertilizers for improving crop yield in an integrated nutrient management system (Maheshwari 2013). Integrated nutrient management system help to minimize chemical input and to enhance nutrient use efficacy by combining chemical and biological sources of plant nutrients in an efficient and environmentally prudent manner (Adesmoye and Kloepper 2009) and also helps to minimize the use of chemical pesticides and fertilizers (Dilantha et al. 2006). In order to successfully utilize PGPRs in agriculture as bioinoculants, it is essential to identify their metabolic, phenotypic and genotypic diversities and their capability for the production of different ranges of antimicrobial metabolites. Conventionally, phenotypic identification methods play an important role but identifying at molecular level becomes much authenticated and reliable. Since the discovery of PCR and DNA sequencing, comparison of the gene sequences of bacterial species have showed that the 16S rRNA gene is highly conserved within a species and among species of the same genus, and hence can be used for identification of bacteria at species level (Olsen and Woese 1993). To understand genotypic and phenotypic diversities of PGPRs and their potential role in plant growth promotion, it is essential to understand their role in the rhizosphere and their interaction with plants, also application as inoculant (Rameshkumar et al. 2012; Maheshwari et al. 2014).

2 Phytohormones Production by PGPRs

Plant growth regulators (PGRs) are organic substances present in extremely small concentrations that affect biochemical, morphological and physiological processes of plants. PGRs act as signal molecules working as chemical messengers and significantly participate in plants as growth regulators (De Salamone et al. 2005; Martínez et al. 2010). Five majors PGRs, viz, auxins, abscisic acid (ABA), cytokinins, 1-Aminocyclopropane-1-carboxylate (ACC) deaminase and gibberellins are usually called phytohormones that have advantageous effects on plant growth and are endogenous in origin of plants (Arshad and Frankenberger 1993). Polyamines and Brassinosteroids are also PGRs produced naturally by tissues. Some synthetic compounds also trigger many physiological responses when they are artificially applied to plant tissues (Galston and Sawhney 1990; Salisbury and Ross 1992). Many bacterial and fungal species synthesize phytohormones and synthesizing ability is broadly distributed among plant- and soil-associated bacteria. Several studies confirmed that the PGPRs can improve plant growth through auxins production (indole acetic acid), ethylene, gibberellins and cytokinins (Bottini et al. 2004; Spaepen et al. 2008).

2.1 Indole-3-Acetic Acid (IAA) Production

Indole-3-acetic acid (IAA) is the most common, well-studied and naturally occurring auxin having the ability to control many aspects of plant growth. Some of them include the vascular tissues differentiation, growth elongation, apical dominance, initiation of lateral root, fruit setting and ripening. Plants produce active IAA produced by de novo synthesis from tryptophan which passes either through oxidative deamination (through indole-3-pyruvic acid formation) or decarboxylation (through tryptamine formulation by using indole-3-acetic acid aldehyde as an intermediate) (Ahemad and Khan 2011) and by releasing IAA from conjugates (Dilfuza 2012). There are different pathways involved in the synthesis of IAA by microbes (Fig. 1) (i) IAA formation via indole-3-pyruvic acid and indole-3-acetic acid aldehyde is present in most of rhizobacteria like *Agrobacterium, Azospirillum, Bradyrhizobium, Rhizobium, Enterobactor, Erwinia herbicola, Pseudomonas, Klebsiella*, etc.; (ii) Conversion of tryptophan into indole-3-acetic aldehyde and produce tryptamine, e.g. *Azospirilla* and Pseudomonads; (iii) Biosynthesis of IAA via indole-3-acetamide formation is reported by *Azospirillum, A. tumefaciensm, E. herbicola, Rhizobium* spp., *Bradyrhizobium*

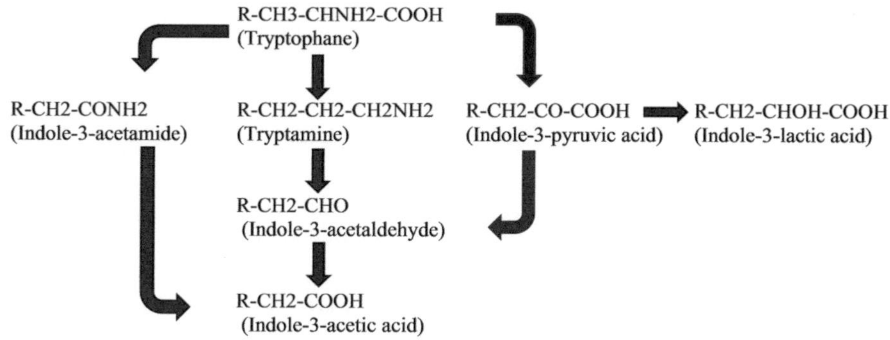

Fig. 1 Biosynthetic pathway of IAA synthesis in bacteria

sp. and Saprophytic Pseudomonads, etc.; (iv) In plant, biosynthesis of IAA via indole-3-acetonitrile is present, in the *Cyanobacteria* and *Alcaligenes faecalis* (v) Tryptophan independent pathway, mostly present in *Cyanobacteria* and *Azospirilla.* However, information of IAA using this pathway is non-significant and its mechanism is unknown. It is well-documented that more than 80 % bacteria isolated from rhizospheric soil of different crops have the capability to produce and release auxin (Loper and Schroth 1986). Among the auxin-producing PGPRs species, *Azospirillum* is the most studied IAA-producers (Dobbelaere et al. 1999). Other IAA-producing bacteria belong to genera *Aeromonas, Azotobacter, Bacillus, Burkholderia, Enterobacter, Pseudomonas* and *Rhizobium* (Swain et al. 2007; Ahmad et al. 2008; Hariprasad and Niranjana 2009; Shoebitz et al. 2009). The formation of different amount of IAA by bacterial strain could be varied because of participation of different biosynthetic pathway, regulatory sequences, genes location and availability of enzymes to convert active, free IAA to fixed form and could also be affected by environmental conditions (Patten and Glick 1996; Ahemad and Khan 2011). Regulation of these different physiological processes by auxin is believed to involve auxin-induced changes in gene expression (Guilfoyle et al. 1998). In addition to IAA, *P. polymyxa* and *Azospirilla* also release other compounds in the rhizosphere that could indirectly contribute to plant growth promotion.

2.2 *Aminocyclopropane-1-Carboxylate (ACC) Deaminase Production*

Ethylene is an important metabolite in regulating normal plant growth and developmental processes (Khalid et al. 2006; Ahemad and Kibret 2014). Ethylene has been recognized as a growth regulator and a good stress hormone (Saleem et al. 2007; Ahemad and Kibret 2014). Its production is due to various environmental factors such as salinity, high temperature and drought, physical impendence, wounding, water logging, metal stress and during disease development (Arteca and Arteca 2007; Belimov et al. 2009; Bhattacharyya and Jha 2012; Ahemad and Kibret 2014). Low level of ethylene has a positive effect but higher levels inhibit normal plant growth. PGPRs with enzyme, ACC deaminase, support growth and development by declining level of ethylene, prompting salt tolerance and decreasing drought stress in plants (Ahemad and Kibret 2014). Presently, bacterial strains containing ACC deaminase enzymes belong to wide range of genera such as *Acinotobacter, Achromobacter, Enterobacter*, *Pseudomonas*, *Azospirillum*, *Agr obacterium, Burkholderia* spp., *Alcaligenes*, *Serratia, Ralstonia*, *Rhizobium*, etc. (Pandey et al. 2005; Shaharoona et al. 2007a, b; Zahir et al. 2009; Kang et al. 2010; Ahemad and Kibret 2014). Such rhizobacteria utilize ethylene precursor ACC and transform it into NH_3 and 2-oxobutanoate (Arshad et al. 2007). Numerous forms of stress, such as effects of phytopathogenic microorganisms (bacteria, fungi and viruses) and resistance to stress from flooding, extreme temperatures, polyaromatic hydrocarbons, heavy metals, high salt concentration, insect predation, radiation, high light intensity, draft and wounding (Lugtenberg and Kamilova 2009; Glick 2012) are overcome due to ACC containing rhizobacteria in plants. ACC deaminase containing PGPRs application show good effects on plant growth and development proving to be good candidates for biofertilizer preparation (Shaharoona et al. 2006).

Ethylene is an effective PGR synthesized by many species of bacteria (Primrose 1979), and serve as a ripening hormone, promotes adventitious roots and root hair formation, induces germination, breaks seed dormancy, enhance plant growth, development and delay senescence. However, higher ethylene concentration after germination proved to be toxic and inhibited root elongation as well as symbiotic N_2 fixation in leguminous plants. One of the mechanisms of growth promotion by PGPRs is by lowering the ethylene level in plants, which is accredited to the activity of the enzyme 1-aminocyclopropane-1-carboxylate (ACC) deaminase, which hydrolyzes ACC, the immediate biosynthesis precursor of ethylene in plants. The product of this hydrolysis, ammonia and α-ketobutyrate, can be utilized by the microbes as nitrogen and carbon source for growth. Therefore, the microorganisms act as a pool for ACC-deaminase and result in lowering of ethylene level in plants, thus preventing some of the precluding deleterious effects of high ethylene concentrations (Glick et al. 1998). PGPR with ACC deaminase activity is attributed to an improved plant growth and yield and thus, are potential candidate for biofertilizer formulation (Shaharoona et al. 2006).

2.3 Cytokinin Production

Cytokinins are good PGRs that control cytokinesis in tissues of crop plants (Skoog et al. 1965). Over 100 years ago, numerous scientists discovered the presence of substances that were capable to prompt cell division in cultured or damaged plant tissue (El-Showk et al. 2013). Letham (1963) stated that zeatin was isolated from *Zea mays*. According to him, it was the first natural cytokinin with pure crystalline structure. Chemical synthesis proved the structure of zeatin to be (E)-4-(hydroxy-3-methyl-but-2-enyl) aminopurine. The most observable effect of cytokinin on plant is stimulation of shoot and root growth and enhancement in cell division (Hayat et al. 2010) and they have been involved in many other important developmental processes in plants, including seed germination, organ formation, shoot meristem formation and maintenance, and leaf senescence (Mok and Mok 2001). Above 30 different growth-promoting cytokinins compounds have been found in plants, plant-associated microorganisms and in in vitro conditions most of microorganisms are capable of releasing cytokinins with different proportions (Hayat et al. 2012). For biosynthesis of cytokinins two pathways have been proposed. Direct pathway, involving development of dimethylallyl pyrophosphate (DMAPP) and N6-isopentenyladenosine monophosphate (i6 AMP) from AMP, followed by formation zeatin-type compounds from hydroxylation of the side-chain and indirect pathway, in which cytokinins are released by turnover of tRNA containing *cis*-zeatin. Cytokinins play significant role during development processes, from germination of seed to plant senescence and regulate different physiological and morphological processes throughout the plant life, including respiration and photosynthesis (Arshad and Frankenberger 1993). The variable effects suggest that cytokinins might have different mechanisms of action depending on the type of tissues, or the impacts of primary and secondary effects caused by the variation in physiological states of the target cells (Salisbury and Ross 1992). Cytokinins phytohormones are usually present in small amounts in biological samples (Vessey 2003) and enhance cell division, root development and root hair formation (Frankenberger and Arshad 1995). Cytokinins are involved in processes such as photosynthesis or chloroplast differentiation. They are also known to induce the opening of stomata, to suppress auxin-induced apical dominance, and to inhibit senescence of plants organs, especially in leaves (Crozier et al. 2001). Plants and plant-associated microorganisms have been found to contain over 30 growth-promoting compounds of the cytokinin group. Nieto and Frankenberger (1991) studied the effect of the cytokinin precursor's adenine (ADE) and isopentyl alcohol (IA), and the cytokinin-producing bacteria *Azotobacter chroococcum* on the morphology and growth of maize under in vitro, greenhouse and field conditions, resultant, found improvement in plant growth.

2.4 Gibberellins Production

Gibberellins (GAs) are other important natural plant growth regulators in higher plants. They are usually derived from gibberellic acid and control seed dormancy, stem proliferation, expansion of leaves and flowering (Javid et al. 2011). GAs were discovered in 1938 and isolated from *Ibberella fujikuroi*, a pathogenic fungus of rice (Miransari and smith 2014). In root nodule symbiosis, GAs plays a significant role. There are more than 80 different gibberellins, but among all GA_3 is the most commonly used form and GA_1 is the most active in plants, which is primarily responsible for stem elongation (Davies 1995). Several pathways are involved for the biosynthesis of gibberellins from geranyl diphosphate. DELLA proteins are involved in the regulation of gibberellins, C-terminal GRAS domain is the core part of structure of DELLA protein (eventually degraded by E3 ubiquitin ligase SCF (GID2/SLY1). Regulation of gibberellins is conducted by this protein (Miransari and Smith 2014). The accumulation of DELLAs in seeds becomes a cause to express the genes involved in the production of F-box proteins. The gibberellins receptor has recently been identified in rice. Gibberellin insensitive dawrf1 (GID1) protein interacts with DELLA proteins followed by their degradation in nucleus and binding with biologically active gibberellins (Willige et al. 2007). Role of gibberellins in plant growth and development is quite evident. The growth of stem is highly dependent on the production of gibberellins. Their low levels in plant metabolism results in shorter height as compared to natural height. In reality shorter and thicker stems are preferred as they can resist stress conditions and give better support; therefore, in grain production extensive use of gibberellin synthesis inhibitors is preferably chosen. On the other hand, they are considered beneficial for seed germination at breaking seed dormancy thereby positively considered for seeds that show resistance for germination. PGPR also produced gibberellic acid (GA) and gibberellins (GAs). Dobbelaere et al. (2003) reviewed that over 89 GAs are known to date and are numbered GA_1 through GA_{89} in approximate order of their discovery. The most widely recognized gibberellin is GA_3 (gibberellic acid), the most active GA in plants is GA_1, which is primarily responsible for stem elongation. GAs also affects reproductive processes in a wide range of plants (Crozier et al. 2001). PGPRs like *Azospirillum* and *Pseudomonas* spp. produce cytokinins and gibberellins (gibberellic acid), in addition to IAA. Different genera of soil bacteria released variety of phytohormones and when inoculated, crops responded positively (Table 1).

3 Approaches to Develop PGPRs

Screening of PGPRs includes traditional as well as modern approaches. Modern approaches of screening these organisms from rhizospheric and non-rhizospheric soils are considered to be potent to improve the results of studying their effects on plant in lab. Soil and crop cultural practices, inoculant formulation and delivery are considered for rhizosphere management (Bowen and Rovira 1999; McSpadden and Fravel 2002). Root-associated traits to enhance the establishment and proliferation of beneficial organisms are being pursued by genetic manipulation of host crops (Smith and Goodman 1999; Mansouri et al. 2002). Multi-strain inocula formulations of PGPR with known functions may enhance the stability in the field (Jetiyanon and Kloepper 2002; Siddiqui and Shaukat 2002). Molecular techniques are playing lead roles in mounting our ability to understand and manage rhizosphere

Table 1 Plant growth regulator release by PGPRs and crop responses

PGPR	PGRs	Crops	Responses	References
Azospirillum brasilense	Indole-3-acetic acid, Gibberellins and cytokinin	Pearl millet	Increased lateral root and root hair	Tien et al. (1979)
Bacillus firmus	Indole-3-acetic acid	Rice	Increased grain yield and phosphate uptake	Datta et al. (1982)
Pseudomonas polymyxa	Indole-3-acetic acid	Wheat grass	Increased growth over uninoculated control	Holl et al.(1988)
Azospirillum spp. and *Bacillus* spp.	Gibberellin	Rice	Increased N^{15} uptake	Kucey (1988)
Azotobacter paspali	IAA and other plant hormones	Canola, tomato and wheat	Increased plant growth	Abbass and Okon (1993)
Bacillus firmus	P-solubilizing and Indole-3-acetic acid	Rice	Increased grain yield and P-uptake of rice in a P-deficient soil	De Freitas et al. (1997)
A. lipoferum	Gibberellin	Maize	Alleviate temporary drought	
Enterobacter cloacae, Pseudomonas putida and *Achromobacter piechaudii*	ACC deaminase	Tomato	Inoculated tomato seed increased plant resistance in 55 days to nine consecutive days of flooding and increased resistance to salinity	Grinchko and Glick (2001)
Bacillus circulans, Bacillus firmus and *Bacillus globisporus*	ACC deaminase	Mustard	Increased root length	Ghosh et al. (2003)
Achromobacter piechaudii	ACC deaminase	Tomato	Increased fresh and dry weight of inoculated plants under saline and water stress conditions	Mayak et al. (2004)
Pseudomonas asplenii	ACC deaminase	Rape seeds	Significant increase in fresh and dry weight and biomass yield	Reed and Glick (2005)

(continued)

Table 1 (continued)

PGPR	PGRs	Crops	Responses	References
Pseudomonas putida	Indole-3-acetic acid	Canola	Two–threefold increases in the length of seedling roots	Ahmad et al. (2005)
Pseudomonas fluorescens	ACC deaminase	Maize	Increased root length and fresh weight under saline conditions	Kausar and Shahzad (2006)
Sphingomonas spp. and *Mycobacterium* spp.	Indole-3-acetic acid	Orchid plant seeds	Increased seed germination rate	Tsavkelova et al. (2007)
Bacillus subtilis	Indole-3-acetic acid	Edible tubercle	Increased root and stem length and root and stem fresh weight	Swain et al. (2007)
Spingobacterium sp. and *Mycobacterium* spp.	Indole-3-acetic acid	Orchid plant seed	Significantly increase rate of germination and stimulate root growth	Tsavkelova et al. (2007)
Pseudomonas fragi	Hydrogen cyanide	Wheat seedlings	Significantly increases the germination percentage, germination rate, plant biomass and nutrient uptake	Selvakumar et al. (2008)
Providencia spp.	Hydrogen cyanide	Wheat	Twofold increase in germination percentage compared to untreated controls	Zarrin et al. (2009)
Bacillus subtilis	Indole-3-acetic acid	Sweet potatoes	Increase in root and stem length, fresh weight of the root and stem, root: stem ratio. and significantly enhanced numbers of sprouts	Martínez et al. (2010)
Pseudomonas fluorescens	ACC deaminase	Groundnut plants	Improved the saline resistance and yield	Siddikee et al. (2010)
Pseudomonas putida UW4	Indole-3-acetic acid and ACC deaminase	Canola	Under saline conditions, protected the seedling of canola from growth inhibition	Siddikee et al. (2010)

(continued)

Table 1 (continued)

PGPR	PGRs	Crops	Responses	References
Providencia spp. and *Pseudomonas aeruginosa*	Hydrogen cyanide	Wheat	Control fungus diseases and enhance defense against phytopathogen	Rana et al. (2011)
Corynebacterium agropyri, Enterobacter gergoviae, Bacillus amyloliquefaciens	Indole-3-acetic acid	Rice	Improved seed germination and seedling establishment	Ng et al. (2012)
Pseudomonas chlororaphis	Siderophore production	Maize	Increased root shoot biomass and seed germination rate	Hayat et al. (2012)

for obtaining improved and potent products (Nelson 2004). Large number of mechanisms has been studied yet for engineering the rhizosphere for improved productivity of crops. This includes manipulation of plant for the modification of rhizosphere. This plays vital role in promoting the nutrient availability to plants, immunity against pathogens and boosting PGPR bacterial growth (Ryan et al. 2008). A study conducted by Sundheim et al. (1998) reported during an in vitro technique that a modified strain of *Pseudomonas* with chitinase gene from *Serratia marcescens* had the potential to control *Fusarium oxysporum* f. sp. *redolens* and *Gaeumannomyces graminis* var. tritici effectively. Recent experiments performed on *Pseudomonas fluorescens* (DAPG-producing PGPR strain) have demonstrated that in rhizosphere different plant species have the ability to support and nourish unique microbial population and genotypes in rhizosphere (Fuente et al. 2006; Landa et al. 2006). DAPG accumulation by *Pseudomonas fluorescens* CHA0 with the expression of DAPG biosynthesis gene phlA has been significantly correlated by Notz et al. (2001). It was observed that the expression in rhizosphere of monocots was greater than that of dicots. Gregorio et al. (2006) noticed that in EDTA-amended soil, inocula with combined application (Triton X-100 and *Sinorhizobium* sp. Pb002) were beneficial for phytoextraction of lead by *Brassica juncea.*

4 Conclusion

PGPR synthesizes and exports phytohormones, also called as plant growth regulators (PGRs), may play regulatory role synthesized in defined organs of the plant that can be translocated to other sites, where it triggers specific biochemical, physiological and morphological responses in plant growth and development. PGRs are organic substances that influence physiological processes of plants at extremely low concentrations and are also active in the tissues where they are produced. Among different PGRs, auxins, gibberellins, cytokinins, ethylene and abscisic acid are well studied. In addition to IAA, abscisic acid (ABA) has also been detected by radioimmunoassay and TLC in supernatants of *Azospirillium* and *Rhizobium* spp. cultures. Role of PGPR in production of phosphatases, β-glucanase, dehydrogenase, antibiotic solubilization of mineral phosphates and other nutrients, stabilization of soil aggregates, improved soil structure and organic matter contents has been recognized. The mechanisms involved have a significant plant growth-promoting potential, retaining more soil organic N and other nutrients in the plant-soil system thereby reducing the need for fertilizer N and P and enhancing the release of nutrients. Another recently identified mechanism for plant growth promotion is due to production of volatiles by PGPR. PGPR release different volatile blends and the differences in these volatile blends stimulate the plant growth. Volatile compounds like 3-hydroxy-2-butanone (acetoin) and 2, 3-butanediol, produced by *Bacillus subtilis* and *B. amyloliquefaciens* stimulated the growth of *Arabidopsis thaliana* by in vitro experiments. The volatile-mediated growth promotion of plant accomplished by PGPR is due to activation of cytokinin-signaling pathways.

References

Abbas Z, Okon Y (1993) Plant growth promotion by *Azotobacter paspali* in the rhizosphere. Soil Biol Biochem 25:1075–1083

Adesemoye AO, Kloepper JW (2009) Plant-microbes interactions in enhanced fertilizer-use efficiency. Appl Microbiol Biotechnol 85:1–12

Ahemad M, Kibret M (2014) Mechanisms and applications of plant growth promoting rhizobacteria: Curent perspective. J King Saud Uni Sci 26(1):1–20

Ahemad M, Khan MS (2011) Toxicological assessment of selective pesticides towards plant growth promoting activities of phosphate solubilizing *Pseudomonas aeruginosa*. Acta Microbiol Immunol Hung 58:169–187

Ahmad F, Ahmad I, Khan MS (2008) Screening of free-living rhizospheric bacteria for their multiple plant growth promoting activities. Microbiol Res 163:173–181

Ahmad F, Ahmad I, Khan MS (2005) Indole acetic acid production by the indigenous isolates of *Azotobacter* and fluorescent *Pseudomonas* in the presence and absence of tryptophan. Turk J Biol 29:29–34

Amara MAT, Dahdoh MSA (1997) Effect of inoculation with plant growth promoting Rhizobacteria (PGPR) on yield and uptake of nutrients by wheat grown on sandy soil. Egyptian J Soil Sci 37:467–484

Arshad M, Saleem M, Hussain S (2007) Perspectives of bacterial ACC-deaminase in phytoremedation. Trends in Biotech 25:356–362

Arshad M, Frankenberger Jr WT (1993) Microbial production of plant growth regulators. In: Blaine, F, Metting Jr (eds) Soil Microbial Ecology, Marcel and Dekker, Inc., New York, pp 307–347

Arteca RN, Arteca JM (2007) Heavy-metal-induced ethylene production in *Arabidopsis thaliana*. J Plant Physiol 164(11):1480–1488

Asghar HN, Zahir ZA, Arshad M, Khaliq A (2002) Relationship between in vitro production of auxins by rhizobacteria and their growth promoting activities in *Brassica juncea*. Biol Fertil Soil 35:231–237

Belimov AA, Dodd IC, Hontzeas N, Theobald JC, Safronova VI, Davies WJ (2009) Rhizosphere bacteria containing ACC-deaminase increase yield of plants grown in drying soil via both local and systemic hormone signaling. New Phytol 181:413–423

Bhattacharyya PN, Jha DK (2012) Plant growth-promoting rhizobacteria (PGPR): emergence in agriculture. World J Microbiol Biotechnol 28:1327–1350

Biswas JC, Ladha JK, Dazzo FB (2000) Rhizobia inoculation improves nutrient uptake and growth of lowland rice. Soil Sci Soc Am J 64:1644–1650

Bottini R, Cassan F, Picolli P (2004) Gibberellin production by bacteria and its involvement in plant growth promotion. Appl Microbiol Biotechnol 65:497–503

Bowen GD, Rovira AD (1999) The rhizosphere and its management to improve plant growth. Adv Agron 66:1–102

Cattelan AJ, Hartel PG, Fuhrmann JJ (1999) Screening for plant growth rhizobacteria to promote early soybean growth. Soil Sci Soc Amer J 63:1670–1680

Chen Y, Mei R, Lu S, Liu L, Kloepper JW (1994) The use of yield increasing bacteria as plant growth promoting rhizobacteria in Chinese agriculture. In: Gupta VK, Utkhede R (eds) Management of Soil Born Diseases. M/S Narosa, New Dehli, India, pp 1–13

Crozier A, Kamiya Y, Bishop G, Yokota T (2001) Biosynthesis of hormones and elicitors molecules. In: Buchanan BB, Grussem W, Jones RL (eds) Biochemistry and molecular biology of plants. American Society of Plant Biologists, Rockville, pp 850–900

Datta M, Banish S, Gupta RK (1982) Studies on the efficacy of a phytohormone producing phosphate solubilizing *Bacillus firmus* in augmenting paddy yield in acid soils of Nagaland. Plant Soil 69:365–373

Davies PJ (1995) The plant hormones: Their nature, occurrence, and functions. In: Davies PJ (ed) Plant Hormones: Physiology. Biochemistry and Molecular Biology, Kluwer, Boston, pp 13–38

De Freitas JR, Banerjee MR, Germida JJ (1997) Phosphate-solubilizing rhizobacteria enhance the growth and yield but not phosphorus uptake of canola (*Brassica napus*). Biol Fertil Soils 36:842–855

De Salamone IEG, Hynes RK, Nelson LM (2005) Role of cytokinins in plant growth promotion by rhizosphere bacteria. In: Siddiqui ZA (ed) PGPR: biocontrol and biofertilization. Springer, The Netherlands, pp 173–195

Dilfuza E (2012) Indole acetic acid production by root associated bacteria and its involvement in plant growth and development In: Keller AH, Fallon MD (eds) Auxins: Structure, biosynthesis and functions, Nova Publishers, pp 103–122

Dilantha F, Nakkeeran S, Yilan Z (2006) Biosynthesis of antibiotics by PGPR and its relation in biocontrol of plant diseases. PGPR Biocontrol Biofert 67–109

Dobbelaere S, Vanderleyden J, Okon Y (2003) Plant growth-promoting effects of diazotrophs in the rhizosphere. Crit Rev Plant Sci 22:107–149

Dobbelaere S, Croonenborghs A, Thys A, Broek AV, Vanderleyden J (1999) Phytostimulatory effect of *Azospirillum brasilense* wild type and mutant strains altered in IAA production on wheat. Plant Soil 212:153–162

El-Showk S, Ruonala R, Helariutta Y (2013) Crossing paths: cytokinin signalling and crosstalk. Development 140:1373–1383

Frankenberger WTJ, Arshad M (1995) Photohormones in soil: microbial production and function. Deker, New York, p 503

Fuente DL, Landa BB, Weler DM (2006) Host crop affects rhizosphere colonization and competiveness of 2, 4-diacetylphloroglucinol producing *Pseudomonas fluorescens*. Phytopathol 96:751–762

Galston AW, Sawhney RK (1990) Polyamines in plant physiology. Plant Physiol 94:406–410

Garcìa JAL, Probanza A, Ramos AB, Barriuso J, Mañero FJG (2004) Effect of inoculation with plant growth promoting rhizobacteria (PGPRs) and *Sinorhizobium fredii* on biological nitrogen fixation, nodulation and growth of *Glycine max* cv. Osumi. Plant Soil 267:143–153

Ghosh S, Penterman JN, Little RD, Chavez R, Glick BR (2003) Three newly isolated plant growth-promoting bacilli facilitate the seedling growth of canola, *Brassica campestris*. Plant Physiol Biochem 41:277–281

Glick BR (2012) Plant growth-promoting bacteria: mechanisms and applications. Hindawi Publishing Corporation, Scientifica

Glick BR, Penrose DM, Li J (1998) A model for lowering plant ethylene concentrations by plant growth promoting rhizobacteria. J Theoretical Biol 190:63–68

Glick BR (1995) The enhancement of plant growth by free-living bacteria. Can J Microbiol 41:109–117

Gregorio DS, Barbafieri M, Lampis S, Sanangelantoni AM, Tasi E, Di Gregorio S, Lampis S, Sanangelantoni AM, Tasi E, Valini G (2006) Combined application of Triton X-10 and *Sinorhizobium* sp. PbO_2 inoculum for the improvement of lead phytoextraction by *Brassica juncea* in EDTA amended soil. Chemosphere 63:293–29

Grichko VP, Glick BR (2001) Amelioration of flooding stress by ACC deaminase containing plant growth-promoting bacteria. Plant Physiol Biochem 39:11–17

Guilfoyle T, Hagen G, Ulmasov T, Murfett J (1998) How does auxin turn on genes? Plant Physiol 118:341–347

Hagen G (1990) The control of gene expression by auxin. In: Davies PJ (ed) Plant hormones and their role in plant growth and development. Kluwer Academic Publishers, Dordrecht, Netherlands, pp 149–163

Hariprasad P, Niranjana SR (2009) Isolation and characterization of phosphate solubilizing rhizobacteria to improve plant health of tomato. Plant Soil 316:13–24

Hayat R, Ali S, Amara U, Khalid R, Ahmed I (2010) Soil beneficial bacteria and their role in plant growth promotion: a review. Ann Microbiol 60:579–598

Hayat R, Ahmed I, Sheirdil RI (2012) An overview of plant growth promoting rhizobacteria (PGPR) for sustainable agriculture. In: Ashraf M, Öztürk M, Ahmad MSA, Aksoy A (eds) Crop Production for Agricultural Improvement, Part 3. Springer, Dordrecht, pp 557–579

Hilali A, Przost D, Broughton WJ, Antoun A (2001) Effect ded Ì inoculation avecdes souches ded *Rhizobium leguminoserum* bv. Trifolii sur la croissance du bledans dues sols du Marco. Can J Microbiol 47:590–593

Holl F B, Chaay CP, Turkingtdn R, Radley RA (1988) Growth response of crested wheatgrass (*Agropyron crisrarum* L.), white clover (*Trifilium repens* L.) and perennial rycgrass (*Lolium perenne* L.) to inoculation with *Bacillus polymyxa*. Soil Biol Biochem 20:19–24

Javid MG, Sorooshzadeh A, Moradi F, Sanavy SAMM, Allahdadi I (2011) The role of phytohormones in alleviating salt stress in crop plants. Aus J Crop Sci 5:726–734

Jetiyanon K, Kloepper JW (2002) Mixtures of plant growth-promoting rhizobacteria for induction of systemic resistance against multiple plant diseases. Biol Control 24:285–291

Kang BG, Kim WT, Yun HS, Chang SC (2010) Use of plant growth-promoting rhizobacteria to control stress responses of plant roots. Plant Biotechnol Rep 4:179–183

Kausar R, Shahzad SM (2006) Effect of ACC-deaminase containing rhizobacteria on growth promotion of maize under salinity stress. J Agri Soci Sci 2:216–218

Khalid A, Arshad M, Zahir ZA (2006) Phytohormones: microbial production and applications. In: Uph N et al (eds) Biological Approaches to Sustainable Soil Systems. Taylor and Francis, Boca Raton, Florida, pp 207–220

Kloepper JW (1993) Plant growth promoting rhizobacteria as biological control agents. In: Metting FB Jr (ed) Soil Microbial Ecology. Dekker, New York, pp 255–274

Kucey RMN (1988) Plant growth-altering effects of *Azospirillum brasilense* and *Bacillus* C-11-25 on two wheat cultivars. J Appl Bacteriol 64:187–196

Landa BB, Mavrodi OV, Schroeder KL, Allende-Molar R, Weller DM (2006) Enrichment and genotypic diversity of *phl*D-containing fluorescent *Pseudomonas* spp. in two soils after a century of wheat and flax monoculture. FEMS Microbiol Ecol 55:351–368

Letham DS (1963) Zeatin, a factor inducing cell division from *Zea mays*. Life Sci 8:569–573

Loper JE, Schroth MN (1986) Influence of bacterial sources of indole-3-acetic acid on root elongation of sugar beet. Phytopathol 76:386–389

Lugtenberg BJ, Kamilova F (2009) Plant-growth-promoting rhizobacteria. Ann Rev Microbiol 63:541–556

Maheshwari DK, Agarwal M, Dheeman S (2014) Trends and prospects of microbial diversity in rhizosphere. In: Maheshwari DK (ed) Bacterial Diversity in Sustainable Agriculture. Springer, Gewerbestrasse, Switzerland, pp 1–22

Maheshwari DK (2013) Bacteria in Agrobiology: Plant Nutrient Management. Springer-Verleg Berlin, Germany, p 352

Mansouri H, Petit A, Oger P, Dessaux Y (2002) Engineered rhizosphere: the trophic bias generated by opine-producing plants is independent of opine type, the soil origin and the plant species. Appl Environ Microbiol 68:2562–2566

Martínez-Viveros O, Jorquera MA, Crowley DE, Gajardo G, Mora ML (2010) Mechanisms and practical considerations involved in plant growth promotion by rhizobacteria. J Soil Sci Plant Nutr 10:293–319

Mayak S, Tirosh T, Glick BR (2004) Plant growth-promoting bacteria confer resistance in tomato plants to salt stress. Plant Physiol Biochem 42:65–572

McSpadden GBB, Fravel DR (2002) Biological control of plant pathogens: research, commercialization and application in the USA. Plant Health Progress

Miransari M, Smith DL (2014) Improving Soybean (*Glycine max* L.) N_2-fixation under salinity stress. Lap Lambert Academic Publishing, Germany

Mok DW, Mok MC (2001) Cytokinin metabolism and action. Annu Rev Plant Mol Biol 52:89–118

Nelson LM (2004) Plant growth promoting rhizobacteria (PGPR): prospects for new inoculants

Ng LC, Sariah M, Sariam O, Radziah O, Abidin MAZ (2012) Rice seed bacterization for promoting germination and seedling growth under aerobic cultivation system. Aust J Crop Sci 6:170–175

Nieto KF, Frankenberger WT Jr (1991) Influence of adenine, isopentyl alcohol and *Azotobacter chroococcum* on the vegetative growth of *Zea mays*. Plant Soil 135:213–221

Notz R, Maurhofer M, Schnider KU, Duffy B, Haas D, Défago G (2001) Biotic factors affecting expression of the 2,4-diacetylphloroglucinol biosynthesis gene phlA in *Pseudomonas fluorescens* biocontrol strain CHA0 in the rhizosphere. Phytopathol 91:873–881

Olsen GJ, Woese CR (1993) Ribosomal RNA: a key to phylogeny. J Bacteriol 7:113–123

Pandey P, Kang SC, Maheshwari DK (2005) Isolation of endophytic plant growth promoting *Burkhoderia* sp. MSSP from root nodules of *Mimosa pudica*. Curr Sci 89:177–180

Patten C, Glick BR (1996) Bacterial biosynthesis of indole-3-acetic acid. Can J Microbiol 42:207–220

Primrose SB (1979) Ethylene and agriculture: the role of the microbe. J Appl Bacteriol 46:1–25

Rameshkumar N, Ayyadurai N, kayalvizhi N, Gunasekaran P (2012) Genotypic and phenotypic diversity of PGPR Fluorescent Pseudomonas isolated from the rhizosphere of sugarcane (*Saccharum officinarum* L.). J Microbiol Biotechnol 22(1):13–24

Rana A, Saharan B, Joshi M, Prasanna R, Kumar K, Nain L (2011) Identification of multi-trait PGPR isolates and evaluating their potential as inoculants for wheat. Ann Microbiol 61:893–900

Reed MLE, Glick BR (2005) Growth of canola (*Brassica napus*) in the presence of plant growth-promoting bacteria and either copper or polycyclic aromatic hydrocarbons. Can J Microbiol 51:1061–1969

Ryan RP, Germaine K, Franks A, Ryan DJ, Dowling DN (2008) Bacterial endophytes: recent developments and applications. FEMS Microbiol Letters 278:1–9

Saleem M, Arshad M, Hussain S, Bhatti AS (2007) Perspective of plant growth promoting rhizobacteria (PGPR) containing ACC deaminase in stress agriculture. J Ind Microbiol Biotechnol 34:635–648

Salisbury FB, Ross C (1992) Plant Physiology, 4th edn. Wadsworth, Belmont, CA

Selvakumar G, Mohan M, Kundu S, Gupta AD, Joshi P, Nazim S, Gupta HS (2008) Cold tolerance and plant growth promotion potential of *Serratia marcescens* strain SRM (MTCC 8708) isolated from flowers of summer squash (*Cucurbita pepo*). Lett Appl Microbiol 46:171–175

Shaharoona B, Arshad M, Khalid A (2007a) Differential response of etiolated pea seedling to 1-aminocyclopropane-1-carboxylate and/or L-methionine utilizing rhizobacteria. J Microbiol 45:15–20

Shaharoona B, Jamro GM, Zahir ZA, Arshad M, Memon KS (2007b) Effectiveness of various *Pseudomonas* spp. and *Burkholderia caryophylli* containing ACC-deaminase for improving growth and yield of wheat (*Triticum aestivum* L.). J Microbiol Biotechnol 17:1300–1307

Shaharoona B, Arshad M, Zahir ZA (2006) Effect of plant growth promoting rhizobacteria containing ACC deaminase on maize (*Zea mays* L.) growth under axenic conditions and on nodulation in mung bean (*Vigna radiata* L.). Lett App Microbiol 42:155–159

Shoebitz M, Ribaudo CM, Pardo MA, Cantore ML, Ciampi L, Curá JA (2009) Plant growth promoting properties of a strain of *Enterobacter ludwigii* isolated from *Lolium perenne* rhizosphere. Soil Biol Biochem 41:1768–1774

Siddiqui IA, Shaukat SS (2002) Rhizobacteria-mediated induction of systemic resistance (ISR) in tomato against *Meloidogyne javanica*. J Phytopathol 150:469–473

Siddikee MA, Chauhan PS, Anandham R, Han GH, Sa T (2010) Isolation, characterization, and use for plant growth promotion under salt stress, of ACC deaminase-producing halotolerant bacteria derived from coastal soil. J Microbiol Biotechn 20(11):1577–1584

Skoog F, Strong FN, Miller CO (1965) Cytokinins Sci 148:532–533

Smith KP, Goodman RM (1999) Host variation for interactions with beneficial plant-associated microbes. Ann Rev Phytopathol 37:473–491

Spaepen S, Dobbelaere S, Croonenborghs A, Vanderleyden J (2008) Effects of *Azospirillum brasilense* indole-3-acetic acid production on inoculated wheat plants. Plant Soil 312:15–23

Sturz AV, Nowak J (2000) Endophytic communities of rhizobacteria and the strategies required to create yield enhancing associations with crops. Appl Soil Ecol 15:183–190

Sundheim L, Poplawsky AR, Elingboe AH (1998) Molecular cloning of two chitinase genes from *Serratia marcescens* and their expression in *Pseudomonas* species. Physiol Mol Plant Pathol 3:483–491

Swain MR, Naskar SK, Ray RC (2007) Indole-3-acetic acid production and effect on sprouting of Yam (*Dioscorea rotundata* L.) minisetts by *Bacillus subtilis* isolated from culturable cow-dung microflora. Pol J Microbiol 56:103–110

Tien TM, Gaskins MH, Hubbell DH (1979) Plant growth substances produced by *Azospirillum brasilense* and their effect on the growth of pearl millet (*Pennisetum americanum* L.). Appl Environ Microbiol 37:1016–1024

Tsavkelova EA, Cherdyntseva TA, Klimova SY, Shestakov AI, Botina SG, Netrusov AI (2007) Orchid-associated bacteria produce indole-3-acetic acid, promote seed germination, and increase their microbial yield in response to exogenous auxin. Arch Microbiol 188:655–664

Vessey JK (2003) Plant growth-promoting rhizobacteria as biofertilizers. Plant Soil 255:571–586

Willige BC, Ghosh S, Nill C, Zourelidou M, Dohmann EMN, Maier A, Schwechheimer C (2007) The della domain of GA insensitive mediates the interaction with the GA insensitive dwarf1a gibberellin receptor of *Arabidopsis*. Plant Cell 19:1209–1220

Zahir ZA, Ghani U, Naveed M, Nadeem SM, Asghar HN (2009) Comparative effectiveness of *Pseudomonas* and *Serratia* sp. containing ACC-deaminase for improving growth and yield of wheat (*Triticum aestivum* L.) under salt-stressed conditions. Arch Microbiol 191:415–424

Zahir ZA, Arshad M, Frankenberger WT (2004) Plant growth promoting rhizobacteria: application and perspectives in agriculture. Adv Agron 81:96–168

Zhang F, Pan B, Smith DL (1997) Application of gibberellic acid to the surface of soybean seed (*Glycine max* (L.) Merr.) and symbiotic nodulation, plant development, final grain and protein yield under short season conditions. Plant Soil 188:329–335

The Importance of Phytohormones and Microbes in Biofertilizers

W.S. Wong, S.N. Tan, L. Ge, X. Chen and J.W.H. Yong

Abstract Plant growth is dependent on meristems where cell proliferations (cell division and growth) give rise to new plant structures and allow the plant to increase in size. We provided scientific linkages and evidence to show that the growth promoting factors in biofertilizers regulating cell proliferation and ultimately modulating plant growth and development are phytohormones. The known biological functions of phytohormones (cytokinins, auxins, gibberellins, etc.) are in tandem with the observed physiological characteristics and crop yield of plants. When light, water and mineral nutrients are not limiting, phytohormones especially cytokinins, in biofertilizers help to drive plant growth by progressing faster through the various plant cell cycle checkpoints leading to the production of more cells. In the soil matrices, PGPRs (Plant Growth Promoting Rhizobacteria) have the ability to promote plant growth via various mechanisms such as nitrogen fixation, phosphorus and zinc solubilization. Some PGPRs secrete phytohormones, especially cytokinins, and can be cultured and developed into a biofertilizer. In the near future, a hybrid approach of combining organic and conventional fertilization

W.S. Wong · J.W.H. Yong (✉)
Singapore University of Technology and Design, 8 Somapah Road,
Singapore 487372, Singapore
e-mail: jyong@sutd.edu.sg

S.N. Tan
Natural Sciences and Science Education Academic Group,
Nanyang Technological University, 1 Nanyang Walk, Singapore 637616, Singapore

L. Ge
Nanyang Environment & Water Research Institute, Nanyang Technological University,
1 CleanTech Loop, CleanTech One, #06-08, Singapore 637141, Singapore

X. Chen
College of Life Science, Zhejiang University, Hangzhou 310058, China

D.K. Maheshwari (ed.), *Bacterial Metabolites in Sustainable Agroecosystem*,
Sustainable Development and Biodiversity 12, DOI 10.1007/978-3-319-24654-3_6

regimes will be the likely scenario as we have achieved a better understanding of plant growth and development through the regulatory controls on the cell proliferation processes by phytohormones and mineral nutrients delivered by fertilizers. The futuristic green biofertilizer should come in the form of granules in which the active plant growth promoting and soil improving substances and/or suitable microbes, with carefully selected mineral nutrients, are embedded in the packing materials giving slow and sustained release over a desired period.

Keywords Phytohormones · Fertilizers · Biofertilizers · Mineral nutrition · Plant growth · Cell cycle · Cytokinins · Auxins · Microbes · PGPR · Rhizobacteria

1 Introduction

1.1 General Plant Growth and Mineral Nutrition

General plant growth and development require 16 chemical elements (Marschner 1995). These essential elements can be classified into groups of non-mineral and mineral elements. Non-mineral elements, hydrogen (H), oxygen (O) and carbon (C), are found in air and water. Plants require these elements as raw ingredients, in the form of carbon dioxide (CO_2) and water (H_2O), to produce their own food via photosynthesis. During photosynthesis, CO_2 and H_2O are converted into glucose and ultimately to complex sugars, starch and cellulose using the energy from the sun light. Starch and complex sugar in turn provide source of carbon, energy and polymeric substrates of the plant for their growth and biosynthesis processes (Rolland et al. 2002).

Unlike non-mineral nutrients, mineral nutrients are obtained from the soil (Bronick and Lal 2005). These nutrients can be classified into macronutrients and micronutrients according to their relative concentration in plant tissues (Shaviv and Mikkelsen 1993; Taiz and Zeiger 2010). Macronutrients can be further divided into primary and secondary nutrients. Amongst all the mineral nutrients, primary nutrients, nitrogen (N), phosphorus (P) and potassium (K), are utilized by plants in the largest amounts for growth and survival and thus these three components are most lacking in nutrient-depleted soil. The secondary macronutrients are calcium (Ca), magnesium (Mg) and sulfur (S). Under natural conditions, the soil contains sufficient amount of secondary macronutrients and hence supplementation through fertilization is not always necessary. Furthermore, large amounts of Ca and Mg are added when lime is applied to reduce soil acidity; while S content in the soil is maintained by those released from the organic matter that underwent slow decomposition.

Micronutrients, boron (B), copper (Cu), iron (Fe), chloride (Cl), manganese (Mn), molybdenum (Mo) and zinc (Zn), on the other hand are required by plants in much smaller (micro) amount (Hänsch and Mendel 2009). They are also commonly termed as trace elements. Like S, these elements are recycled from decomposed organic matter. The above-mentioned elements in the form of micronutrient and macronutrients are mandatory for plant growth and development. However, their availability for plant uptake is determined not only by the amount present in the soil; soil composition and physicochemical properties (e.g. soil texture, soil structure and soil pH) also determine the extent to which the nutrients are bioavailable to the plants.

1.2 Soil Properties and Nutrient Availability

The soil is a complex physical, chemical and biological substrate (Bronick and Lal 2005; Berendsen et al. 2012). It is the most common medium in which many plants grow, and thus good soil condition is the prerequisite for promoting favorable plant growth which subsequently affects crop yield.

The properties contributing to the soil's function and ability to support plant life are the soil structure, texture and pH. Many ecological and plant physiological processes are heavily influenced by these factors: nutrient cycling, erosion, root penetration and gas exchange are some of the processes more directly related to plant growth.

Soil texture refers to the proportion of sand, silt, clay and organic matter in the soil, which is influenced by the geographical location and seasons. Soil texture influences nutrient and water retention in the soil which in turn benefits the plants (Bronick and Lal 2005). Clay and organic soil have much better nutrients and water retention capacity than sandy soil. In soils with poor nutrient and water retention capacity, leaching and loss of soil nutrients into groundwater occur as nutrients drain away, along with the water that is not being retained. This results in less nutrients being available for plant uptake. When the soil has more clay and organic matter, then water might be retained for too long causing the soil to become waterlogged. Under waterlogged conditions, the oxygen content in the soil depletes, plant roots might also rot due to prolonged soaking in water and aerobic respiration at the roots ceases. Production of nitrates, N source for plants, is also inhibited by the anaerobic condition of waterlogged soil (Crawford and Glass 1998). Hence, soil that contains optimum portions of sand, silt, clay and organic matter is ideal for farming and agricultural use.

Apart from the soil composition, soil structure, i.e. aggregation of soil particles, is also an important property of productive soil. It is the key factor that determines the functioning of the soil and enables the soil to support plant life on top of moderating environmental quality with respect to soil carbon sequestration and water quality (Bronick and Lal 2005). The pore spaces created between the particles in the aggregates affect water and air movement within the rhizospheres, nutrient availability for plant root growth and microbial activity. Thus, favorable soil structure helps to improve soil fertility, agronomic productivity and enhance soil porosity while lowering erodibility of the soil (Bronick and Lal 2005).

Soil pH, a measure of the acidity or alkalinity of the soil, is another important property that directly affects the availability of nutrients for plant uptake. At low pH, macronutrients tend to be less available; while at high pH, micronutrients tend to be less available. Under most circumstances, most soil has low pH level due to the release of hydrogen ion from the reaction between soil water and carbon dioxide produced during organic matter decomposition. Lime application can raise the soil pH level to the ideal range of 6.0–6.5. This slightly acidic pH range promotes root growth, weathering of rocks that releases minerals (Ca, K, Mg and Mn), and increases the solubility of carbonates, sulfates and phosphates (Taiz and Zeiger 2010). Beneficial plant bacterial activities, such as microbial nitrogen fixation and conversion of sulfur to forms suitable for plant uptake, also become more prevalent. Furthermore, the added lime also contributes to the pool of Ca and Mg for plant use and enhances the soil structure which subsequently promotes water and air movement.

Although nutrients occur naturally in the soil, some nutrients should be added to the soil such as lime or fertilizer to sustain plant growth and especially under situations where there are significant biomass removal periodically (Shaviv and Mikkelsen 1993; Chen 2006).

1.3 Fertilizers

Sustainable agriculture ideally should produce good crop yields with minimal impact on important ecological factors such as soil fertility (Tilman 1998; Mäder et al. 2002; Chen 2006). Mäder et al. (2002) defined fertile soil as a soil that provides essential nutrients for crop plant growth, supports a diverse and active biotic community, exhibits a typical soil structure, and allows for an undisturbed decomposition. Such "ideal" fertile soil is, however, yet to be achieved widely in the current green revolution that practices high-intensity agriculture. Intensive agriculture, often referred to as conventional agriculture, has successfully increased crop yields to meet the demands of the growing global population, but it leads to serious environmental costs (Tilman 1998). These costs include contamination of groundwater, release of greenhouse gases, change in the natural soil structure, loss of crop genetic diversity, eutrophication of water bodies and aquatic ecosystems and alteration of aquatic food webs.

Over the years, an alternative agriculture practice known as organic farming, is steadily gaining wide acceptance and practice. Agriculture products from organic farming are also marketed globally. According to the survey conducted by Foundation Ecology and Agriculture (SÖL), 4.0 % (as of 2004) of the agricultural land in Asia is managed organically; a rapid rise from the 0.33 % in 2001 (Hsieh 2005). The survey also revealed that a significant percentage of agricultural land in other continents were also managed organically, e.g. Oceania: 42 %; Latin American: 24 %; North America: 6 %; Europe: 23 %; Africa: 1 % (as of 2004). The 21-year-long organic farming study conducted by Mäder et al. (2002) showed that this method is also environment friendly. Nutrient inputs (N, P, K) and energy required to produce a crop dry matter unit was significantly reduced to 34, 51, 20 and 56 %, respectively, as compared to the conventional practice. There was, however, 20 % less crop yield. But how sustainable is organic farming? Until today, the answer still remains elusive.

The success of both farming methods is highly dependent on soil fertility, i.e. mineral nutrients. Mineral nutrients from the soil are dissolved in water and absorbed through plant roots. These nutrients may occur naturally in the soil but the bioavailable amount may not be sufficient to support healthy and robust plant growth. Furthermore, farming (especially the conventional method) may also deplete the soil of nutrients, especially primary macronutrients—N, P and K. To overcome this nutritional limitation, farmers and gardeners add nutrients externally through the application of fertilizers to compensate for the shortage of mineral nutrients in the soil. Globally, fertilizer usage has been increasing steadily over the years; typically, 40–70 % of a food production company's operating cost is spent on fertilizer usage.

In general, fertilizers refer to the substances added to the soil to increase its fertility. While most fertilizers are applied to the soil, some are formulated to be sprayed on leaves and the other aerial plant parts. Fertilizers can be derived from either organic sources or can be chemically synthesized. Regardless of the route of application and how they are derived, fertilizers supply the plants with nutrients that are generally absorbed in the form of inorganic ions (Taiz and Zeiger 2010). Nevertheless, there is new evidence to demonstrate that plants are able to absorb proteins directly and without involving other organisms (Paungfoo-Lonhienne et al. 2008).

Fertilizers play a pivotal role in regulating the growth of crop plants, and thus the reliability of food supply. The application of fertilizers is important for plant growth via cell proliferation (cell division and enlargement; detailed mechanisms will be discussed in Sects. 2 and 3), and the periodic replenishment of essential nutrients, especially the primary macronutrients, N, P, K, which are most likely to be depleted in heavily utilized soil, and other trace micronutrient, is essential to maintain soil fertility.

N, P and K are the three major mineral nutrients essential for plant growth and development (Orhan et al. 2006). Among these three nutrients, N is the mineral nutrient needed in greatest abundance by plants (Crawford 1995). N is the key constituent of molecules such as amino acids, amides, proteins, nucleic acids, nucleotides, coenzymes that are essential for various biological functions and comprises about 1.5 % of a plant's dry weight (Taiz and Zeiger 2010). In addition to serving as a nutrient and as an osmolyte, N also functions as a signal that reprograms N and C metabolism and influences root and shoot growth (Crawford 1995; Wang et al. 2007). N level in the soil is often lowered due to plant uptake, leaching and microbial denitrification, resulting in the dependence on N fertilizers to sustain the productivity of any modern intensive agriculture (Crawford and Glass 1998).

After N, P is the second most frequently limiting macronutrient for plant growth (Schachtman et al. 1998). Making up about 0.2 % of a plant dry weight, P is a component of key molecules such as nucleic acids, phospholipids, sugar phosphates, coenzymes and ATP (Schachtman et al. 1998; Taiz and Zeiger 2010). Phosphorus in the form of orthophosphate (Pi), is also required in the regulation of metabolic pathways and enzymatic reactions (Theodorou and Plaxton 1993).

Potassium, K, generally constitutes 1 % of a plant dry weight and thus it is the nutrient to be absorbed in second largest amount, after N. It is essential as a cofactor for enzyme activities involved in nutrient absorption, respiration, transpiration and photosynthesis. Unlike N and P, K does not become a part of endogenous organic compounds but remain as a cation in the plant tissues. It is also the crucial for the establishment of cell turgor and maintenance of cell electroneutrality (Taiz and Zeiger 2010).

A wide variety of fertilizers are available commercially and they can be classified into chemical fertilizers and organic/green fertilizers. Another class of fertilizers, biofertilizers, is also gaining worldwide attention, due to the awareness of the detrimental effects of chemical fertilizers imposed on the environment globally and of the improved knowledge on the relationships between plants and microorganisms occurring in the soil (Malusá and Vassilev 2014). Thus, agricultural fertilizers currently available in the market can be classified traditionally into three broad categories, namely chemical fertilizers, organic fertilizers and biofertilizers.

Chemical fertilizers are chemically synthesized compounds that contain specific nutrients, macro and/or micronutrients. This group of fertilizers provides the plants with nutrients in inorganic forms. In this review, we refer chemical fertilizers specifically to synthetic fertilizers containing N, P and K in various ratios in terms of weight percentage, otherwise also known as NPK fertilizers or inorganic fertilizers (Shaviv and Mikkelsen 1993).

Organic fertilizers are fertilizers generally derived from natural sources such as plant and animal matter. The commonly known examples are composts (decomposed plant materials) and manure (animal excrement). Meat and bone meal is another form of organic material being used as organic fertilizers (Jeng et al. 2006). In this review, we consider composts and manures as organic fertilizers.

Biofertilizers are the new and emerging entities in the realm of agricultural fertilizers or "Biostimulants". Biofertilizers are defined as fertilizers that enhance plant growth via the activities of microorganisms, i.e. conversion of nutritionally important elements/compounds from the "unavailable state" to (bio)available form(s) and production of active ingredients, particularly phytohormones (e.g. cytokinins, auxins, gibberellins etc.), amino acids and proteins. To date, the definition of biofertilizers is still unclear and remains highly debatable. No consensus has been agreed upon on the inclusion of indirect microbial activities, such as biocontrol properties targeting pathogens and conferring resistance against pathogens, as plant growth enhancing properties (Malusá and Vassilev 2014). Interestingly, there are also some people who do not consider biofertilizers as organic fertilizers. For the purpose of this review, we would, however, like to define biofertilizers as organic products containing biomass-based structural matrix, e.g. composts, humic acid and fulvic acid, with different types of useful (natural) microorganisms that enhance plant growth through their biological activities. Thus, in this review, our holistic classification of organic fertilizers is to include biofertilizers, unless otherwise stated.

In the following sections, we will discuss briefly on the advantages and disadvantages of chemical and organic fertilizers, while biofertilizers will be discussed in greater depth in Sect. 5.

1.3.1 Chemical Fertilizers

Chemical fertilizers are NPK-based formulation that has been widely used for over 100 years. Chemical fertilizers, e.g. urea, are the preferred choice as they are deemed to be highly effective and can be transported more economically. They supply the plants with mineral nutrients in the form of organic/inorganic salts that are easily and readily taken up by the plants via the roots, giving rise to immediate or quick plant growth improvement. Most importantly, chemical fertilizers supplement the soil with nutrients in desired ratios of N:P:K at a low cost.

Despite the advantages, a century-long usage of chemical fertilizers has seemed to reveal that these fertilizers are "losing" its growth promoting efficacy and increasingly, creating environmental problems. It has been reported that large fractions of N fertilizers applied to agriculture systems are lost as N_2, trace gases and nitrate leachate (Adesemoye et al. 2010), as the soil structure degrades and loses its mineral and water retention capacity. Larger amount of chemical fertilizers is thus required to achieve the same effect. The rampant use of chemical fertilizers in turn leads to eutrophication of water resources, pollution and contamination of soil, further degradation of soil structure, reduced soil fertility and reduced fertilizer efficiency (Vitousek et al. 1997; Tilman 1998; Mahdi et al. 2010; Xiang et al. 2012). Prolonged supply of high amount of N to plants also causes plant tissues softening, causing the plant to be more sensitive to pests and diseases (Chen 2006).

Chemical fertilizers also pose adverse effects on the biological properties of the soil. Its continuous application usually causes the soil to become acidic and repels earthworms, beneficial entity in fertile soil. Acidic soil also alters microbial species composition and diversity in the rhizosphere and results in the destruction and hindrance of beneficial microbial activities such as organic matter decomposition and symbiotic interaction with plants. Also, biologically inactive soil contains less organic matter and do not release as much nutrients as biologically active soil (Chandramohan et al. 2013). Over time, soil treated only with chemical fertilizers will lose its organic matter and disrupt the interactions with the living organisms, namely earthworms and microorganisms, which contribute immensely to soil "health" (Tilman 1998).

1.3.2 Organic Fertilizers

It is increasingly evident that the intensive agricultural methods employed as part of the conventional agriculture have been proven to be unsustainable (Tilman 1998; Mäder et al. 2002). Small but growing cohorts of farmers/growers have recognized this issue and have turned to utilize an alternate farming method, i.e. organic agriculture. Organic agriculture is defined by the FAO/WHO as "a holistic production management system which promotes and enhances agro-ecosystem health, including biodiversity, biological cycles and soil biological activity. It emphasizes the use of sustainable management practices in all aspects of the farm management and taking into account of the regional conditions and locally adapted systems. This is accomplished by using, wherever possible, agronomic, biological and mechanical methods, as opposed to using synthetic materials, to fulfill any specific function within the system" (FAO/WHO 2015). In comparison to conventional agriculture, organic agriculture is deemed to be an effective farming system with self-sustainability features.

Theoretically, organic agriculture is self-sustainable but it still requires the input of external fertilizers to replenish the nutrients that are quickly sequestered by the growing plants. To adhere to its organic practices, organic agriculture supply their soil with N by more natural means such as growing cover crops, mainly leguminous species that have the ability to fix atmospheric N. It makes use of organic fertilizers instead of chemical fertilizers. Organic fertilizers, also termed as green fertilizers, however, work on a different basis as compared to chemical fertilizers.

Organic fertilizers, including biofertilizers, supplement the soil with nutrients, but usually at a much lower concentration and slower release rate as compared to chemical fertilizers. N is released slowly from organic fertilizers (compost) due to its slow mineralization rate (Hernández et al. 2010) and its availability is dependent on soil properties (Fricke and Vogtmann 1993). P is contributed at a percentage of 20–40 % (Fricke and Vogtmann 1993) of its total low P content (vermicompost: 0.014 ± 0.0009 %; compost: 0.015 ± 0.0009 % (Hernández et al. 2010)). K is, however, contributed at an exceptionally higher percentage of 85 % (Fricke and Vogtmann 1993).

Generally, organic fertilizers have lower mineral nutrient contents which are often not well characterized and quantified, and may vary between production batches and methods when compared with chemical fertilizers (Shaviv and Mikkelsen 1993; Mäder et al. 2002; Chen 2006; Mahdi et al. 2010). Thus, within the plant industry, there is certainly a need to develop/produce organic fertilizers in a reproducible way in order to gain wider acceptance by farmers utilizing the conventional farming approach, i.e. quality control and assurance for organic fertilizers as a reliable commercial product. Furthermore, the rate of nutrient release in some poorly prepared organic fertilizers may not meet the needs of the vigorously growing plants, accustomed to conventional chemical fertilizers. Hence, larger amount of organic fertilizers has to be applied under certain situations. This in turn incurs more cost to the growers as compared with using the traditional chemical fertilizers.

Organic fertilizers are, however, desirable over chemical fertilizers due to its ability to improve the soil structure via the enhancement of soil biological diversity. As discussed earlier (Sect. 1.2), soil properties play a very vital role in determining plant growth and nutrient availability for plant uptake. Thus, organic fertilizers are more valued for their soil-improving qualities and to a lesser extent for their mineral nutrient, mainly N and P, contribution.

In comparison to chemical fertilizers, the environmentally friendly organic fertilizers (composts and manure) do appear to be the solution to attain sustainability in modern agriculture. However, due to our current limited understanding about the growth stimulating mechanism(s) of organic fertilizers on plant growth, we will not be able to achieve optimal and sustainable agriculture yield by relying solely on organic fertilizers at present. We believe that the combined use of chemical and/or organic fertilizer with the active ingredients obtained from biofertilizers, would help the present day farmers to achieve sustainable farming with maximal yield (Shafi et al. 2012; Qin et al. 2015; Song et al. 2015).

In order to understand how organic fertilizers and phytohormones work in tandem to govern plant growth and development, and the other associated useful plant performance characteristics such as conferring resistance against pathogens, the fundamentals of plant growth mechanisms and cell proliferation regulations by various substrates (sucrose, phytohormones, availability of nutrients, etc.) will be revisited and discussed in depth and in relation to fertilizers.

2 Plant Growth Mechanisms

Plant growth is dependent on meristems, groups of dividing cells that give rise to new plant structures (Steeves and Sussex 1989; Coen and Meyerowitz 1991; Wolters and Jürgens 2009) and enable the plant to increase in size continuously throughout its lifetime (Huala and Sussex 1993). In order to be competent and ready to divide, proliferative cells in the plant meristems have to undergo four distinctive phases of the cell cycle: postmitotic interphase (G1), DNA synthetic phase (S), postsynthetic interphase (G2) and mitosis (M phase); and subdivision phases of mitosis, cytokinesis and G1-phase (G0); all of which are governed by a series of checkpoints (Francis and Sorrell 2001) regulated by cyclin-dependent kinases (CDKs) (Den Boer and Murray 2000; Inzé 2003; Dewitte and Murray 2003) (Fig. 1). Cyclins, CDK inhibitors, retinoblastoma proteins, E2F/DP transcription factors, histones, sucrose and phytohormones (specifically cytokinins, auxins, gibberellin and abscisic acid) are some of the other factors involved in the regulation of the cell cycles (Inzé 2003; Souza et al. 2010).

Meristems have the ability to divide and differentiate but can only give rise to certain structures; for example, the root meristems give rise to roots while the shoot meristems may result in leaves, flowers, axillary buds and internodes (Huala and Sussex 1993). Primary shoot apical meristem (SAM) that arises during embryogenesis together with additional meristems formed after seed germination plays

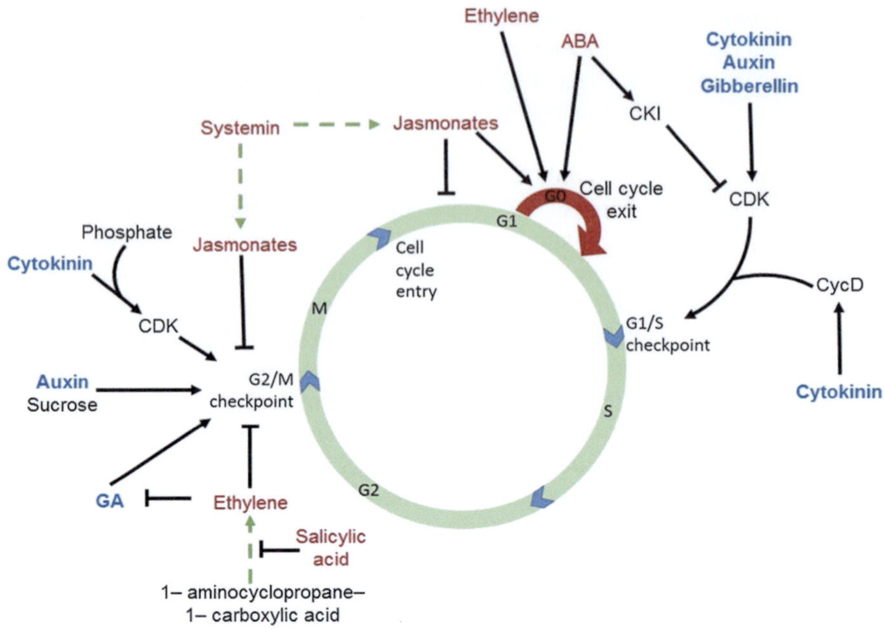

Fig. 1 Schematic diagram of the plant cell cycle and regulatory actions of phytohormones. Cell cycle has four distinct phases, G1 (postmitotic interphase), S (DNA synthetic phase), G2 (postsynthetic interphase) and M (mitosis); and subdivision phases of mitosis, cytokinesis and G1-phase (G0); all of which are governed by a series of checkpoints, mainly between G1 and S (G1/S checkpoint) and between G2 and M (G2/M checkpoint). The plant cell cycle is predominantly regulated by phytohormones while other factors exerting varying degrees of controls under different environmental (abiotic) and biotic circumstances in which the whole plant is exposed to. Generally, auxins, cytokinins and gibberellin play stimulatory roles; while abscisic acid (ABA), ethylene and jasmonates inhibit the progression of cell cycle. Systemin, a plant peptide hormone, down regulates cell cycle by promoting the biosynthesis of jasmonates, while salicylic acid up-regulates cell cycle progression by inhibiting the biosynthesis of ethylene

a crucial role in giving the plants its various forms (Steeves and Sussex 1989). Primary SAM provides the plant with the main axis while the other meristems determine the development of the shoot branches, and the brunches temporal and spatial development determines the complexity of the branching pattern (Shimizu-Sato and Mori 2001). Plant forms although plastic in nature, influenced by environmental cues, are still genetically governed and thus retain their species-specific forms (Shimizu-Sato and Mori 2001).

Axillary meristems are typically located on the adaxial (upper) region where the primary organ, such as leaf axil and stems join (Grbić and Bleecker 2000; Long and Barton 2000; Shimizu-Sato and Mori 2001). Being secondary meristems, axillary meristems are crucial for the continuous development of the plant morphology, such as lateral plant growth (Bennett and Leyser 2006). Lateral plant growth is, however, governed by a mechanism known as apical dominance. Apical dominance, broadly defined as the inhibitory control of the shoot apex over the outgrowth of lateral buds, is one of the mechanisms that ensure the plants survival with a reservoir of meristems to replace damaged primary shoot (Shimizu-Sato and Mori 2001). Apical dominance, however, can be released by development programs, hormonal and environmental cues (Turnbull et al. 1997; Bangerth el al. 2000; Shimizu-Sato and Mori 2001; Schmülling 2002). Increased cytokinin concentrations and changes in phytohormone ratio(s) (e.g. auxins:cytokinins) are some of the hormonal cues that enable the plant to overcome apical dominance.

3 Plant Growth Regulation

Plant growth is regulated by various abiotic and biotic factors such as temperature, light intensity, water availability, soil compositions and characteristics, nutrient availability, phytohormones (plant hormones) availability, interactions with the immediate organisms (microorganisms, fungi, other plants) in its surroundings, and other factors (for reviews, see Steeves and Sussex 1989; Mok 1994; Rolland et al. 2002; Van Loon 2007; Wolters and Jürgens 2009).

In this review, we will focus our discussion on sucrose and phytohormones availability in relation to fertilizer types and usage.

3.1 Regulation by Sucrose Availability

Photosynthesis is the fundamental process in plants and sugars are produced in the process. Sugars are essential for the plant growth as they are the source of carbon, energy and polymer substrates for biosynthesis (Rolland et al. 2002). They are transported via the plants phloem from the sites of photosynthesis (usually the leaves) to the various sink organs such as roots, flowers, developing fruits and seeds, mainly in the form of sucrose. Sucrose concentration plays an important role in regulating various plant growth processes. Low sucrose concentration stimulates leaf photosynthetic activities, nutrient mobilization and export from the sink organs, while high sucrose concentration inhibits photosynthetic activities but stimulate growth and storage in the sink organs (Wang and Ruan 2013). Before sucrose can be utilized for metabolism and biosynthesis, they are normally converted to simpler forms like glucose and fructose by invertase or UDP-glucose and fructose by sucrose synthase (Wang and Ruan 2013). In addition to their essential roles as source of carbon, energy and building blocks for plant growth, sucrose and its cleavage compound hexose have important hormone-like functions as signaling molecules that regulate specific gene expression (Rolland et al. 2002; Wang and Ruan 2013).

A close correlation was observed between the supply of sucrose and the expressions of cyclins, specifically d-type cyclins (CycD2 and CycD3), that induce the cell to progress beyond the G1 phase and become committed to complete the full plant cell cycle (Riou-Khamlichi et al. 2000) (Fig. 1). Expression of CycD2 only requires sucrose while CycD3 expression requires the presence of sucrose and phytohormones, specifically cytokinins and auxins (Koning 1994; Riou-Khamlichi et al. 2000). In other words, sucrose is essential in the upstream regulation prior to hormonal regulation of CycD3 expression. Sucrose is also required for the activation of mitotic entry by activating the transcription of key components that drives the G2 to M transition (Skylar et al. 2011) (Fig. 1). Thus, it is evident that sucrose could be involved in the reactivation of cell from the state of growth arrest (Souza et al. 2010). Apart from driving plant cell cycle transition indirectly, sucrose is also required for general plant growth such as tuberization in potato (Šimko 1994), to induce formation of adventitious roots in *Arabidopsis* seedlings (Takahashi et al. 2003), to induce flowering (Roldán et al. 1999) and various other growth processes (for a review, see Gibson 2005).

3.2 Regulation by Phytohormones Availability

Phytohormones are naturally occurring substances known to be crucial for regulating various aspects of physiology and development throughout the lifecycle (for reviews, see Bleecker and Kende 2000; Mok and Mok 2001; Pimenta Lange and Lange 2006; Matsubayashi and Sakagami 2006; Wolters and Jürgens 2009; Zhao 2010; Pacifici et al. 2015). Some of the growth regulatory functions include cell division and expansion, cell elongation, stem elongation, inhibition, root growth, activation of bud growth, branch development, promoting or delay in leaf senescence and chlorophyll production. Regulatory functions of cytokinins and auxins will be discussed in detail in the following sections. Other classes of phytohormones, such as gibberellins, ethylene, abscisic acid and strigolactones, are listed in Table 1 and will not be discussed in detail.

3.2.1 Cytokinins

The vast majority of naturally occurring cytokinins are a group of adenine compounds with an isoprene, modified isoprene, or aromatic side chain attached to the N^6 amino group. Refer to Fig. 2a for the basic structure of cytokinins and Fig. 2b for the representative cytokinins. Cytokinins occur in both free and tRNA-bound forms, and are essential in regulating various physiological processes in plants (Letham and Palni 1983; Haberer and Kieber 2002; Stirk and Van Staden 2010). The biosynthetic gene *ipt*, which encodes the enzyme isopentenyltransferase, is responsible for the synthesis and expression of cytokinins (Kamínek et al. 1997). This enzyme is produced in the roots and shoots (Chen et al. 1985), with the root apical meristems being the major site of synthesis. Isopentenyltransferase is essential in the first step of cytokinin synthesis. It transfers the isopentenyl moiety from

Table 1 List of hormonal and non-hormonal plant growth promoters and their known biological functions

	Classes of plant growth promoters	Molecular structure	Biological functions	References
Hormonal	**Cytokinins** *trans*-zeatin (*t*Z) *trans*-zeatin riboside (ZR) Kinetin (K) N^6-[2-isopentyl] adenine (iP) N^6-isopentyladenosine (iPR) Dihydrozeatin	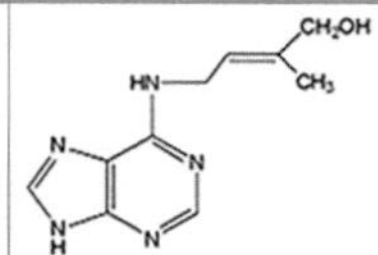 *t*Z 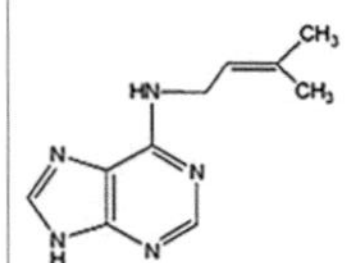 iP	Activate cell division and regulate plant growth from the cellular level through to the tissue, organ and whole plant level	Letham and Palni (1983), Frank and Schmülling (1999), Francis and Sorrell (2001), Schmülling (2002), Sakakibara (2006)
			Upregulate CycD3 at the G1 checkpoint and the phosphoregulation of CDK at the G2/M checkpoint	Frank and Schmülling (1999), Francis and Sorrell (2001)
			Involved in formation of embryo vasculature, leaf expansion, branching, chloroplast development, root growth, seed germination and delay senescence	Letham and Palni (1983), Mok (1994), Schmülling (2002), Howell et al. (2003)
			Involved in the initiation and outgrowth of axillary buds, release from shoot apical dominance	Turnbull et al. (1997), Bangerth et al. (2000), Shimizu-Sato and Mori (2001), Yong et al. (2014)
			Function as signals	Leopold and Kawase (1964), Gersani and Kende (1982), Mauk and Noodén (1992), Yong et al. (2000), Emery and Atkins (2002), Ma et al. (2002), Schmülling (2002), Frugier et al. (2008), Stirk and Van Staden (2010); Yong et al. (2014)
	Auxins Indole-3-acetic acid (IAA) Indole-3-butyric acid (IBA) Naphthaleneacetic acid (NAA)	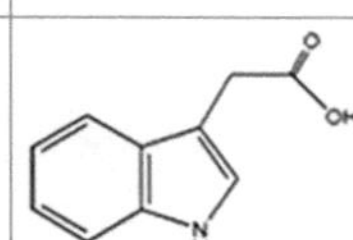 IAA	Promote cell expansion/elongation, root growth, regulate cell cycle and inhibit growth of axillary buds	Zhao (2010)

(continued)

Table 1 (continued)

	Classes of plant growth promoters	Molecular structure	Biological functions	References
			Induce the expression of CycD3 and CDKs	Wang and Ruan (2013)
			Interact with other phytohormones to regulate various plant growth processes	Shimizu-Sato and Mori (2001), Haberer and Kieber (2002), Wang and Ruan (2013), Zhou et al. (2013), Zhao et al. (2015)
			Regulate synthesis of other phytohormones	Al-Babili and Bouwmeester (2015)
			Involved in organ primordia initiation and the initiation and maintenance of organ founder cell populations	Wolters and Jürgens (2009)
	Gibberellin (GA) GA1 GA3 GA4 GA7	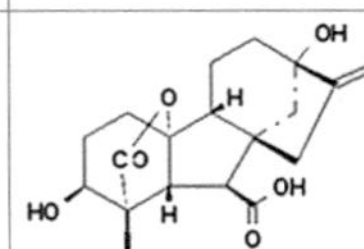 GA1	Regulate plant growth, promote stem elongation, leaf expansion, trichome development, flower formation, sex expression, fruit setting, seed development and germination	Pimenta Lange and Lange (2006), Wolters and Jürgens (2009), Nelson et al. (2012), Pacifici et al. (2015)
			Transcriptional activation of G2/M checkpoint proteins that promote cell division	Sauter et al. (1995)
			Work in antagonism with ABA in the control of seed germination	Nelson et al. (2012)
			Inhibit fruit ripening by decreasing respiratory activity and delaying anthocyanin synthesis and chlorophyll degradation	Martínez et al. (1994)

(continued)

Table 1 (continued)

Classes of plant growth promoters	Molecular structure	Biological functions	References
Ethylene	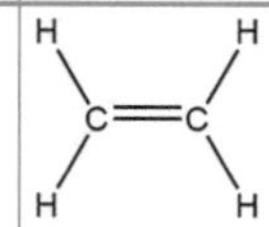	Inhibits cell proliferation	Herbert et al. (2001)
		Involved in stress response and activation of defense responses and resistance against certain pathogens	Klessig et al. (2000), Wolters and Jürgens (2009)
		Regulator of plant senescence	Bleecker and Kende (2000)
Abscisic Acid (ABA)	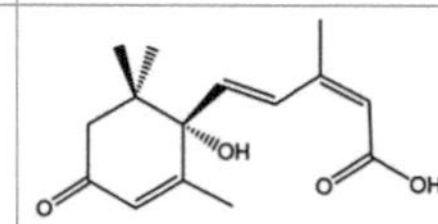	Inhibit cell proliferation (cell division); prevent cells from entering S phase	Koning (1994), Francis and Sorrell (2001), Dewitte and Murray (2003)
		Inhibit bud growth	Shimizu-Sato and Mori (2001)
		Involved in stress response (closing of stomata)	Wolters and Jürgens (2009), Yoshida et al. (2010)
		Interact with auxin to regulate root development	Zhao et al. (2015)
		Downregulates GA signaling	Wolters and Jürgens (2009)
Strigolactones	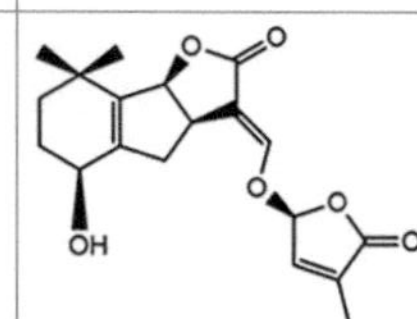 Strigol	Carotenoid-derived phytohormones	Kretzschmar et al. (2012), Al-Babili and Bouwmeester (2015)
		Stimulate growth of arbuscular mycorrhizal (AM) fungi and seed germination of root-parasitic plants	Umehara (2011), Kretzschmar et al. (2012), Nelson et al. (2012)
		Coordinate root and shoot development in response to phosphorus starvation	Koltai (2011), Smith (2013)
		Interact strongly with the auxin signaling pathway to modulate auxin-dependent secondary growth, e.g. increasing stem thickness	Agusti et al. (2011)

(continued)

Table 1 (continued)

	Classes of plant growth promoters	Molecular structure	Biological functions	References
	Brassinosteroids	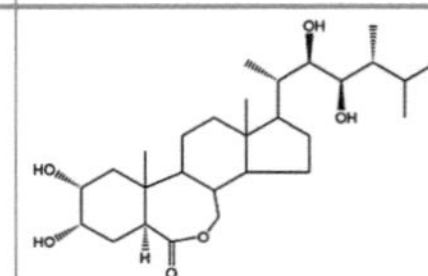	Regulate growth, xylem differentiation, disease resistance, stress tolerance and control of photomorphogenesis bending	Dewitte and Murray (2003), Müssig et al. (2006), Wolters and Jürgens (2009), Li et al. (2013), Pacifici et al. (2015)
			Required for the induction of a protein, ARGOS-LIKE (ARL), responsible for cell expansion during leaf growth	Wolters and Jürgens (2009)
			Required as signal for genes activation involved in a wide range of physiological processes	Müssig et al. (2006)
	Salicylic Acid	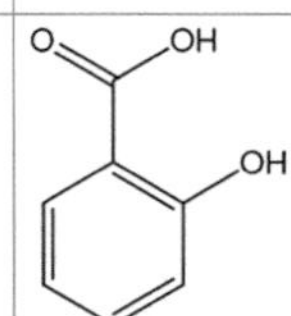	Involved in plant defense responses	Doares et al. (1995), Klessig et al. (2000), An and Mou (2011)
			Involved in cell growth, seed germination and establishment, flower formation, stomatal responses, thermotolerance, thermogenesis and nodulation	An and Mou (2011)
			Inhibits biosynthesis of ethylene	Raskin (1992)
			Induce production of plant phenolics	Raskin (1992)
	Jasmonates	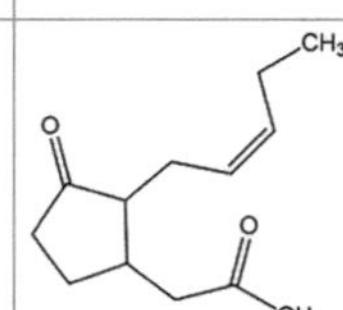	Prevent DNA synthesis during the G1 phase and inhibit M phase progression	Dewitte and Murray (2003)

(continued)

Table 1 (continued)

	Classes of plant growth promoters	Molecular structure	Biological functions	References
			Involved in stress response by mediating the activation of defense response and resistance genes against certain pathogens and insects	Creelman and Mullet (1997), Klessig et al. (2000)
			Modulate growth processes, e.g. fruit ripening, production of viable pollen, root growth and tendril coiling	Creelman and Mullet (1997)
			Induce tuberization in potatoes	Pelacho and Mingo-Castel (1991)
			Regulation of genes encoding vegetative storage proteins	Creelman and Mullet (1997)
			Repress genes encoding proteins involved in photosynthesis	Creelman and Mullet (1997)
	Plant peptide hormones Systemin		Function as signals in the regulation of various growth processes, e.g. meristem organization, root growth regulation, leaf shape regulation, nodule development, callus growth, pollen tube growth and organ abscission	Matsubayashi and Sakagami (2006), Matsubayashi (2014)
			Involved in plant defense responses	Ryan (2000), Lindsey et al. (2002), Ryan and Pearce (2003), Matsubayashi and Sakagami (2006), Matsubayashi (2014)

(continued)

Table 1 (continued)

	Classes of plant growth promoters	Molecular structure	Biological functions	References
	Polyamines Putrescine Spermidine Spermine	 Putrescine	Required for cell cycle progression	Weiger and Hermann (2014)
			Regulate fruit growth and ripening, leaf senescence, reproductive organ development and tuberization	Martin-Tanguy (2001)
			Involved in stress response	Groppa and Benavides (2008)
	Nitric oxide (NO)	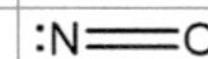	Function as signaling compound in regulatory pathways	Ya'acov (1996), Nelson et al. (2012)
			Stimulatory and inhibitory effects on plant growth	Ya'acov (1996), Durner and Klessig (1999)
			Regulate expression of genes related to metabolism, cellular detoxification, iron homeostasis, signaling, flowering, lignin biosynthesis and transport	Lamotte et al. (2005)
			Involved in plant defense gene activation upon pathogen attack	Durner et al. (1998), Klessig et al. (2000)
			Modulate the synthesis of salicylic acid, jasmonate and ethylene	Lamotte et al. (2005)

(continued)

Table 1 (continued)

	Classes of plant growth promoters	Molecular structure	Biological functions	References
	Karrikins	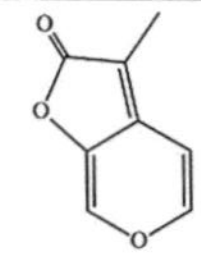 Karrikinolide	Stimulate seed germination for certain plant species	Nelson et al. (2009, 2012), Scaffidi et al. (2012)
			Enhance light responses during germination and seedling development	Nelson et al. (2010)
Non-hormonal	**Humic substances** Humic acid Fulvic acid		Enhance plant growth and enhance root development	Rauthan and Schnitzer (1981), Piccolo et al. (1992), Atiyeh et al. (2002)
			Involved in cell respiration, oxidative phosphorylation, protein synthesis, photosynthesis and enzymatic reactions	Atiyeh et al. (2002)
			Improve nutrient uptake, e.g. iron and zinc	Atiyeh et al. (2002), Chen et al. (2004)
			Improve soil structure and accumulation of organic matter	Piccolo et al. (1992), Atiyeh et al. (2002), Piccolo (2002)

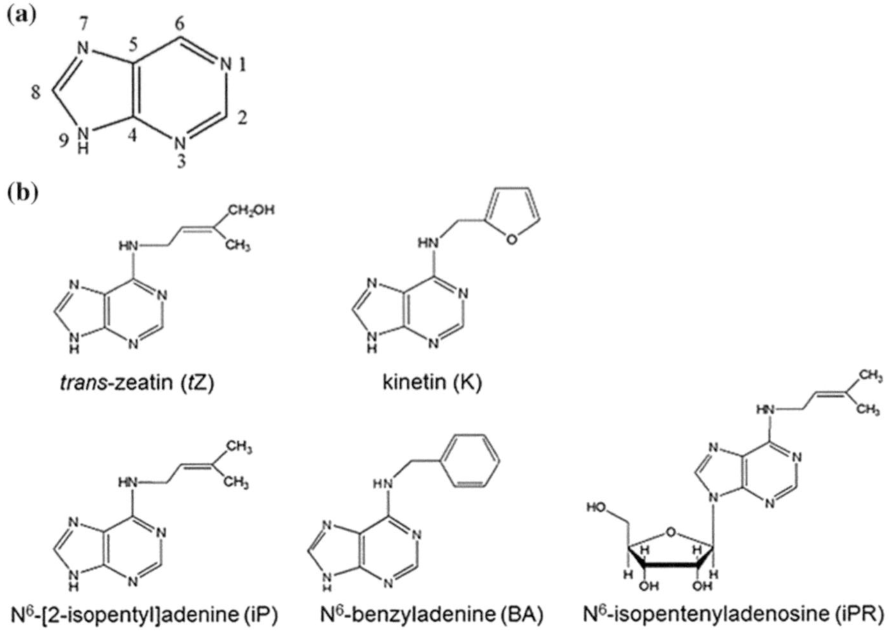

Fig. 2 Chemical structures of cytokinins and its basic structure. **a** Adenine structure and the numbering system for the nomenclature of cytokinins. **b** Representative cytokinins: *trans*-zeatin (*tz*), kinetin (K), N^6-[2-isopentyl]adenine (iP), N^6-benzyladenine (BA) and N^6-isopentyladenosine (iPR). Note that the isoprene, modified isoprene, or aromatic side chains are attached to the N^6 amino group of adenine compounds

dimethylallyl diphosphate (DMAPP) to ATP/ADP, which is more efficiently utilized by the plant isopentenyltransferase compared to AMP (Kakimoto 2001). In contrast, bacterial cytokinin synthesis, which shares a similar pathway, are able to start the first step by transferring isopentenyl moiety from 1-hydroxy-2-methyl-2-(E)-butenyl 4-diphosphate (HMBDP) to AMP (Ueda et al. 2012) (Fig. 3).

Cytokinins are classified according to their side-chain configurations as either isoprenoid or aromatic cytokinins (Stirk and Van Staden 2010) with the latter being the rarer form (Kakimoto 2003). Naturally occurring isoprenoid cytokinins are either isopentenyladenine (iP)-type, which carries an isopentenyl N^6 side chain, or zeatin-type, which carries hydroxylated isopentenyl N^6 side chain (Kakimoto 2003). Zeatin-type cytokinins can occur in *cis* or *trans* configuration, depending on the hydroxylation of the methyl group on the side chain; and *trans*-zeatin and its derivatives have higher biological activity than the *cis* forms. Zeatin-type cytokinins are also the main constituents in plants (Mok et al. 2000). Cytokinins also occur in different forms, namely in the form of free base, riboside, or ribotide (or nucleotide); with the free base form being biologically active and the riboside form being the form of transportation via the xylem system. The riboside type of cytokinins are later converted to their active form by another enzyme at the shoot (Sakakibara 2006).

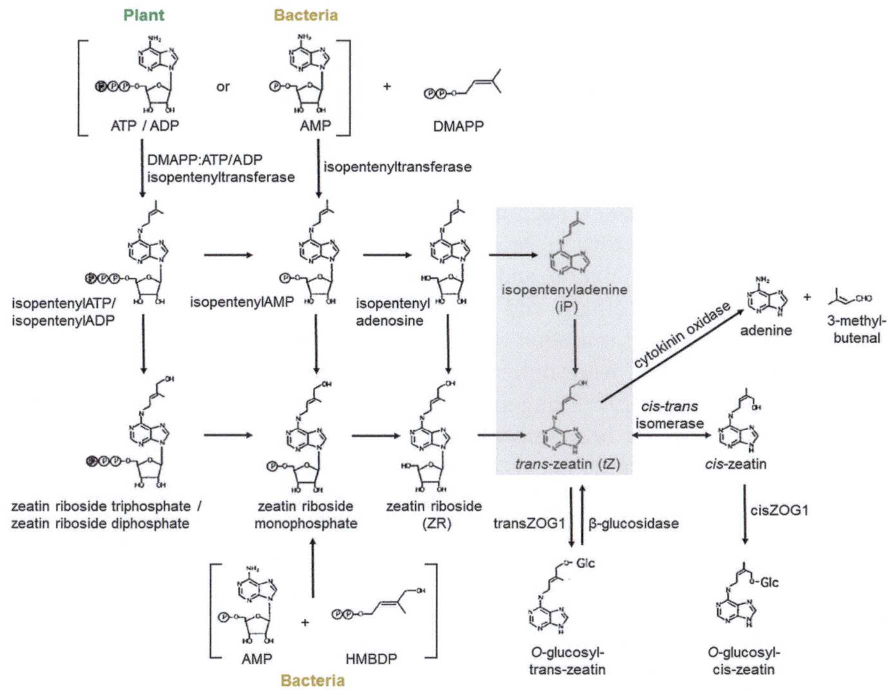

Fig. 3 A model for cytokinins biosynthesis and metabolic pathway in plants and bacteria. In plants, the isopentenyl moiety from dimethylallyl diphosphate (DMAPP) is transferred to ATP/ADP while the bacterial pathways start off with AMP. Bacteria cytokinin biosynthesis may also start by transferring isopentenyl moiety from 1-hydroxy-2-methyl-2-(E)-butenyl 4-diphosphate (HMBDP) to AMP. Active cytokinins are shaded (adapted from: Haberer and Kieber (2002), Kakimoto (2003), Sakakibara (2006), Tarkowski et al. (2009) and Frébort et al. (2011))

The biological concentrations of these phytohormones are closely regulated by the rates of biosynthesis, metabolism, inactivation and degradation with homeostasis under the influence of both internal and external factors (Sakakibara 2006). Any exchange or interconversion of cytokinins between plants and its various external components can potentially influence and even disrupt cytokinin homeostasis in the plant and alter its growth pattern as low concentrations of cytokinins are needed to elicit a physiological response (Letham and Palni 1983; Stirk and Van Staden 2010).

Role of Cytokinins in Plant Growth

Cytokinins are crucial for activating cell division and to regulate plant growth from the cellular level through to the tissue, organ and whole plant level (Letham and Palni 1983; Francis and Sorrell 2001; Schmülling 2002; Sakakibara 2006).

On the cellular level, cytokinins upregulate plant d-type cyclin (CycD3) at the G1 checkpoint and the phosphoregulation of the CDK at the G2/M checkpoint, thereby inducing a continuum of cell cycle activation that leads to plant growth (Francis and Sorrell 2001). The main cytokinin driver of the cell cycle is zeatin, which peaks its concentrations at the end of S phase, during the G2/M phase transition and in the late G1 phase.

Developmental processes such as formation of embryo vasculature, nutritional signaling, leaf expansion, branching, chlorophyll production, root growth, promotion of seed germination and delay of senescence are also heavily influenced by cytokinins (Letham and Palni 1983; Mok 1994; Schmülling 2002; Howell et al. 2003). The initiation and outgrowth of axillary buds, released from shoot apical dominance, were reported to be well correlated with the cytokinins levels (Turnbull et al. 1997; Bangerth et al. 2000; Shimizu-Sato and Mori 2001; Yong et al. 2014). It has been known that plants with reduced endogenous cytokinins have distinct morphological and developmental alterations such as shorter shoot internodes, delayed flowering, fewer flowers and reduced leaf surface area with smaller vasculature, smaller shoot apical meristems with reduced cell division, enhanced root growth and a larger root meristem (Schmülling 2002). Thus, any change in the levels of endogenous cytokinins could alter the regulation of the above-mentioned physiological processes and result in the disruption of normal plant growth (Letham and Palni 1983; Schmülling 2002). However, it is also important to note that different classes of phytohormones interact in a synergistic way for regulation of physiological processes and optimum plant growth. The roles of these phytohormones will be discussed in their respective sections.

Cytokinins have been reported to function as local and long-range chemical signals in plants. They are transported via the xylem and phloem (Hwang and Sakakibara 2006) and the transpiration stream from the root tips to aerial plant parts (Yong et al. 2000; Schmülling 2002; Stirk and Van Staden 2010; Yong et al. 2014). Studies conducted by Ma et al. (2002) showed that cytokinins synthesized in the embryo function as local signal for increased meristematic activity. Reallocation of nutrients, minerals and nonmetabolizable substances are also initiated with an increase in cytokinins concentrations in leaves (Leopold and Kawase 1964; Gersani and Kende 1982; Mauk and Noodén 1992), a phenomenon termed as cytokinin-induced nutrient mobilization. It has also been suggested by Frugier et al. (2008) that cytokinins may function as the central signal for controlling lateral organ differentiation. Their study revealed that a local increase in cytokinin concentrations within the roots induces nodule organogenesis while repressing lateral root formation (Fig. 4). Cytokinins functioning as long-range biochemical signals help to coordinate root–shoot development (Schmülling 2002; Stirk and Van Staden 2010), communicate root biotic interactions (e.g. with *Rhizobium*, Yong et al. 2014) and environmental stresses such as nutritional status, low temperatures, salinity and drought to the shoots (Goicoechea et al. 1996; Yong et al. 2000; Emery and Atkins 2002; Schmülling 2002), a phenomenon termed as root-to-shoot signaling.

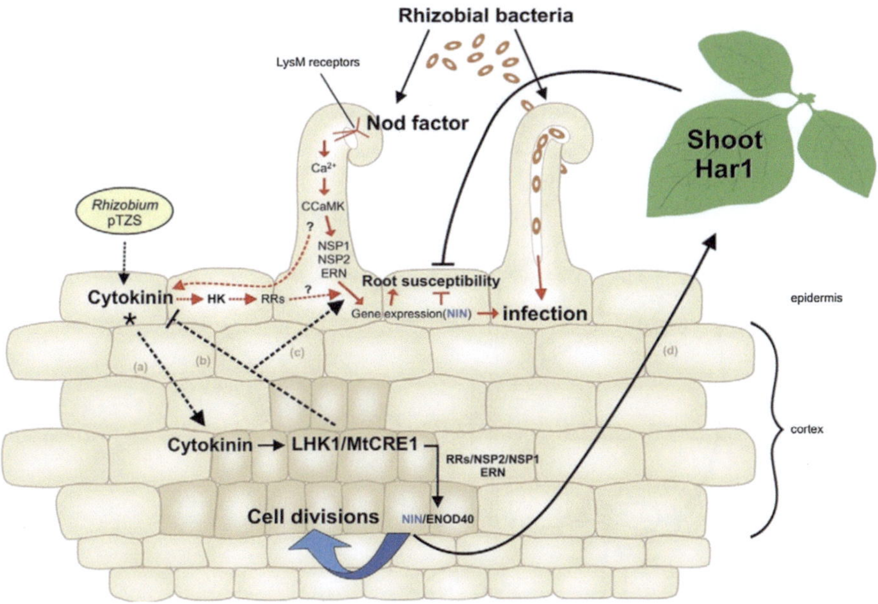

Fig. 4 Proposed role of cytokinin in nodulation and infection events. **a** Nod factor perception by lysin motif (LysM)-containing receptors elicits calcium signaling through a calcium- and calmodulin-dependent kinase (CCaMK). This in turn leads to localized biosynthesis and/or activation of cytokinin signaling by an as yet unknown mechanism. Part of this signaling cascade can be bypassed by bacterially produced cytokinins (*Rhizobium* pTZS) or cytokinin-like molecules, enabling Nod factor-independent nodulation. Epidermally produced cytokinin might be translocated to the cortex by diffusion and/or by selective transport from cell to cell. Alternatively, an intermediate messenger (*) might travel to the cortex to elicit de novo localized cytokinin signaling. **b** Cytokinin perception by LHK1 or ortholog MtCRE1, and signaling through cytokinin response regulators (RRs) leads to initiation of nodule organogenesis (cell divisions). This requires transcription factors such as NSP1, NSP2 and ERN, as well as downstream functions, such as NIN and ENOD40. In the epidermis, NIN is required for infection thread formation but also negatively regulates root susceptibility to rhizobial signaling. Hypothetically, cytokinin might participate in this process by signaling through an unknown histidine kinase receptor(s) (HK). In this scenario, cytokinin signaling contributes to, but is not fully responsible for, reprogramming of gene expression, possibly by regulating the activity and/or localization of transcriptional factors, such as NSP2, which is known to relocate from the nuclear envelope to the nucleus upon Nod factor signaling. Cytokinin might also be involved in both local and systemic feedback regulation of infection. In *Lotus*, LHK1 is not required for initiation and progression of infection events, but it participates in negative regulation of root susceptibility to infection. **c** In *M. truncatula*, both nodule inception and infection thread progression, but not initiation, are tightly linked to MtCRE1 function. **d** Cytokinin might also participate in systemic autoregulatory feedback mechanisms, possibly involving HAR1, to restrict nodule number (reprinted from Trends in Plant Science, 13, F. Frugier, S. Kosuta, J.D. Murray, M. Crespi, K. Szczyglowski, Cytokinin: secret agent of symbiosis, 115–120, Copyright (2008), with permission from Elsevier)

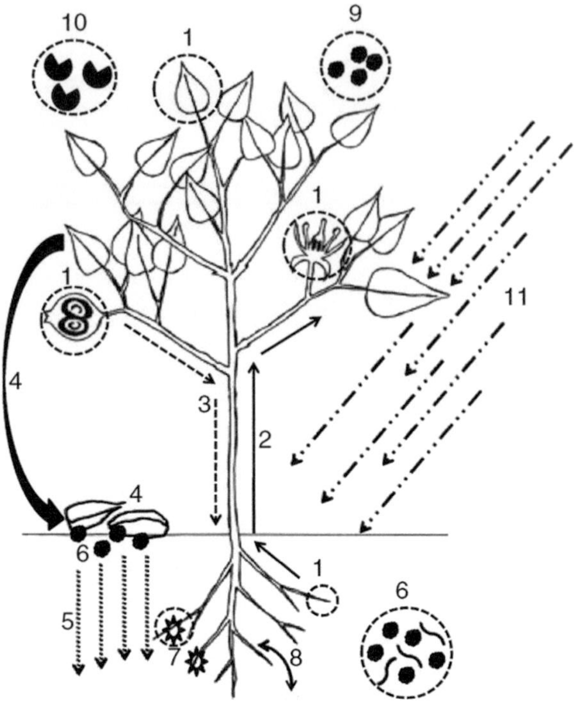

Fig. 5 Diagrammatic scheme showing the movement of cytokinins through the environment. *Dotted circles* indicate sites of cytokinins biosynthesis. *1* sites of cytokinin biosynthesis in vascular plants including roots, flowers and fruits; *2* cytokinin translocation via the xylem from the roots; *3* cytokinin translocation via the phloem from aerial plant organs; *4* cytokinins derived from degradation of leaf litter; *5* movement of cytokinins through the soil due to rainwater and irrigation; *6* free-living microorganisms (bacteria, fungi, Cyanophyta and Chlorophyta) in the rhizosphere; *7* microorganisms, including nematodes in a symbiotic relationship (either beneficial or parasitic) with the host plant's roots; *8* cytokinins released into the soil by root exudates and cytokinin uptake by the roots; *9* air-borne microorganisms (bacteria and fungi) in a parasitic relationship with the host plant; *10* insects infecting the host plant to form galls; *11* agricultural input including irrigation with water that contains cytokinins and application of natural (e.g. seaweed concentrates) and synthetic cytokinins for crop improvement (reprinted from Plant Growth Regulation, 62, 2010, 101–116, Flow of cytokinins through the environment, W.A. Stirk and J. Van Staden, Fig. 1, with kind permission from Springer Science and Business Media)

Flow of Cytokinins Through the Environment

Cytokinins are widely distributed throughout the plant kingdom (Stirk and Van Staden 2010), and are widely available and highly fluid within the environment (Fig. 5). In the following section, we will discuss the various plant-related sources of cytokinins and their distribution in the environment.

Sources of Cytokinins

Cytokinins can be derived from various sources and the most prominent source would be from plants as they are called phytohormones. Cytokinins have been known to be released directly into the soil from plant roots (Van Staden 1976). Studies conducted by Arthur et al. (2001) showed that tomato seedlings metabolize the cytokinins taken up from their external environment before releasing them back to the external environment. Plant parts, such as cotyledons, flowers and leaf litter, are also sources of cytokinins (Letham and Palni 1983; Stirk and Van Staden 2010). Hence, root exudates and any plant parts that contain cytokinins are potential sources that may contribute to the pool of cytokinins available in the environment for the uptake by other plants.

Microalgae of both prokaryotic and eukaryotic nature, namely Cyanophyta and Chlorophyta, respectively, are also natural sources of cytokinins. Ördög et al. (2004), Stirk et al. (2003, 2009) and Burkiewicz (1987) have detected cytokinins and cytokinins-like activity in isolated microalgae samples. Stirk et al. (1999) had successfully proven the presence of cytokinins, specifically iP, in microalgae with the use of GC-MS. Also, studies on *Chlorella* had expressed highest biosynthetic rates for iPR, iP and cZ in samples harvested 8 h into the light period compared to samples harvested 8 h into the dark period (Stirk et al. 2011). These studies indicated that microalgae are capable of synthesizing cytokinins which might be released into the soil during cell decomposition, thereby further contributing to the pool of cytokinins available for plant uptake (Stirk and Van Staden 2010).

To date, various studies have proven that certain bacteria are important sources of cytokinins (Philip and Torrey 1972; Upadhyaya et al. 1991; Arkhipova et al. 2005; Kudoyarova et al. 2014) and harbor the potential for the discovery and extractions of cytokinins from nature for plant industry applications. Interestingly, there are novel cytokinins that are yet to be discovered and characterized from these microbial sources. Genetic studies have successfully identified biosynthetic gene responsible for the expression of cytokinins in various bacteria (Powell and Morris 1986; Crespi et al. 1992; Binns 1994; for a review, see Taylor et al. 2003). It has also been reported that bacteria enhance or promote plant cytokinin production. *Agrobacterium tumefaciens*, which induces crown galls in plants, was reported to be capable of transferring and integrating part of their Ti-plasmid DNA into the host plants' genome (Sakakibara et al. 2005). The integrated bacterial genome which encodes an enzyme, adenosine phosphate-isopentenyltransferase, confers the host plant with the ability to synthesize cytokinin via an alternative biosynthesis pathway leading to increased plant cytokinin production. Unlike plants, bacteria produce iP-type cytokinins. These iP-type cytokinins could be taken up by plants, gets converted to Z (Mok and Mok 2001) and subsequently to ZR. ZR is then transported within the xylem to target sites where it gets cleaved into the bioactive form (Z) to drive growth (active cell cycle) using NPK (raw materials) to achieve optimal growth. Studies conducted by Ueda et al. (2012), however, proved that *A. tumefaciens* was capable of efficient biosynthesis of *tZ* during tumor formation in infected galls. Thus, there is a possibility that other bacteria capable of synthesizing zeatin-type cytokinins remains to be discovered.

Apart from bacteria, other studies have also detected cytokinins production in various mycorrhizal fungi. Crafts and Miller (1974) had successfully obtained crystalized Z and ZR from the media in which *Rhizopogon roseolus* (Corda) Hollos, a fungus, had been cultured; presenting definite evidence for the production of cytokinins by the various mycorrhizal fungi screened. Studies conducted by Barea and Azcón-Aguilar (1982), Ng et al. (1982) and Kraigher et al. (1991) had also successfully detected substances with cytokinin-like activity and cytokinins in growth cultures of the various mycorrhizal fungi screened. Thus, it is possible that many cytokinin-producing fungi are contributing to the pool of cytokinins in the environment and these fungal species have yet to be identified.

Parasitic nematodes, root-knot nematodes (*Meloidogyne* species) and cyst nematodes (*Heterodera* species), the common cause of gall and cyst formation in plant roots (Stirk and Van Staden 2010) are known to be capable of exuding cytokinin-like substances (Bird and Loveys 1980). De Meutter et al. (2003) had detected the presence of cytokinins, specifically BA, iP and Z, from in vitro grown nematode exudates or lysates in concentrations high enough to be biologically active. The secretion of cytokinins together with host plant cytokinins had been reported to trigger a change in the nematodes' surface lipophilicity (Akhkha et al. 2002), which might be an infection strategy against the host plant's defense responses. It is also likely that the cytokinins help to establish and/or maintain the feeding cells as a nutrient sink (De Meutter et al. 2003) as roots with overexpressed cytokinins oxidase have reduced gall formation (Lohar et al. 2004).

Similar to nematodes, insect larvae had been reported to synthesize and secrete cytokinins into the plant tissues (Mapes and Davies 2001). Results from the experiments conducted by Van Staden and Bennett (1991) and McDermott et al. (1996) revealed an elevation in cytokinin concentrations in plant tissues that were infected with insect larvae compared to healthy tissue. Elevated concentration of up to 53 times was recorded in the cytokinin profile comparison between the insect larvae with the gall structure and the plant stem tissue. This observation suggested that the insect larvae are capable of synthesizing and secreting their own cytokinins (Mapes and Davies 2001).

Although the roles of cytokinins exuded by nematode and insect larvae are unclear, it is likely that high concentrations of cytokinins are secreted in the gall so that the feeding site remains as an active nutrient sink (Stirk and Van Staden 2010) as elevated concentration of cytokinins may function as local signal for the reallocation of nutrients and photoassimilates; similar to the cytokinin-induced nutrient mobilization phenomenon observed in plants. Thus, both nematodes and insect larvae are also potential sources of cytokinins which could contribute to the environmental cytokinins pool as exudates or lysates.

3.2.2 Auxins

Apart from cytokinins, auxins are another important class of phytohormones that play crucial roles in regulating various plant growth processes (for a review, see Zhao 2010). Auxins are synthesized at the root (Pacifici et al. 2015), in the shoot apex and young expanding leaves (Al-Babili and Bouwmeester 2015). From the site of biosynthesis, auxins are transported basipetally via auxin transporters, which in turn create an auxin gradient across the plant (Al-Babili and Bouwmeester 2015). Peaks of auxin concentration and gradient then act as positional information for auxin activity and maintenance of correct cell division, polarity and fates at the root apex (Sabatini et al. 1999; Al-Babili and Bouwmeester 2015). Auxins also regulate the synthesis of other hormones such as strigolactones (Al-Babili and Bouwmeester 2015), trigger organ primordia initiation and play a role in the initiation and maintenance of the organ founder cell populations (Wolters and Jürgens 2009).

At the cellular level, auxins induce the expression of CycD3 and CDKs, both play crucial roles governing the various checkpoints of the cell cycle, especially in the transition of G1/S phase (Wang and Ruan 2013). Auxin signaling is also required in the later phase of G2/M transition to complete the mitosis process (Wang and Ruan 2013). The mechanisms on how auxins are involved in the initiation of the various cell cycle stages have been reviewed by Wang and Ruan (2013).

Auxins are known to inhibit axillary bud growth, however, the mechanisms of axillary bud outgrowth are dependent on the ratio of cytokinins to auxin rather than the absolute concentration levels of either hormone (Shimizu-Sato and Mori 2001). Direct application of auxin to axillary buds, however, cannot prevent bud growth (Shimizu-Sato and Mori 2001). Apart from regulating axillary bud growth, cytokinin to auxin ratios also determine the development of roots and shoots. A balanced ratio of the two hormones keep the cells in undifferentiated state, while low cytokinin to auxin ratios promote root development and high ratios promote shoot development (Haberer and Kieber 2002). Apart from interacting with cytokinins, interactions between auxins and other phytohormones have been reported as well. Examples include interaction with ABA to regulate root growth (Zhao et al. 2015), with sugar for cell division and expansion regulation (Wang and Ruan 2013) and with brassinosteroids to regulate differential growth (Zhou et al. 2013).

Like cytokinins, auxins can be found in various fertilizer sources and these included vermicomposts (Zhang et al. 2015) and humic acids (Canellas et al. 2002). Hayat et al. (2010) provided a useful listing of bacteria (*Azospirillum*, *Azobacter*, *Bacillus*, *Kluyvera*, *Paenibacillus*, *Pseudomonas*, *Rhizobacteria*, *Rhizobium*) that produce auxins although not all the research described in the review paper provided unequivocal evidence for the occurrence of auxins using mass spectrometery. Interestingly, Patten and Glick (2002) provided scientific evidence for a direct role of auxins produced by *Pseudomonas putida* in regulating mung bean root development.

4 Non-hormonal Plant Growth Promoters (Humic Substances)

Humic substances are the major components of natural organic matter found in soil, water and organic deposits such as sediments, peats, coals, leaf litters and composts (for reviews, see Piccolo et al. 1992; Piccolo 2002; International Humic

Substances Society 2007). The bioactive components of humic substances are humic acid and fulvic acid. Fulvic acid is essentially a polymerized humic acid and thus this section will focus on humic substances in general as both humic acid and fulvic acid share similar if not the same chemical characteristics. Both humic and fulvic acids are available commercially in the forms of pellets.

Studies have often shown positive effects of humic substances on seed germination, root initiation and total plant biomass (Rauthan and Schnitzer 1981; Chen et al. 2004). This is achieved via their ability to improve soil structure and their hormone-like activity on plants (Piccolo et al. 1992; Atiyeh et al. 2002). In plants, humic acids are involved in cell respiration, oxidative phosphorylation, protein synthesis, photosynthesis and various enzymatic reactions (Atiyeh et al. 2002) by improving iron, and possibly zinc nutrition (Chen et al. 2004). At the roots, humic acids enhance root initiation, root hair proliferation and mineral nutrient uptake by increasing the permeability of membranes of the root cells (Atiyeh et al. 2002).

Piccolo (2002) stated that the beneficial effects which humic substances have on the physical properties of soil and their role in the soil environment are significantly greater than that attributed to their contributions to sustaining plant growth, and have provided a comprehensive review on how humic substances contribute to the soil properties and environment. Piccolo's review (2002) provided insights that the hydrophobic nature of humic components protects compounds that are easily degradable and enhance their persistence in soil. This contributes to the accumulation of organic matter which harbors beneficial effects on the rhizosphere and plant growth. Review by Bronick and Lal (2005) also corroborated this view that humic acids help improve the soil condition by increasing the aggregate stability of the soil structure which results in better plant growth and higher yield. Furthermore, humic acid, being a weak acid, could function as a buffer that keeps the soil at the optimal pH for both plant and microbial growth.

5 Biofertilizers

With the ever growing concerns in environmental-related issues and increasing efforts to promote more environmentally friendly farming practices in conventional farms and plantations, the usage of biofertilizers is gaining global acceptance. Unlike conventional chemically synthesized fertilizers that contain N, P and K, biofertilizers are biomass-based structural matrix (e.g. compost, humic acid, etc.) that contain live or latent cells of microorganisms that have the ability to augment nutrients for plant assimilation through microbial processes such as atmospheric nitrogen fixation, phosphate solubilization, cellulolytic degradation and production of phytohormones (Vessey 2003; Van Loon 2007; Lugtenberg and Kamilova 2009; Mishra et al. 2013; Ahemad and Kibret 2014; Owen et al. 2014). It has been reported that the application of biofertilizers to the seeds and the soil, has helped to increase nutrient availability for plant uptake, increased and/or improved plant growth parameters and increased crop yield up to 10–20 %

without any adverse effect on the environment (Bhattacharjee and Dey 2014). Thus, biofertilizers are plausible means to tap onto the natural nutrient cycle without posing any threat on the environment.

The use of biofertilizers in our modern farming practice should be encouraged so as to reduce the adverse effects of long-term chemical fertilizers usage. There is currently a wide range of biofertilizers available commercially (please refer to Table 2 for the mode of action of the various types of biofertilizers and their known microorganisms) and we will focus our discussion on PGPR and vermicomposts.

5.1 Plant Growth Promoting Rhizobacteria (PGPR)

PGPR are bacteria found within the rhizosphere and have the ability to promote plant growth (for reviews, see Kloepper et al. 1989; Vessey 2003; Hayat et al. 2010; Ahemad and Kibret 2014) via various mechanisms such as nitrogen fixation, phosphorus and zinc solubilization, which help to enhance the availability of plant nutrients for absorption (Çakmakçi et al. 2006; Mahdi et al. 2010). The use of PGPR has been reported to increase plant uptake of nitrogen from fertilizer (Adesemoye et al. 2010), and aid to sustain soil productivity and environmental health by reducing dependence on chemical fertilizers (Shaviv and Mikkelsen 1993). PGPR are also referred to as biocontrol agents due to their ability to reduce the incidence or severity of plant diseases (Beattie 2006). Applications of PGPR have been investigated in various plants and crops such as maize, wheat, oat, barley, peas, canola, soy, potatoes, tomatoes, lentils, radicchio, cucumber and chickpea (Gray and Smith 2005; Gopalakrishnan et al. 2015).

While some would consider bacteria localized on the epidermis of plant leaves to be PGPR (Maksimov et al. 2011), we consider PGPR to be the bacteria found within the rhizosphere, free-living or in association with plant roots. However, most PGPR are bacteria that form close association with the plants on the root surface (rhizoplane) or penetrate into the radicular tissues of the root. Most bacterial growth usually occurs at the junctions between epidermal cells and areas where side roots appear (Lugtenberg and Kamilova 2009). Some researchers speculated that PGPR must colonize the root surface efficiently, compete well against other microbes present within the same rhizosphere for nutrients secreted by the root and for sites that can be occupied on the root before being able to exert beneficial effects on the plants (Lugtenberg and Kamilova 2009).

In general, PGPR can affect plant growth in two different ways, directly or indirectly. Direct effects include the various positive influences that PGPR have on plant growth which occur in the absence of pathogens (Lugtenberg and Kamilova 2009). Minimizing or preventing deleterious effects of plant pathogenic organisms via production of antagonistic substances or induction of plant resistance against pathogens is referred to as indirect effects (Glick 1995). It is difficult to classify the effects of PGPR on plant growth into the two distinct groups as a direct effect

Table 2 Different types of biofertilizers and the known microorganisms

Biofertilizer	Mode of action	Microorganism	References
Symbiotic nitrogen fixers	Fix nitrogen only after colonizing host plants' roots	*Rhizobium* spp., *Azospirillum* spp	Steenhoudt and Vanderleyden (2000), Vessey (2003), Lugtenberg and Kamilova (2009), Mehboob et al. (2009), Mishra et al. (2013), Sivasakthivelan and Saranraj (2013)
Nonsymbiotic nitrogen fixer	Fix nitrogen without any symbiotic associations with plants	Cyanobacteria, *Azotobacter*	Johnstone (1967), Obreht et al. (1993), Zachmann and Molina (1993), Stirk et al. (1999), Mishra et al. (2013), Ahemad and Kibret (2014)
Phosphate solubilizers	Convert insoluble organic and inorganic bound phosphates into available forms for plant uptake and utilization May synthesize and release pathogen-suppressing metabolites (siderophores), phytohormones (IAA) and lytic enzymes	*Pseudomonas fluorescens*, *Thiobacillus* spp., *Aspergillus niger*, *Trichoderma* spp., *Paecilomyce* spp, *Bacillus* spp., *Rhizobium* spp., *Burkholderia* spp., *Achromobacter* spp., *Agrobacterium* spp., *Microccocus* spp., *Aereobacter* spp., *Flavobacterium* spp. and *Erwinia* spp.	Handelsman and Stabb (1996), Vassilev et al. (2006), Berendsen et al. (2012), Mishra et al. (2013), Bhattacharjee and Dey (2014)
Phosphate mobilizers	Form symbiotic associations with the plants at the roots and enhance uptake of phosphorous and other nutrients	Fungi	Torelli et al. (2000), Mishra et al. (2013)

(continued)

Table 2 (continued)

Biofertilizer	Mode of action	Microorganism	References
Zinc solubilizer	Convert zinc in the soil into suitable form for plant uptake and utilization	*Pseudomonas* spp., *Bacillus* spp., *Rhizobium* spp.	Mishra et al. (2013)
		Bacillus subtilis, *Thiobacillus thioxidans* and *Saccharomyces* sp.	
Plant growth promoting rhizobacteria (PGPR)	Fix nitrogen, solubilize phosphorus and zinc and enhance availability and uptake of plant nutrients	*Bacillus* spp., *Pseudomonas* spp., *Azospirillum* spp.	Phillips and Torrey (1972), Kloepper et al. (1989), Upadhyaya et al. (1991), Döbereiner (1992), Glick (1995), Glick and Bashan (1997), Okon et al. (1998), Ryu et al. (2003), Vessey (2003), Kravchenko et al. (2004), Gray and Smith (2005), Beattie (2006), Çakmakçi et al. (2006), Crowley (2006), Glick et al. (2007), Van Loon (2007), Lugtenberg and Kamilova (2009), Adesemoye et al. (2010), Hayat et al. (2010), Mahdi et al. (2010), Maksimov et al. (2011), Ahemad and Kibret (2014), Talaat et al. (2015)
	Confer biocontrol against plant diseases. Secrete bacterial phytohormones, e.g. cytokinins		
	Release volatiles and cofactor pyrrolquinoline quinone (PQQ) that stimulate plant growth		
	Reduce stress response in plants		
	Produce siderophores that help enhance plant iron uptake and remediate soil pollutants		

Adapted from Vessey (2003), Hayat et al. (2010), Mishra et al. (2013), Bhattacharjee and Dey (2014) might lead to an indirect influence. For example, production of phytohormones helps to enhance growth (direct effect) but may also induce disease resistance (indirect effect). Thus, our review will not classify the resultant effects into direct or indirect groups.

PGPR share similar functions with the other groups of biofertilizers. The common functions include converting atmospheric nitrogen (Döbereiner 1992) and facilitating the uptake of nutrients such as phosphorus and zinc via solubilizing inaccessible forms trapped in insoluble compounds. Unlike the other groups of biofertilizers that mainly exert one type of positive effect, PGPR enhance plant growth in many more ways. One of the most prominent enhancements is the secretion of bacterial phytohormones (Glick 1995), specifically cytokinins (Philip and Torrey 1972; Upadhyaya et al. 1991), which promotes plant growth. The phytohormones, specifically auxins which the bacteria synthesize using the tryptophan present in root exudates (Kravchenko et al. 2004), also promote better root system formation, thereby enhancing water and nutrient absorption (Patten and Glick 2002). These in turn help the plants to pass through the pathogen-sensitive early development stage more rapidly (Maksimov et al. 2011). This characteristic is especially important as studies with added inorganic nitrogen (to increase nitrogen fixation) suggested that plant growth promotion is caused by the production of plant growth factors such as phytohormones rather than nitrogen fixation (Okon et al. 1998).

Apart from providing the plants with phytohormones, PGPR are also known to stimulate plant growth by releasing volatiles and cofactor pyrrolquinoline quinone (PQQ) (Ryu et al. 2003; Lugtenberg and Kamilova 2009). Volatiles are reported to increase photosynthetic efficiency and chlorophyll content in *Arabidopsis thaliana* through the modulation of endogenous signaling of glucose and abscisic acid sensing (Zhang et al. 2008). PQQ on the other hand functions as antioxidants and cofactor of enzymes involved in antifungal activity and induction of systemic resistance (Lugtenberg and Kamilova 2009; Choi et al. 2008).

Other major substances known to be synthesized by PGPR and are beneficial to the plants include antibiotics, siderophores and hydrolytic enzymes. PGPR antibiotics are oligopeptides that inhibit cell wall synthesis in pathogens at the cell wall synthesis initiation stage. The antibiotics disrupt the functions of ribosomes and inhibit the formation of initiation complex on small subunit of ribosomes (Maksimov et al. 2011). PGPR antibiotics are said to be effective against Gram-positive and Gram-negative bacteria and pathogenic fungi (Maksimov et al. 2011).

Microbial siderophores synthesis by PGPR is induced by low ferric ion level in the environment. Siderophores have high affinity to ferric ions and have the ability to solubilize and extract ferric ions from mineral or organic complexes (Wandersman and Delepelaire 2004). Thus, this increases the pool of iron available for plant assimilation. It was reported that microbial siderophores help to enhance iron uptake by plants that are able to recognize the bacterial ferric-siderophore complex (Masalha et al. 2000; Katiyar and Goel 2004; Dimkpa et al. 2009). Iron uptake by plants in the presence of other metals such as nickel and cadmium is also enhanced by siderophores (Burd et al. 1998; Dimkpa et al. 2008). By enhancing plant iron uptake, pathogens are deprived of the iron that is much needed for their growth and development and thus reducing the occurrence of plant diseases (Maksimov et al. 2011). Calcium assimilation by plants is also enhanced by siderophores.

The production of bacterial hydrolytic enzymes, e.g. chitinases, glucanases, proteases, lipases, that lyse fungal cells, volatile compounds and their toxins are also ways which PGPR help reduce and/or prevent pathogenic diseases (Neeraja et al. 2010; Maksimov et al. 2011) and suppress nematode populations within the rhizosphere (Youssef and Eissa 2014).

PGPR also help facilitate plant growth and development by reducing the stress response within plants via decreased ethylene levels. Ethylene precursor 1-aminocyclopropane-1-carboxylate (ACC) is converted into 2-oxobutanoate and ammonia by bacterial ACC deaminase (Glick et al. 2007), relieving the inhibition of root growth (Van Loon 2007) while rendering the plants to be more resistant against stress due to heavy metals (Ca^{2+} and Ni^{2+}), salt, draught and phytopathogenic bacteria (Glick and Bashan 1997; Lugtenberg and Kamilova 2009; Van Loon 2007). Studies by Talaat et al. (2015) also provide concrete evidence on the application of suitable microorganisms to help the plants gain tolerance against salinity stress via alteration of nutrient acquisition and accumulation of compatible solutes.

It is also noteworthy that PGPR play the role of rhizoremediator by degrading soil pollutants (Lugtenberg and Kamilova 2009). Siderophores produced by PGPR are able to aid in the bioremediation of soil pollutants (Crowley 2006) by isolating and degrading heavy metals and toxic organic matters such as metabolites of pathogenic origins (Maksimov et al. 2011) reducing the occurrence of contaminated crops that may pose adverse effects when consumed. Cleaner soil also allows more microorganisms and organisms (such as earthworms) to flourish, conferring beneficial effects on the plants.

Another way that plants benefit from the association with PGPR is the activation of defense mechanism—induced systemic resistance (ISR) against pathogens (Van Loon et al. 1998). Exudates produced by PGPR are able to stimulate ISR by activating components such as lipoxygenases, lipid peroxidases and reactive oxygen species (Maksimov et al. 2011) conferring protection against diseases caused by different organisms (Lugtenberg and Kamilova 2009) by reducing the rate of disease development in terms of severity or number of diseased plants (Van Loon 2007). ISR activation is dependent on jasmonic acid and ethylene signaling (Van Loon 2007). Systemic acquired resistance (SAR) which also enhances resistance against diseases is, however, induced by pathogens and dependent on salicylic acid (SA) signaling (Van Loon 2007). It is important to note that PGPR that elicit ISR in one plant species may not do so in another due to interaction specificity between rhizobacteria and plants (Van Loon 2007).

5.2 Vermicomposts

Another form of biofertilizer that is gaining widespread acceptance globally is vermicompost and vermicompost tea, a leachate of the vermicompost. Vermicompost is the highly valued compost produced by earthworms (for a review, see Edwards et al. 2010). Vermicomposting not only helps to reduce organic wastes in volume, but also turning them into humus-like substance that is finer than compost and generally contains high concentration of mineral matter. This makes vermicompost a very good fertilizer that is up to 70 times more efficient than conventional manure (Červená et al. 2013). The earthworm activity also stimulates and increases the diversity of microbial activity. Typically vermicompost is applied at low concentration to the plant growth medium or as soil drench or foliar spray.

Vermicomposts have been reported to have beneficial effects on plant growth such as improved seed germination, enhanced seedling growth and development, and increased plant productivity (Atiyeh et al. 2002; for a review, see Edwards et al. 2010). It enhances plant growth by improving the physical structure and moisture retention capacity of the soils (Arancon et al. 2004) while supplying the plants with N in stable form (Chaoui et al. 2003) and phytohormones or phytohormone-like compounds produced by the microorganisms present within.

Our group has been actively characterizing the phytohormones in vermicomposts and their leachate (vermicompost tea). Recently, a new method has been successfully established by our group for the analysis of phytohormones present within vermicompost (Zhang et al. 2015) and quantitative evidence of the various growth regulating factors, such as phytohormones, i.e. cytokinins, auxins, gibberellins and brassinosteroids, present in vermicompost tea and leachate have been provided by Zhang et al. (2014) and Aremu et al. (2015). Hopefully, the phytohormone screening approach developed for vermicomposts can be extended to all types of organic fertilizers. Aremu et al. (2015) have also provided insightful discussion on the importance of different phytohormones on their roles in regulating plant growth and development. Results from these studies indicated that vermicomposts harbor a rich diversity of plant growth promoting factors, specifically phytohormones. The origins of these "subterranean" phytohormones are likely to be linked to the symbiotic microbes living in the gut of the earthworms. There is also a possibility that vermicomposts may contain other factors that are beneficial for the plants that have yet to be detected. The beneficial effects of vermicomposts can also be attributed to the presence of humic acids or growth regulators associated with humic acids as demonstrated by Arancon et al. (2004) and Canellas et al. (2002).

Synergistic relationship between vermicompost and PGPR had been reported to improve plant growth, reduce plant mortality and increase microbial biomass (Sahni et al. 2008; Song et al. 2015). This could be due to the reason that vermicompost contains humus which allows PGPR to thrive well and multiply in population. Thus, farming practice can turn to a new biofertilization regime which utilizes both the vermicomposts and PGPR to reap the full synergistic benefits of these natural resources that are beneficial for the plants and to maintain good soil and environmental health.

6 How Biofertilizers Work in Tandem with Microorganisms and Phytohormones to Influence Plant Growth?

An extensive review of published literature (Gharib et al. 2008; Datta et al. 2009; Edwards et al. 2010; Hayat et al. 2010; Mahdi et al. 2010; Shafi et al. 2012; Ahemad and Kibret 2014; Bhattacharjee and Dey 2014; Qin et al. 2015; Sarma et al. 2015; Song et al. 2015) and our extensive field observations have shown that organic fertilizers, despite having low NPK value, can sometimes produce the same growth promoting effect and/or achieving comparable yields, when compared to plants grown using conventional chemical fertilizers with high NPK ratios (e.g. 10–21) (for reviews, see Shaviv and Mikkelsen 1993; Chen 2006). Hence, there must be some growth-promoting factors present in organic fertilizers and these positive factors are certainly not the NPK mineral nutrients per se, that are driving plant growth and development. These salient and positive growth enhancement observations had been noted by many farmers/growers in the plant industry and a plausible scientific explanation remains elusive. In this review, we provided scientific evidence that the growth promoting factors in biofertilizers modulating plant growth and development are phytohormones, and that the known biological functions of phytohormones are in tandem with the observed physiological characteristics and crop yield (Fig. 6).

At the whole plant level and in relation to the plant–soil continuum, the interactions between the whole plant and the microorganisms present in the soil can be best illustrated by Fig. 7. The soil and the entire subterranean root system form a diverse and intimate association of "biological networks and entities" comprising of plant roots, microbes (bacteria, fungi) and many very small organisms (nematodes, earthworms, etc.). Amidst the complex array of biological networks and entities is the soil matrix and water medium where multitudes of biological activities (e.g. microbial biochemical activities like enzyme production, plant exudations [allelopathic] and uptake, ingestion by earthworms, etc.) and interactions are taking place. It is therefore conceivable that the soil matrix and water medium contain many naturally produced substances, biological metabolites and these include the phytohormones and their precursors (see Sect. 5). From a holistic perspective, one may view the entire plant subterranean root system as a "receiver" of the multitudes of biochemical signals and this information allows the plants to "sense" the prevailing soil conditions for water, nutrient and phytohormone availability. The selected biochemical signals are "received" at the root tips, "assimilated" and sent to the various plant parts. These signals induce most responses at the actively growing areas within the plants. The actively growing areas within any plant are the plant meristems found mainly in the aboveground shoot system: shoot apices, axillary buds, flower buds and the root tips (belowground). The growth rates of these meristems are governed by the various phytohormonal chemical signals arriving there, from the roots. Most of these phytohormonal chemical signals have their origins in the subterranean soil and they normally travel with the

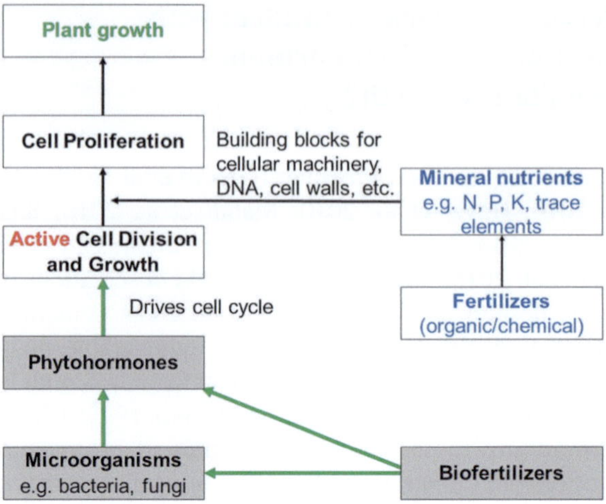

Fig. 6 The contribution of chemical, organic and biofertilizers to plant growth and development through cell proliferation. Chemical and organic fertilizers provide nutrients, e.g. N, P, K, that are essential for the basic cellular structural machinery and the biosynthesis of DNA, enzymes, cells walls, etc. These processes are active when the meristematic cells are dividing, i.e. cell undergoes the entire cell cycle process, and new cells are produced. Under normal conditions, plants grow at fairly predictable pace due in part to cell cycle regulation governed predominantly by the availability of resources (mineral nutrients, water), suitable environmental conditions (adequate sunlight, optimal temperature) and phytohormones. Thus, the addition of mineral nutrients via chemical fertilization will not necessarily increase the rate of plant growth per se, when there are other limitations imposed on the plant. The application of biofertilizers supplies the plants with phytohormones (in addition to those synthesized endogenously by the plants) that help the plant meristems to overcome the various cell cycle checkpoints' "restrictions" and to facilitate active cell proliferation. The calibrated and integrated usage of different fertilizers (both chemical and organic) to supply the plants with ample nutrients for their cellular structural needs and appropriate phytohormonal signals to proceed through cell cycle checkpoints will eventually lead to active plant growth

transpirational water flow through the xylem and onto many plant parts. Closer to the meristems where there may not be any functional conduits leading from the main transport tubes to the meristems, these chemical signals may travel via the phloem and/or through cell-to-cell linkages via diffusion and onto the meristems. Nevertheless, the xylem represents the main conduit for the root-to-shoot transmission of the phytohormonal signals for plant growth and development, and there are ample scientific evidence to support this growth regulatory mechanism (Yong et al. 2000, 2014; for reviews, see Schmülling 2002; Stirk and Van Staden 2010).

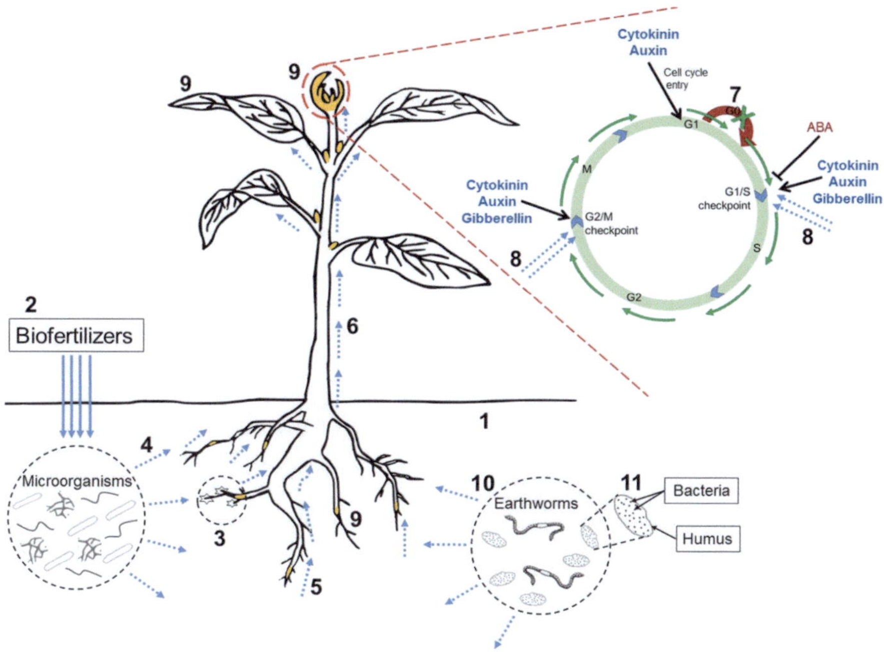

Fig. 7 Schematic diagram to describe how biofertilizers work in tandem with microorganisms and phytohormones to influence plant growth. *1* Soil provides the plants with general nutrients like N, P and K, that serve as cellular structural materials and building blocks for components such as nucleic acids, ATP and enzymes; and cofactors for enzyme activities. Fertilizers, either conventional or organic in origin, are added to restore the "nutrient loss" from biomass removal and/or through leaching or weathering processes. *2* Application of biofertilizers introduces microorganisms (e.g. bacteria and fungi) into the rhizosphere. *3* Microorganisms may be free-living or in symbiosis with the plants (at the roots). *4* Certain microorganisms produce various useful enzymes for improving soil availability of certain nutrients, and other useful substances including phytohormones or possibly, their precursors (mainly cytokinins and auxins). *5* The phytohormones (and/or their precursors), present in the soil, are then taken up by the plant, via the roots, and *6* transported to/or near the sites of active growth, i.e. shoot apices and axillary meristems, through the xylem. *7* The phytohormones help the cells to bypass the G0 phase and *8* proceed beyond the checkpoints (G1/S and G2/M) that results in active cell divisions, producing more cells at the meristems, and leading eventually to plant growth. *9* Phytohormones transported to the other parts of the plants also regulate various biological processes such as the opening of stomata, chloroplast production, release from apical dominance, flower development and root development. *10* Earthworms produce vermicompost that contains phytohormones (through earthworms' intestinal microbial activities) that are released into the soil and can be taken up by the plants. *11* Vermicompost also contains humus that allows beneficial bacteria (either of earthworm gut origin and/or soil origin) to thrive and multiply, and thereby increasing the bioactivity of the soil. Humus also help to improve the soil structure that allow better water and air movement within the rhizospheres and thereby increasing soil fertility

6.1 Improving Current Fertilizer Regime in Light of the Linkages Between Plant Growth and Microorganisms-Derived Phytohormones

For the general plant industry at present, the sole application of each type of fertilizer (either chemical or organic) appears to be the current trend adopted by many growers/farmers. However, there is a gradual and progressive shift by some of the conventional farmers/growers toward using organic additives (organic fertilizers and biofertilizers) to supplement their chemically based NPK fertilizer applications in the farms/plantations. This is because many conventional farmers/growers have realized that the long-term and prolonged usage of chemical fertilizers on their lands had somewhat led to a drop in plant growth promoting efficacy and subsequently lower yield. Due to the complexity of these biotic and abiotic processes (some of which are still unclear) and interactions, it was often impossible to give farmers/growers adequate scientific explanations. Some research suggested a plausible link between lower yields (despite increasing chemical fertilization) and diminishing microbial activities ("poor soil health"). Such ambiguity paves the way for more intensive research toward understanding the role of soil microbes (and their associated phytohormonal content), in relation to plant growth and fertilizer formulations. Nevertheless, various scientific studies had been conducted on the use of a combination of fertilizers, chemical and organic (Shafi et al. 2012), organic and biofertilizers (Gharib et al. 2008; Sarma et al. 2015), vermicompost and PGPR (Song et al. 2015) and even a combination of all three types (chemical, organic and biofertilizers; Datta et al. 2009). The studies have proven that the integrated use of different fertilizers is highly beneficial in terms of crop yield and environmental friendliness.

Some conventional plant industries have started using humic acid (commercially available in pellet forms) and/or biocharcoal, in their farming practices. Humic acid pellets are known to improve the soil properties, which are likely to be degraded by prolonged chemical fertilizer application. Other conventional farmers/growers have used vermicomposts which not only improve the soil composition and structure but also introduce beneficial microorganisms. These microorganisms in turn provide the necessary phytohormones needed to support plant growth and development. While organic fertilizers and biofertilizers contain phytohormones (varying levels) and are environmentally friendly, they may fail to deliver stable and predictable growth stimulation comparable to chemical fertilizers as it is difficult to manage the batch–batch variation in microbial activities and missing mineral nutrients. Moving forward, plant industry that uses organic fertilizer or biofertilizers could consider fortifying their fertilizers with selected micro- and macromineral nutrients so as to supplement the low amount of nutrients, especially NPK and other trace elements, to improve or to maintain the yield.

6.2 The Novel Futuristic Green Biofertilizer—With Microbial Phytohormones

In our current plant industry and fertilization practices, the shortage of mineral nutrients can be assessed rapidly and easily overcome by adding fertilizers. However, the apparent shortage of phytohormones in promoting plant growth and development has never been recognized as an important tenet of whole plant nutrition. Ironically, the importance of phytohormones in promoting in vitro plant growth and development in the plant tissue culture industry is widely recognized and supplementing phytohormones within a desired mineral nutrient formulation is the standard practice (for reviews, see George and Sherrington 1984; George 1993). Unknown to many farmers/growers from the scientific perspective of plant cell cycle regulation through phytohormones, the use of organic fertilizers and biofertilizers (with their microbes and naturally occurring phytohormones and especially cytokinins) by either an organic farm (routinely) or as a periodic supplement to their conventional farming methods, help to increase the levels of growth promoting phytohormones in the subterranean root environment. With the new understanding about the pivotal role of phytohormones in regulating plant cell proliferation when mineral nutrients are sufficient, we believe that plant industry productivity or yield can be further enhanced with the supplementation (or natural occurrence) of phytohormones like cytokinins, auxins, gibberellins, etc., within the current fertilizer formulation, and possibly through microbial avenues with varying levels of controlled release technologies (Bashan et al. 2014).

Moving forward, we propose a new practice of introducing phytohormones, specifically cytokinins, in current agricultural/horticultural plant nutrition methodologies. Cytokinins appear to be one of the limiting factors in regulating plant growth due to its scarcity and fluidity in the fragile subterranean environment, as discussed in earlier sections. Thus, we believe that agricultural/horticultural yield/productivity can be greatly enhanced with the supplementation of cytokinins and other phytohormones in fertilizers. The preferred sources of cytokinins in any futuristic green biofertilizer should preferably be "natural" and originating from microorganisms (e.g. bacteria like *Azospirillum* and *Rhizobium*) and natural sources (coconuts, macroalgae, or seaweeds) that are widely available to the growers/farmers (for reviews, see Letham and Palni 1983; Stirk et al. 2003; Ördög et al. 2004; Yong et al. 2009). Microbial production of cytokinins has been well documented (Phillips and Torrey 1972; Ng et al. 1982; Burkiewicz 1987; Kraigher et al. 1991; Upadhyaya et al. 1991; Arkhipova et al. 2005; Kudoyarova et al. 2014) and could be cultured economically in large quantity using bioreactors and formulated for agricultural/horticultural use (Fig. 8). Thus, cytokinin-producing bacteria harbors the greatest potential for large-scale cytokinin production and can be developed into the next generation of green fertilizer for agricultural/horticultural applications.

The new generation of green biofertilizer is likely to come in the form of granules, and/or coated by natural/hydrophobic polymers or as matrices in which the

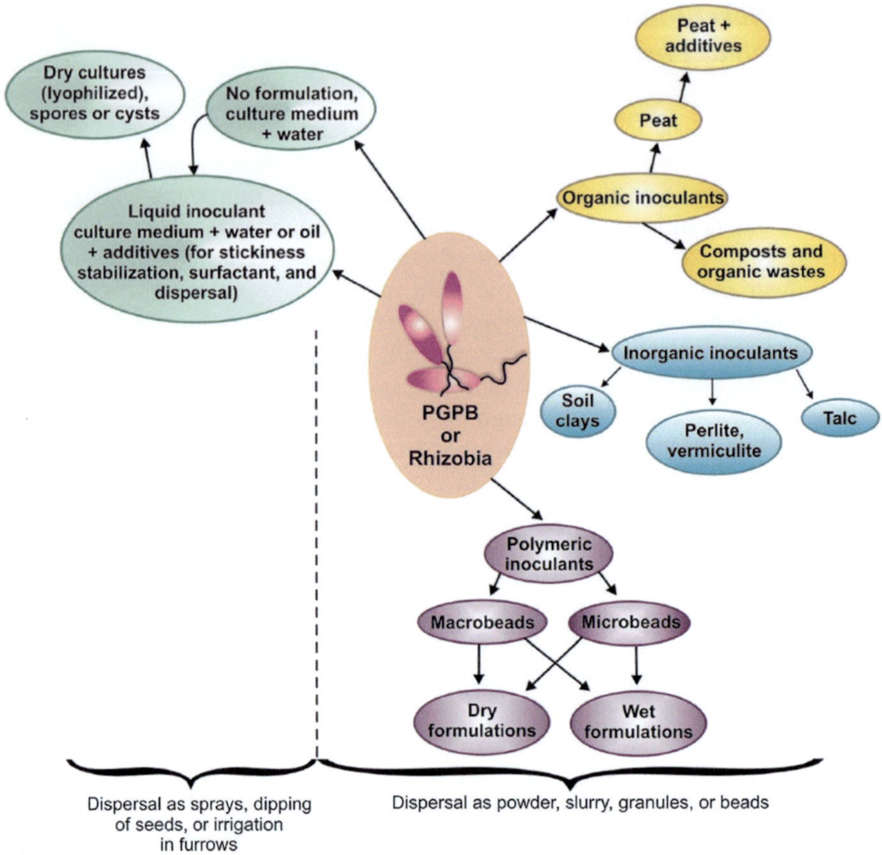

Fig. 8 Formulations of inoculants as biofertilizers for agricultural and environmental uses. (reprinted from Plant and Soil, 378, 2014, 1–33, Advances in plant growth promoting bacterial inoculant technology: formulations and practical perspectives (1998–2013), Y. Bashan, L.E. de-Bashan, S.R. Prabhu, Juan-Pablo Hernandez, Fig. 2, with kind permission from Springer Science and Business Media)

active plant growth promoting (e.g. cytokinins, auxins, gibberellins, etc.) and soil improving substances (e.g. humic acid), and/or suitable microbes, with carefully selected mineral nutrients, are embedded in packing materials (e.g. alginate) that restricts the rapid dissolution of the fertilizer, and consequently providing the plants with a sustained source of phytohormones, mineral nutrients, amino acids, proteins, etc. (Fig. 9). The current slow and controlled release or "stabilized" fertilizers is a useful design template to develop the futuristic fertilizer (for reviews, see Shaviv and Mikkelsen 1993; Bashan et al. 2014). The selection of specific mineral nutrients to be embedded in the novel futuristic green fertilizer should be formulated only after a proper chemical analysis of the targeted soil type or locality had been carried out. In situations where certain macro- and

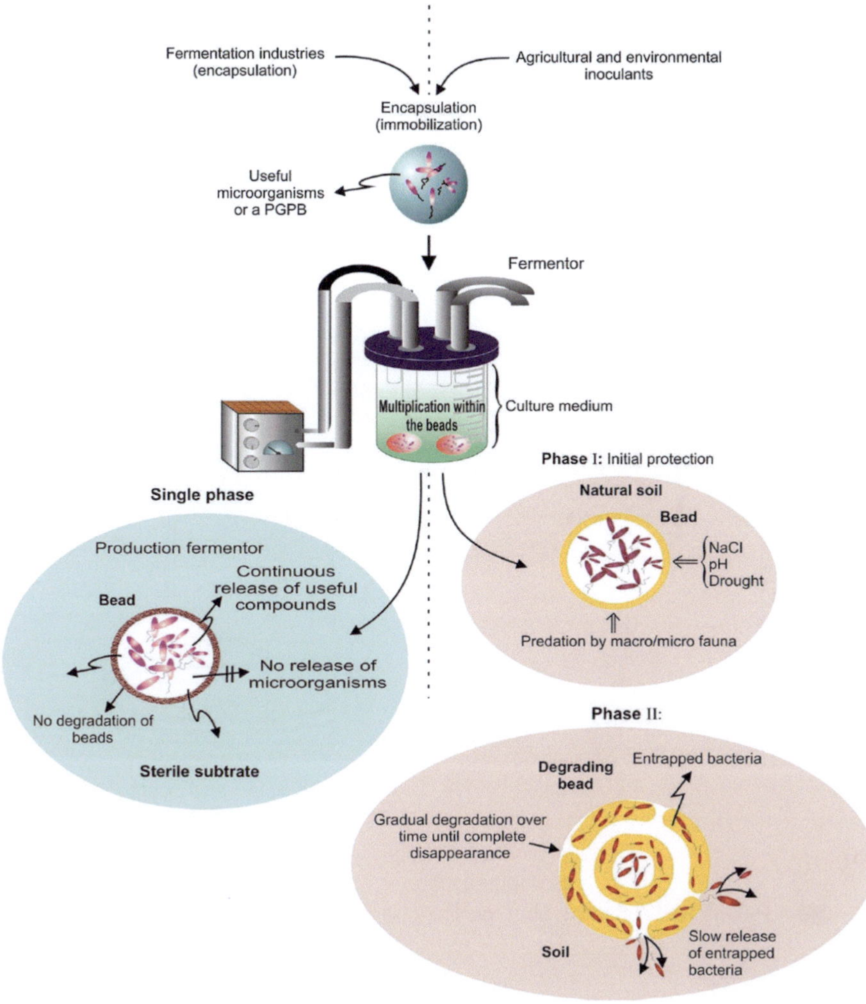

Fig. 9 Encapsulating bacteria for use in industrial fermentation and as an inoculant for agricultural/environmental use (reprinted from Plant and Soil, 378, 2014, 1–33, Advances in plant growth promoting bacterial inoculant technology: formulations and practical perspectives (1998–2013), Y. Bashan, L.E. de-Bashan, S.R. Prabhu, Juan-Pablo Hernandez, Fig. 3, with kind permission from Springer Science and Business Media)

micronutrients may be biologically unavailable, suitable and effective microbial populations for solubilizing the chemically fixed phosphorus and to improve the availability of other macro and micronutrients, could be added onto these granules. When there are threats to plant health, the novel fertilizer may be formulated to have some levels of biocontrol properties against certain pathogens. Some other organic compounds such as humic acids and vermicomposts with favorable soil structure improving properties (such as water and mineral nutrient retention) and soil microbial activity-enhancing properties, could also be added in order to augment the overall performance of the novel futuristic green biofertilizer. Bashan et al. (2014) provided interesting and useful methodologies toward designing varying levels of controlled release centered on either a seed (Fig. 10a) or a microbial pellet (Fig. 10b). These useful microbial-inoculant design conceptual templates

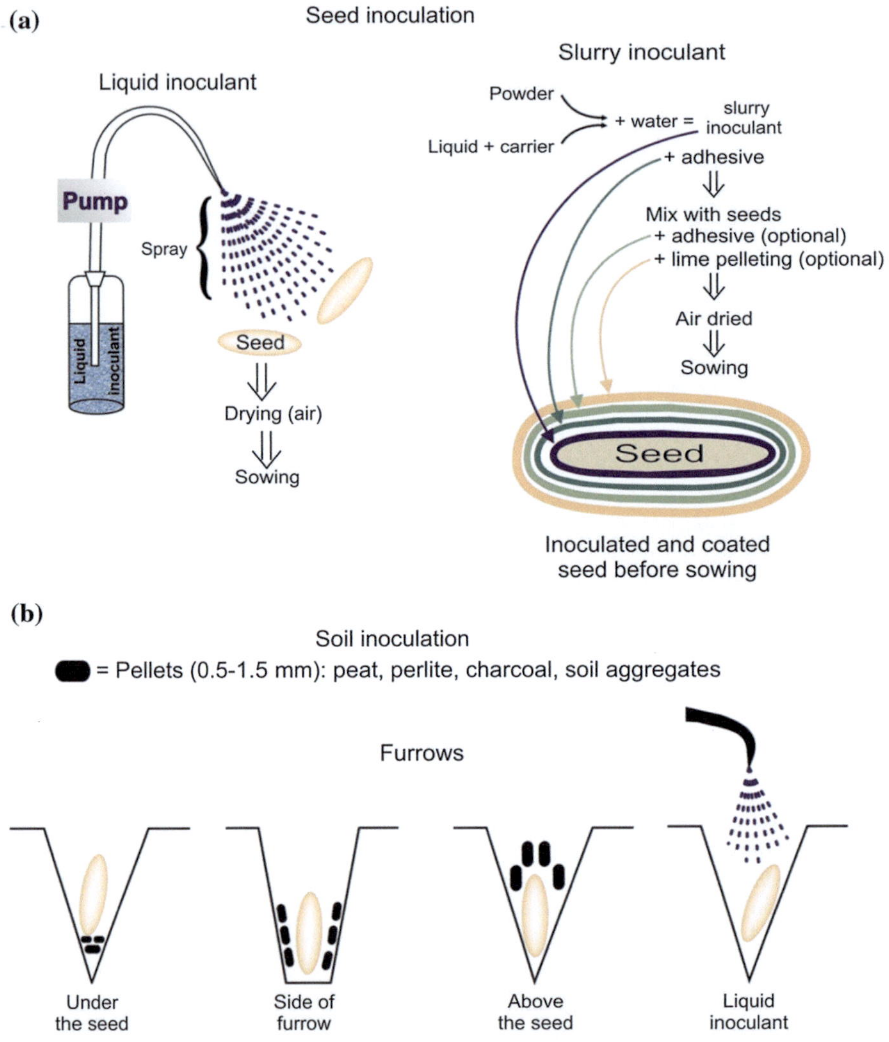

Fig. 10 Schematic representation of the various strategies available for **a** multilayered seed inoculation and **b** microbial pellets for soil inoculation (reprinted from Plant and Soil, 378, 2014, 1–33, Advances in plant growth promoting bacterial inoculant technology: formulations and practical perspectives (1998–2013), Y. Bashan, L.E. de-Bashan, S.R. Prabhu, Juan-Pablo Hernandez, Fig. 4, with kind permission from Springer Science and Business Media)

can be broaden further to include the futuristic green fertilizer granule design that may carry some useful substances (see Sects. 3.2 and 5) and perhaps, concomitantly embedding two different microbes with different biological functions in the soil. With above-mentioned benefits, together with a reasonable storage life (up to 3–6 months), the novel futuristic green biofertilizer in the form of granules is likely to be favored by the conventional farmers/growers.

The ideal fertilizers should have the following three characteristics:

- A single application should supply enough mineral nutrients and phytohormones throughout the entire growing season to meet plant demand for optimum growth;
- Maximal plant growth stimulation thus allowing the largest financial return for the cost of input;
- Minimum detrimental ecological effects on the soil, water and atmospheric environment.

7 Conclusion

Plant growth is dependent on meristems where cell proliferations give rise to new plant structures and allow the plant to increase in size. We provided scientific linkages and evidence to show that the growth promoting factors in biofertilizers regulating cell proliferation and ultimately modulating plant growth and development are phytohormones. The known biological functions of phytohormones (cytokinins, auxins, gibberellins, etc.) are in tandem with the observed physiological characteristics and crop yield of plants. When light, water and mineral nutrients are not limiting, phytohormones, especially cytokinins, in biofertilizers help to drive plant growth by progressing faster through the various plant cell cycle checkpoints leading to the production of more cells.

There is an enormous diversity of microbes found within the soil matrices of the subterranean environment where the plant root system exists. Within the rhizosphere, there is a group of PGPR bacteria that has the ability to promote plant growth via various mechanisms such as nitrogen fixation, phosphorus and zinc solubilization. Another prominent ability of some PGPR is the secretion of bacterial phytohormones, like cytokinins and auxins and other useful substances. Phytohormone-producing PGPR harbors the greatest potential for large-scale phytohormone (especially cytokinins) production and can be developed into the next generation of green fertilizer with microbial phytohormones and/or microbial inoculant for agricultural/horticultural applications.

Evidently, the above literature reviews and discussions reveal that biofertilizers bring about great advantages and improvements to our modern and intensive conventional agriculture practice. In addition, although not discussed in depth in this review, certain biofertilizers may also confer natural biocontrol property that would be useful for disease management. The long-term and fundamental sustainable criterion for any futuristic farmland/plantation is essentially soil health. Good soil provides the foundation for healthy plant growth with minimal external mineral nutrient addition. However, despite its environmental friendliness, any fertilizer regime that solely relies upon biofertilizer may not be feasible in terms of crop productivity per unit area when compared to conventional agriculture practice. Large amount of biofertilizers would be required as they generally contain lower mineral nutrient content, variable elemental composition and/or releasing nutrients at a much slower rate that is unable to sustain maximum plant growth over a limited time period (Chen 2006). Thus, for those involved solely in organic farming practices, we suggest that their organic fertilizers be fortified with selected macro- and micronutrients when there is a drop in horticultural productivity/crop yield. Conversely, for those involved in conventional farming practices, adding organic and biofertilizers periodically is the remedy to reduce chemical fertilizer usage while maintaining their expected yields. Under certain circumstances, selective biofertilizer application to support conventional farming practices is considered the best way to restore the effective microbial populations in order to solubilize chemically fixed phosphorus and to improve the availability of other macro- and micronutrients for plant uptake.

In the near future, we envisage that a hybrid approach of combining organic and conventional fertilization regimes will be widely accepted throughout the global plant industry. This is evident from the 15-year-long study conducted by Qin et al. (2015) which demonstrated that the combined application of organic fertilizers (manure was investigated in that study), together with chemical fertilizer are of great importance to improving agricultural economy as well as sustaining soil heath and quality. Moving forward, the new generation of green biofertilizer should come in the form of granules in which the active plant growth promoting (e.g. cytokinins, auxins, gibberellins, etc.) and soil improving substances (e.g. humic acid) and/or suitable microbes ("inoculants"), with carefully selected mineral nutrients, are embedded in the packing materials giving slow and sustained release over a desired period. The futuristic green fertilizers should provide the plants with a sustained source of phytohormones and mineral nutrients. In situations where certain macro- and micronutrients may be unavailable, suitable and effective microbial populations for solubilizing the chemically fixed phosphorus and to improve the availability of other macro- and micronutrients could be added onto these green fertilizer granules.

References

Adesemoye AO, Torbert HA, Kloepper JW (2010) Increased plant uptake of nitrogen from ^{15}N-depleted fertilizer using plant growth-promoting rhizobacteria. Appl Soil Ecol 46:54–58

Agusti J, Herold S, Schwarz M, Sanchez P, Ljung K, Dun EA, Brewer PB, Beveridge CA, Sieberer T, Sehr EM, Greb T (2011) Strigolactone signaling is required for auxin-dependent stimulation of secondary growth in plants. Proc Natl Acad Sci USA 108:20242–20247

Ahemad M, Kibret M (2014) Mechanisms and applications of plant growth promoting rhizobacteria: current perspectives. J King Saud Univ—Sci 26:1–20
Akhkha A, Kusel J, Kennedy M, Curtis R (2002) Effects of phytohormones on the surfaces of plant-parasitic nematodes. Parasitology 125:165–175
Al-Babili S, Bouwmeester HJ (2015) Strigolactones, a novel carotenoid-derived plant hormone. Annu Rev Pl Biol 66:161–186
An C, Mou Z (2011) Salicylic acid and its function in plant immunity. J Integr Plant Biol 53:412–428
Arancon NQ, Edwards CA, Bierman P, Welch C, Metzger JD (2004) Influences of vermicomposts on field strawberries: 1. Effects on growth and yields. Bioresour Technol 93:145–153
Aremu AO, Strik WA, Kulkarni MG, Tarkowská D, Turečková V, Gruz J, Šubrtová M, Pěnčík A, Novák O, Doležal K, Strnad M, Van Staden M (2015) Evidence of phytohormones and phenolic acids variability in garden-waste-derived vermicompost leachate, a well-known plant growth stimulant. Plant Growth Regul 75:483–492
Arkhipova TN, Veselov SU, Melentiev AI, Martynenko EV, Kudoyarova GR (2005) Ability of bacterium *Bacillus subtilis* to produce cytokinins and to influence the growth and endogenous hormone content of lettuce plants. Plant Soil 272:201–209
Arthur GD, Jäger AK, Van Staden J (2001) Uptake of [$_3$H]DHZ by tomato seedlings. S Afr J Bot 67:661–666
Atiyeh RM, Lee S, Edwards CA, Arancon NQ, Metzger JD (2002) The influence of humic acids derived from earthworm-processed organic wastes on plant growth. Bioresour Technol 84:7–14
Bangerth F, Li CJ, Gruber J (2000) Mutual interaction of auxin and cytokinins in regulating correlative dominance. Plant Growth Regul 32:205–217
Barea JM, Azcón-Aguilar C (1982) Production of plant growth-regulating substances by the vesicular-arbuscular mycorrhizal fungus *Glomus mosseae*. Appl Environ Microbiol 43:810–813
Bashan Y, de-Bashan LE, Prabhu SR, Hernandez JP (2014) Advances in plant growth-promoting bacterial inoculant technology: formulations and practical perspectives (1998–2013). Plant Soil 378:1–33
Beattie GA (2006) Plant-associated bacteria: survey, molecular phylogeny, genomics and recent advances. In: Gnanamanickam LJ (ed) Plant-associated bacteria. Springer, Netherlands, pp 1–56
Bennett T, Leyser O (2006) Something on the side: axillary meristems and plant development. Plant Mol Biol 60:843–854
Berendsen RL, Pieterse CMJ, Bakker PAHM (2012) The rhizosphere microbiome and plant health. Trends Plant Sci 17:478–486
Bhattacharjee R, Dey U (2014) Biofertilizer, a way towards organic agriculture: a review. Afr J Microbiol Res 8:2332–2343
Binns AN (1994) Cytokinin accumulation and action: biochemical, genetic, and molecular approaches. Annu Rev Plant Biol 45:173–196
Bird AF, Loveys BR (1980) The involvement of cytokinins in a host–parasite relationship between the tomato (*Lycopersicon esculentum*) and a nematode (*Meloidogyne javanica*). Parasitology 80:497–505
Bleecker AB, Kende H (2000) Ethylene: a gaseous signal molecule in plants. Annu Rev Cell Dev Biol 16:1–18
Bronick CJ, Lal R (2005) Soil structure and management: a review. Geoderma 124:3–22
Burd GI, Dixon DG, Glick BR (1998) A plant growth-promoting bacterium that decreases nickel toxicity in seedlings. Appl Environ Microbiol 64:3663–3668
Burkiewicz K (1987) Active substances in the media after algae cultivation. Acta Physiol Plant 9:211–217
Çakmakçi R, Dönmez F, Aydın A, Şahin F (2006) Growth promotion of plants by plant growth-promoting rhizobacteria under greenhouse and two different field soil conditions. Soil Biol Biochem 38:1482–1487

Canellas LP, Olivares FL, Okorokova-Façanha AL, Façanha AR (2002) Humic acids isolated from earthworm compost enhance root elongation, lateral root emergence, and plasma membrane H^+ -ATPase activity in maize roots. Plant Physiol 130:1951–1957

Červená K, Lyčková B, Kučerová R (2013) Vermicompost—an effective natural fertilizer. In: Proceedings in EIIC-The 2nd electronic international interdisciplinary conference. http://www.eiic.cz/archive/?vid=1&aid=3&kid=20201-110&q=f1. Accessed 13 Nov 2014

Chandramohan A, Sivasankar V, Ravichandran C, Sakthivel R (2013) A probe on the status of microorganisms in the air, soil and solid waste samples of Ariyamangalam dumping site at Tiruchirappalli district, South India. In: Velu RK (ed) Microbiological research in agroecosystem management. Springer, India, pp 1–9

Chaoui HI, Zibilske LM, Ohno T (2003) Effects of earthworm casts and compost on soil microbial activity and plant nutrient availability. Soil Biol Biochem 35:295–302

Chen CM, Ert JR, Leisner SM, Chang CC (1985) Localization of cytokinin biosynthetic sites in pea plants and carrot roots. Plant Physiol 78:510–513

Chen Y, Clapp CE, Magen H (2004) Mechanisms of plant growth stimulation by humic substances: the role of organo-iron complexes. Soil Sci Plant Nutr 50:1089–1095

Chen JH (2006) The combined use of chemical and organic fertilizers and/or biofertilizers for crop growth and soil fertility. In: International workshop on sustained management of the soil-rhizosphere system for efficient crop production and fertilizer use, vol 16. Land Development Department Bangkok, Thailand, p 20

Choi O, Kim J, Kim JG, Jeong Y, Moon JS, Park CS, Hwang I (2008) Pyrroloquinoline quinone is a plant growth promotion factor produced by *Pseudomonas fluorescens* B16. Plant Physiol 146:657–668

Coen ES, Meyerowitz EM (1991) The war of the whorls: genetic interactions controlling flower development. Nature 353:31–37

Crafts CB, Miller CO (1974) Detection and identification of cytokinins produced by mycorrhizal fungi. Plant Physiol 54:586–588

Crawford NM (1995) Nitrate: nutrient and signal for plant growth. Plant Cell 7:859–868

Crawford NM, Glass AD (1998) Molecular and physiological aspects of nitrate uptake in plants. Trends Plant Sci 3:389–395

Creelman RA, Mullet JE (1997) Biosynthesis and action of jasmonates in plants. Annu Rev Plant Biol 48:355–381

Crespi M, Messens E, Caplan AB, Van Montagu M, Desomer J (1992) Fasciation induction by the phytopathogen *Rhodococcus fascians* depends upon a linear plasmid encoding a cytokinin synthase gene. EMBO J 11:795–804

Crowley DE (2006) Microbial siderophores in the plant rhizosphere. In: Barton LL, Abadia J (eds) Iron nutrition in plants and rhizospheric microorganisms. Springer, Netherlands, pp 169–198

Datta JK, Banerjee A, Sikdar MS, Gupta S, Mondal NK (2009) Impact of combined exposure of chemical, fertilizer, bio-fertilizer and compost on growth, physiology and productivity of *Brassica campestries* in old alluvial soil. J Environ Biol 30:797–800

De Meutter J, Tytgat T, Witters E, Gheysen G, Van Onckelen H, Gheysen G (2003) Identification of cytokinins produced by the plant parasitic nematodes *Heterodera schachtii* and *Meloidogyne incognita*. Mol Plant Pathol 4:271–277

Den Boer BG, Murray JA (2000) Triggering the cell cycle in plants. Trends Cell Biol 10:245–250

Dewitte W, Murray JA (2003) The plant cell cycle. Annu Rev Plant Biol 54:235–264

Dimkpa C, Svatoš A, Merten D, Büchel G, Kothe E (2008) Hydroxamate siderophores produced by *Streptomyces acidiscabies* E13 bind nickel and promote growth in cowpea (*Vigna unguiculata* L.) under nickel stress. Can J Microbiol 54:163–172

Dimkpa CO, Merten D, Svatoš A, Büchel G, Kothe E (2009) Siderophores mediate reduced and increased uptake of cadmium by *Streptomyces tendae* F4 and sunflower (*Helianthus annuus*), respectively. J Appl Microbiol 107:1687–1696

Doares SH, Narvaez-Vasquez J, Conconi A, Ryan CA (1995) Salicylic acid inhibits synthesis of proteinase inhibitors in tomato leaves induced by systemin and jasmonic acid. Plant Physiol 108:1741–1746
Döbereiner J (1992) History and new perspectives of diazotrophs in association with non-leguminous plants. Symbiosis 13:1–13
Durner J, Klessig DF (1999) Nitric oxide as a signal in plants. Curr Opin Plant Biol 2:369–374
Durner J, Wendehenne D, Klessig DF (1998) Defense gene induction in tobacco by nitric oxide, cyclic GMP, and cyclic ADP-ribose. Proc Natl Acad Sci USA 95:10328–10333
Edwards CA, Arancon NQ, Sherman R (2010) Vermiculture technology: earthworms, organic wastes, and environmental management. CRC Press, Boca Raton
Emery RJN, Atkins CA (2002) Roots and cytokinins. In: Waisel Y, Eshel A, Kafkafi U (eds) Plant roots—the hidden half. Marcel Dekker Inc, New York, pp 417–434
FAO/WHO (2015) What is organic agriculture. http://www.fao.org/organicag/oa-faq/oa-faq1/en/. Accessed 3 Jan 2015
Francis D, Sorrell DA (2001) The interface between the cell cycle and plant growth regulators: a mini review. Plant Growth Regul 33:1–12
Frank M, Schmülling T (1999) Cytokinin cycles cells. Trends Plant Sci 4:243–244
Frébort I, Kowalska M, Hluska T, Frébortová J, Galuszka P (2011) Evolution of cytokinin biosynthesis and degradation. J Exp Bot 62:2431–2452
Fricke K, Vogtmann H (1993) Quality of source separated compost. BioCycle 34:64–70
Frugier F, Kosuta S, Murray JD, Crespi M, Szczyglowski K (2008) Cytokinin: secret agent of symbiosis. Trends Plant Sci 13:115–120
George EF (1993) Plant propagation by tissue culture, part 1: the technology, 2nd edn. Exegetics Ltd, Edington
George EF, Sherrington PD (1984) Plant propagation by tissue culture-handbook and directory of commercial laboratories. Exegetics Ltd, Edington
Gersani M, Kende H (1982) Studies on cytokinin-stimulated translocation in isolated bean leaves. J Plant Growth Regul 1:161–171
Gharib FA, Moussa LA, Massoud ON (2008) Effect of compost and bio-fertilizers on growth, yield and essential oil of sweet marjoram (*Majorana hortensis*) plant. Int J Agr Biol 10:381–382
Gibson SI (2005) Control of plant development and gene expression by sugar signalling. Curr Opin Plant Biol 8:93–102
Glick BR (1995) The enhancement of plant growth by free-living bacteria. Can J Microbiol 41:109–117
Glick BR, Bashan Y (1997) Genetic manipulation of plant growth-promoting bacteria to enhance biocontrol of phytopathogens. Biotechnol Adv 15:353–378
Glick BR, Cheng Z, Czarny J, Duan J (2007) Promotion of plant growth by ACC deaminase-producing soil bacteria. Eur J Plant Pathol 119:329–339
Goicoechea N, Antolin MC, Strnad M, Sánchez-Díaz M (1996) Root cytokinins, acid phosphatase and nodule activity in drought-stressed mycorrhizal or nitrogen-fixing alfalfa plants. J Exp Bot 47:683–686
Gopalakrishnan S, Srinivas V, Alekhya G, Prakash B, Kudapa H, Rathore A, Varshney RK (2015) The extent of grain yield and plant growth enhancement by plant growth-promoting broad-spectrum *Streptomyces* sp. in chickpea. SpringerPlus 4:31. doi:10.1186/s40064-015-0811-3
Gray EJ, Smith DL (2005) Intracellular and extracellular PGPR: commonalities and distinctions in the plant–bacterium signaling processes. Soil Biol Biochem 37:395–412
Grbić V, Bleecker AB (2000) Axillary meristem development in *Arabidopsis thaliana*. Plant J 21:215–223
Groppa MD, Benavides MP (2008) Polyamines and abiotic stress: recent advances. Amino Acids 34:35–45

Haberer G, Kieber JJ (2002) Cytokinins. New insights into a classic phytohormone. Plant Physiol 128:354–362
Handelsman J, Stabb EV (1996) Biocontrol of soilborne plant pathogens. Plant Cell 8:1855–1869
Hänsch R, Mendel RR (2009) Physiological functions of mineral micronutrients (Cu, Zn, Mn, Fe, Ni, Mo, B, Cl). Curr Opin Plant Biol 12:259–266
Hayat R, Ali S, Amara U, Khalid R, Ahmed I (2010) Soil beneficial bacteria and their role in plant growth promotion: a review. Ann Microbiol 60:579–598
Herbert RJ, Vilhar B, Evett C, Orchard CB, Rogers HJ, Davies MS, Francis D (2001) Ethylene induces cell death at particular phases of the cell cycle in the tobacco TBY-2 cell line. J Exp Bot 52:1615–1623
Hernández A, Castillo H, Ojeda D, Arras A, López J, Sánchez E (2010) Effect of vermicompost and compost on lettuce production. Chil J Agr Res 70:583–589
Hsieh SC (2005) Organic farming for sustainable agriculture in Asia with special reference to Taiwan experience. Research Institute of Tropical Agriculture and International Cooperation, National Pingtung University of Science and Technology, Taiwan. http://www.agnet.org/htmlarea_file/library/20110801133519/eb558.pdf. Accessed 23 Oct 2014
Howell SH, Lall S, Che P (2003) Cytokinins and shoot development. Trends Plant Sci 8:453–459
Huala E, Sussex IM (1993) Determination and cell interactions in reproductive meristems. Pl Cell 5:1157–1165
Hwang I, Sakakibara H (2006) Cytokinin biosynthesis and perception. Physiol Plant 126:528–538
International Humic Substances Society (2007) What are humic substances? http://www.humicsubstances.org/whatarehs.html. Accessed 24 Oct 2014
Inzé D (2003) Why should we study the plant cell cycle? J Exp Bot 54:1125–1126
Jeng AS, Haraldsen TK, Grønlund A, Pedersen PA (2006) Meat and bone meal as nitrogen and phosphorus fertilizer to cereals and rye grass. Nutr Cycl Agroecosys 76:183–191
Johnstone DB (1967) Isolation of *Azotobacter insignis* from fresh water. Ecology 48:671–672
Kakimoto T (2001) Identification of plant cytokinin biosynthetic enzymes as dimethylallyl diphosphate: ATP/ADP isopentenyltransferases. Plant Cell Physiol 42:677–685
Kakimoto T (2003) Biosynthesis of cytokinins. J Plant Res 116:233–239
Kamínek M, Motyka V, Vaňková R (1997) Regulation of cytokinins content in plant cells. Physiol Plant 101:689–700
Katiyar V, Goel R (2004) Siderophore mediated plant growth promotion at low temperature by mutant of fluorescent pseudomonad. Plant Growth Regul 42:239–244
Klessig DF, Durner J, Noad R, Navarre DA, Wendehenne D, Kumar D, Zhou JM, Shah J, Zhang S, Kachroo P, Trifa Y, Pontier D, Lam E, Silva H (2000) Nitric oxide and salicylic acid signaling in plant defense. Proc Natl Acad Sci USA 97:8849–8855
Kloepper JW, Lifshitz R, Zablotowicz RM (1989) Free-living bacterial inocula for enhancing crop productivity. Trends Biotechnol 7:39–44
Koltai H (2011) Strigolactones are regulators of root development. New Phytol 190:545–549
Koning RE (1994) Cell cycle. Plant Physiology Information Website. http://plantphys.info/plant_physiology/cellcycle.shtml. Accessed 24 Oct 2014
Kraigher H, Grayling A, Wang TL, Hanke DE (1991) Cytokinin production by two ectomycorrhizal fungi in liquid culture. Phytochem 30:2249–2254
Kravchenko LV, Azarova TS, Makarova NM, Tikhonovich IA (2004) The effect of tryptophan present in plant root exudates on the phytostimulating activity of rhizobacteria. Microbiology 73:156–158
Kretzschmar T, Kohlen W, Sasse J, Borghi L, Schlegel M, Bachelier JB, Reinhardt D, Bours R, Bouwmeester HJ, Martinoia E (2012) A petunia ABC protein controls strigolactone-dependent symbiotic signalling and branching. Nature 483:341–344
Kudoyarova GR, Melentiev AI, Martynenko EV, Timergalina LN, Arkhipova TN, Shendel GV, Kuz'mina LY, Dodd IC, Veselov SY (2014) Cytokinin producing bacteria stimulate amino acid deposition by wheat roots. Plant Physiol Biochem 83:285–291

Lamotte O, Courtois C, Barnavon L, Pugin A, Wendehenne D (2005) Nitric oxide in plants: the biosynthesis and cell signalling properties of a fascinating molecule. Planta 221:1–4
Leopold AC, Kawase M (1964) Benzyladenine effects on bean leaf growth and senescence. Am J Bot 51:294–298
Letham DS, Palni LMS (1983) The biosynthesis and metabolism of cytokinins. Ann Rev Plant Physiol 34:163–197
Li P, Chen L, Zhou Y, Xia X, Shi K, Chen Z, Yu J (2013) Brassinosteroids-induced systemic stress tolerance was associated with increased transcripts of several defence-related genes in the phloem in *Cucumis sativus*. PLoS ONE 8:e66582. doi:10.1371/journal.pone.0066582
Lindsey K, Casson S, Chilley P (2002) Peptides: new signalling molecules in plants. Trends Plant Sci 7:78–83
Lohar DP, Schaff JE, Laskey JG, Kieber JJ, Bilyeu KD, Bird DM (2004) Cytokinins play opposite roles in lateral root formation, and nematode and rhizobial symbioses. Plant J 38:203–214
Long J, Barton MK (2000) Initiation of axillary and floral meristems in *Arabidopsis*. Dev Biol 218:341–353
Lugtenberg B, Kamilova F (2009) Plant-growth-promoting rhizobacteria. Annu Rev Microbiol 63:541–556
Ma QH, Lin ZB, Fu DZ (2002) Increased seed cytokinin levels in transgenic tobacco influence embryo and seedling development. Funct Plant Biol 29:1107–1113
Mäder P, Fliessbach A, Dubois D, Gunst L, Fried P, Niggli U (2002) Soil fertility and biodiversity in organic farming. Science 296:1694–1697
Mahdi SS, Hassan GI, Samoon SA, Rather HA, Dar SA, Zehra B (2010) Bio-fertilizers in organic agriculture. J Phytol 2:42–54
Maksimov IV, Abizgil'Dina RR, Pusenkova LI (2011) Plant growth promoting rhizobacteria as alternative to chemical crop protectors from pathogens (review). Appl Biochem Microbiol 47:333–345
Malusá E, Vassilev N (2014) A contribution to set a legal framework for biofertilisers. Appl Microbiol Biot 98:6599–6607
Mapes CC, Davies PJ (2001) Cytokinins in the ball gall of *Solidago altissima* and in the gall forming larvae of *Eurosta solidaginis*. New Phytol 151:203–212
Marschner H (1995) Mineral nutrition of higher plants, 2nd edn. Academic Press, London
Martínez GA, Chaves AR, Añón MC (1994) Effect of gibberellic acid on ripening of strawberry fruits (*Fragaria annanassa* Duch.). J Plant Growth Regul 13:87–91
Martin-Tanguy J (2001) Metabolism and function of polyamines in plants: recent development (new approaches). Plant Growth Regul 34:135–148
Masalha J, Kosegarten H, Elmaci Ö, Mengel K (2000) The central role of microbial activity for iron acquisition in maize and sunflower. Biol Fert Soils 30:433–439
Matsubayashi Y (2014) Posttranslationally modified small-peptide signals in plants. Annu Rev Plant Biol 65:385–413
Matsubayashi Y, Sakagami Y (2006) Peptide hormones in plants. Annu Rev Plant Biol 57:649–674
Mauk CS, Noodén LD (1992) Regulation of mineral redistribution in pod-bearing soybean explants. J Exp Bot 43:1429–1440
McDermott J, Meilan R, Thornburg R (1996) Plant–insect interactions: the hackberry nipple gall. World Wide Web J Biol 2 http://epress.com/w3jbio/vol2/mcdermott/mcdermott.html
Mehboob I, Naveed M, Ahmad Z (2009) Rhizobial association with non-legumes: Mechanisms and applications. Crit Rev Plant Sci 28:432–456
Mishra DJ, Singh R, Mishra UK, Kumar SS (2013) Role of bio-fertilizer in organic agriculture: a review. Res J Recent Sci 2:39–41
Mok MC (1994) Cytokinins and plant development–an overview. In: Mok DWS, Mok MC (eds) Cytokinins: chemistry, activity and function. CRC Press, Boca Raton, pp 155–166
Mok DW, Mok MC (2001) Cytokinin metabolism and action. Annu Rev Plant Biol 52:89–118

Mok MC, Martin RC, Mok DW (2000) Cytokinins: biosynthesis, metabolism and perception. In Vitro Cellular Dev-Plant 36:102–107

Müssig C, Lisso J, Coll-Garcia D, Altmann T (2006) Molecular analysis of brassinosteroid action. Plant Biol 8:291–296

Neeraja C, Anil K, Purushotham P, Suma K, Sarma PVSRN, Moerschbacher BM, Podile AR (2010) Biotechnological approaches to develop bacterial chitinases as a bioshield against fungal diseases of plants. Critl Rev Biotechnol 30:231–241

Nelson DC, Riseborough JA, Flematti GR, Stevens J, Ghisalberti EL, Dixon KW, Smith SM (2009) Karrikins discovered in smoke trigger *Arabidopsis* seed germination by a mechanism requiring gibberellic acid synthesis and light. Plant Physiol 149:863–873

Nelson DC, Flematti GR, Riseborough JA, Ghisalberti EL, Dixon KW, Smith SM (2010) Karrikins enhance light responses during germination and seedling development in *Arabidopsis thaliana*. Proc Natl Acad Sci USA 107:7095–7100

Nelson DC, Flematti GR, Ghisalberti EL, Dixon KW, Smith SM (2012) Regulation of seed germination and seedling growth by chemical signals from burning vegetation. Annu Rev Plant Biol 63:107–130

Ng PP, Cole ALJ, Jameson PE, McWha JA (1982) Cytokinin production by ectomycorrhizal fungi. New Phytol 91:57–62

Obreht Z, Kerby NW, Gantar M, Rowell P (1993) Effects of root-associated N_2-fixing cyanobacteria on the growth and nitrogen content of wheat (*Triticum vulgare* L.) seedlings. Biol Fert Soils 15:68–72

Okon Y, Bloemberg GV, Lugtenberg BJJ (1998) Biotechnology of biofertilization and phytostimulation. In: Altman A (ed) Agricultural biotechnology. Marcel Dekker, New York, pp 327–349

Ördög V, Stirk WA, Van Staden J, Novák O, Strnad M (2004) Endogenous cytokinins in three genera of microalgae from the chlorophyta. J Phycol 40:88–95

Orhan E, Esitken A, Ercisli S, Turan M, Sahin F (2006) Effects of plant growth promoting rhizobacteria (PGPR) on yield, growth and nutrient contents in organically growing raspberry. Sci Hortic 111:38–43

Owen D, Williams AP, Griffith GW, Withers PJA (2014) Use of commercial bio-inoculants to increase agricultural production through improved phosphorus acquisition. Appl Soil Ecol 86:41–54

Pacifici E, Polverari L, Sabatini S (2015) Plant hormone cross-talk: the pivot of root growth. J Exp Bot 66:1113–1121

Patten CL, Glick BR (2002) Role of *Pseudomonas putida* indoleacetic acid in development of the host plant root system. Appl Environ Microb 68:3795–3801

Paungfoo-Lonhienne C, Lonhienne TG, Rentsch D, Robinson N, Christie M, Webb RI, Gamage HK, Carroll BJ, Schenk PM, Schmidt S (2008) Plants can use protein as a nitrogen source without assistance from other organisms. Proc Natl Acad Sci USA 105:4524–4529

Pelacho AM, Mingo-Castel AM (1991) Jasmonic acid induces tuberization of potato stolons cultured *in vitro*. Plant Physiol 97:1253–1255

Phillips D, Torrey J (1972) Studies on cytokinin production by *Rhizobium*. Plant Physiol 49:11–15

Piccolo A (2002) The supramolecular structure of humic substances: a novel understanding of humus chemistry and implications in soil science. Adv Agron 75:57–134

Piccolo A, Nardi S, Concheri G (1992) Structural characteristics of humic substances as related to nitrate uptake and growth regulation in plant systems. Soil Biol Biochem 24:373–380

Pimenta Lange MJ, Lange T (2006) Gibberellin biosynthesis and the regulation of plant development. Plant Biol 8:281–290

Powell GK, Morris RO (1986) Nucleotide sequence and expression of a *Pseudomonas savastanoi* cytokinin biosynthetic gene: homotogy with *Agrobacterium tumefaciens* tmr and tzs loci. Nucleic Acids Res 14:2555–2565

Qin H, Lu K, Strong PJ, Xu Q, Wu Q, Xu Z, Xu J, Wang H (2015) Long-term fertilizer application effects on the soil, root arbuscular mycorrhizal fungi and community composition in rotation agriculture. Appl Soil Ecol 89:35–43

Rauthan BS, Schnitzer M (1981) Effects of a soil fulvic acid on the growth and nutrient content of cucumber (*Cucumis sativus*) plants. Plant Soil 63:491–495

Raskin I (1992) Role of salicylic acid in plants. Annu Rev Plant Biol 43:439–463

Riou-Khamlichi C, Menges M, Healy JS, Murray JA (2000) Sugar control of the plant cell cycle: differential regulation of *Arabidopsis* D-type cyclin gene expression. Mol Cell Biol 20:4513–4521

Roldán M, Gómez-Mena C, Ruiz-García L, Salinas J, Martínez-Zapater JM (1999) Sucrose availability on the aerial part of the plant promotes morphogenesis and flowering of *Arabidopsis* in the dark. Plant J 20:581–590

Rolland F, Moore B, Sheen J (2002) Sugar sensing and signaling in plants. Plant Cell 14:S185–S205

Ryan CA (2000) The systemin signaling pathway: differential activation of plant defensive genes. Biochim Biophys Acta 1477:112–121

Ryan CA, Pearce G (2003) Systemins: a functionally defined family of peptide signals that regulate defensive genes in Solanaceae species. Proc Natl Acad Sci USA 100:14577–14580

Ryu CM, Farag MA, Hu CH, Reddy MS, Wei HX, Paré PW, Kloepper JW (2003) Bacterial volatiles promote growth in *Arabidopsis*. Proc Natl Acad Sci USA 100:4927–4932

Sabatini S, Beis D, Wolkenfelt H, Murfett J, Guilfoyle T, Malamy J, Benfey P, Leyser O, Bechtold N, Weisbeek P, Scheres B (1999) An auxin-dependent distal organizer of pattern and polarity in the *Arabidopsis* root. Cell 99:463–472

Sahni S, Sarma BK, Singh DP, Singh HB, Singh KP (2008) Vermicompost enhances performance of plant growth-promoting rhizobacteria in *Cicer arietinum* rhizosphere against *Sclerotium rolfsii*. Crop Prot 27:369–376

Sakakibara H (2006) Cytokinins: activity, biosynthesis, and translocation. Annu Rev Plant Biol 57:431–449

Sakakibara H, Kasahara H, Ueda N, Kojima M, Takei K, Hishiyama S, Asami T, Okada K, Kamiya Y, Yamaya T, Yamaguchi S (2005) *Agrobacterium tumefaciens* increases cytokinin production in plastids by modifying the biosynthetic pathway in the host plant. Proc Natl Acad Sci USA 102:9972–9977

Sarma I, Phookan DB, Boruah S (2015) Influence of manures and biofertilizers on carrot (*Daucus carota* L.) cv. *Early Nantes* growth, yield and quality. EFAJ 10:25–27

Sauter M, Mekhedov SL, Kende H (1995) Gibberellin promotes histone H1 kinase activity and the expression of cdc2 and cyclin genes during the induction of rapid growth in deepwater rice internodes. Plant J 7:623–632

Scaffidi A, Waters MT, Bond CS, Dixon KW, Smith SM, Ghisalberti EL, Flematti GR (2012) Exploring the molecular mechanism of karrikins and strigolactones. Bioorg Med Chem Lett 22:3743–3746

Schachtman DP, Reid RJ, Ayling SM (1998) Phosphorus uptake by plants: from soil to cell. Plant Physiol 116:447–453

Schmülling T (2002) New insights into the functions of cytokinins in plant development. J Plant Growth Regul 21:40–49

Shafi M, Shah A, Bakht J, Shah M, Mohammad W (2012) Integrated effect of inorganic and organic nitrogen sources on soil fertility and productivity of maize. J Plant Nutr 35:524–537

Shaviv A, Mikkelsen RL (1993) Controlled-release fertilizers to increase efficiency of nutrient use and minimize environmental degradation—a review. Fert Res 35:1–12

Shimizu-Sato S, Mori H (2001) Control of outgrowth and dormancy in axillary buds. Plant Physiol 127:1405–1413

Smith SM (2013) Plant biology: Witchcraft and destruction. Nature 504:384–385

Šimko I (1994) Sucrose application causes hormonal changes associated with potato tuber induction. J Plant Growth Regul 13:73–77

Sivasakthivelan P, Saranraj P (2013) *Azospirillum* and its formulation. Int J Microl Res 4:275–287

Skylar A, Sung F, Hong F, Chory J, Wu X (2011) Metabolic sugar signal promotes *Arabidopsis* meristematic proliferation via G2. Dev Biol 351:82–89

Song X, Liu M, Wu D, Griffiths BS, Jiao J, Li H, Hu F (2015) Interaction matters: synergy between vermicompost and PGPR agents improves soil quality, crop quality and crop yield in the field. Appl Soil Ecol 89:25–34

Souza BM, Molfetta-Machado JB, Freschi L, Figueira A, Purgatto E, Buckeridge MS, Van Sluys M, Mercier H (2010) Axillary bud development in pineapple nodal segments correlates with changes on cell cycle gene expression, hormone level, and sucrose and glutamate contents. In Vitro Cell Dev-Plant 46:281–288

Steenhoudt O, Vanderleyden J (2000) *Azospirillum*, a free-living nitrogen-fixing bacterium closely associated with grasses: genetic, biochemical and ecological aspects. FEMS Microbiol Rev 24:487–506

Steeves TA, Sussex IM (1989) Patterns in Plant Development, 2nd edn. Cambridge University Press, London

Stirk WA, Van Staden J (2010) Flow of cytokinins through the environment. Plant Growth Regul 62:101–116

Stirk WA, Ördög V, Van Staden J (1999) Identification of the cytokinin isopentenyladenine in a strain of *Arthronema africanum* (Cyanobacteria). J Phycol 35:89–92

Stirk WA, Novák O, Strnad M, Van Staden J (2003) Cytokinins in macroalgae. Plant Growth Regul 41:13–24

Stirk WA, Novák O, Hradecká V, Pěnčík A, Rolčík J, Strnad M, Van Staden J (2009) Endogenous cytokinins, auxins and abscisic acid in *Ulva fasciata* (Chlorophyta) and *Dictyota humifusa* (Phaeophyta): towards understanding their biosynthesis and homoeostasis. Eur J Phycol 44:231–240

Stirk WA, Van Staden J, Novák O, Doležal K, Strnad M, Dobrev P, Sipos G, Ördög V, Bálint P (2011) Changes in endogenous cytokinin concentrations in *Chlorella* (Chlorophyceae) in relation to light and the cell cycle. J Phycol 47:291–301

Taiz L, Zeiger E (2010) Plant physiology, 5th edn. Sinauer Associates Inc, Sunderland

Takahashi F, Sato-Nara K, Kobayashi K, Suzuki M, Suzuki H (2003) Sugar-induced adventitious roots in *Arabidopsis* seedlings. J Plant Res 116:83–91

Talaat NB, Ghoniem AE, Abdelhamid MT, Shawky BT (2015) Effective microorganisms improve growth performance, alter nutrients acquisition and induce compatible solutes accumulation in common bean (*Phaseolus vulgaris* L.) plants subjected to salinity stress. Plant Growth Regul 75:281–295

Tarkowski P, Ge L, Yong JWH, Tan SN (2009) Analytical methods for cytokinins. Trend Anal Chem 28:323–335

Taylor NJ, Stirk WA, Van Staden J (2003) The elusive cytokinin biosynthetic pathway. S Afr J Bot 69:269–281

Theodorou ME, Plaxton WC (1993) Metabolic adaptations of plant respiration to nutritional phosphate deprivation. Plant Physiol 101:339–344

Tilman D (1998) The greening of the green revolution. Nature 396:211–212

Torelli A, Trotta A, Acerbi L, Arcidiacono G, Berta G, Branca C (2000) IAA and ZR content in leek (*Allium porrum* L.), as influenced by P nutrition and arbuscular mycorrhizae, in relation to plant development. Plant Soil 226:29–35

Turnbull CG, Raymond MA, Dodd IC, Morris SE (1997) Rapid increases in cytokinin concentration in lateral buds of chickpea (*Cicer arietinum* L.) during release of apical dominance. Planta 202:271–276

Ueda N, Kojima M, Suzuki K, Sakakibara H (2012) *Agrobacterium tumefaciens* tumor morphology root plastid localization and preferential usage of hydroxylated prenyl donor is important for efficient gall formation. Plant Physiol 159:1064–1072

Umehara M (2011) Strigolactone, a key regulator of nutrient allocation in plants. Plant Biotechnol 28:429–437

Upadhyaya NM, Letham DS, Parker CW, Hocart CH, Dart PJ (1991) Do *Rhizobia* produce cytokinins? Biochem Int 24:123–130

Van Loon LC (2007) Plant responses to plant growth-promoting rhizobacteria. Eur J Plant Pathol 119:243–254

Van Loon LC, Bakker PAHM, Pieterse CMJ (1998) Systemic resistance induced by rhizosphere bacteria. Annu Rev Phytopathol 36:453–483

Van Staden J (1976) The release of cytokinins by maize roots. Plant Sci Lett 7:279–283

Van Staden J, Bennett PH (1991) Gall formation in crofton weed. Differences between normal stem tissue and gall tissue with respect to cytokinin levels and requirement for in vitro culture. S Afr J Bot 57:246–248

Vassilev N, Vassileva M, Nikolaeva I (2006) Simultaneous P-solubilizing and biocontrol activity of microorganisms: potentials and future trends. Appl Microbiol Biot 71:137–144

Vessey JK (2003) Plant growth promoting rhizobacteria as biofertilizers. Plant Soil 255:571–586

Vitousek PM, Aber JD, Howarth RW, Likens GE, Matson PA, Schindler DW, Schlesinger WH, Tilman DG (1997) Human alteration of the global nitrogen cycle: sources and consequences. Ecol Appl 7:737–750

Wandersman C, Delepelaire P (2004) Bacterial iron sources: from siderophores to hemophores. Annu Rev Microbiol 58:611–647

Wang L, Ruan YL (2013) Regulation of cell division and expansion by sugar and auxin signaling. Front Plant Sci 4:1–9

Wang R, Xing X, Crawford N (2007) Nitrite acts as a transcriptome signal at micromolar concentrations in *Arabidopsis* roots. Plant Physiol 145:1735–1745

Weiger TM, Hermann A (2014) Cell proliferation, potassium channels, polyamines and their interactions: a mini review. Amino Acids 46:681–688

Wolters H, Jürgens G (2009) Survival of the flexible: hormonal growth control and adaptation in plant development. Nat Rev Genet 10:305–317

Xiang W, Zhao L, Xu X, Qin Y, Yu G (2012) Mutual information between beneficial microorganisms and the roots of host plants determined the bio-functions of biofertilizers. Am J Plant Sci 3:1115–1120

Ya'acov YL (1996) Nitric oxide in biological systems. Plant Growth Regul 18:155–159

Yong JWH, Wong SC, Letham DS, Hocart CH, Farquhar GD (2000) Effects of elevated [CO_2] and nitrogen nutrition on cytokinins in the xylem sap and leaves of cotton. Plant Physiol 124:767–780

Yong JWH, Ge L, Ng YF, Tan SN (2009) The chemical composition and biological properties of coconut water. Molecules 14:5144–5164

Yong JWH, Letham DS, Wong SC, Farquhar GD (2014) *Rhizobium*-induced elevation in xylem cytokinin delivery in pigeonpea induces changes in shoot development and leaf physiology. Funct Plant Biol 41:1323–1335

Yoshida T, Fujita Y, Sayama H, Kidokoro S, Maruyama K, Mizoi J, Shinozaki K, Yamaguchi-Shinozaki K (2010) AREB1, AREB2, and ABF3 are master transcription factors that cooperatively regulate ABRE-dependent ABA signaling involved in drought stress tolerance and require ABA for full activation. Pl J 61:672–685

Youssef MMA, Eissa MFM (2014) Biofertilizers and their role in management of plant parasitic nematodes. A review. EJBPR 5:1–6

Zachmann JE, Molina JAE (1993) Presence of culturable bacteria in cocoons of the earthworm *Eisenia fetida*. Appl Environ Microbiol 59:1904–1910

Zhang H, Xie X, Kim MS, Kornyeyev DA, Holaday S, Paré PW (2008) Soil bacteria augment *Arabidopsis* photosynthesis by decreasing glucose sensing and abscisic acid levels in planta. Plant J 56:264–273

Zhang H, Tan SN, Wong WS, Ng CYL, Teo CH, Ge L, Chen X, Yong JWH (2014) Mass spectrometric evidence for the occurrence of plant growth promoting cytokinins in vermicompost tea. Biol Fert Soils 50:401–403

Zhang H, Tan SN, Teo CH, Yew YR, Ge L, Chen X, Yong JWH (2015) Analysis of phytohormones in vermicompost using a novel combinative sample preparation strategy of ultrasound-assisted extraction and solid-phase extraction coupled with liquid chromatography–tandem mass spectrometry. Talanta 139:189–197

Zhao Y (2010) Auxin biosynthesis and its role in plant development. Annu Rev Plant Biol 61:49–64

Zhao FY, Cai FX, Gao HJ, Zhang SY, Wang K, Liu T, Wang X (2015) ABA plays essential roles in regulating root growth by interacting with auxin and MAPK signaling pathways and cell-cycle machinery in rice seedlings. Plant Growth Regul 75:535–547

Zhou XY, Song L, Xue HW (2013) Brassinosteroids regulate the differential growth of *Arabidopsis* hypocotyls through auxin signaling components IAA19 and ARF7. Mol Plant 6:887–904

Phytohormone-Producing PGPR for Sustainable Agriculture

Dinesh Kumar Maheshwari, Shrivardhan Dheeman and Mohit Agarwal

Abstract Looking into account, the effective microorganisms (EM) discussed in green revolution are able to enhance plant growth and crop productivity. These act as fertilizers without causing any hazard on edaphic profile and ecological sustainability. In recent scenario, these microorganisms as PGPR are known to produce phytohormones and cover tremendous role in sustainable agriculture. Major six classes of phytohormone including natural, semi-synthetic and synthetic check out seed dormancy of several crop plants and allow to germinate in short period and further induce plant growth in sustainable manner, trigger plant immunity, maintain stress tolerance and aid plant maturity for fruiting and seedling. Under this review, discussion lies on all the aspects covering role and significance of auxin and other phytohormone-producing PGPR important to agriculture in near future.

Keywords PGPR · Phytohormone · Cytokinin · Auxin · Salicylic acid · Plant immunity · Crop productivity

1 Introduction

The ecology of root vicinity is called as rhizosphere, which harbors diversified microorganisms having various interactions, under two broad means as symbiotic (formation of nodule) and non-symbiotic. Based on the mode of interactions in rhizosphere, plant growth promoting rhizobacteria are termed as extracellular PGPR (ePGPR) and intracellular PGPR (iPGPR) as stated by Martinez-Viveros

---

D.K. Maheshwari (✉) · S. Dheeman · M. Agarwal
Department of Botany and Microbiology, Gurukul Kangri University, Haridwar, UK, India
e-mail: maheshwaridk@gmail.com

D.K. Maheshwari (ed.), *Bacterial Metabolites in Sustainable Agroecosystem*, Sustainable Development and Biodiversity 12, DOI 10.1007/978-3-319-24654-3_7

et al. (2010). Due to their importance in agro-ecosystem, these are now established to improve soil health and profile so as to increase crop productivity.

In the beginning of twentieth century, Starling was first to define the term phytohormone as *organic substance which are synthesized in minute quantities in one part of the plant body and transported to another part where these influence specific physiological processes*. Phytohormones are structurally unrelated small molecule in nature, regulating plant growth and development i.e. auxin, abscisic acid, cytokinin, gibberellin and ethylene. However, recently, many semi-synthetic and synthetic phytohormones have been identified: brassinosteroids, jasmonate, salicylic acid, nitric oxide, strigolactones, etc. (Santner and Estelle 2009). Phytohormone producing bacteria are gaining full swing in whole globe under the means of exploitation due to synergy between bacteria and plant led to sustainable agriculture (Narula et al. 2006). Growth and development of plant is predominantly influenced by mineral nutrients, hormones and other secreting metabolites. In fact, almost all the communication in plant cells is brought by plant hormones produced by plant cells or by rhizobacteria.

The most commonly occurring phytohormone is auxin (indole acetic acid). IAA produced in shoot apical meristem of plant and found throughout the plant body. It occurs in the form of free auxins (diffusible auxins), which is released out from plant tissues, released out from tissues only after hydrolysis, autolysis and enzymolysis. Production of IAA is widespread among rhizospheric bacteria (Table 1). Different IAA biosynthesis pathways are used by these bacteria and sometimes a single bacterial strain exhibit more than one pathway (Patten and Glick 1996). There are many chemically synthesized phytohormones such as indole-3-butyric acid (IBA), 2-methyl-4-chlorophenoxy acetic acid (MCPA), indole-3-propionic acid (IPA), 2,4-dichlorophenoxy acetic acid (2,4-D), etc., able to trigger various physiological processes (Table 2).

Plant growth regulatory hormone generally called gibberellins (GAs) forms a large family of plant growth substances with distinct functions during the life cycle of higher plants. Gibberellins are involved in a number of developmental and physiological processes (Crozier et al. 2000) including seed germination, seedling emergence, stem and leaf growth, floral induction and flower or/and fruit growth (King and Evans 2003; Sponsel 2003), regulation of vegetative and reproductive (bud) dormancy and delay of senescence (Bottini and Luna 1993; Fulchieri et al. 1993; Reinoso et al. 2002). Gibberellins in combination with other phytohormones, are directly effective in promotion of shoot elongation in plants (Crozier et al. 2000). Very few bacteria produce GA during their cultivation on artificial culture medium.

Cytokinin mediates the responses to variable extrinsic factors, such as light conditions in the shoot and availability of nutrients and water in the root and also play role in the response to biotic and abiotic stresses. Together, these activities contribute to the fine-tuning of quantitative growth regulation in plants (Werner and Schmülling 2009; Gupta and Rashotte 2012). Cytokinin concentration in plant cells depends on biosynthesis immobilization from extracellular sources,

Table 1 Various indigenous genera-producing IAA and their influence on different crops

Name of genera	Accession number (NCBI)	IAA production (μg/ml)	Yield (%) increase	References
Pseudomonas aeruginosa GRC_1		31.00	42.6 (G/P)	Aeron et al. (2010)
P. aeruginosa PS2		36.00	38.8 (G/P)	Aeron et al. (2010)
P. aeruginosa PSII		30.00	39.2 (G/P)	Aeron et al. (2010)
P. aeruginosa LES4	HQ123431	42.00	41.2 (G/P)	Aeron et al. (2010)
P. aeruginosa PRS4	AB666551	40.00	40.8 (G/P)	Aeron et al. (2010)
P. aeruginosa PS15		41.00	47.3 (G/P)	Aeron et al. (2010)
P. aeruginosa PSI		36.00	45.74	Aeron et al. (2011)
Mesorhizobium loti MP6		24	34.28	Chandra et al. (2007)
Bacillus sp. BPR7	JN208240	17		Kumar et al. (2012)
Bradyrhizobium sp. BMP17	AB665550	40		Maheshwari et al. (2014)
Sinorhizobium meliloti PP3		80		Pandey and Maheshwari (2007)
S. meliloti MSSP		100		Pandey and Maheshwari (2007)

Author's Lab
G/P Growth per plant over control

metabolic inter-conversions, inactivation and degradation. Increased cytokinin concentration results either from their uptake or biosynthesis. Accumulated cytokinins are capable of inducing cytokinin oxidase which consequently decreases cytokinin levels. This seems to be the mechanism of re-establishment and maintenance of cytokinin homeostasis required for further development of physiological events induced by transient cytokinin accumulation (Kaminek et al. 1997).

Abscisic acid (ABA) is produced in a very low concentration which influences physiological processes such as respiration rate, metabolism and root abundance. ABA is involved in protection against drought, salt stress and toxic metals. It also induces stomatal closure of leaf. Rhizospheric bacteria capable of producing ABA are experimentally poorly underpinned. ABA has effective role in synthesis and inhibition of cytokinin (Miernyk 1979), and increases plant growth by managing with cytokinin concentration (Spaepen et al. 2009). It also alleviates plant stress by increasing rhizosphere in terms of root abundance (Maheshwari 2011; Boiero et al. 2007).

Plants use ethylene in gaseous form to regulate myriad developmental processes and stress responses. Ethylene production by infected plants is an early resistance response leading to activation of plant defense pathways. However, plant pathogens are also capable of producing ethylene, which might have an effect not only on the plant but also on the pathogen as well. Therefore, ethylene plays a dual role in plant–pathogen interactions by affecting the plant as well as the pathogen (Chagué et al. 2006). Ethylene regulates seed germination, root

Table 2 Involvement of various genes in biosynthetic pathways of IAA

Bacteria	Gene	Enzymatic activity	Pathway	References
Azospirillum brasilense Yu62	*ald*A	Aldehyde dehydrogenase	IPA	Xie et al. (2005)
A. brasilense Sp245	*ipd*C	Indole pyruvate decarboxylase	IPA	Costacurta et al. (1994)
A. brasilense Sp7	*his*C1	Aromatic amino acid	IPA	Castro-Guerrero et al. (2012)
Rhizobium sp. NGR234	*y4w*E	Aminotransferase	IPA	Kittell et al. (1989)
	*y4w*E			
Enterobacter cloacae FERM BP-1529	*ipd*C	Indole pyruvate decarboxylase	IPA	Koga et al. (1991)
Pseudomonas fluorescens Psd	*iaa*M	Tryptophan monooxygenase	IAM	Kochar et al. (2011)
Ralstonia solanacearum	*iaa*M	Tryptophan monooxygenase	IAM	Salanoubat et al. (2002), Kurosawa et al. (2009)
	*iaa*H	Indole acetamide hydrolase		
Erwinia chrysanthemi 3937	*iaa*M	Tryptophan monooxygenase	IAM	Yang et al. (2007)
	*iaa*H	Indole acetamide hydrolase		
Streptomyces En-1	*iaa*M	Tryptophan monooxygenase	IAM	Lin and Xu (2013)
	*iaa*H	Indole acetamide hydrolase		
P. fluorescens EBC191	*nit*	Indoleacetonitrilase	IAN	Kiziak et al. (2005)
	*nth*AB	Nitrile hydrolase		
Bacillus amyloliquefaciens FZB42	*yhc*X	Indoleacetonitrilase	IAN	Idris et al. (2007)
Bacillus sp. OxB-1	*oxd*	Phenylacetaldoxime dehydratase	IAOX	Kato et al. (2005)
Rhodococcus globerulus A-4	*oxd*RG	Phenylacetaldoxime dehydratase	IAOX	Kato et al. (2005)
	*Nha*1	Nitrile hydratase		
Pseudomonas sp. K-9	*oxd*K	Phenylacetaldoxime dehydratase	IAOX	Kato and Asano (2006)
	*Nha*1	Nitrile hydratase		
Rhodococcus erythropolis JCM 3201	*oxd*K	Phenylacetaldoxime dehydratase	IAOX	Kato et al. (2005)
	*Nha*1	Nitrile hydratase		
Rhodococcus rhodochrous J-1	*oxd*K	Phenylacetaldoxime dehydratase	IAOX	Kato et al. (2005)
	*Nha*1	Nitrile hydratase		
Rhodococcus sp. AK32	*oxd*	Phenylacetonitrilase	IAOX	Kato et al. (2005)

(continued)

Table 2 (continued)

Bacteria	Gene	Enzymatic activity	Pathway	References
Brevibacterium butanicum	*oxd*	Phenylacetaldoxime dehydratase	IAOX	Kato et al. (2005)
	*Nha*1	Nitrile hydratase		
Corynebacterium sp. C5	*oxd*	Phenylacetonitrilase	IAOX	Kato et al. (2005)

IPA Indole-3-pyruvic acid; *IAM* Indole-3-acetamide pathway; *IAN* Indole-3-acetonitrile pathway; *IAOX* Indole acetaldoxime pathway

initiation, flower development, fruit ripening, senescence and responses to biotic and abiotic stresses. It thus plays a key role in responses to the environment that have a direct bearing on a plant's fitness for adaptation and reproduction (Lin et al. 2009).

Beneficial group of bacteria presently dominating by auxin-producing and plant growth promoting rhizobacteria profoundly increases seed germination, root development and water utilization by plants. These rhizobacteria can arouse plant growth directly or indirectly by changing microbial balance in the rhizosphere in service of beneficial microorganisms. They can subdue a broad spectrum of bacterial, fungal and nematode diseases. PGPR can also deliver protection against viral diseases (Siddiqui 2006).

More recently, there has been a resurgence of interest in environmental friendly, sustainable and organic agricultural practices (Esitken et al. 2005). Bio-inoculants or microbial inoculants are agricultural amendments that use beneficial microorganisms to promote plant health by providing phytohormone production in rhizosphere on colonization. Microbial inoculants can induce systemic acquired resistance (SAR) of crop species to several common crop diseases (provides resistance against pathogens).

In this review, present state of knowledge is being discussed for better understanding the nature of beneficial bacterial physiology responsible to deliver phytohormone in root vicinity and interaction with plant for their growth promotion and disease management.

2 Classification, Biochemistry and Biosynthesis of Phytohormones

Phytohormone production by soil bacteria has significant influence on plant growth and performance (Smaill et al. 2010). In fact such hormones are crucial signaling molecules that coordinate all aspects of plant growth, development and defense mechanism. Production of the phytohormone particularly auxin (IAA) is widespread among bacteria that inhabit the rhizosphere of plants.

Phytohormones are commonly classified as semi-synthetic and synthetic hormone including few herbicides. These are grouped into five classes based on structural similarity (biochemistry) and physiological effect on plant (or plant part). Growth regulators of other synthetic hormones are not grouped into these classes; they may occur naturally, chemically synthesized, or organically (biochemically/microbiologically) synthesized in bacterial cells which further may harvest through several criteria and strategies. In each class of phytohormone including chemically synthesized, one have their pragmatic effect on plant for their growth regulation/promotion and health management. Naturally occurring bacteria are able to do such kind of action (production of phytohormone) which signifies their benefits to plant and soil by residing in rhizospheric habitat. Exhaustive information about the biosynthesis of phytohormone auxins, gibberellins, cytokinins, ethylene and abscisic acid, as well as plant growth regulators such as polyamines and nitric oxide in *Azospirillum* spp. have been observed (Cassan et al. 2014). High level of auxin, gibberellins and salicylic acid in chemically defined media was produced by *Bacillus amyloliquefaciens.* Co-inoculation of this strain with *Bradyrhizobium japonicum* enhances soybean nodulation (Masciarelli et al. 2014). Phytohormonal basis for the plant growth promoting action of naturally occurring biostimulators attributed for enhancing the growth of agricultural and horticultural crops (Kurepin et al. 2014). PGPR treatment improves seedling growth and quality of cabbage and increases GA, SA and IAA in plants raised by such group of bacteria (Turan et al. 2014).

2.1 Auxins

Several decades before, the term "Auxin" was introduced into the identification of scientific community (Went and Thirmann 1937). In recent scenario, understanding of IAA in plant growth promotion has been truly spectacular. Undoubtedly, IAA has wide approach in enhancement of plant growth and health promotion. Five bacteria producing IAA in pure culture include members of genera *Bacillus*, *Microbacterium*, *Methylophages*, *Agromyces* and *Paenibacillus* which have considerable impact on root elongation of tropical rice plant (Bal et al. 2013). Rhizospheric halotolerant IAA-producing bacteria *Kocuria turfanensis* were able to promote growth of *A. hypogoea* both in nonsaline and saline soils (Goswami et al. 2014).

Auxins are versatile in nature which exhibit differential physiological action. They belong to five major groups namely indole acids, naphthalene acids, chlorophenoxy acids, benzoic acids and picolinic acids and their derivatives. The first group belongs to indole propionic acids and indole butyric acid; second group comprises Nephthaleneacetic acid and β-naphthoxyacetic acid; third group has 2,4-dichlorophenoxyacetic acid, 2,4,5-trichlorophenoxyacetic acid and 2-methyl-4-chlorophenoxyacetic acid; whereas, 2,4,6-trichlorobenzoic acid and 2-methoxy-3-6-dichlorobenzoic acid categorized in fourth group and 4-amino-3,5,6-trichloropicoline acid in last group.

It is interesting to note that diverse bacterial genera produce "Auxin." Recently, a number of studies have clearly shown that IAA can be a signaling molecule in microorganisms, in both IAA-producing and IAA-non-producing species. These findings raise new intriguing questions on the role of IAA in bacteria and their interaction with plants. Such phytohormones of bacterial origin directly affect plants physiology particularly in root colonization strategies adopted by bacteria during plant–microbe interaction. IAA acts as signaling molecule in bacteria, therefore, facilitates positive outcome on the plant, which ranges from phytostimulation to plant immunity (Cheynier et al. 2013).

High degree of similarity between IAA biosynthesis pathway in plants and bacteria has been observed. Tryptophan has been identified as a main precursor of IAA biosynthesis pathway in bacteria. Basically, two precursors of IAA formation are presumed, either tryptophan or a tryptophan precursor. Various intermediate products namely indole-3-pyruvate, indole-3-acetaldoxime, indole-3-acetaldehyde (IAAld), indole-3-acetonitrile (IAN), indole-3-acetamide (IAM), or tryptamine (TAM) are involved in IAA biosynthesis.

IAA pathways are usually classified based on these intermittent compounds. Independently, indole-3-acetamide pathways are characterized in bacterial genera *Agrobacterium*, *Pseudomonas*, *Pantoea*, *Rhizobium*, *Bradyrhizobium* (Theunis et al. 2004). Conversion of tryptophan into IAA is accomplished by two steps: first step involved IAM that is produced by enzymatic action (tryptophan-2-monooxygenase) and during second step, IAA is obtained by enzymatic hydrolysis of IAM by IAM hydrolase (Bar and Okon 1993; Prinsen et al. 1993).

The other pathway is named as indole-3-pyruvate (IPyA) pathway which occurs in some pathogenic bacteria including species of *Pantoea* and few beneficial genera such as *Rhizobium* and *Bradyrhizobium*. Initially, tryptophan is converted into IPyA by enzymatic transformation and further decarboxylated into indole-3-acetaldehyde (IAAld) by indole-3-pyruvate decarboxylase (IPDC). In the terminal step, IAAld is oxidized into IAA. In tryptamine (TAM) pathway, TAM is directly converted to IAAld by amine oxidase and further decarboxylation brought about with indole-3-pyruvate decarboxylase lead to the formation of IAA (Hartmann et al. 1983) as identified in *Bacillus* spp. (Perley and Stowe 1966).

In tryptophan side chain oxidase (TSO) pathway, tryptophan is converted into IAAld by IPyA and oxidized to IAA simultaneously as in *Pseudomonas fluorescens* CHA0 (Oberhänsli et al. 1991). Conversion of indole-3-acetamide via nitrilase is another pathway where indole-3-acetonitrile is produced by tryptophan via indole-3-acetaldoxime (Patten and Glick 1996). The diagrammatic representation is shown in Fig. 1.

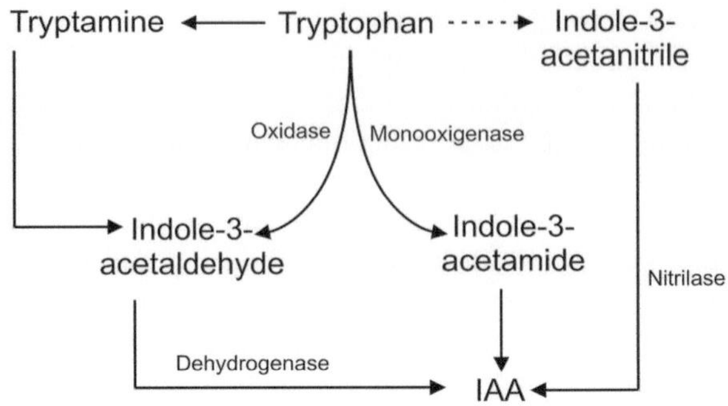

Fig. 1 Auxin (IAA) biosynthesis—an overview via different routes (*Dashed line* represents the unknown pathways and *solid lines* represent the pathways of enzymatic action)

2.2 Gibberellic Acid

The gibberellic acid (GA) production was first noticed in fungal genera *Giberella fujikuroii* and later discovered in higher plant species. The hormone induces stem elongation, early flowering/budding, breaks seed dormancy and delay senescence in plants. Free GAs, conjugated GAs and bound GAs are three major states of GAs. Natural GAs are the conjugates of β-D-glucose. About 76 derivatives of glucosyl esters of GAs are presently identified and more than 6 glucoside derivatives namely GA_1-glucoside, GA_3, GA_8, GA_{26}, GA_{27} and GA_{29} have been characterized. GA is tetracyclic diterpenoid compound and their biosynthetic pathways are quite complex (Richman et al. 1999). Gibberellin biosynthetic pathways comprise three stages according to the nature of the enzymes involved.

During first stage, isopentenyl pyrophosphate (IPP) and ent-kaurene are synthesized from mevalonic acid (Graebe 1987; Macmillan et al. 1997). Synthesized IPP is further converted into dimethylallyl pyrophosphate (DMAPP) which later on converted into geranyl-geranyl pyrophosphate (GGPP), IPP-isomerase and GGPP-synthase, localized in plastids of higher plants (Dogbo and Camara 1987). GGPP is further cyclized into ent-copalyl pyrophosphate (ent-CCP) and finally leads to ent-kaurene by CPP synthase and ent-kaurene synthase (Fig. 2). In second stage, the ent-kaurene so formed is converted into GA12. Ent-kaurene is oxidized into six steps to GA12 via ent-kaurenol, ent-kaurenal, ent-kaurenoic acid, ent-7a-hydroxykaurenoic acid and GA12-aldehyde. Microsomal NADPH-dependent cytochrome P-450 monooxygenases catalyze this intermediate in endoplasmic reticulum (Graebe 1987). 7-oxidation, 12a-hydroxylation and 13-hydroxylation are certain biosynthetic steps catalyzed by both particulate monooxygenases and soluble dioxygenases, which occasionally occur together within the same species or same tissues with few exceptions (Lange and Graebe 1993; Bearder 1983).

The final step involves the oxidation of GA12-aldehyde by 2-oxoglutarate-dependent dioxygenases to form GA12. In fact, GA 20-oxidase catalyzes the whole series of oxidation reactions carried out at carbon-20, leading to either C20-GAs (GA25), or after loss of C20 to form C19-GAs. Later, 3b-hydroxylation activates C19-GAs to plant hormone and subsequently inactivated by

Fig. 2 Gibberellin biosynthesis

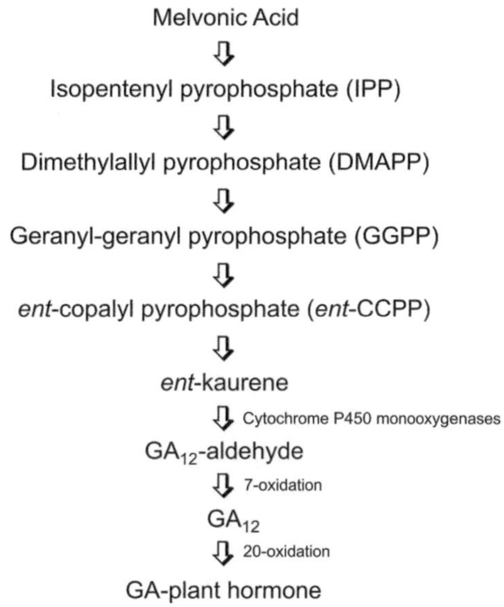

2b-hydroxylation. C20-GAs are also 2b and 3b-hydroxylated but the resulting products are GA13 and GA43 (Fig. 2).

Among bacteria, characterization of GA was first reported in *Rhizobium meliloti* (Atzorn et al. 1988). The presence of GA_1, GA_4, GA_9 and GA_{20} was also demonstrated in gnotobiotic cultures. Apart from *Rhizobium* spp., production of gibberellins was also observed in other bacterial genera *Acetobacter diazotrophicus*, *Herbaspirillum seropedicae* (Bastian et al. 1998) and *Bacillus* spp. (Gutiérrez-Mañero et al. 1998). Involvement of GA production in enhanced growth of *Pinus pinea,* is inoculated with *Bacillus licheniformis* and *B. pumilus* was reported by Probanza et al. (2002). Other than *Bacillus,* endophytic *Sphingomonas* is recently reported to enhance growth of tomato (Khan et al. 2014). Kang et al. (2014) reported that *Pseudomonas putida* modulates stress physiology of soybean and enhances its growth under saline conditions.

2.3 Cytokinin

Cytokinins play distinguish role in cell division, leaf expansion, delay senescence and induce seed germination (Mok 1994). Naturally occurring cytokinins, N6-(D2-isopentenyl) adenine (i6Ade) and Zeatin (trans-zeatin), contain hydroxylated side chain. Broadly, direct and indirect pathways have been proposed for cytokinin biosynthesis (Fig. 3).

During direct pathway of cytokinin biosynthesis, N6-isopentenyladenosine monophosphate formed from AMP and dimethylallyl pyrophosphate (DMAPP). Its

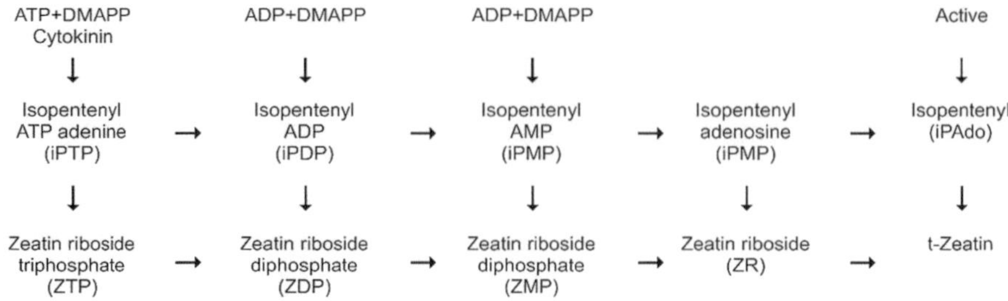

Fig. 3 Cytokinin biosynthesis

side chain hydrolyzed to form zeatin-type compounds, whereas indirect pathway involved release of cytokinin by turnover of tRNA containing cis-zeatin (Chen 1997).

Cytokinin production in *Agrobacterium* and *Pseudomonas* spp. was observed by Akiyoshi et al. (1987). Besides these genera, members of *Methylobacterium* spp. are capable to influence plant growth promotion by production of cytokinin (Lee et al. 2006; Madhaiyan et al. 2006). Earlier, Timmusk et al. (1999) observed that rhizobacteria of wheat produce cytokinin. Different bacterial genera *Proteus*, *Klebsiella*, *Bacillus*, *Escherichia*, *Pseudomonas* and *Xanthomonas* have ability to produce cytokinins (García de Salamone et al. 2001; Karadeniz et al. 2006).

2.4 Ethylene

In general, ethylene is known as hydrocarbon gas (C_2H_4) that acts as plant hormone. Resurgence and current research focus on its role in fruit ripening, inhibition of seedling growth, increase in the membrane permeability and root gravitropism.

For its biosynthesis, S-adenosyl-methionine (S-AdoMet) and ACC are the main precursors. S-AdoMet is used as a substrate for many biochemical pathways including polyamines in plants (Martin-Tanguy 2001). Initially, ethylene biosynthesis occurs by conversion of S-AdoMet to ACC by enzyme ACC synthase (S-adenosyl-L-methionine methylthioadenosine-lyase). In addition to ACC, ACC synthase (ACS) produces 5′-methylthioadenosine (MTA) which later on converted to methionine using a modified methionine cycle (Bleecker and Kende 2000) (Fig. 4).

Ethylene production has been observed in almost all seed-bearing plants. Various plant parameters such as seed germination, tissue differentiation, formation of root and shoots primordial, root elongation, lateral bud formation, flowering initiation, anthocyanin synthesis, flower opening and senescence, fruit ripening and degreening, production of aroma, leaf and fruit abscission and response of plant to biotic and abiotic stresses as the major processes influenced due to ethylene-influenced regulation that affects diverse developmental processes and stress responses (McKeon and Yang 1984; Abeles et al. 2012). Recently, few reports of bacterial species that produce plant growth modulating volatiles have been published. Blom et al. (2011) suggested the effects of bacterial volatiles highly dependent on the

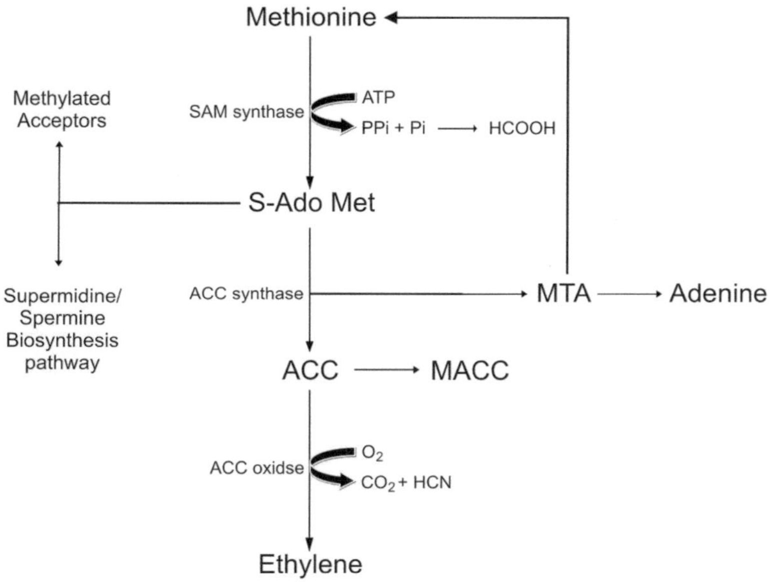

Fig. 4 Ethylene biosynthesis

cultivation medium and the inoculum quantity. However, less bacterial genera are identified to produce ethylene. Long back, Freebairn and Buddenhagen (1964) considered that ethylene might be involved in the apparent early ripening of banana fruits, which is characteristic of infection by *P. solanacearum*. Infected banana fruits indeed produce more ethylene than healthy fruits of comparable age and interestingly, no ethylene was detectable in uninfected green fruit.

On the other hand, IAA can activate the transcription of ACC synthase (Kim et al. 1992; Kende 1993). It may also inhibit IAA transport and signal transduction (Swarup et al. 2007), when ACC deaminase-containing bacteria lower the ethylene concentration in plant roots, relieve the ethylene repression of auxin response synthesis and indirectly increase plant growth. Thus, ACC deaminase-containing bacteria decrease ethylene inhibition, permitting IAA stimulation without the negative effects of increasing ACC synthase and plant ethylene levels. In addition, it acts as signaling molecule in plant protection against pathogens. Ethylene production was reported to act as a virulence factor for bacterial pathogens, e.g. *P. syringae* (Weingart and Volksch 1997; Weingart et al. 2001).

2.5 Abscisic Acid

Abscisic acid (ABA) is a naturally occurring growth inhibitor for leaf abscission and has a significant role in seed dormancy. This hormone is a sesquiterpenoid

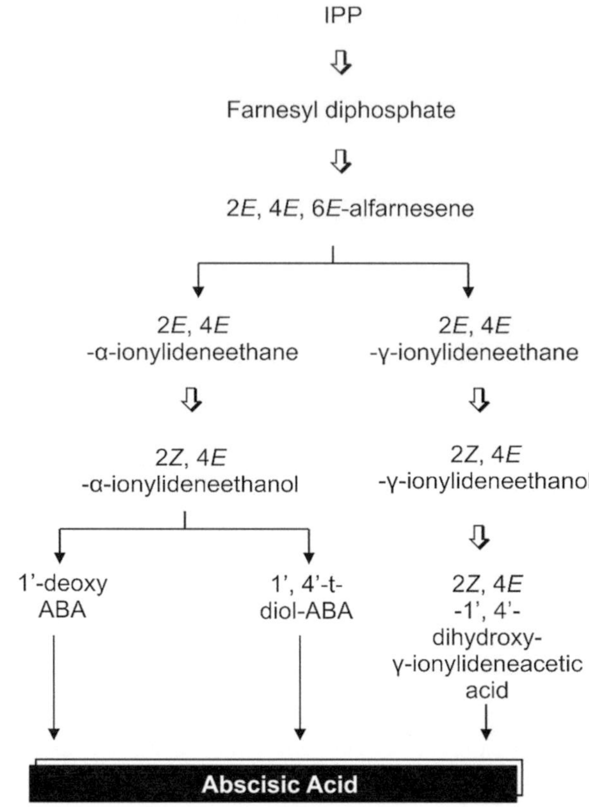

Fig. 5 Abscisic acid biosynthesis

(15-C) compound related to monoterpenes, diterpenes (including GAs), carotenoids and triterpenes. The structural feature of ABA and related compound/molecule contains a free carboxyl group, cyclohexane ring with double bond in α or β- position and C-2 double bond in cis geometry. The hormone regulates physiological processes such as stress adaptation and seed maturation. ABA is synthesized by two distinct pathways (Oritani and Kiyota 2003; Schwartz and Zeevaart 2010). Direct pathway occurs in phytopathogenic fungi in which IPP is synthesized from mevalonate pathway (MVA) (Newman and Chappell 1999), while indirect pathway is present in higher plants wherein Methylerythritol phosphate (MEP) is the source of IPP (Fig. 5).

ABA is produced by bacterial genera *Azospirillum brasilense* (Cohen et al. 2008) and *Bradyrhizobium japonicum* (Boiero et al. 2007). Available literature revealed that the effect of inoculation with ABA-producing bacteria on plant growth is under infancy. Since, ABA inhibits the synthesis of cytokinins (Miernyk 1979), it is therefore speculated that ABA increases plant growth by interfering with the cytokinin pool (Spaepen et al. 2009) and also alleviates plant stress by increasing the root/shoot ratio (Watts et al. 1981). Recent studies have reported regulation of endogenous ABA produced by PGPR of *Oryza sativa* (Belimov et al. 2014).

3 Applications

Phytohormone-producing PGPR are the free-living and associative community of bacteria in rhizosphere which encourage beneficial effects on plant health and growth, suppress disease-causing microbes and accelerate nutrient availability and assimilation. In virtue of metabolite production, survival of disease-causing organisms in its niche (rhizosphere) is quite less. Phytohormone production also corroborates plant immunity to withstand against density-dependent and density-independent stress in rhizosphere. Density-dependent stress includes the parasitic (biotic) mode of interaction brought by pathogenic fungi, bacteria and viruses, while density-independent stress mainly occurs due to abiotic factors such as temperature, pH, water, salinity, etc. Thus, PGPR are the potential candidates to protect plant by colonizing within the rhizosphere and producing antimicrobial metabolites antagonistic in nature. Phytohormones produced by such community of bacteria provide plant health and immunity by regulatory hormones (Pieterse et al. 2012).

3.1 Seed Germination, Seedling Emergence and Elongation

Consideration of phytohormone to maintain seed dormancy is circumstantial evidence and ABA is involved in regulating the onset dormancy and its state. Interaction of abscisic acid (ABA), gibberellins (GA), ethylene (ET), brassinosteroids (BR), auxin and cytokinin influences the regulation of interconnected molecular processes that control dormancy release and seed germination in dicots (Kucera et al. 2005). ABA promotes dormancy induction and maintenance, whereas GA induces progression from release through seed germination. Environmental signals regulate this balance by modifying expression of biosynthetic and catabolic enzymes include both positive and negative regulators that are mainly feedback, regulate to enhance, or attenuate the response. The net result is a slightly heterogeneous response, thereby providing more temporal options for successful seed germination (Finkelstein et al. 2008).

The benefits derived from plant–PGPR interactions are improvements of seed germination rate, root development, shoot and root weights, yield, leaf area, chlorophyll content, hydraulic activity, protein content and nutrient uptake—including phosphorus and nitrogen. PGPRs promote plant growth and development using any one, or more, of these mechanisms as elaborated in Sect. 3.4. Interestingly, PGPR may lower the plants ethylene concentration. Inhibition of seedling root length and lowering of ethylene levels in plants are through the synthesis of the enzyme 1-aminocyclopropane- 1-carboxylate (ACC) deaminase (Glick et al. 2007; Saraf et al. 2011).

Auxins such as indole acetic acid (IAA) and indole acetamide (IAM) influence root development, tissue differentiation and responses to light and gravity

(Adesemoye and Kloepper (2009). Bhatia et al. (2008) reported that IAA containing fluorescent pseudomonads increase seed germination, growth promotion and suppression of charcoal rot disease in oil seed crops. Jagadeesh et al. (2001) tested the influence of deleterious bacteria and PGPR on germination and growth of tomato in vitro. Deleterious bacteria inhibited seed germination, but PGPR (*Pseudomonas* sp. RDV 108) significantly suppressed the growth of deleterious bacteria and increased seed germination, root and shoot length in plants. Çakmakçı et al. (2007) Nitrogen fixing and phytohormone-secreting bacterial inoculant improved growth of spinach (Çakmakçı et al. 2007). Inoculation with PGPR increased shoot fresh weight, leaf area and plant height as compared with the non-inoculated control. Recently, a study done by Singh et al. (2010) showed *P. aeruginosa* PN1, which produce IAA, cyanogen, siderophore and cellulolytic enzymes when inoculated as seed dressing, resulted in increase biomass, root and shoot length in chir-pine seedlings. Increase in length of root and shoot enhanced due to significance of PGPR-mediated IAA which may prominently involve in growth promotion of several pulses and oil seed crop (Maheshwari 2008). Such observation was an agreement for stating that phytohormone-producing PGPRs have positive effect in early and increased seed germination, seed vigor index and increase in biomass with no side effect on plants.

3.2 Somatic Embryogenesis Initiation and Enhancement

Development of somatic cells into zygotic embryos is called somatic embryogenesis (SE). The combination of auxin and cytokinin induces callus formation. Auxin regulates stem cell formation during SE (Su et al. 2009). On the other hand, auxin and cytokinin regulate many processes that are critical to plant growth, development and environmental responsiveness (Jones et al. 2010). Initiation in response to auxins and cytokinins is complex due to strong interactions between these two classes of growth regulators (Hooker and Nabors 1977). Bai et al. (2013) reported that ethylene level decreased progressively during SE initiation, whereas treatment with the metabolic precursor of ethylene, 1-aminocyclopropane-1-carboxylic acid (ACC), or mutation of ethylene-overproduction1 (ETO1) disrupted SE induction. Somatic embryo production was increased more by the presence of exogenous GA_3 in the differentiation than that of induction medium. These results indicate that GA is beneficial for both embryo induction and formation. The level of endogenous gibberellins is presumably sufficient for callus induction and growth. Various PGRs involve in SE/callus induction and development signifies their role in plant development and their early growth. For example, Ruduś et al. (2002) studied the influence of exogenous GA_3 and paclobutrazol, an inhibitor of gibberellin biosynthesis, on growth of callus and SE in petiole-derived tissue cultures of *Medicago sativa* L. resulting increase in the weight of callus and number of somatic embryos. Gutiérrez-Mañero et al. (2001) studied that PGPR *B. pumilus* and *B. licheniformis* isolated from the rhizosphere of alder

(*Alnus glutinosa* [L.] Gaertn.) induce seedlings of *Quercus* species. The promotion and elongation induced by the PGPR could be mediated by bacterial GAs. Earlier, Phillips and Torrey (1970) found hormonal interactions between soybean roots and the *Rhizobium* initiating root nodule proliferation. Recently, Pallai et al. (2012) observed the ability of various strains of *P. fluorescens* that produce cytokinins involved in enhancement of roots elongation and seedling growth.

3.3 Defense Mechanism (Plant Immunity)

Various groups of pathogenic microorganisms such as fungi, bacteria, viruses, nematodes, etc. cause disease in plants. Despite of attack, plants tend to protect themselves against disease. Plant defense mechanisms (immune system) are usually multifaceted and operative against diverse array of pathogens. On the other hand, plants also utilize physical and chemical barriers to avoid pathogen entry and pathogenesis. These consist of molecular, biochemical and morphological changes, such as oxidative burst, expression of defense-related genes, production of antimicrobial compounds and/or programed cell death, lignification of tissues, thickening of cell wall, etc. (van Loon et al. 2006).

Besides other metabolites, phytohormones auxins, gibberellins (GA), abscisic acid (ABA), cytokinins (CK), salicylic acid (SA), ethylene (ET), jasmonates (JA), brassinosteroids (BR) and peptide hormones play important roles in defense mechanisms. Infection of plants with diverse pathogens altered the level of various phytohormones (Robert-Seilaniantz et al. 2006; Adie et al. 2007). Microbial pathogens have also developed the ability to manipulate the defense-related regulatory network mimicking plants by producing phytohormones resulting into hormonal imbalance causing failure of defense responses (Robert-Seilaniantz et al. 2006). Beneficial PGPR able to produce hormones involve in strengthening of induced systemic resistance (ISR) and systemic acquired resistance (SAR) of complex regulatory networks where multiple hormonal pathways interact and influence plant defense responses (van Loon et al. 1998; Pieterse et al. 2014).

In rhizosphere, PGPR antagonize pathogens through competition for nutrients, production of antibiotics and secretion of lytic enzymes (Maheshwari 2013). PGPR reduce the activity of pathogenic microorganisms not only through microbial antagonism, but also by activating the plant to better defend itself. This phenomenon was termed 'induced systemic resistance' (ISR). When plant get infected with pathogens, the activation of certain *PR* genes in some, though not all, the systemic resistance is induced by the rhizobacteria, which is similar to pathogen-induced systemic acquired resistance (SAR). In both, exogenous and endogenous productions of plant growth hormone are essential to maintain and develop disease resistance. In fact, plant's root secretes plethora of organic compounds creating a favorable niche for diverse microbial populations. The means of disease suppression by PGPRs include siderophore-mediated competition for iron, antibiosis, production of lytic enzymes and ISR, which are added advantages. The signal molecules elicit defense mechanisms in plants by activating quiescent defense genes which are present in healthy plants (Vidhyasekaran 1988a, b).

3.3.1 Induced Systemic Resistance (ISR)

ISR is triggered by PGPR without causing any adverse effect on plant system. Some lipopeptide-producing bacteria induce defense responses in plants. For example, *B. subtilis* S499 produces biosurfactant, viz., fengycins and surfactins, which in turn provide an ISR-mediated protective effect on tomato plant against *Botrytis cinerea* (Ongena et al. 2007). Varnier et al. (2009) also showed that rhamnolipids and other metabolites trigger defense responses in plants.

Bacteria-produced salicylic acid (SA) contributes to the induction of systemic resistance. Involvement of SA as a precursor of pyochelin, its role for pyochelin in ISR cannot be ruled out (Delaney et al. 1994; De Meyer and Hofte 1997). While in initial stage of SA production, it triggers resistance in iron-chelating conditions. JA has also been implicated as a signal in several defensive responses (Wasternack and Parthier 1997).

Several PGPB initiate and carried SA-dependent pathway in rhizosphere exogenously. For example, *Burkholderia phytofirmans* PsJN triggers ISR against *Botrytis cinerea* on grapevine (Ait Barka et al. 2002). Several *Pseudomonas* spp. are able to induce ISR in a wide range of plants against different pathogens (van Loon 2007). ISR is associated with an increase in sensitivity to the related hormone rather than an increase in production. This might lead to the activation of a partially different set of defense gene (Hase et al. 2003). SA, JA and ET, involved in ISR although 2,4-diacetylphloroglucinol (DAPG) are known for its antibiotic property. DAPG has dual nature as hormone and antibiotic-like substance produced by *Pseudomonas* and *Bacillus* spp. leads to physiological changes that subsequently exit ISR (Weller et al. 2012).

3.3.2 Systemic Acquired Resistance (SAR)

Three phytohormones—SA, JA and ET are known to play major role in regulating plant defense responses against various pathogens, pests and abiotic stresses. SA plays a crucial role in plant defense and is generally involved in the activation of defense responses against biotrophic and hemi-biotrophic pathogens as well as the establishment of SAR (Grant and Jones 2002). SAR devised to plant by PGPR and hormones SA, JA and 2,6-dicholoro-isonicotinic acid (INA) or benzo (1,2,3) thiadiazole-7-carbothioic acid *S*-methyl ester (BTH) required for the establishment of SAR coordination through accumulation of SA of JA in whole plant. Any disruption in the plant's ability to accumulate SA resulted in the loss of *PR* gene expression and decrease in SAR response. JA-signaling mutants sgt1b, opr3 and jin1 failed to develop SAR upon leaf infiltration with an avirulent strain of the pathogen *Pseudomonas syringae* pv. tomato, suggesting that JAs play a role in SAR as well (Cui et al. 2005). SAR is characterized by the increased expression of a large number of PR genes both in local and systemic tissues. Antimicrobial properties of PR proteins function in defense response. SAR results from the corrected effect of many PR proteins rather than specific PR proteins. SAR preceded by an increase in SA concentration which changes in redox status and the induction of defense gene expression (Mou et al. 2003; Durrant and Dong 2004).

The classical form of SAR can be triggered by exposing the plant to virulent, avirulent and non-pathogenic microbes or phytohormones. Depending upon plant and elicitor, a set period of time is required for the establishment of SAR that corresponds to the time required for the coordinated accumulation of PR (Vallad and Goodman 2004).

3.4 Plant Growth Promotion and Crop Productivity

PGPR emerged as biostimulators on account of their ability of efficient phytohormone production, which in turn contributes in plant growth and promotion (Maheshwari 2010). Plant growth and development is regulated by an array of structurally unrelated collection of plant hormones (Santner and Estelle 2009). On the other hand, lowering of plant ethylene levels by the ACC deaminase (Glick et al. 2007) is also a fortifying mechanism to provide drought tolerance to plant or remain healthy in adverse conditions. Other signal molecules are also involved in plant–microbe interactions in the form of nucleic acids, protein, lipid and polysaccharides (Halverson and Stacey 1986). Bacteria interact with plants and bacterial auxins cause interference with plant developmental processes regulated by auxin (Spaepen and Vanderleyden 2011) and affect gene expression in some microorganisms. Therefore, IAA acts as a reciprocal signaling molecule in microbe–plant interactions.

It is interesting to note that plant growth promotion is facilitated by PGPR via diverse mechanisms, due to the production and degradation of the major groups of plant hormones; although plant root exudates have many potential substrates for rhizobacterial growth including plant hormones or their precursors. Rhizobacterial mediation of plant hormone status not only showed local effects on root elongation and architecture, mediating water and nutrient capture, but also affect plant root-to-shoot hormonal signaling that regulates leaf growth and gas exchange. Combining rhizobacterial traits (or species) influences plant hormones and status, thereby, modifying root architecture (to capture existing soil resources) to make additional resources available (e.g. nitrogen fixation, phosphate solubilization) which may enhance the sustainability of crops (Dodd et al. 2010). Hence, the hormones play central role in the ability of plants to adapt to the changing environments, by mediating growth, development, nutrient allocation and source/sink transitions (Mordukhova et al. 1991; Gupta et al. 1999; García et al. 2001; Peleg and Blumwald 2011) leading to sustainable growth and development.

4 Conclusion

Phytohormones produced by PGPR are major signaling molecule employed in enhancement of crops production. IAA in majority and other hormones such as ABA, CK, ET, etc. (natural, semi-synthetic and synthetic) proved beneficial by stabilizing plant immunity, biocontrol and crop productivity. The role of phytohormone in seed dormancy, seedling emergence and elongation as well as somatic embryogenesis, initiation and enhancement bring immense need to manage increasing food production to account sustainable agriculture. Phytohormone is exploiting endogenously and exogenously in the maintenance of several physiological traits of plants. It has been revealed that some PGPR secrete novel signaling molecules that also promote plant growth. The use of rhizobacterial signaling in promoting plant growth offers a new window of opportunity especially to provide novel biological products for enhancing plant growth and development in sustainable manner.

Acknowledgments Thanks are due to UCOST (Dehradun) and UGC (New Delhi) for providing financial support in the form of research project to DKM.

References

Abeles FB, Morgan PW, Saltveit Jr ME (2012) Ethylene in plant biology. Academic press

Adesemoye AO, Kloepper JW (2009) Plant–microbes interactions in enhanced fertilizer-use efficiency. Appl Microbiol Biotechnol 85(1):1–12

Adie BA, Pérez-Pérez J, Pérez-Pérez MM, Godoy M, Sánchez-Serrano JJ, Schmelz EA, Solano R (2007) ABA is an essential signal for plant resistance to pathogens affecting JA biosynthesis and the activation of defenses in *Arabidopsis*. Plant Cell Online 19(5):1665–1681

Aeron A, Pandey P, Maheshwari DK (2010) Differential response of sesame under influence of indigenous and non-indigenous rhizosphere competent fluorescent pseudomonads. Curr Sci 99(2):166–168

Aeron A, Dubey RC, Maheshwari DK, Pandey P, Bajpai VK, Kang SC (2011) Multifarious activity of bioformulated *Pseudomonas fluorescens* PS1 and biocontrol of *Sclerotinia sclerotiorum* in Indian rapeseed (*Brassica campestris* L.). Eur J Plant Pathol 131(1):81–93

Akiyoshi DE, Regier DA, Gordon MP (1987) Cytokinin production by *Agrobacterium* and *Pseudomonas* spp. J Bacteriol 169:4242–4248

Atzorn R, Crozier A, Wheeler CT, Sandberg G (1988) Production of gibberellins and indole-3-acetic acid by *Rhizobium phaseoli* in relation to nodulation of *Phaseolus vulgaris* roots. Planta 175(4):532–538

Bai B, Su YH, Yuan J, Zhang XS (2013) Induction of somatic embryos in arabidopsis requires local YUCCA expression mediated by the down-regulation of ethylene biosynthesis. Mol Plant 6:1247–1260

Bal HB, Nayak L, Das S, Adhya TK (2013) Isolation of ACC deaminase producing PGPR from rice rhizosphere and evaluating their plant growth promoting activity under salt stress. Plant Soil 366(1–2):93–105

Bar T, Okon Y (1993) Tryptophan conversion to indole-3-acetic acid via indole-3-acetamide in *Azospirillum brasilense* Sp7. Canadian J Microbiol 39(1):81–86

Barka EA, Gognies S, Nowak J, Audran JC, Belarbi A (2002) Inhibitory effect of endophyte bacteria on *Botrytis cinerea* and its influence to promote the grapevine growth. Biol Cont 24(2):135–142

Bastian F, Cohen A, Piccoli P, Luna V, Bottini R, Baraldi R (1998) Production of indole-3-acetic acid and gibberellins A1 and A3 by *Acetobacter diazotrophicus* and *Herbaspirillum seropedicae* in chemically-defined culture media. Plant Growth Regul 24(1):7–11

Bearder JR (1983) In vivo diterpenoid biosynthesis in *Gibberella fujikuroi*: the pathway after ent-kaurene In: Crozier A (ed) The biochemistry and physiology of gibberellins, vol 1, pp 251–387

Belimov AA, Dodd IC, Safronova VI, Dumova VA, Shaposhnikov AI, Ladatko AG, Davies WJ (2014) Abscisic acid metabolizing rhizobacteria decrease ABA concentrations in planta and alter plant growth. Plant Physiol Biochem 74:84–91

Bhatia S, Maheshwari DK, Dubey RC, Arora DS, Bajpai VK, Kang SC (2008) Beneficial effects of fluorescent *pseudomonads* on seed germination, growth promotion, and suppression of charcoal rot in groundnut (*Arachis hypogea* L.). J Microbiol Biotechnol 18:1578–1583

Bleecker AB, Kende H (2000) Ethylene: a gaseous signal molecule in plants. Ann Rev Cell Developmental Biology 16(1):1–18

Blom D, Fabbri C, Eberl L, Weisskopf L (2011) Volatile-mediated killing of *Arabidopsis thaliana* by bacteria is mainly due to hydrogen cyanide. Applied Environ Microbiol 77(3):1000–1008

Boiero L, Perrig D, Masciarelli O, Penna C, Cassan F, Luna V (2007) Phytohormone production by three strains of *Bradyrhizobium japonicum* and possible physiological and technological implications. Appl Microbiol Biotechnol 74:874–880

Bottini R, Luna V (1993) Bud dormancy in deciduous fruit trees. Curr Top Plant Physiol 1:147–159

Çakmakçı R, Erat M, Erdoğan Ü, Dönmez MF (2007) The influence of plant growth–promoting rhizobacteria on growth and enzyme activities in wheat and spinach plants. J Plant Nutr Soil Sci 170(2):288–295

Cassan F, Vanderleyden J, Spaepen S (2014) Physiological and agronomical aspects of phytohormone production by model plant-growth promoting rhizobacteria (PGPR) belonging to the genus *Azospirillum*. J Plant Growth Regul 33(2):440–459

Castro-Guerrero J, Romero A, Aguilar JJ, Xiqui ML, Sandoval JO, Baca BE (2012) The hisC1 gene, encoding aromatic amino acid aminotransferase-1 in *Azospirillum brasilense* Sp7, expressed in wheat. Plant Soil 356:139–150

Chagué V, Danit LV, Siewers V, Schulze-Gronover C, Tudzynski P, Tudzynski B, Sharon A (2006) Ethylene sensing and gene activation in *Botrytis cinerea*: a missing link in ethylene regulation of fungus-plant interactions? MPMI 19:33–42

Chandra S, Choure K, Dubey RC, Maheshwari DK (2007) Rhizosphere competent *Mesorhizobium loti* MP6 induces root hair curling, inhibits *Sclerotinia sclerotiorum* and enhances growth of Indian mustard (*Brassica campestris*). Brazilian J Microbiol 38(1):124–130

Chen CM (1997) Cytokinin biosynthesis and interconversion. Physiol Plant 101(4):665–673

Cheynier V, Comte G, Davies KM, Lattanzio V, Martens S (2013) Plant phenolics: recent advances on their biosynthesis, genetics, and ecophysiology. Plant Physiol Biochem 72:1–20

Cohen AC, Bottini R, Piccoli PN (2008) *Azospirillum brasilense* Sp 245 produces ABA in chemically-defined culture medium and increases ABA content in arabidopsis plants. Plant Growth Regul 54(2):97–103

Costacurta A, Keijers V, Vanderleyden J (1994) Molecular cloning and sequence analysis of an *Azospirilium brasilense* indole-3-pyruvate decarboxylase gene. Mol Gen Genet 243:463

Crozier A, Kamiya Y, Bishop G, Yokota T (2000) Biosynthesis of hormones and elicitor molecules. In: Buchanan BB, Gruissem W, Jones RL (eds) Biochemistry and molecular biology of plants. American Society of Plant Physiology, Rockville, pp 850–929

Cui J, Bahrami AK, Pringle EG, Hernandez-Guzman G, Bender CL, Pierce NE, Ausubel FM (2005) *Pseudomonas syringae* manipulates systemic plant defenses against pathogens and herbivores. Proc Natl Acad Sci USA 102:1791–1796

De Meyer G, Hofte M (1997) Induction of systemic resistance by the rhizobacterium *Pseudomonas aeruginosa* 7NSK2 is a salicylic acid dependent phenomenon in tobacco. In: Proceedings of the IOBC/EFPP molecular approaches in biological control, 5th, Delemont

Delaney TP, Uknes S, Vernooij B, Friedrich L, Weymann K, Negrotto D, Gaffney T, Gut-Rella M, Kessmann H, Ward E, Ryals J (1994) A central role of salicylic acid in plant disease resistance. Science 266:1247–1250

Dodd IC, Zinovkina NY, Safronova VI, Belimov AA (2010) Rhizobacterial mediation of plant hormone status. Annal Appl Biology

Dogbo O, Camara B (1987) Purification of isopentenyl pyrophosphate isomerase and geranylgeranyl pyrophosphate synthase from Capsicum chromoplasts by affinity chromatography. Biochimica et Biophysica Acta (BBA)-Lipids and Lipid Metabolism 920(2):140–148

Durrant WE, Dong X (2004) Systemic acquired resistance. Annu Rev Phytopathol 42:185–209

Esitken A, Ercisli S, Karlidag H, Sahin F, Libek A, Kaufmane E, Sasnauskas A (2005) Potential use of plant growth promoting rhizobacteria (PGPR) in organic apricot production. In: Proceedings of the international scientific conference: environmentally friendly fruit growing. Polli, Estonia, 7–9 Sept 2005. Tartu University Press, pp 90–97

Finkelstein R, Reeves W, Ariizumi T, Steber C (2008) Molecular aspects of seed dormancy. Annu Rev Plant Biol 59:387–415

Freebairn HT, Buddenhagen IW (1964) Ethylene production by *Pseudomonas solanacearum*

Fulchieri M, Lucangeli C, Bottini R (1993) Inoculation with *Azospirillum lipoferum* affects growth and gibberellin status of corn seedling roots. Plant Cell Physiol 34:1305–1309

García de Salamone IE, Hynes RK, Nelson LM (2001) Cytokinin production by plant growth promoting rhizobacteria and selected mutants. Canadian J Microbiol 47(5):404–411

Glick BR, Todorovic B, Czarny J, Cheng Z, Duan J, McConkey B (2007) Promotion of plant growth by bacterial ACC deaminase. Crit Rev Plant Sci 26(5–6):227–242

Goswami D, Pithwa S, Dhandhukia P, Thakker JN (2014) Delineating *Kocuria turfanensis* 2M4 as a credible PGPR: a novel IAA-producing bacteria isolated from saline desert. J Plant Inter 9(1):566–576

Graebe JE (1987) Gibberellin biosynthesis and control. Ann Rev Plant Physiol 38(1):419–465

Grant MR, Jones JDG (2002) Hormone (Dis) harmony moulds plant health and disease. Science 324:750–752

Gupta S, Rashotte AM (2012) Down-stream components of cytokinin signaling and the role of cytokinin throughout the plant. Plant Cell Rep 31(5):801–812

Gupta VP, Sharma DD, Rekha M, Chandrashekar DS (1999) Integration of *Trichoderma pseudokoningii* with agrochemicals for disease management and plant development in mulberry. Arch Phytopathol Plant Protec 32(6):521–529

Gutiérrez-Mañero FJ, Ramos-Solano B, Probanza A, Mehouachi J, Tadeo RF, Talon M (2001) The plant-growth-promoting rhizobacteria Bacillus pumilus and Bacillus licheniformis produce high amounts of physiologically active gibberellins. Physiol Plant 111(2):206–211

Gutiérrez-Mañero FJ, Lucas JA, Barrientos ML, Ramos B, Probanza A, Acero N (1998) Effect of the rhizobacteria of *Vicia villosa* Roth. on nitrogen fixation. In: Biological nitrogen fixation for the 21st century. Springer Netherlands, pp 652–652

Halverson LJ, Stacey G (1986) Signal exchange in plant-microbe interactions. Microbiol Rev 50(2):193–225

Hartmann A, Singh M, Klingmüller W (1983) Isolation and characterization of *Azospirillum* mutants excreting high amounts of indoleacetic acid. Canadian J Microbiol 29(8):916–923

Hase S, Van Pelt JA, Van Loon LC, Pieterse CM (2003) Colonization of Arabidopsis roots by *Pseudomonas fluorescens* primes the plant to produce higher levels of ethylene upon pathogen infection. Physiol Mol Plant Pathol 62(4):219–226

Hooker MP, Nabors MW (1977) Callus initiation, growth, and organogenesis in sugarbeet (Beta vulgaris L.). J Plant Physiol 84(3):237–246

Idris EE, Iglesias DJ, Talon M, Borriss R (2007) Tryptophan dependent production of indole-3-acetic acid (IAA) affects level of plant growth promotion by *Bacillus amyloliquefaciens* FZB42. Mol Plant-Microbe Interact 20:619–626

Jagadeesh KS, Kulkarni JH, Krisharaj PU (2001) Evaluation of role of fluorescent siderophore in the biological control of bacterial wilt in tomato using Tn5 mutants of fluorescent *Pseudomonas* sp. Curr Sci 81:882–883

Jones B, Gunnerås SA, Petersson SV, Tarkowski P, Graham N, May S, Dolezal K, Sandberg G, Ljung K (2010) Cytokinin regulation of auxin synthesis in *Arabidopsis* involves a homeostatic feedback loop regulated via auxin and cytokinin signal transduction. Plant Cell 22:2956–2969

Kaminek M, Motyka V, Vankova R (1997) Regulation of cytokinin content in plant cells. Physiol Plant 101:689–700

Kang SM, Radhakrishnan R, Khan AL, Kim MJ, Park JM, Kim BR, Shin DH, Lee IJ (2014) Gibberellin secreting rhizobacterium, *Pseudomonas putida* H-2-3 modulates the hormonal and stress physiology of soybean to improve the plant growth under saline and drought conditions. Plant Physiol Biochem 84:115–124

Karadeniz A, Topcuoğlu ŞF, Inan S (2006) Auxin, gibberellin, cytokinin and abscisic acid production in some bacteria. World J Microbiol Biotechnol 22(10):1061–1064

Kato Y, Asano Y (2006) Molecular and enzymatic analysis of the "aldoxime–nitrile pathway" in the glutaronitrile degrader *Pseudomonas* sp. K-9. Appl Microbiol Biotechnol 70:92–101

Kato Y, Yoshida S, Asano Y (2005) Polymerase chain reaction for identification of aldoxime dehydratase in aldoxime-or nitrile-degrading microorganisms. FEMS Microbiol Lett 246:243–249

Kende H (1993) Ethylene biosynthesis. Annual RevPlant. Biol 44(1):283–307

Khan AL, Waqas M, Kang SM, Al-Harrasi A, Hussain J, Al-Rawahi A, Al-Khiziri S, Ullah I, Ali L, Jung H, Lee IJ (2014) Bacterial endophyte *Sphingomonas* sp. LK11 produces gibberellins and IAA and promotes tomato plant growth. J Microbiol 52(8):689–695

Kim WT, Silverstone A, Yip WK, Dong JG, Yang SF (1992) Induction of 1-aminocyclopropane-1-carboxylate synthase mRNA by auxin in mung bean hypocotyls and cultured apple shoots. Plant Physiol 98(2):465–471

King RW, Evans LT (2003) Gibberellins and flowering of grasses and cereals: prising open the lid of the "Florigen" black box. Ann Rev Plant Physiol Plant Mol Biol 54:307–328

Kittell BL, Helinski DR, Ditta GS (1989) Aromatic aminotransferase activity and indoleacetic acid production in *Rhizobium meliloti*. J Bacteriol 171:5458–5466

Kiziak C, Conradt D, Stolz A, Mattes R, Klein J (2005) Nitrilase from *Pseudomonas fluorescens* EBC191: cloning and heterologous expression of the gene and biochemical characterization of the recombinant enzyme. Microbiology 151:3639–3648

Kochar M, Upadhyay A, Srivastava S (2011) Indole-3-acetic acid biosynthesis in the biocontrol strain *Pseudomonas fluorescens* psd and plant growth regulation by hormone overexpression. Res Microbiol 162:426–435

Koga J, Adachi T, Hidaka H (1991) Molecular cloning of the gene for indolepyruvate decarboxylase from *Enterobacter Cloacae*. Mol Gen Genet 226:10–16

Kucera B, Cohn MA, Leubner-Metzger G (2005) Plant hormone interactions during seed dormancy release and germination. Seed Science Research 15(4):281–307

Kumar P, Dubey RC, Maheshwari DK (2012) *Bacillus* strains isolated from rhizosphere showed plant growth promoting and antagonistic activity against phytopathogens. Microbiol Res 167(8):493–499

Kurepin LV, Zaman M, Pharis RP (2014) phytohormone basis for the plant growth promoting action of naturally occurring biostimulators. J Sci Food Agri 94(9):1715–1722

Kurosawa N, Hirata T, Suzuki H (2009) Characterization of putative tryptophan monooxygenase from *Ralstonia solanasearum*. J Biochem 146:23–32

Lange T, Graebe JE (1993) 16 enzymes of gibberellin synthesis. Methods Plant Biochem 9:403

Lee HS, Madhaiyan M, Kim CW, Choi SJ, Chung KY, Sa TM (2006) Physiological enhancement of early growth of rice seedlings (*Oryza sativa* L.) by production of phytohormone of N_2-fixing methylotrophic isolates. Biol Fertil Soil 42(5):402–408

Lin L, Xu X (2013) Indole-3-acetic acid production by endophytic *Streptomyces* sp. En-1 isolated from medicinal plants. Curr Microbiol 67:209–217

Lin Z, Zhong S, Grierson D (2009) Recent advances in ethylene research. J Exp Botany 60:3311–3336

MacMillan J, Ward DA, Phillips AL, Sanchez-Beltran MJ, Gaskin P, Lange T, Hedden P (1997) Gibberellin biosynthesis from gibberellin A12-aldehyde in endosperm and embryos of Marah macrocarpus. Plant Physiol 113(4):1369–1377

Madhaiyan M, Poonguzhali S, Ryu J, Sa T (2006) Regulation of ethylene levels in canola (*Brassica campestris*) by 1-aminocyclopropane-1-carboxylate deaminase-containing *Methylobacterium fujisawaense*. Planta 224(2):268–278

Maheshwari DK (2008) Potential microorganisms for sustainable agriculture: a techno-commercial perspective. IK International Pvt Ltd

Maheshwari DK (2010) Plant growth and health promoting rhizobacteria. Springer-Verlag Berlin Heidelberg

Maheshwari DK (2011) Bacteria in agrobiology: crop ecosystem. Springer, Heidelberg

Maheshwari DK (2013) Bacteria in agrobiology: disease management. Springer-Verlag Berlin Heidelberg

Maheshwari DK, Aeron A, Dubey RC, Agarwal M, Dheeman S, Shukla S (2014) Multifaceted beneficial associations with *Pseudomonas* and rhizobia on growth promotion of *Mucuna pruriens* L. J Pure Appl Microbiol 8(6):4657–4667

Martinez-Viveros O, Jorquera MA, Crowley DE, Gajardo G, Mora ML (2010) Mechanisms and practical considerations involved in plant growth promotion by rhizobacteria. J Soil Sci Plant Nutr 10:293–319

Martin-Tanguy J (2001) Metabolism and function of polyamines in plants: recent development (new approaches). Plant Growth Regul 34(1):135–148

Masciarelli O, Llanes A, Luna V (2014) A new PGPR co-noculated with *Bradyrhizobium japonicum* enhances soybean nodulation. Microbiol Res 169(7–8):609–615

McKeon TA, Yang SF (1984) A comparison of the conversion of 1-amino-2-ethylcyclopropane-1-carboxylic acid stereoisomers to 1-butene by pea epicotyls and by a cell free system. Planta 160:84–87

Miernyk JA (1979) Abscisic acid inhibition of kinetin nucleotide formation in germinating lettuce seeds. Physiol Plant 45:63–66

Mok MC (1994) Cytokinins and plant development: an overview—Cytokinins: chemistry, activity, and function. CRC Press, Boca Raton, FL, pp 155–166

Mordukhova EA, Skvortsova NP, Kochetkov VV, Dubeikovskii AN, Boronin AM (1991) Synthesis of the phytohormone indole-3-acetic acid by rhizosphere bacteria of the genus *Pseudomonas*. Microbiology 60:494–500

Mou Z, Fan W, Dong X (2003) Inducers of plant systemic acquired resistance regulate NPR1 function through redox changes. Cell 113(7):935–944

Narula N, Deubel A, Gans W, Behl RK, Merbach W (2006) Paranodules and colonization of wheat roots by phytohormone producing bacteria in soil. Plant Soil Environ 52:119–129

Newman JD, Chappell J (1999) Isoprenoid biosynthesis in plants: carbon partitioning within the cytoplasmic pathway. Crit Rev Biochem Mol Biol 34(2):95–106

Oberhänsli T, Défago G, Haas D (1991) Indole-3-acetic acid (IAA) synthesis in the biocontrol strain CHA0 of *Pseudomonas fluorescens*: role of tryptophan side chain oxidase. J Gen Microbiol 137(10):2273–2279

Ongena M, Jourdan E, Adam A, Paquot M, Brans A, Joris B, Arpigny J, Thonart P (2007) Surfactin and fengycin lipopeptides of *Bacillus subtilis* as elicitors of induced systemic resistance in plants. Environ Microbiol 9(4):1084–1090

Oritani T, Kiyota H (2003) Biosynthesis and metabolism of abscisic acid and related compounds. Nat Prod Rep 20(4):414–425

Pallai R, Hynes RK, Verma B, Nelson LM (2012) Phytohormone production and colonization of canola (*Brassica napus* L.) roots by *Pseudomonas fluorescens* 6-8 under gnotobiotic conditions. Canadian J Microbiol 58(2):170–178

Pandey P, Maheshwari DK (2007) Two-species microbial consortium for growth promotion of *Cajanus cajan*. Curr Sci 92(8):1137–1142

Patten CL, Glick BR (1996) Bacterial biosynthesis of indole-3-acetic acid. Canadian J Microbiol 42:207–220

Peleg Z, Blumwald E (2011) Hormone balance and abiotic stress tolerance in crop plants. Curr Opin Plant Biol 14(3):290–295

Perley JE, Stowe BB (1966) The production of tryptamine from tryptophan by *Bacillus cereus* (KVT). Biochem J 100:169–174

Phillips DA, Torrey JG (1970) Cytokinin production by *Rhizobium japonicum*. Physiol Plant 23(6):1057–1063

Pieterse CM, Van der Does D, Zamioudis C, Leon-Reyes A, Van Wees SC (2012) Hormonal modulation of plant immunity. Annu Rev Cell Develop Biol 28:489–521

Pieterse CM, Zamioudis C, Berendsen RL, Weller DM, Van Wees SC, Bakker PA (2014) Induced systemic resistance by beneficial microbes. Phytopathology 52(1):347

Prinsen E, Costacurta A, Michiels K, Vanderleyden J, Van Onckelen H (1993) *Azospirillum brasilense* indole-3-acetic acid biosynthesis: evidence for a non-tryptophan dependent pathway. MPMI 6:609

Probanza A, García JL, Palomino MR, Ramos B, Mañero FG (2002) *Pinus pinea* L. seedling growth and bacterial rhizosphere structure after inoculation with PGPR *Bacillus* (*B. licheniformis* CECT 5106 and *B. pumilus* CECT 5105). Applied Soil Ecol 20(2):75–84

Reinoso H, Dauría C, Luna V, Pharis R, Bottini R (2002) Dormancy in peach (Prunus persica L.) flower buds VI. Effects of gibberellins and an acylcyclohexanedione (Cimectacarb) on bud morphogenesis in field experiments with orchard trees and on cuttings. Can J Bot 80:656–663

Richman AS, Gijzen M, Starratt AN, Yang Z, Brandle JE (1999) Diterpene synthesis in Stevia rebaudiana: recruitment and up-regulation of key enzymes from the gibberellin biosynthetic pathway. Plant J 19(4):411–421

Robert-Seilaniantz A, Shan L, Zhou JM, Tang X (2006) The *Pseudomonas syringae* pv. tomato DC3000 type III effector HopF2 has a putative myristoylation site required for its avirulence and virulence functions. Mol Plant Microbe Interact 19(2):130–138

Ruduś I, Kępczyńska E, Kępczyński J (2002) Regulation of *Medicago sativa* L. somatic embryogenesis by gibberellins. Plant Growth Regul 36:91–95

Salanoubat M, Genin S, Artiguenave F, Gouzy J, Mangenot S, Arlat M, Billault A, Brottier P, Camus J, Cattolico L (2002) Genome sequence of the plant pathogen *Ralstonia solanacearum*. Nature 415:497–502

Santner A, Estelle M (2009) Recent advances and emerging trends in plant hormone signalling. Nature 459:1071–1078

Saraf M, Jha CK, Patel D (2011) The role of ACC deaminase producing PGPR in sustainable agriculture. In: Mahreshwari DK (ed) Plant growth and health promoting bacteria. Springer-Berlin Heidelberg, pp 365–385

Schwartz SH, Zeevaart JA (2010) Abscisic acid biosynthesis and metabolism. Plant Hormones 137–155

Siddiqui ZA (2006) PGPR: Prospective biocontrol agents of plant pathogens. In: PGPR: biocontrol and biofertilization, pp 111–142

Singh N, Kumar S, Bajpai VK, Dubey RC, Maheshwari DK, Kang SC (2010) Biological control of *Macrophomina phaseolina* by chemotactic fluorescent *Pseudomonas aeruginosa* PN1 and its plant growth promotory activity in chir-pine. Crop Protec 29(10):1142–1147

Smaill SJ, Leckie AC, Clinton PW, Hickson AC (2010) Plantation management induces long-term alterations to bacterial phytohormone production and activity in bulk soil. Appl Soil Ecol 45:310–314

Spaepen S, Vanderleyden J (2011) Auxin and plant-microbe interactions. Perspectives in Biology Cold Spring Harbor Press 3(4):a001438

Spaepen S, Das F, Luyten E, Michiels J, Vanderleyden J (2009) Indole-3-acetic acid- regulated genes in *Rhizobium etli* CNPAF512. FEMS Microbiol Lett 291:195–200

Sponsel VM (2003) Gibberellins. In: Henry HL, Norman AW (eds) Encyclopedia of hormones, vol 2. Academic, pp 29–40

Su YH, Zhao XY, Liu YB, Zhang CL, O'Neill SD, Zhang XS (2009) Auxin-induced WUS expression is essential for embryonic stem cell renewal during somatic embryogenesis in *Arabidopsis*. Plant J 59:448–460

Swarup R, Perry P, Hagenbeek D, Van Der Straeten D, Beemster GT, Sandberg G, Bhalerao R, Ljung K, Bennett MJ (2007) Ethylene upregulates auxin biosynthesis in *Arabidopsis* seedlings to enhance inhibition of root cell elongation. Plant Cell Online 19(7):2186–2196

Theunis M, Kobayashi H, Broughton W, Prinsen E (2004) Flavonoids, NodD1, NodD2, and nod-box NB15 modulate expression of the y4wEFG locus that is required for indole-3-acetic acid synthesis in Rhizobium sp. strain NGR234. Mol Plant Microbe Interact 17:1153–1161

Timmusk S, Nicander B, Granhall U, Tillberg E (1999) Cytokinin production by Paenibacillus polymyxa. Soil Biol Biochem 31(13):1847–1852

Turan M, Ekinci M, Yildirim E, Gunes A, Karagoz K, Kotan R, Dursun A (2014) Plant growth-promoting rhizobacteria improved growth, nutrient and hormone content of cabbage (*Brassica oleracea*) seedlings. Turkish J Agri Forestry 38(3):327–333

Vallad GE, Goodman RM (2004) Systemic acquired resistance and induced systemic resistance in conventional agriculture. Crop Sci 44(6):1920–1934

van Loon LC (2007) Plant responses to plant growth-promoting rhizobacteria. Euro J Plant Pathol 119(3):243–254

van Loon LC, Bakker PAHM, Pieterse CMJ (1998) Systemic resistance induced by rhizosphere bacteria. Annual Rev Phytopathol 36(1):453–483

van Loon LC, Rep M, Pieterse CMJ (2006) Significance of inducible defense-related proteins in infected plants. Annu Rev Phytopathol 44:135–162

Varnier AL, Sanchez L, Vatsa P, Boudesocque L, García-Brugger A, Rabenoelina F, Sorokin A, Renault J, Kauffmann S, Pugin A, Clement C, Baillieul F, Dorey S (2009) Bacterial rhamnolipids are novel MAMPs conferring resistance to *Botrytis cinerea* in grapevine. Plant, Cell Environ 32(2):178–193

Vidhyasekaran P (1988a) Physiology of disease resistance in plants, vol 2. CRC Press, Inc

Vidhyasekaran P (1988b) Host enzymes and disease resistance. Physiol Dis Resist Plants 2:5–18

Wasternack C, Parthier B (1997) Jasmonate-signalled plant gene expression. Trends Plant Sci 2:302–307

Watts S, Rodriguez JL, Evans SE, Davies WJ (1981) Root and shoot growth of plants treated with abscisic acid. Ann Bot 47(5):595–602

Weingart H, Volksch B (1997) Ethylene production by *Pseudomonas syringae* pathovars in vitro and in planta. Applied Environ Microbiol 63(1):156–161

Weingart H, Ullrich H, Geider K, Völksch B (2001) The role of ethylene production in virulence of *Pseudomonas syringae* pvs. glycinea and phaseolicola. Phytopathology 91(5):511–518

Weller DM, Mavrodi DV, van Pelt JA, Pieterse CM, van Loon, LC, Bakker, PA (2012) Induced systemic resistance in *Arabidopsis thaliana* against *Pseudomonas syringae* pv. tomato by 2, 4-diacetylphloroglucinol-producing *Pseudomonas fluorescens*. Phytopathol 102(4):403–412

Went FW, Thimann KV (1937) Phytohormones. Macmillan: New York

Werner T, Schmülling T (2009) Cytokinin action in plant development. Curr Opin Plant Biol. doi:10.1016/j.pbi.2009.07.002

Xie B, Xu K, Zhao HX, Chen SF (2005) Isolation of transposon mutants from *Azospirillum brasilense* Yu62 and characterization of genes involved in indole-3-acetic acid biosynthesis. FEMS Microbiol Lett 248:57–63

Yang S, Zhang Q, Guo J, Charkowski AO, Glick BR, Ibekwe AM et al (2007) Global effect of indole-3-acetic acid biosynthesis on multiple virulence factors of *Erwinia chrysanthemi* 3937. Appl Environ Microbiol 73:1079–1088

Indole-3-Acetic Acid and 1-Aminocyclopropane-1-Carboxylate Deaminase: Bacterial Traits Required in Rhizosphere, Rhizoplane and/or Endophytic Competence by Beneficial Bacteria

Hassan Etesami, Hossein Ali Alikhani and Hossein Mirseyed Hosseini

Abstract Majority of plants harbor a diverse community of bacteria, which can positively affect host plant growth. Plant-associated bacteria have various plant growth-promoting (PGP) traits. Rhizobacteria are PGP bacteria within rhizosphere that can enhance plant growth by a wide variety of mechanisms like production of phytohormones, siderophore, 1-aminocyclopropane-1-carboxylate (ACC) deaminase and volatile organic compounds, phosphate solubilization, biological nitrogen fixation, rhizosphere engineering, quorum sensing signal interference and inhibition of biofilm formation, exhibiting antifungal activity, induction of systemic resistance, promoting beneficial plant–microbe symbioses and interference with pathogen toxin production. In recent years, interest in the use of plant growth-promoting rhizobacteria (PGPRs) to promote plant growth has increased. The use of PGPRs has steadily increased in agriculture and offers an attractive alternative to replace chemical fertilizers, pesticides and supplements. To act as PGPRs, any bacteria should be able to colonize and survive in the rhizosphere of plants. A competent colonization is essential for PGP effects produced by the bacteria and the important first step in the interaction of bacteria with plants. The purpose of this review was to give an overview on the most important PGP traits involved in plant more colonization. It seems that PGP traits of production of IAA and ACC deaminase may be required for endophytic and rhizosphere competence by PGPRs. In addition, this review indicates that the selected bacterial isolates based on their IAA and ACC deaminase-producing traits have the potential for more colonization of plants. Such bacteria may be used for a sustainable crop management under field conditions. Bacterial IAA together with ACC deaminase increase root

H. Etesami (✉) · H.A. Alikhani · H. Mirseyed Hosseini
Department of Soil Sciences, University of Tehran, Tehran, Iran
e-mail: hassanetesami@ut.ac.ir

D.K. Maheshwari (ed.), *Bacterial Metabolites in Sustainable Agroecosystem*,
Sustainable Development and Biodiversity 12, DOI 10.1007/978-3-319-24654-3_8

surface area and length, and thereby provide the plant to have greater access to soil nutrients under different environmental conditions including stress situations. Therefore, proper screening of PGPRs can be useful for future agricultural applications, providing higher production yields, reduced input costs and negative environmental impact due to the use of chemical fertilizers.

Keywords Colonization · PGPR · IAA · ACC deaminase · Plant growth-promoting traits · Rhizosphere

1 Introduction

Food security is one of the fundamental needs that can never be ignored by any society. The extensive increases in both environmental damage due to unsuitable agricultural practices and human population pressure have the unlucky consequence that global food production may soon become inadequate to feed all of the world's people. To supplement the nutritional need, it is therefore essential that agriculture becomes intensive and sustainable. In addition, the agricultural productivity must significantly increase without destroying environment within the next few decades. The development of such a global system for sustainable food production is one of the greatest challenges faced by the humans. To this end, agricultural practice is moving toward a more sustainable and environmentally friendly approach. This includes both the use of transgenic plants and plant growth-promoting rhizobacteria (PGPRs) as a part of conventional agricultural practice (Glick 2012). In both managed and natural ecosystems, PGPRs play a key role in supporting and enhancing plant health and growth (Maheshwari 2010). These bacteria are of interest for application in agriculture as biofertilizers and pesticides (biocontrol), as well as for phytoremediation applications (Bhattacharjee et al. 2008; Berg 2009; Lugtenberg and Kamilova 2009; Weyens et al. 2009). Rhizobacteria colonize plant roots and enhance plant growth through a variety of mechanisms. Based on the area of colonization, these bacteria can be grouped into associative bacteria that include rhizosphere (in the vicinity of root) rhizoplane (on the surface of root) and endophytic bacteria. Plant-associated bacteria isolated from rhizoplane and phylloplane surfaces are known as epiphytes (Andrews and Harris 2000), whereas those isolated from the interior of tissues, which they inhabit without causing harm to the host, are called endophytes (Petrini et al. 1989; Azevedo et al. 2000; Sturz et al. 2000), with some bacterial populations fluctuating between endophytic and epiphytic colonization (Hallmann 1997). There are three basic categories of microbial interactions based on ecology, namely neutral, negative and positive interactions generally exist between rhizobacteria and plants (Whipps 2001). Most of the rhizobacteria are commensals in which the bacteria establish an innocuous interaction with the host plants exhibiting no visible effect on the growth and overall physiology of the host (Beattie 2006). In negative interactions, the phytopathogenic rhizobacteria produce phytotoxic substances such as hydrogen cyanide (HCN) or ethylene, thus negatively influence on the growth and physiology of the plants (Khalid et al. 2005). In contrast to these deleterious bacteria, some PGPRs isolate can promote plant growth and development either directly or/and indirectly. Direct stimulation includes biological nitrogen fixation, producing phytohormones like auxins, cytokinins and gibberellins, solubilizing minerals like phosphorus and iron, while indirect stimulation is basically related to biocontrol, including antibiotic production, production of siderophores and enzymes and induction of systemic resistance, chelation of available Fe in the rhizosphere, synthesis of extracellular enzymes to hydrolyze the fungal cell wall and competition for niches within the rhizosphere (Zahir et al. 2004; van Loon 2007; Akhtar and Siddiqui 2008; Castro et al. 2009). Associative bacteria as well as endophytic bacteria use the same mechanisms to influence plant growth (Lugtenberg and Kamilova 2009). Since the extensive use of chemical based components can cause unanticipated environmental impacts (including nutrient imbalance, substantial economic loss to the farmers and reducing the population of beneficial microorganisms, disruption and degradation of agroecosystem and decreased soil fertility) and impart pesticide resistance in pests (Ayala and Rao 2002), interest in the use of PGPRs to promote plant growth has been increased in recent years. Based on their ability to stimulate plant growth, it is imperative to develop microbial inoculants for use in agricultural production. Depending on their mode of action and effects, these products can be used as biofertilizers (direct mechanisms) and biocontrol agents (indirect mechanisms). This application can help to minimize dependence on chemical fertilizers, which have adverse effects on the environment, finally leading to have sustainable agriculture and environment (Fig. 1).

PGPRs may use more than one of these mechanisms to enhance plant growth, as experimental evidence suggests that the plant growth stimulation is the net result of multiple mechanisms that may be activated simultaneously (Martinez-Viveros et al. 2010). Despite their different mechanisms of action, their use has not been developed to its full potential due to inconsistencies in their performance and their commercialization has been limited to a few developed countries. In many cases, PGPRs fail to induce the desired effects when applied in the field. This might be due to insufficient rhizosphere and plant colonization, which is as an important step required for exhibiting beneficial effects (Lugtenberg et al. 2001). In addition, the variability in the performance of PGPRs under in vitro and field conditions may be due to various environmental factors that may affect their growth and exert their effects on plant. The environmental factors include climate, weather conditions, soil characteristics or the composition or activity of the indigenous microbial flora of the soil (Chanway and Holl 1993; Zhender et al. 1999). To achieve the maximum growth-promoting interaction between PGPRs and plant, it is important to discover how the rhizobacteria exerting their effects on plant and whether the effects are altered by various environmental factors, including the presence of other microorganisms (Bent et al. 2001). One possible approach is to investigate soil microbial diversity for PGPRs having combination of plant growth-promoting (PGP) activities and well adapted to particular soil environment. Regardless of the mechanism of plant growth promotion, to be more effective in

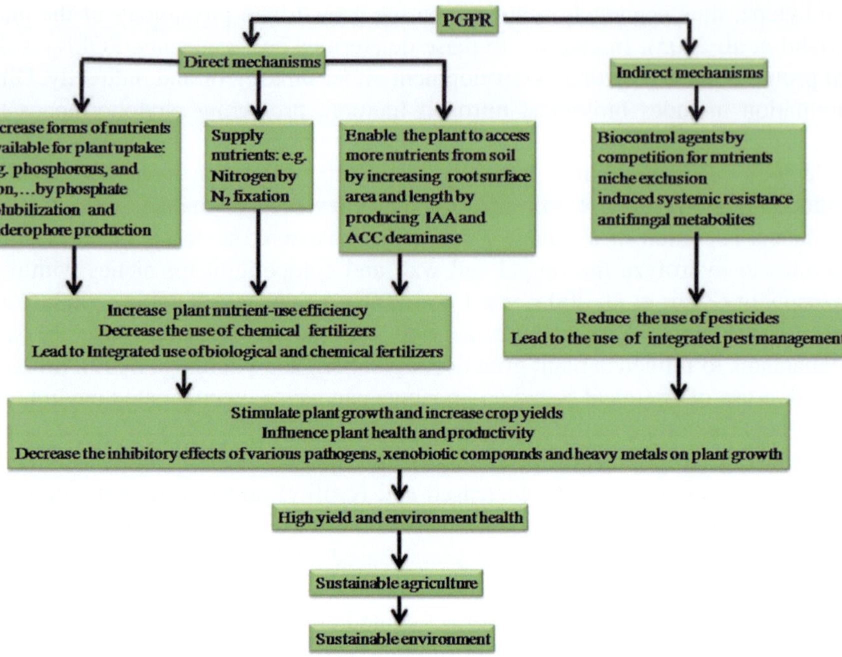

Fig. 1 The role of PGPRs using different mechanisms of action in sustainable agriculture and environment

the rhizosphere, PGPRs must maintain a critical population density for a longer period (Compant et al. 2005). In addition to these traits, PGP bacterial isolates must be rhizosphere/endophytic competence, able to survive and colonize in the rhizosphere soil (Cattelan et al. 1999; Chandra et al. 2007; Martínez-Viveros1 et al. 2010). Therefore, not only mechanisms responsible for plant growth promotion have to be investigated, but also a thorough understanding of all steps involved in plant colonization by PGPRs is required to improve the efficiency and reliability of inoculant isolates. PGP traits can be assessed under laboratory conditions and allow the selection of strains that could lead to increased plant growth (Yanni et al. 1997). Naturally, plants select PGPRs that are competitively fit to occupy compatible niches without causing pathological stress on them. Plant is restricting or directing the development of the attracted organisms in a way to keep control of these guests by excreting quite selective mixtures of substances that provide selective conditions for rhizosphere microorganisms. Furthermore, rhizosphere is a quite heavily populated microhabitat, which is characterized by competition and even predation among the inhabitants. Therefore, soil microorganisms do experience the rhizosphere environment as microhabitat of great opportunities but also of big challenges. The use of epiphytic and rhizosphere bacteria in agricultural production depends on our knowledge of the bacteria–plant interaction and our ability to maintain, manipulate and modify beneficial bacterial populations under field conditions (Hallmann 1997). The interactions that occur between plants and their associated microorganisms have long been of interest, as knowledge of these processes could lead to the development of novel agricultural applications. However, when screening bacteria for PGP agents, it is better to screen them for the most promising isolates having suitable colonization and PGP traits. In most researches, it has been seen that following incubation, bacterial flora are taken at random from Petri plates or morphological representatives are selected for further study. However, this type of selection may remove some superior bacteria of PGP traits and with high colonization ability. Gram reaction test and other phenotypic characteristics could not definitively determine the classification for the isolates. Therefore, it is essential to study all the bacteria isolated in an economic way. On the other hand, if we test all strains isolated from plants for all PGP traits, this process will take a long time and will be costly. Several methods have been used to demonstrate that root colonization is taking place, including use of fluorescence techniques, antibiotic-resistant mutants and marker genes, such as *LUX* and *GUS*. However, these methods are relatively expensive and time-consuming (Silva et al. 2003). Hence, we were interested in reviewing the previous studies for finding the most important PGP traits in selection of the isolates with more colonization and PGPR potentiality. The studies show IAA can be as a microbial metabolic and signaling molecule in microorganisms, in both IAA-producing and IAA-non-producing species (in plant–bacteria interactions). In addition, the role of bacterial IAA together with 1-aminocyclopropane-1-carboxylate (ACC) deaminase in different bacteria–plant interactions highlights the fact that bacteria use this phytohormone (together with ACC deaminase) to interact with plants as part of their colonization strategy, including phytostimulation and circumvention of basal plant defense mechanisms. It may be suggested that plants select endophytic and rhizosphere bacteria with these traits or that these bacteria harbor other traits that allow them to more effectively reach and establish themselves in rhizoplane and the inner plant tissue. This chapter will focus on the effect of IAA and ACC deaminase-producing bacteria and will provide an insight into plant–bacteria interactions.

2 Plant Growth-Promoting Rhizobacteria (PGPRs)

A diverse group of free-living soil bacteria capable of stimulating plant growth by a number of different mechanisms is known as plant growth-promoting rhizobacteria (PGPRs) (Klopper et al. 1989; Glick 1995) or yield increasing bacteria (YIB) (Tang 1994). The interactions between bacteria and plants may be beneficial, harmful, or neutral for the plant and sometimes the effect of a particular bacterium may vary as the soil conditions change (Lynch 1990). The mechanisms by which these PGPRs increase plant phytohormones, increasing the local availability of nutrients, or facilitating the uptake of nutrients by plants. They also may decrease heavy metal toxicity, antagonize plant pathogens and even induce systemic resistance in the plant against pathogens. This section will focus on plant growth promotion by PGPRs directly. There are several ways in which PGPRs can directly facilitate plant proliferation (Glick 1995) and they can be distinguished based on the modes of action of PGPRs.

2.1 Providing Nutrients for Plants

Under such conditions, PGPRs can provide the nutrients in soil, which is lacking, such as nitrogen by atmospheric nitrogen (N_2) fixation. Nitrogen (N) is the most vital nutrient for plant growth and productivity. Although, there is about 78 % N_2 in the atmosphere, soil nitrogen is mostly in organic forms and unavailable for plants. The atmospheric N_2 is converted into plant-utilizable forms by biological N_2 fixation (BNF) which changes nitrogen to ammonia by nitrogen-fixing PGPRs using a complex enzyme system known as nitrogenase (Kim and Rees 1994).

2.2 Increasing Nutrients Availability to Plants

A large proportion of nutrients are unavailable for the root uptake by plants, because the nutrients in soils are generally bound to inorganic and organic soil constituents, or alternatively present as insoluble precipitates. Therefore, in these conditions, PGPRs enhance the availability of these nutrients to growing plants by influencing solubility or uptake conditions (such as enhancing the solubility of phosphorus and iron). For example, phosphorus (P) is precipitated after addition to soil, thus becoming less available to plants (Gyaneshwar et al. 2002; Kuklinsky-Sobral et al. 2004). Despite large reservoir of P, the amount of available forms to plants is generally low. This low availability of phosphorous to plants is because the majority of soil P is found in insoluble forms, while the plants absorb it only in two soluble forms, the monobasic ($H_2PO_4^-$) and the diabasic (HPO_4^{-2}) ions (Bhattacharyya and Jha 2012). A considerable amount of phosphorus applied to soil as fertilizers is rapidly fixed into less available forms through complexation with aluminum or iron in acidic soils or with calcium in calcareous soils before plant roots have a chance to absorb it in orthophosphate form (Malboobi et al. 2009). Another PGP activity of PGPRs consists in solubilization of inorganic insoluble phosphates, transforming them into bioavailable forms. Phosphate-solubilizing bacteria (PSB) have been reported for promoting plant growth and increasing yield (Altomare et al. 1999; Barea et al. 2002; Amir et al. 2005; Canbolat et al. 2006; Khan et al. 2009). Secretion of organic acids (production of gluconic acid), proton release or production of chelating substances, exchange reactions and phosphatase enzymes are common mechanisms that facilitate the conversion of insoluble forms of phosphorous to plant accessible forms (Rodriguez and Fraga 1999; Chung et al. 2005; Zaidi et al. 2009; Gulati et al. 2010; Singh and Satyanarayana 2011). Bacteria producing trace element-chelating organic acids, such as citric, oxalic, or acetic acid have been shown to mobilize various elements in soil (Abou-Shanab et al. 2006; Li et al. 2009). Increased trace element uptake in various plants after inoculation with acid producers or PSB has been reported (Ma et al. 2011a). In aerobic conditions, iron exists primarily as ferric state (Fe^{3+}) and is largely unavailable to plants and microorganisms. Iron bioavailability is also low at neutral pH, as it is mostly in the form of insoluble Fe (III) hydroxides. Siderophores are iron-chelating secondary metabolites, which some PGPRs release under iron-limiting conditions. Siderophore production is widespread among bacteria, which can solubilize and sequester iron, making the nutrient more available to plants. All siderophores possess higher affinity for Fe (III) than for Fe (II) or any other trace element ion (Hider and Kong 2010). In general, soil microorganisms are known to affect the nutrients mobility and availability to the plant, through acidification and redox changes, or by producing iron chelators and siderophores (Burd et al. 2000; Guan et al. 2001; Abou-Shanab et al. 2003).

2.3 Enhancing Plant Greater Access to Soil Nutrients

Nutrient presence in soil and its solubility may be high, but still plants do not have any access to it due to limitations in root growth or activities. Because essential plant nutrients are taken up from the soil by roots (Mills and Jones 1996), good root growth is considered as a prerequisite for enhanced plant development. Therefore, PGPRs enhance the access of plants to the nutrient and more uptake of it by increasing the root growth (such as production of IAA and ACC deaminase). For example, applied N can be lost through nitrate leaching (Biswas et al. 2000). Previous reports have suggested positive impacts of bacteria on N uptake involving non-legume biological fixation (Boddey et al. 1995; Kennedy et al. 1997; Biswas et al. 2000a; Dobbelaere et al. 2001; Saubidet et al. 2002; Wu et al. 2005; Aseri et al. 2008). Many PGPRs cause stimulation of root growth (Biswas et al. 2000, Lucy et al. 2004), sometimes via production of phytohormones by the plant or the bacteria (Lucy et al. 2004; Shaharooma et al. 2008). If promotion of root growth by PGPRs could be achieved with high frequency in the field, PGPR may be potential tools for increasing nutrient uptake (Adesemoye et al. 2009). In general, bacterial IAA and ACC deaminase increase root surface area and length and thereby provides the plant greater access to soil nutrients and water uptake (Vessey 2003; Ryan et al. 2008).

3 Plant–Bacteria Interactions

Plant–bacteria interactions may occur at phyllosphere, endosphere and rhizosphere. Very important and intensive interactions are expected to take place among the plant environment, soil and microflora (Bringhurst et al. 2001). The term rhizospheric effect designs the fact that bacterial density is higher in the rhizosphere in comparison with non-rhizosphere soil (Foster and Rovira 1978). Although all parts of the plant are colonized by microorganisms, the rhizosphere represents the main source of bacteria with plant-beneficial activities. Biochemical interactions and exchanges of signal molecules between plants and soil microbes have been described and reviewed (Pinton et al. 2007). The plant–bacteria interactions in the rhizosphere are responsible for increasing plant health and soil fertility (Khan et al. 2006). Both aboveground and underground parts of the plants constitute an excellent ecosystem for bacterial activity and development (Bonaterra et al. 2003). The relationship between the PGPRs and their host can be categorized into two basic levels of complexity: (i) rhizospheric and (ii) endophytic. In rhizospheric relationship, the PGPRs can colonize the rhizosphere, the surface of the root or even the superficial intercellular spaces of plant roots (McCully 2001). In endophytic relationship, PGPRs reside within the apoplastic spaces inside the host plants. However, the degree of intimacy between the PGPRs and host plant can vary depending on where and how the PGPRs colonize the plant. PGPRs present in the rhizosphere play important roles in ecological fitness of their host plant. Exploring these bacteria by figuring out their possible relationships with plants, has started a new and fascinating area of investigations in the rhizosphere research. Understanding the interaction between consortium of microbial inoculants and plant systems will pave way to link more benefits from bacterial inoculants for improving plant growth and yield (Raja et al. 2006). Theoretically, the composition of microbes, which colonize the rhizosphere, can be a result of a positive or negative selection procedure or both. In many rhizospheric relationships, the PGPRs are known to colonize the plant root (Andrews and Harris 2000) and exert beneficial effects on plant growth and development by a wide variety of mechanisms.

4 Rhizosphere, Rhizoplane and Endophytic Bacterial Colonization

Root colonization includes the ability of bacteria to establish on or in the plant root, to propagate, survive and disperse along the growing root in presence of the native microflora (Whipps 2001; Lugtenberg et al. 2002; Kamilova et al. 2005; Babalola and Glick 2012). Colonization of bacteria in rhizosphere or on plant surface is a complex process which involves relationship between several bacterial traits and genes due to multistep process. Migration toward plant roots, attachment, distribution along the root as well as growth and survival of the population have all been identified as colonization determinants and have widely been studied in symbiotic, pathogenic and associative plant–microbe interactions. For endophytic bacteria, one additional step is required that is entry into root and formation of microcolonies inter- or intracellularly. Each trait may vary for different associative and endophytic bacteria (Lugtenberg and Dekkers 1999; Benizri et al. 2001;

Rodríguez-Navarro et al. 2007; Compant et al. 2010). The primary colonizers of the bacterial population are strongly influenced by the substances secreted as the root exudates and bacteria benefit from these derive nutrients (Bais et al. 2001; Dakora and Phillips 2002; Walker et al. 2003). Bacteria move toward rhizosphere in response to root exudates, which are rich in amino acids, sugars, organic acids, purines/pyrimidines, vitamins and other metabolic products. In addition to providing nutritional substances, plants start cross talk by secreting some signals which cause colonization by some bacteria while inhibits the other (Bais et al. 2006; Compant et al. 2011). Rhizospheric and/or rhizoplane and endophytic competence are a necessary prerequisite for rhizobacteria to be PGPRs (Compant et al. 2005). The root competence plays a major role in antagonistic activities of some bacteria and is very much essential to deliver the beneficial bacteria at the right place and time on the root, as poor root colonization may result in decreased biocontrol activity (Schippers et al. 1987; Weller 1988; Lugtenberg et al. 1999). Indeed, population size was reported in many works as correlated to the efficiency of biocontrol activity against plant pathogens (Bull et al. 1991). As endophytic PGPRs colonize an ecological niche similar to certain plant pathogens, they are likely candidates for biocontrol agents (Adhikari et al. 2001; Arora et al. 2001; Lacava et al. 2007). Most PGPRs with their efficient PGP potential fail to increase plant yield under field trials in agricultural soils at most of the times. Attempts to exploit PGPRs as biocontrol inoculants, biofertilizers, phytostimulants, or inoculants for bioremediation had limited success so far. This has been attributed to their incompetence to successfully colonize the rhizosphere. In field soil, environmental conditions and competition or displacement by the numerous microorganisms present in the rhizosphere limit colonization (Elliot and Lynch 1984; Thomas et al. 2008). A major factor contributing to inconsistent results from field experiments seems to be variable ecological performance (Somers et al. 2004). Many factors as nature of colonizing organism (bacterial traits), composition of root exudates, bacterial quorum sensing effects, the PGPRs environment, seasonal changes, plant tissue (Bacilio-Jimenez 2003; Mocali et al. 2003), plant species and cultivar, soil type (Kinkel et al. 2000; Fromin et al. 2001; Gnanamanickam 2006; Saleem et al. 2007), sufficient population density, root colonizing ability, PGP ability of the bacteria (Lugtenberg and Dekkers 1999), interaction with other beneficial or pathogenic microorganisms (Araújo et al. 2001; Araújo et al. 2002) and several other biotic and abiotic factors can be involved in rhizosphere and rhizoplane competence by PGPRs (Benizri et al. 2001; Gnanamanickam 2006; Saleem et al. 2007). Further, the phenomenon of chemotaxis, the nature of bacteria flagella (through motility), lipopolysaccharides (LPS) and exopolysaccharides structure, the outer membrane protein *Opr*F and to a lesser extent, presence of pili, all are important for competitive root colonization which determine the colonization of the roots by PGPRs (Lugtenberg and Bloemberg 2004; Fujishige et al. 2006; Böhm et al. 2007). Approaches aiming to enhance PGPRs root colonization have focused on the effect of abiotic factors (Howie et al. 1987) and biotic factors (Notz et al. 2001): host genotype (Baldani and Dobereiner 1980; Smith and Goodman 1999; Adams and Kloepper 2002; Arnold and Lutzoni 2007) and microbial genotypes

(Landa et al. 2002, 2003). Bacteria residing in the rhizosphere of plants may gain access into the root interior and establish endophytic populations. The endophytic colonization of host plant by bacteria reflects on their ability to selectively adapt themselves to these specific ecological niches resulting in an intimate association without any apparent harm to the plant (Sturz and Nowak 2000; Compant et al. 2005a). Exploitation of endophyte–plant interactions can result in the promotion of plant health and can play a significant role in low-input sustainable agriculture applications for both food and non-food crops. An understanding of the mechanisms enabling these endophytic bacteria to interact with plants will be essential to fully achieve the biotechnological potential of efficient plant–bacterial partnerships for a range of applications (Senthilkumar et al. 2011). Successful establishment of the introduced bacteria depends on proper PGPRs selection that must be tailored to the soil and crop combination. There has been considerable confusion over the precise effects of PGPRs, which confounds scientific studies aimed at quantifying their contribution to plant growth. This is largely due to poor understanding of the interactions between PGPRs and their plant hosts and the resident microorganisms, as well as a paucity of information on how environmental factors influence processes that contribute to plant growth promotion (Martínez-Viveros et al. 2010). Therefore, before the deliberate use of PGPRs as biofertilizers or biocontrol agents, it is necessary to know some key parameters such as root colonization capacity, location of infection and degree of persistence of the inoculum (Wiehe and Hoflich 1995). These parameters must be studied under the most realistic conditions possible. The intimacy between plants and environment in rhizosphere is essential for better acquisition of water and nutrients by plants as well beneficial interactions of plants with soil-borne microorganisms (Ryan et al. 2009). Therefore, in this section we will focus on PGP attributes of ACC deaminase and IAA as useful traits in more colonization of rhizosphere, rhizoplane and subsequent endosphere and promoting plant growth (root system) and subsequently more uptake of water and nutrients. For instance, we reported that plant growth promotion observed in rice was more pronounced with endosphere-competent *Pseudomonas fluorescens* as compared to a non-endosphere-competent isolate. This isolate produced both ACC deaminase and IAA (Etesami et al. 2014a). In general, the understanding of colonization processes is important to better predict how bacteria interact with plants and whether they are likely to establish themselves in the plant environment after field application.

5 Indole-3-Acetic Acid (IAA)

A member of the group of phytohormones, IAA is usually considered to be the most important native auxin which influences division, extension and differentiation of plant cells and tissues, stimulate seed and tuber germination, increase the rate of xylem and root development, control processes of vegetative growth and initiate lateral and adventitious roots. Auxins can mediate responses to light and gravity, florescence, fructification of plants and affect photosynthesis, pigment formation, biosynthesis of various metabolites and resistance to stressful conditions (Tsavkelova et al. 2006). Microbial production of IAA has been known for a long time (Yamada 1993; Costacurta and Vanderleyden 1995; Ludwig-Muller 2004). This property is best documented for bacteria that interact with plants because bacterial IAA can cause interference with many plant developmental processes regulated by this hormone. Many important plant–microbial interactions focus on the production of IAA detected in many pathogenic, symbiotic and free-living bacterial species (Costacurta and Vanderleyden 1995; Tsavkelova et al. 2006). Production of IAA is widespread among a wide range of soil bacteria (estimated to be ~80 % of all soil bacteria) (Khalid et al. 2004), including in streptomycetes, methylobacteria, cyanobacteria and archaea. At present, IAA-producing PGPRs are the most well-studied phytohormone producers (Tsavkelova et al. 2006; Spaepen et al. 2007). These PGP rhizobacteria produce IAA from L-Tryptophan (L-Trp) by different pathways, although it can also be synthesized via L-Trp-independent processes, though in lower quantities (Spaepen et al. 2007). Among PGPRs species, *Azospirillum* is one of the best studied IAA producers (Dobbelaere et al. 1999) and it is generally agreed that IAA production is the major factor responsible for the stimulation of root system development and growth promotion by this bacterium (Spaepen et al. 2007; van Loon 2007). Other IAA-producing bacteria belonging to *Aeromonas* (Halda-Alija 2003), *Azotobacter* (Ahmad et al. 2008), *Bacillus* (Swain et al. 2007), *Burkholderia* (Halda-Alija 2003), *Enterobacter* (Shoebitz et al. 2009), *Pseudomonas* (Hariprasad and Niranjana 2009), *Variovorax* (Belimov et al. 2005; Jiang et al. 2012) and *Rhizobium* (Ghosh et al. 2008) genera have been isolated from different rhizosphere soils. Inoculation with IAA-producing PGPRs has been used to stimulate seed germination, to accelerate root growth and modify the architecture of the root system and increase the root biomass. The ability to synthesize IAA is responsible for symbiotic associations and pathogenesis as well (Patten and Glick 1996; Khalid et al. 2004). A positive correlation between IAA production and growth-promoting activity of diverse PGPRs has been also reported in some plants (Asghar et al. 2002; Khalid et al. 2004; Etesami et al. 2013, 2014b). The root exudates and root-associated microflora are environmentally controlled sources of the IAA influx into the rhizosphere (Kravchenko et al. 1994; Muller et al. 1989; Benizri et al. 1998; Siciliano et al. 1998; Patten and Glick 2002; Badri and Vivanco 2009). Different IAA concentrations have diverse effects on the physiology of plants with plant responses being a function of the type of plant, the particular tissue involved, and its developmental stage. The actual effective range of IAA concentrations varies according to plant species and the sensitivity of the plant tissue to IAA; levels below this range have no effect, whereas higher concentrations inhibit growth (Peck and Kende 1995). For example, Evans et al. (1994) found that only exogenous concentrations between 10^{-10} and 10^{-12} M stimulated primary root elongation in *Arabidopsis thaliana* seedlings. Moreover, the endogenous pool of IAA in the plant is affected by soil microorganisms able to synthesize this phytohormone, and also the impact of bacterial IAA on plant development ranges from positive to negative effects according to the amount of IAA available to the plant and to the sensitivity of the host plant to the phytohormone. In addition, the level of IAA synthesized by the plant itself may be important in determining whether bacterial IAA will stimulate or suppress plant growth. In plant roots, endogenous IAA may be suboptimal or optimal for growth (Pilet and Saugy 1987) and additional IAA from bacteria could alter the such amount resulting in plant growth promotion or inhibition, respectively (Martínez-Morales et al. 2003; Spaepen et al. 2007). IAA biosynthesis in bacteria is affected by a number of factors including environmental stress, pH, osmotic and matrix stress, carbon starvation and the composition of the root exudates. However, due to the diversity of IAA expression and regulation according to the biosynthetic pathways and bacterial species, all of these factors cannot easily be integrated into a comprehensive regulatory scheme of IAA biosynthesis in bacteria (Spaepen et al. 2007). In general, the production of IAA seems to be one of the most prevalent PGP traits among PGPRs.

5.1 IAA and Stimulation of Plant Growth

Plant-associated bacteria can promote plant growth through modulating the level of plant hormones (Glick 1995; Lee et al. 2004; Dodd et al. 2010). Plants respond properly to environmental changes and adapt their physiology by changing hormones (IAA) levels (De Salamone et al. 2005). The ability of bacteria to produce IAA in the rhizosphere depends on the availability of biochemical precursors and uptake of microbial IAA by plant. However, the total amount of IAA produced by the plant and the bacteria should be optimum to promote plant growth. On the other hand, the production of high levels of IAA is often a main trait of plant pathogens (Rezzonico et al. 1998). Based on the integrated IAA levels produced by plant and PGPRs, a detailed examination of action mechanisms of IAA-producing bacteria in the presence and absence of ACC deaminase activity is described below (Fig. 2).

5.1.1 Stimulation of Plant Growth in the Optimal Levels of IAA Without ACC Deaminase Activity

Plants typically exude a large fraction of their photosynthetically fixed carbon through their roots. Depending on the plant species and environmental conditions, the exudated substrates can account for up to 40 % of the dry matter produced by plants. Root exudates generally contain large amounts of sugars, organic acids and amino acids (L-Trp), vitamins, nucleotides, enzymes and other plant metabolites including IAA, which represent an important source of nutrients for microorganisms in the rhizosphere. They also participate in early colonization by inducing chemotactic response of rhizospheric bacteria. Presence of these compounds is

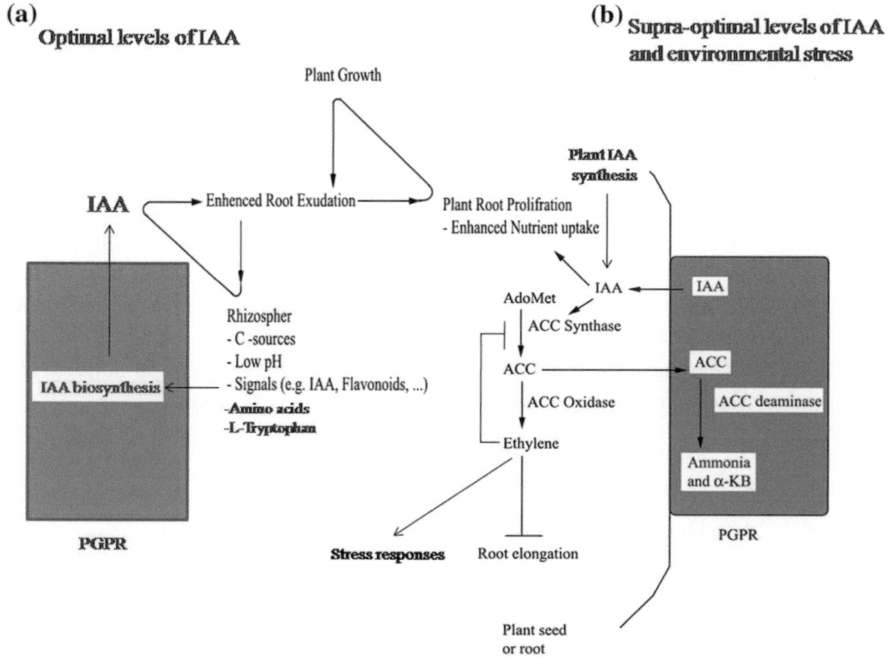

Fig. 2 A possible mechanism of how action mechanisms of IAA-producing bacteria in the presence and the absence of ACC deaminase activity. **a** Stimulation of plant growth in the optimal levels of IAA without ACC deaminase activity. In this case, IAA does not act to stimulate the synthesis of ethylene in the plant. **b** Stimulation of plant growth in the supra-optimal levels of IAA with ACC deaminase activity. In these conditions, IAA acts to stimulate the synthesis of ethylene in the plant. IAA induce the transcription of the plant enzyme ACC synthase that catalyzes the formation of ACC. AdoMet is converted to ACC by the enzyme ACC synthase; ACC is converted to ethylene by ACC oxidase. ACC synthesized in plant tissues by ACC synthase is exuded from plant roots and be taken up by ACC deaminase-producing PGPR. Subsequently, the PGPR hydrolyze ACC to ammonia and α-ketobutyrate. This ACC hydrolysis maintains ACC concentrations low in PGPR and permits continuous ACC transfer from plant roots to bacteria. Otherwise, ethylene can be produced from ACC and then cause stress responses including root elongation. Here, in the absence of ACC deaminase, root-produced ethylene inhibits transcription of IAA response factors, thereby limiting IAA stimulated plant growth as well as IAA promotion of ACC synthase transcription. In the presence of ACC deaminase, ethylene levels are decreased and the obstruction of IAA response factor transcription is alleviated thereby facilitating plant growth. Abbreviations: *ACC* 1-aminocyclopropane-1-carboxylate; *IAA* indole-3-acetic acid; *S*-AdoMet, *S*-adenosy-*L*-methionine. (Modified from Glick (2013) and Lambrecht et al. (2000))

the main reason why the numbers of bacteria in rhizosphere are 10–1000 times higher than in the bulk soil (Glick 2013). Plant-derived IAA presence or adequate amount of IAA precursor molecules in the rhizosphere could be adequate for IAA-producing bacteria to enhance the expression of the *ipd*C gene, involved in IAA biosynthesis (Lambrecht et al. 1999, 2000). An important molecule that can alter the level of IAA synthesis is the amino acid L-Trp, identified as the main precursor for IAA and thus expected to play a role in modulating the level of IAA biosynthesis. In the rhizosphere, L-Trp is originated from degrading root and microbial cells and from root exudates (Spaepen et al. 2007). In the plant root exudates, PGPRs synthesize and secrete IAA, responding to L-Trp and other small molecules. This IAA, together with endogenous plant-synthesized IAA, can stimulate plant cell proliferation and/or plant tissue elongation (increase of root growth and root length), resulting in greater root surface area. This would enable the plant to access more nutrients from soil (Jacobson et al. 1994; Boiero et al. 2007; Ortiz-Castro et al. 2009) and in turn release more exudates. This IAA can also loosen plant cell walls promoting an increase of root exudation that provides additional nutrients to support the growth of rhizosphere bacteria (James et al. 2002; Chi et al. 2005). The release of more nutrients in turn increases microbial activity and subsequently IAA, and this process continues in a cycle (Fig. 2a).

5.1.2 Stimulation of Plant Growth in the Supra-Optimal Levels of IAA with ACC Deaminase Activity

The majority of substrates for microbial activity in the rhizosphere are derived from the plant. As mentioned above in response to the presence of L-Trp and other small molecules in the plant root exudates, PGPRs synthesize and secrete IAA, some of which is taken up by the plant. The IAA produced from different pathways can induce the transcript of the plant enzyme ACC synthase that catalyzes the formation of ACC. In this case, IAA acts as a stimulator of ethylene in the plant. Along with other small molecule components of root exudates, some of the plant ACC are exuded from seeds, roots, or leaves and may be taken up by the bacteria associated with these tissues, and later cleaved by ACC deaminase (Penrose and Glick 2003) and it can be used as nitrogen (Jacobson et al. 1994; Glick et al. 1995) and carbon sources (Belimov et al. 2005). The cleavage of exuded ACC by bacterial ACC deaminase is eventually acting as a sink for ACC. Moreover, because of lowering either the endogenous or the IAA-stimulated ACC level, the amount of ethylene that could potentially form in the plant is reduced. Subsequently, by lowering plant ethylene levels, ACC deaminase-containing PGPRs can reduce ethylene inhibition in plant growth following a wide range of abiotic and biotic stresses. As a result, plants that grow in association with ACC deaminase-containing PGPRs generally have longer roots and shoots and are more resistant to growth inhibition by a variety of ethylene-inducing stresses. According to Glick (2013) as plant ethylene levels increase, the ethylene that is produced in response inhibits IAA signal transduction, thereby limiting the extent that IAA can activate ACC synthase transcription (Pierik et al. 2006; Prayitno et al. 2006; Czarny et al. 2007; Glick et al. 2007; Stearns et al. 2012). With PGPRs that both secrete IAA and synthesize ACC deaminase, plant ethylene levels do not become elevated to the same extent as when plants interact with bacteria that secrete IAA but do not synthesize ACC deaminase. In the presence of ACC deaminase, there is much less ethylene and subsequent ethylene feedback inhibition of IAA signal transduction, so that the bacterial IAA can continue to both promote plant growth and increase ACC synthase transcription. However, in this case, a large portion of the additional ACC that is synthesized is cleaved by the bacterial ACC deaminase. The net result of this cross talk between IAA and ACC deaminase is that by lowering plant ethylene levels, ACC deaminase facilitates the stimulation of plant growth by IAA (Fig. 2b).

There are some studies showing IAA and ACC deaminase work in concert to stimulate root elongation. The IAA level affecting the root system ranges from 10^{-13} to 10^{-5} M, depending on the type of root formations (primary or lateral roots, root hairs) and on the plant species (Meuwley and Pilet 1991; Taiz and Zeiger 1998; Dobbelaere et al. 1999). For example, root tissues are more sensitive to fluctuating concentrations of IAA than other plant tissues (Tanimoto 2005). The synthesis of high quantities of IAA by PGPRs has been shown to inhibit the growth of roots rather than to promote it. Primary root growth is stimulated by application of relatively low levels of IAA, typically between 10^{-9} and 10^{-12} M (Alvarez et al. 1989; Meuwley and Pilet 1991; Pilet and Saugy 1987), and is inhibited by higher IAA concentrations, likely by IAA-induced ethylene (Fig. 2b) (Peck and Kende 1995). Production of IAA by *Pseudomonas putida* GR12-2 plays a major role in the root development of canola (*Brassica rapa*) root system as evidenced by the production of roots 35–50 % shorter by an IAA-deficient mutant (Patten and Glick 2002). On the contrary, inoculation of mung bean cuttings with the mutant *aux*1 of the same strain, which overproduces IAA, yielded a greater number of shorter roots compared with controls (Mayak et al. 1999). Treatment of plants with low concentrations (up to 10^{-8} M) of exogenous IAA can enhance nodulation on *Medicago* and *Phaseolus vulgaris*, whereas higher concentrations inhibit nodulation (van Noorden et al. 2006). The combined effect of IAA on growth promotion and inhibition of root elongation by ethylene may be the explanation (Jackson 1991). The bacterial IAA from the plant stimulates the activity of ACC synthase, resulting in increased synthesis of ACC (Jackson 1991), and a rise in ethylene which, in turn, inhibited root elongation (Riov and Yang 1989). Therefore, the production of IAA by itself does not account for the capacity of PGPRs (Xie et al. 1996) in promoting growth. IAA secreted by a bacterium may promote root growth through direct stimulation of plant cell elongation or cell division or indirectly influencing bacterial ACC deaminase activity (Glick 1998; Shah et al. 1998). ACC deaminase hydrolyzes plant ACC and thus prevents the production of plant growth-inhibiting levels of ethylene (inhibitor of root growth) inside the plant because of lack of precursor ACC (Glick 1998, 2005). Mutants of PGPRs that do not produce ACC deaminase have lost the ability to stimulate root elongation (Li et al. 2000), because most IAA knock-out mutants are still able to promote plant growth, IAA biosynthesis alone is not responsible for the overall observed effect (Xie et al. 1996; Dobbelaere et al. 1999, 2003). It is possible that IAA and ACC deaminase work in concert to stimulate root elongation (Jacobson et al. 1994; Li and Glick 2001). In the additive hypothesis, it was suggested that multiple mechanisms, such as IAA biosynthesis, together with ACC deaminase activity, are responsible for the increase in plant growth promotion and yield (Bashan and Holguin 1997; Bhusan et al. 2013). In addition, some PGP traits do not work independently to each other as exemplified by IAA biosynthesis and ACC deaminase activity. Although bacterial IAA production by some ACC deaminase-containing PGPRs (Glick 1998; Glick et al. 2007a) may stimulate root growth, the creation of bacterial mutants with severely diminished ACC deaminase activity abolished their root growth-promoting effect (Glick et al. 1994; Belimov et al. 2007, 2009). Nevertheless, in vitro application of bacterial mutants with decreased ACC deaminase activity resulted in plants with longer root hairs (Contesto et al. 2008) compared to those inoculated with wild-type ACC deaminase-producing PGPRs. ACC deaminase-containing PGPRs did not affect lateral root development or root architecture in *A. thaliana* (Contesto et al. 2008), *Cucumis sativus* (Gamalero et al. 2008) and *P. sativum* (Jiang et al. 2012). In general, it may be suggested that IAA and ACC deaminase-containing PGPRs can lead to better growth of plants than PGPRs producing ACC deaminase or IAA alone. For example, IAA and ACC deaminase-producing *Variovox paradoxus* 5C-2 stimulated root hair elongation of tomato and pea (*Pisum sativum*) in vitro by producing IAA and decreasing ACC concentrations via ACC deaminase activity (Belimov et al. 2005, 2009a; Belimov 2012; Jiang et al. 2012).

6 IAA as a Signaling Molecule in Bacteria

IAA is important in plant–bacteria interactions and may be involved at different levels in plant–bacteria interactions (Costacurta and Vanderleyden 1995; Bashan and Holguin 1998; Patten and Glick 2002; Molina-Favero et al. 2008). IAA acts as a signaling molecule in microorganisms including bacteria (Bianco et al. 2006; Liu and Nester 2006; Yang et al. 2007; Yuan et al. 2008; Spaepen et al. 2009) because it affects gene expression in some microorganisms. Extensive communication occurs between plants and bacteria during different stages of plant development in which signaling molecules from the two partners play an important role. Bacteria are capable to detect the plant host and initiate their colonization strategies in the rhizosphere by producing growth-regulating substances such as IAA. On the other hand, plants are able to recognize microbe-derived compounds and adjust their defense and growth responses according to the type of microorganism encountered. This molecular dialog will determine the final outcome of the relationship, ranging from pathogenesis to symbiosis, usually through highly coordinated cellular processes (Bais et al. 2004). IAA like quorum sensing molecules may play a role in plant–bacterial signaling (Loper and Schroth 1986; Idris et al. 2007; Phi et al. 2008; Van Puyvelde et al. 2011). For example, IAA triggers a broad gene expression response in *Azospirillum brasilense* (Van Puyvelde et al. 2011) and IAA synthesis is controlled by a positive feedback transcriptional mechanism (Vande Broek et al. 1999). In addition to the hypothesis that bacterial IAA contributes to evade the host defense by derepressing the IAA signaling in the plant, IAA also have a direct effect on bacterial survival and its resistance to plant defense (Remans et al. 2006). Evidence has been accumulating that some microorganisms, independent of their ability to produce IAA, make use of IAA as a signaling molecule steering microbial behavior. These results led to the speculation that signaling by indole may have a role in adaptation of bacterial cells to a nutrient-poor environment where amino acid catabolism is an important energy source (Wang et al. 2001). Other targets of indole mediated signaling were found signifying a role for indole signaling in biofilm formation (Domka et al. 2006). Other evidence has accumulated indicating that classic plant signals such as IAA can be produced by microorganisms to efficiently colonize the root and control root system architecture (Randy et al. 2009). Many studies have shown that bacterial IAA is known as an effector's molecule in plant–bacteria interactions, both in pathogenesis and phytostimulation. It has been shown that bacterial IAA biosynthesis contributes to colonization capacity and fitness on the host. A low IAA-producing mutant of *P. fluorescens* HP72 is reduced in colonization ability on bent grass roots as compared with the wild-type (Suzuki et al. 2003). It is logical to postulate that bacteria use IAA as part of their colonization strategy by stimulating proliferation of plant tissues and thus enhanced colonization surface and exudation of nutrients for bacterial growth. Some similarity exists between IAA signaling in bacteria–plant interactions, in which IAA is produced by both partners, and signaling by bacterial quorum sensing molecules in bacteria–host interactions (Spaepen et al. 2007). However, the ecological significance of IAA production by bacteria would be more conclusive if it could be established that bacterial IAA production occurs while bacteria colonize the root system. As both the plant and the bacteria synthesize and secrete IAA, it is difficult to address the contribution of one particular hormone responsible for the effects observed (Spaepen et al. 2007). Nevertheless, it seems bacterial IAA, together with endogenous plant-synthesized IAA may have significantly affected plants and bacterial colonization as mentioned above (Fig. 2).

7 Bacterial IAA in Endophytic and Rhizosphere Colonization

The IAA-producing PGPRs can stimulate root growth and seed germination, modify the architecture of the root system, enhance root exudates and eventually increase the root biomass. These bacteria can facilitate more colonization of endophytic and rhizosphere PGPRs. Enhanced root system and exudates in turn have many other effects as shown in Table 1.

7.1 IAA in Endophytic Bacterial Colonization

Endophytic bacteria can be defined as those bacteria that colonize the internal tissue of the plant showing no external sign of infection or negative effect on their host (Ryan et al. 2008). These bacteria significantly affect plant growth by different mechanisms, which is similar to those used by associative bacteria

Table 1 Effects of resulting from increasing root exudates and root system by IAA-producing PGPRs

Increasing root exudates	Increasing root system
• Affecting growth and metabolism of biocontrol agents	• Enhancing the plant access to nutrients
• Altering the diversity and activity of plant-associated microbes	• Increasing plant growth
• Serving as important nutrients, attractants and deterrents	• Increasing root exudates
• Mobilizing nutrients (toxic/essential ions) such as phosphorus and micronutrient and/or metal immobilization	
• Complexation of toxic and essential ions and increase their mobility for plant uptake	
• A major driving force for microbial root colonization	
• Prolonging metabolic activity	
• Extending colonization persistence	
• Influencing on overall biological control performance	
• Effecting on the physical and chemical properties of the soil and on the indigenous microflora	
• Uptake of nutrient ions by the plant	
• Supporting higher populations of microflora	

(Lugtenberg and Kamilova 2009). Numerous endophytes are actively involved in the synthesis of IAA in pure culture and in plants and increased root growth and root length, resulting in greater root surface area that enables the plant to access more nutrients from soil (Jacobson et al. 1994; Boiero et al. 2007). Production of pectinase and cellulase (pectinolytic activity) are common features of endophytic bacteria (Elbeltagy et al. 2000) responsible for plant invasion by them (Teaumroong et al. 2001). Endophytic bacteria may colonize root tissues and spread actively in aerial parts of plants through expressing moderate amount of degradative enzymes (pectinases and cellulases) (Adriano-Anaya et al. 2006). Utilization of previously mentioned enzymatic activities for colonization by PGPRs has been revealed as one of the efficient methods to get entry into the host plant. Endoglucanase is one of the major determinants for the colonization of endorhizosphere, which was evident from the observation that *Azoarcus* strain lacking endoglucanse was not effective in colonizing the rice plants. The endoglucanase loosens larger cellulose fibers, which may help entering into the plant. However, in our studies, most of the root and rhizosphere isolates produced pectinases and cellulases and some of the isolates were not positive for activity of cellulases and pectinases (Etesami et al. 2014b). In addition, genes encoding plant cell wall degrading enzymes have not been found in endophytic bacteria *Herbaspirillum seropedicae* strain SmR1 (Pedrosa et al. 2011). Previous studies that have shown invasion can happen through lesions particularly occurring on the lateral or adventitious roots. This is through root hairs and between undamaged epidermal cells fissures at the lateral root base and by cortical, intercellular crack entry (Chaintreuil et al. 2000; Sevilla et al. 2001; James et al. 2002). Chi et al. (2005) demonstrated that the colonization of *gfp*-tagged rhizobia in crop plants begins with surface colonization of the rhizoplane at lateral root emergence, followed by endophytic colonization within roots and then ascending endophytic migration into the stem base, leaf sheath and leaves where they develop high populations. *Azospirillum* may also colonize endophytically through wounds and cracks of the plant root (Reinhold-Hurek and Hurek 2011). The colonization of the interior of plant roots by microbial endophytes appears as a most attractive goal, because their plant nutrient resources can be explored even more effectively without the tough competition with the high number of other microbes colonizing the root surface and environment (Rosenblueth and Martinez-Romero 2006; Schulz et al. 2006). However, in this case, the efficient interaction with the plant host gets even more important. The success of invasion and survival within the host also requires that bacteria overcome plant defense responses prompted after microbial recognition, a process in which surface polysaccharides, antioxidant systems, ethylene biosynthesis inhibitors and virulence genes are involved (Soto et al. 2006). However, it can be speculated that IAA production trait is part of the strategy used by IAA synthesizing bacteria to bypass the plant defense system. It has been observed previously that IAA interfere with parts of the host defense system. IAA is able to block several pathogenesis-related (PR) enzymes, including β-glucanase (Mohnen et al. 1985; Jouanneau et al. 1991; Lim and Kim 1995) and chitinase (Shinshi et al. 1987) at the mRNA level. The link between plant defense and IAA signaling gives an extra dimension to the role of bacterial IAA in colonization ability (Spaepen et al. 2007). The capacity to synthesize IAA is common among endophytic bacteria. Most of endophytic diazotroph isolates (62.75 %) in the study conducted by Teaumroong et al. (2001) also produced a significant amount of IAA. Endophytic bacterial isolates from *Thai* rice also showed a high N_2-fixation potential and were able to produce PGP substances such as IAA (Teaumroong et al. 2001). This suggests that the ability of IAA production may help IAA-producing or IAA-non-producing bacteria (with and without pectinolytic activity) invade inside plant roots. In such a process, IAA which is a plant hormone with no apparent function in bacterial cells could improve the fitness of the plant–bacterium interaction. Brandl and Lindow (1998) have studied the contribution of IAA for bacterial epiphytic fitness, and their observations were supported by the investigations of other workers (Glick 1995; Dobbelaere et al. 1999; Verma et al. 2001). Since the first step of bacteria invasion in plant root comprises of the attachment of isolates onto epidermal cells of the root surface, where root hair zone shows one of the major sites of primary colonization (mainly on the basal region of emerging hairs), it is possible that IAA-producing bacteria by increased root system can colonize plant roots better than other bacteria (Katherine et al. 2008; Prieto et al. 2011). In addition, IAA levels weaken plant defense mechanisms making colonization easier. Bacterial IAA can loosen plant cell walls and as a result promotes an increase in root exudation that provides additional nutrients to support the growth of rhizosphere bacteria (James et al. 2002; Chi et al. 2005). Since endophytic microbial communities originate from the soil and rhizosphere (Hallmann 1997; Sturz et al. 2000; Elvira-Recuenco and Van Vuurde 2000), bacterial IAA can attract more rhizosphere bacteria by increasing root exudation. Bacterial IAA stimulates the development of the root system of the host plant (De Salamone et al. 2005) and IAA-producing isolates can improve the fitness of plant–microbe interactions (Brandl and Lindow 1998; De Salamone et al. 2005). Mendes et al. (2007) showed most of the IAA-producing isolates were found among the stem endophytes, followed by root endophytes and rhizosphere isolates. Previous studies indicate higher frequency of IAA-producing bacteria in root compared to rhizosphere (Kuklinsky-Sobral et al. 2004; Mendes et al. 2007; Etesami et al. 2014b). The observation that the frequency of IAA-producing bacteria is higher in the roots than in the rhizosphere of plants suggests that plants select for endophytic bacteria with this trait or that IAA-producing bacteria harbor other traits that allow them to more effectively reach and establish themselves in the inner plant tissue (Mendes et al. 2007). IAA of microbial origin in the interior of plants could induce a physiological response in the host plant. Therefore, screening of the endophytes for their in vitro potential of IAA production could provide a reliable base for selection of effective PGP bacteria (Patten and Glick 2002; Etesami et al. 2015). In general, IAA-producing bacteria by increasing root system and root exudates can have effective role in colonization themselves or other bacteria inside or on plants, explained separately in the following sections.

7.1.1 IAA and Root Exudates

One of the main effects of bacterial IAA is the enhancement of lateral and adventitious rooting leading to improved nutrient uptake and root exudation that in turn stimulates bacterial proliferation on the roots (Tien et al. 1979; Fallik et al. 1988; Xie et al. 1996; Okon and Vanderleyden 1997; Dobbelaere et al. 1999; Lambrecht et al. 2000; Steenhoudt and Vanderleyden 2000; Himanen et al. 2002; Tsavkelova et al. 2007). Rhizosphere and rhizoplane colonization and after that endophytic colonization has been described to be linked to root exudation (Lugtenberg and Dekkers 1999). Carbon fixed by plant photosynthesis is known to be partly translocated into the root zone and released as root exudates (Bais et al. 2006). Various carbohydrates, amino acids (L-Trp), organic acids, as well as other compounds, which provide a source of nutrients for root-associated bacteria, are released in the rhizosphere (Jones 1998; Walker et al. 2003). Microorganisms are known to be chemoattracted and move toward exudates, allowing them to colonize and multiply both in the rhizosphere and in the rhizoplane (Lugtenberg and Kamilova 2009). It is known that bacterial IAA can loosen plant cell walls and as a result promotes an increasing amount of root exudation that provides additional nutrients to support the growth of rhizosphere bacteria (James et al. 2002; Chi et al. 2005). Many compounds present in the root exudates stimulate a positive chemotactic response in bacteria (Somers et al. 2004; Kumar et al. 2007a). Being a major driving force for microbial root colonization, plant root exudation could stimulate microbial colonization on the roots. In addition, greater exudation or nutrient availability may prolong metabolic activity, extend colonization persistence and enhance expression of certain traits (Pielach et al. 2008). Overall, bacterial IAA increases root surface area and length and thereby provides the plant with greater access to soil nutrients. In addition, IAA stimulates overproduction of root hairs and lateral roots in plants and release of saccharides from plant cell walls during the elongation (Davies 2004). Saccharides are a source of nutrients for microorganisms and can increase the colonization ability of plant-associated bacteria (Lindow and Brandl 2003). Failure of PGPRs to produce a desired effect after seeds inoculation is frequently associated with their failure to colonize plant roots (Benizri et al. 2001). The host plants may provide a satisfactory environment for bacteria to proliferate and produce excessive amounts of IAA, thus weakening the plant and promoting root colonization. Since bacterial attachment to plant surfaces begins with attraction by seedling root exudates (Begonia and Kremer 1994; Bellis and Ercolani 2001), bacterial IAA can increase colonization by loosening plant cell walls and as a result facilitating an increasing amount of root exudation. IAA may also regulate root exudation through changing plasmalemma permeability (Brandl and Lindow 1998). It was hypothesized that the production of rhizobacterial IAA contributes to circumvent the plant defense system by depressing auxin signaling (Spaepen et al. 2007). The expression of IAA biosynthesis genes in bacteria colonizing the plant root zone testifies to the importance of IAA production for this colonization (Rothballer et al. 2005). As reviewed by Spaepen et al. (2007), regardless of their ability to produce IAA, bacteria can use the phytohormone as a signaling molecule to trigger the expression of genes related to survival under stress. Therefore, IAA can be involved both in the establishment of plant–bacteria associations and in the regulation of their functioning under changing environmental conditions. Since endophytic microbial communities originate from the soil and rhizosphere (Hallmann 1997; Sturz et al. 2000; Elvira-Recuenco and Van Vuurde 2000), bacterial IAA can attract more rhizosphere bacteria and as a result endophytic bacteria by increasing more amount of root exudation. As the amount of photosynthates secreted as root exudates varies with the type of soil and the availability of nutrients (Kraffczyk et al. 1984; Paterson and Sim 2000), the effect of bacterial IAA in the amount of root exudation and subsequently root colonization can also be different under changing conditions.

7.1.2 IAA and Root System

Bacterial IAA plays a major role in promotion of root elongation when a bacterium is associated with its host plant (Dangar and Basu 1987; Lynch 1990; Arshad and Frankenberger 1991; Glick 1995; García de Salamone et al. 2001; Gutiérrez-Mañero et al. 2001; Persello-Cartieaux et al. 2003; Dobbelaere et al. 2003; Vivas et al. 2005). Promotion of root growth is one of the major markers by which the beneficial effect of PGPRs is measured (Glick 1995). Almost all endophytic bacteria were also found in the rhizosphere, thus supporting the hypothesis that there is a continuum of root-associated bacteria from the rhizosphere to rhizoplane to epidermis and cortex (Kloepper and Beauchamp 1992; Quadt-Hallman et al. 1997). This might explain the close relationship between endophytic and rhizosphere colonizing bacteria. Except for bacteria transmitted through seeds, potential endophytes must first colonize the root surface prior to entering the plant. Potential internal colonists find their host by chemotaxis, electrotaxis, or accidental encounter. Lipopolysaccharides, flagella, pili and twitching motility (Dörr et al. 1998; Böhm et al. 2007) have been shown to affect endophytic colonization and bacterial mobility within host plants. Motility of beneficial associative PGPRs has been described for several bacteria such as *Alcaligenes faecalis*, *A. brasilense* and *P. fluorescens* (Bashan 1986; You et al. 1995). In addition, the secretion of cell wall degrading enzymes is involved in bacterial penetration (Lodewyckx et al. 2002) and spreading within the plant. The penetration process does not necessarily involve active mechanisms and thus all rhizosphere bacteria can be expected to be endophytic at one stage of their life (Hardoim et al. 2008). Entry into a plant tissue can also be via the stomata, lenticels, wounds (including broken trichomes), areas of emergence of lateral roots and emerging radicles. However, the main entry for endophytic bacteria appears to be through wounds that naturally occur because of plant growth or through root hairs and at epidermal conjunctions (Reinhold-Hurek and Hurek 1998). Several authors have reported extensive colonization of the secondary root emergence zone (site of root branches) by bacterial endophytes (Hallmann 1997). The fact that colonization is especially abundant in root tissue may reflect the fact that the root is the primary site where endophytes gain entry into plants. A criterion for some endophytes to colonize the plant is thus must find their way through cracks formed at the emergence of lateral roots or at the zone of elongation and differentiation of the root. During the colonization process, migration of bacteria toward roots is dependent on active motility of bacteria and passive movement of bacteria in percolating water, on vectors, or via carrying and deposition by elongating root tips (Parke 1991; Walker et al. 2002; Bowen and Rovira 1991). Percolating water may enhance root colonization due to the transport and spread of bacteria. Root elongation and expansion can also be involved in transporting bacteria down the root. IAA together with ACC deaminase activity can help transport bacteria by increasing root elongation. In addition, there are many independent evidences using microbiological and molecular techniques indicating that roots stimulate soil microbial communities selectively creating unique rhizosphere communities (Duineveld et al. 1998; Marschner et al. 2001; Rengel and Marschner 2005). IAA by increasing root system may help this selection. In view of function of bacterial IAA in increased root system, it is proposed that IAA-producing bacteria can provide more number of active sites and access to colonization for other PGPRs. For example, the presence of PGPRs in the root vicinity could improve ability of rhizobia to compete with indigenous populations for nodulation. Parmar and Dadarwal (2000) reported that increase in root growth provides more number of active sites and access to nodulation for rhizobia in chickpea.

7.2 IAA in Epiphytic Bacterial Colonization

The biosynthesis of IAA is widespread among bacterial colonizers of the phyllosphere (Fett et al. 1987; Glickmann et al. 1998; Lindow et al. 1998; Brandl et al. 2001). Because IAA is involved in many aspects of plant development, it is of great importance that bacteria which colonize plant surfaces have the ability to synthesize an IAA matching that found in plants. Many studies reported the contribution of IAA for bacterial epiphytic fitness (Glick 1995; Patten and Glick 1996; Bastián et al. 1998; Brandl and Lindow 1998; Dobbelaere et al. 1999; Verma et al. 2001). It is hypothesized that the secretion of IAA may modify the microhabitat of epiphytic bacteria by increasing nutrient leakage from plant cells; enhanced nutrient availability may better enable IAA-producing bacteria to colonize the phyllosphere and may contribute to their epiphytic fitness (Brandl et al. 1996). In competition experiments, an IAA-producing strain of *Pantoea agglomerans* reached twice the population size of an isogenic IAA-deficient mutant on pear flowers in the field and on bean plants in the greenhouse (Brandl and Lindow 1998). This increase in the ratio of the population size of the parental strain over that of the IAA-deficient mutant occurred only during periods of active colonization of the plants. IAA production in *P. agglomerans* was also associated with increased fitness during periods of drought stress on plants (Manulis et al. 1998). IAA stimulates the release of saccharides from the plant cell wall (Goldberg 1980; Vanderhoff and Dute 1981; Fry 1989). Because bacteria on plants are frequently nutrient limited (the nutrient concentration including glucose and other sugars on leaves ranges from 3 to 20 mg L^{-1}) (Chet et al. 1973; Fokkema and Lorbeer 1974), it is hypothesized that the greater epiphytic fitness of IAA-producing strains resulted from enhanced nutrient availability caused by increased leakage of saccharides from plant cells in their vicinity. Brandl et al. (1996) showed a similar release of nutrients from plant cells in response to IAA produced by epiphytic bacteria on plants, which convene upon a selective advantage. Brandl and Lindow (1998) conducted the epiphytic fitness of strains *Erwinia herbicola* 299R and 299XYLE, an isogenic IAA-deficient mutant of strain 299R, evaluated in greenhouse and field studies by analysis of changes in the ratio of the population sizes of these two strains after inoculation as mixtures onto plants. Populations of the parental strain increased to approximately twice those of the IAA-deficient mutant strain after co-inoculation in a proportion of 1:1 onto bean plants in the greenhouse and onto pear flowers in field studies. They showed that IAA synthesis could contribute to the growth of strain 299R on plant surfaces. Their results clearly indicate that a benefit of IAA production occurs primarily when cells can exploit resources in the phyllosphere for further growth. Work performed with the non-pathogenic *E. herbicola* 299R strain showed that *ipd*C transcription increased 32-fold *in planta* on leaves of bean and tobacco and 1000-fold on pears flowers (Brandl and Lindow 1997). Studies involving with wild-type and *ipd*C mutant have demonstrated that IAA production contributed to epiphytic fitness of the bacteria on bean plants and pear blossoms, because the *ipd*C mutants exhibited a tenfold reduced fitness when compared to wild-type strain (Brandl and Lindow 1998). This change in the proportion of IAA-producing to IAA-deficient strains in mixed populations on leaves appears also to reflect a plant specific benefit of IAA production, since no difference in the growth of these two strains was noted in culture. They concluded that this benefit may be mediated by the increased leakage of nutrients from plant cells in the vicinity of IAA-producing bacteria colonizing the plant surface. Another example is *E. herbicola*, a common colonist on plant surfaces such as leaves and buds. *E. herbicola* produces IAA through L-Trp-independent pathways. IAA can increase colonization of plant surfaces by this epiphyte (Brandl and Lindow 1996; Lindow and Brandl 2003). Earlier, Varvaro and Martella (1993) have shown that IAA-deficient mutants of *Pseudomonas syringae* pv. *savastanoi*, obtained by selection for resistance to α-methyltryptophan, reduced in their ability to colonize and survive on olive leaf surfaces. They also tested the importance of IAA production in bacterial colonization of bean leaves with the brown spot pathogen *P. syringae* pv. *syringae* and an IAA-deficient mutant derived by insertional mutagenesis (Mazzola and White 1994). Their results showed IAA biosynthesis is not essential for bacterial growth and survival, since IAA-deficient mutants as well as their IAA-producing parental strain grew in vitro (Brandl and Lindow 1996; Smidt and Kosuge 1978). Increased transcriptional activity of *ipd*C during the growth of *E. herbicola* 299R on plant surfaces provides some evidence for the bacterial production of IAA in the phyllosphere (Brandl and Lindow 1997, 1998). Their results thus indicate that bacterial IAA synthesis can affect the normal physiology of plant cells. Exogenously applied IAA can stimulate the release of large quantities of monosaccharides and oligosaccharides from the plant cell wall (Fry 1989; Goldberg 1980). Therefore, IAA-producing bacteria may modify their microhabitat or the microhabitat of other bacteria by increasing nutrient leakage from plant cells; enhanced nutrient availability may better enable them to colonize the phyllosphere and may contribute to their epiphytic fitness.

8 IAA and Solubilization of Phosphorous

After nitrogen, the essential mineral element that most frequently limits the growth of plants is phosphorus (P), which only is taken up in monobasic ($H_2PO_4^-$) or dibasic (HPO_4^{2-}) soluble forms (Glass 1989). Although soils generally contain a large amount of total P but only a small fraction is available for plant uptake (Khan et al. 2006). Substantial amounts of phosphate fertilizers are applied to agricultural soils due to relative immobility of phosphate and its very low concentration in soil solutions. This results in an accumulation of large quantities of total phosphorus in the soil, of which 20–80 % is in organic form (Richardson 1994). However, plants are well adapted to uptake of P from low concentration soil solution (Jungk 2001). Therefore, it is presumed that the supply and availability of P to the root surface is influenced by the root and microbial processes. The plant-associated microorganisms improve the plant nutrient acquisition by mobilizing nutrients and making it available to plant roots. An example is the P-solubilizing bacteria, which dissolves various sparingly soluble P sources such as $Ca_3(PO_4)_2$ (Rodriguez et al. 2004) and $Zn_3(PO_4)_2$ (Saravanan et al. 2007) through lowering pH of the rhizosphere soil and making P available for plant uptake. The increased plant growth and P uptake have been reported on the inoculations of P-solubilizing *Pseudomonas* sp. in wheat (Babana and Antoun 2006), *Pantoea* J49 in peanut (Taurian et al. 2010) and *Psychrobacter* sp. SRS8 in *Ricinus communis* and *Helianthus annuus* (Ma et al. 2010). Furthermore, presence of high levels of heavy metals in soil interferes with P uptake and lead to plant growth retardation (Zaidi et al. 2006). Under metal stressed conditions, most metal-resistant PGPRs (specially ACC deaminase-producing bacteria) can either convert these insoluble phosphates into available forms through acidification, chelation, exchange reactions and release of organic acids (Chung et al. 2005) or mineralize organic phosphates by secreting extracellular phosphatases (Gyaneshwar et al. 2002; van der Heijden et al. 2008). As mentioned above, PGPRs stimulate the plant growth directly through increase in nutrition acquisition, such as phosphate solubilization, or more generally by rendering the inaccessible nutrients available to the plants (Persello-Cartieaux et al. 2003). Bacterial IAA can increase the root exudates and root system through soil pH and nutrient status. Exudation of organic acids from root results in acidification of the rhizosphere (Amir and Pineau 2003; Dakora and Philips 2002; Jones et al. 2003). The organic acids play an important role in the complexation of toxic and essential ions and increase their mobility for plant uptake. An acidic pH is typical for the rhizosphere environment due to proton extrusion through membranes of root cells (Spaepen et al. 2007). The acidification can also contribute to plant growth by mobilizing nutrients such as phosphorus and micronutrient. Acidification of the surrounding soil can occur with the release of protons and organic acids from the seed and root and uptake of nutrient ions by the plant (Hartman et al. 2009). In addition, phosphorous deficiency in many plants enhances the production and release of phenolic and carboxylate compounds (Hartman et al. 2009). Altered root morphology of inoculated plants may enhance phosphorus uptake. Furthermore, root hair abundance and length are also positively correlated with increased uptake of relatively immobile elements such as phosphorus. Datta et al. (1982) reported that a P-solubilizing and IAA-producing strain of *Bacillus firmus* increased the grain yield and P uptake of rice in a P-deficient soil amended with rock phosphate. In general, in view of function of bacterial IAA in increasing root exudates and root surface area (Dobbelaere et al. 1999; Lambrecht et al. 2000; Steenhoudt and Vanderleyden 2000) (Fig. 2), it may be suggested that IAA-producing bacteria can also solubilize insoluble phosphates similar to phosphate-solubilizing bacteria (Fig. 3).

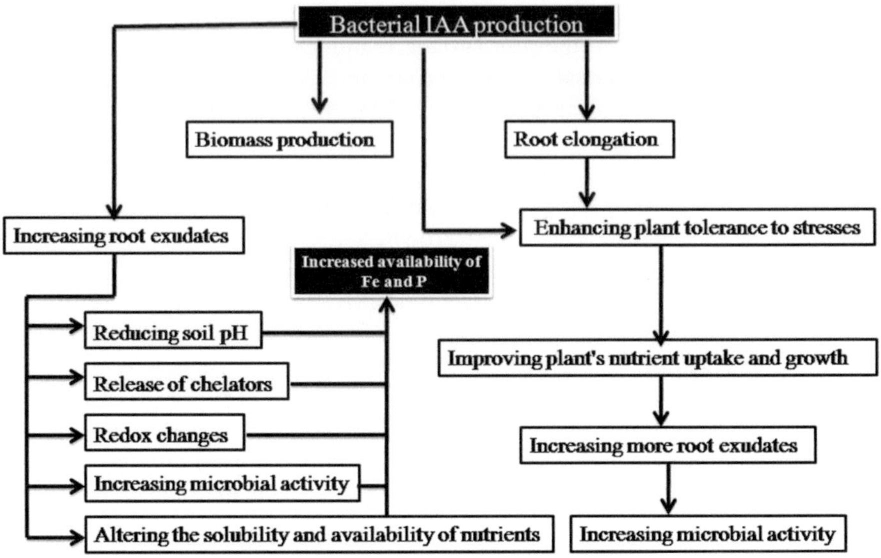

Fig. 3 Functions of bacterial IAA in obviating some of the roles of siderophore-producing bacteria and phosphate-solubilizing bacteria

9 IAA and Availability of Iron

Iron is a necessary cofactor for many enzymatic reactions. Under aerobic conditions, iron exists predominantly as Fe^{3+} and reacts to form highly insoluble hydroxides and oxyhydroxides that are basically unavailable to plants and microorganisms. High soil pH reduces while acidic soil conditions increase Fe availability. As pH increases by one unit, activity of Fe^{3+} decreases by 1000-fold. Under reducing conditions, addition of H^+ or other reductants, Fe solubility increases. Under such situations, Fe can be adsorbed on soil as an exchangeable ion. To acquire sufficient iron, plants under iron stress release phytosiderophores or protons and chelators (phenolics, carboxylates) to acquire iron (Hartman et al. 2009). Poorly soluble inorganic nutrients can be made available through the secretion of organic acids. Most plant-associated bacteria can produce iron chelators called siderophores in response to low iron levels in the rhizosphere. Several examples of increased Fe uptake in plants with concurrent stimulation of plant growth as a result of PGPRs inoculations have been reported (Burd et al. 2000; Carrillo-Castañeda et al. 2003; Barzanti et al. 2007). Exudation of organic acids from root has resulted in acidification of the rhizosphere (Dakora and Philips 2002). Acidification of rhizosphere through organic acids can contribute to plant growth by mobilizing nutrients such as P and Fe. In addition, organic acids are capable of chelating Fe^{3+} and making it available to plant roots. Some of the compounds in root exudates are able to form Fe complexes that improve availability. Carbohydrates, amino acids, organic acids, phenolics and secondary metabolites (low-molecular-weight compounds), proteins and mucilage (high-molecular-weight components) are typically the dominant soluble reduced carbon compounds in rhizodeposits (Lynch and Whipps 1990; Farrar et al. 2003; Wen et al. 2007; Badri and Vivanco 2009). Because of function of IAA in secreting root exudates and increasing rooting system and, since these exudates are involved in acidifying rhizosphere and in providing a reducing conditions required for converting Fe^{3+} to Fe^{2+}, it may be suggested IAA-producing bacteria can also solubilize insoluble Fe sources and induce plant growth and iron uptake in a similar manner to siderophore-producing bacteria (Fig. 3). For example, protons and electrons are secreted within carbon compounds as undissociated acids or compounds with reducing capabilities. Oxygen consumption, due to respiration by the root (increase of root system due to bacterial IAA) and associated microflora (increase of microflora activity due to production of more root exudates), can also result in steep redox gradients in the rhizosphere (Hartman et al. 2009). Because in the aerobic environment, iron occurs principally as Fe^{3+} and is likely to form insoluble hydroxides and oxyhydroxides, thus it is generally inaccessible to both plants and microorganisms (Rajkumar et al. 2010).

10 IAA in Phytopathogenesis

Production of IAA is common among plant-associated bacteria, which may be beneficial or detrimental to the plant health. For example, IAA production by *P. putida* GR12-2 has been found to improve the root proliferation resulting in increased root surface area, which helps in rise of nutrient and water uptake from soil (Patten and Glick 2002). On the other hand, in some reports, IAA production has been found necessary for pathogenesis (Vandeputte et al. 2005; Yang et al. 2007). Plant–microbe interactions were determined by different IAA biosynthesis pathways. For instance, the beneficial plant-associated bacteria synthesize IAA via the indole-3-pyruvate (IPyA) pathway, whereas pathogenic bacteria mainly use the indole-3-acetamide (IAM) pathway (Patten and Glick 1996, 2002; Manulis et al. 1998; Hardoim et al. 2008). For example, in phytopathogenic bacteria, such as *Agrobacterium tumefaciens* and pathovars of *P. syringae*, IAA is synthesized from L-Trp via the intermediate IAM pathway and has been connected to the induction of plant tumors (Glickmann et al. 1998; Patten and Glick 2002; Buell et al. 2003). The production of phytohormones such as IAA and cytokinins in free-living cultures is an indication of many phytophatogenic gall forming bacteria such as *P. agglomerans*, *P. savastanoi* pv. *savastanoi*, *P. syringae* pv. *syringae*, *Ralstonia solanacearum* and *Rhodococcus fascians* (Morris 1995; Vandeputte et al. 2005). In many bacterial pathogens, the *hrp*-gene encoded type III secretion system that directly translocates effector proteins into the eukaryotic host cells is fundamental to pathogenesis and the development of disease symptoms (Jin et al. 2003; He et al. 2004). In *P. syringae*, the presence of a functional *Hrp* promoter upstream of the *iaa*L gene involved in IAA biosynthesis further supports the role for IAA production in virulence (Fouts et al. 2002). The results of Navarro et al. (2006) suggest that decreasing plant IAA signaling can increase resistance to bacterial pathogens. A possible mechanism is the expression of IAA-repressed plant defense genes. They further showed that exogenous application of IAA enhances susceptibility to bacterial pathogens. These findings allow us to hypothesize that bacterial IAA production may contribute to circumvent the host defense system by deactivating repressor gene of IAA signaling. In this way, IAA biosynthesis may play an important role in bacterial resistance and colonization on the plant (Remans et al. 2006). For disease development, the first step is to infect the plant host and obtain nutrients to support the pathogen's growth and survival. In *E. herbicola*, the presence of IAA increases the ability of the bacterium to colonize on plant surfaces (Brandl and Lindow 1996) and the loss of IAA production decreases the colony size and population growth (Lindow and Brandl 2003). For example, a twofold population increase relative to IAA-deficient strains in pear flowers and bean plants was reported in IAA-producing *P. agglomerans* (Brandl and Lindow 1996). It has also been suggested that bacteria synthesize IAA to stimulate the root hairs production and lateral roots in plants relating to release saccharides from plant cell walls during the elongation (Davies 2004). Saccharides are carbohydrates that can be a source of nutrients for microorganisms, increase the colonization ability of a bacterium (Lindow and Brandl 2003) and facilitate bacterial colonization of plant surfaces (Bender et al. 1999). In addition, IAA production has been demonstrated to be a virulence factor in some pathogens (Yamada 1993). Many microorganisms produce IAA in order to perturb host physiological processes for their own benefits (Costacurta and Vanderleyden 1995; Yamada 1993). Exogenous application of IAA produced by pathogens enhances susceptibility to bacterial pathogens. In their interaction with plants, these microorganisms can interfere with plant development by disturbing the IAA balance in plants. This is best documented for phytopathogenic bacteria like *Agrobacterium* spp. and *P. savastanoi* pv. *savastanoi*, causing tumors and galls, respectively (Jameson 2000; Mole et al. 2007), and PGPR such as *Azospirillum* spp. that have impact on plant root development (Persello-Cartieaux et al. 2003; Spaepen et al. 2007). As many bacterial pathogens are known to produce IAA, it can be speculated that this property is part of the strategy used by the pathogen to bypass the plant defense system. The same could apply for IAA-producing PGPRs. Rhizobacteria may affect plant hosts by mechanisms similar to phytopathogenic bacteria through production of enzymes, phytotoxins, or phytohormones (Loper and Schroth 1986; Schippers et al. 1987). Nevertheless, biotrophic phytopathogens and plant-beneficial bacteria are coming closer to each other when taking an IAA perspective. Obviously, as we try to comprehend the challenges in one direction (phytopathology) new and fascinating questions raises in another direction (phytostimulation). In general, the function of bacterial IAA in pathogenesis and disease development is not entirely clear.

11 IAA in Rhizobium–Legume Symbiosis

The IAA produced by PGPRs is involved in plant–bacteria interactions and can affect plant growth promotion and root nodulation. They are involved in many processes of nodule formation by rhizobia in legume plants, such as founder cell specification, nodule initiation and differentiation (IAA accumulation), nodule numbers, vascular bundle formation and cell division and differentiation. These three later events are more necessary for nodule formation. Mutants of the bacterium *Bradyrhizobium elkanii* that had a decreased level of IAA synthesis induced fewer nodules on soybean roots than did the wild-type strain (Fukuhara et al. 1994). Nitrogen fixation capacity in the former nodules was also increased (Camerini et al. 2008). In addition, inoculation of *Medicago truncatula* with IAA-overproducing strain resulted in better plant growth under phosphorus deficiency because of the release of organic acids by the bacterium (Bianco and Defez 2010). In co-inoculation studies with *Azospirillum* and *Rhizobium*, earlier and faster nodulation and higher crop yields were observed (Okon and Itzigsohn 1995; Burdman et al. 1996). However, using an *Azospirillum ipd*C mutant, producing 10 % of IAA produced by the wild-type strain, the increase in nodulation and nitrogen fixation was not observed, showing that bacterial IAA production is important in symbiosis (Remans et al. 2008). An extensive overlap of changes in protein level could be observed in *M. truncatula* in response to IAA treatment and *Sinorhizobium meliloti* inoculation, probably because of regulation of these proteins by IAA during the early stages of nodulation (van Noorden et al. 2007). It was demonstrated that the *nod* inducers, the flavonoids, also stimulate the production of IAA by *Rhizobium* (Prinsen et al. 1991). In fact, *A. brasilense* caused a significant increase in the *nod*-inducing activity of crude alfalfa root exudates. IAA could be important for maintaining a functional root nodule (Badenochjones et al. 1983). However, the origin of IAA in the nodules is still not clear. It has been suggested that elevated levels of IAA in nodules are derived from the prokaryotic microsymbiont because a mutant of *Bradyrhizobium japonicum* that produces 30-fold more IAA than the wild-type strain has higher nodulation efficiency (Kaneshiro and Kwolek 1985). Bacteroids of plants inoculated with mutant *B. japonicum* strains produce high amounts of IAA in comparison with wild-type bacteroids, suggesting that increased IAA biosynthesis in nodules is of prokaryotic origin. It is therefore likely that IAA transport regulation is part of the process leading to nodule initiation (Hunter 1989; Kaneshiro and Kwolek 1985). In addition, rhizobia can also indirectly influence the IAA homeostasis by interfering with plant IAA transport (Badenochjones et al. 1983; Ghosh and Basu 2006). Many studies indicate that changes in IAA balance in the host plant are a prerequisite for nodule organogenesis (Mathesius et al. 1998). An IAA-producing *S. meliloti* strain showed increased tolerance to several stresses, and *M. truncatula* plants inoculated with this strain have a higher IAA content in nodules and roots and are better resistant to salt stress (Bianco and Defez 2009). The link between Nod factors as symbiotic signaling molecules and rhizobial IAA production points to a role for IAA in the *Rhizobium*–legume symbiosis (Theunis 2005). Nevertheless, the exact role of IAA in the different stages of Rhizobium–plant symbiosis remains unclear.

12 IAA in Actinorhizal Symbioses Formation

The term actinorhiza refers both to the filamentous bacteria *Frankia*, an actinobacteria, and to the root location of nitrogen-fixing nodules. Actinorhizal symbioses result from the interaction between *Frankia* and plants belonging to eight angiosperm families collectively called actinorhizal plants (Benson and Silvester 1993). This symbiotic interaction results in the formation of a actinorhizal nodule on the root system, where the bacteria are hosted and fix nitrogen (Obertello et al. 2003). Unlike legume nodules, actinorhizal nodules are structurally and developmentally related to lateral roots (Pawlowski and Bisseling 1996). *Frankia* like many soil bacteria has been known to produce auxins since long ago. For instance, IAA and phenylacetic acid (PAA) are found at relatively high concentration (10^{-5}–10^{-6} M) in the supernatant of various *Frankia* strains in pure culture (Wheeler et al. 1979; Hammad et al. 2003). A specific IAA response might occur in infected cells allowing the infection to proceed. The infection threads are encompassed by the plant cell membrane and a new cell wall-like structure composed mainly of pectin derivatives (Lalonde and Knowles 1975). IAA is known to regulate genes involved in cell wall remodeling and pectin biosynthesis and methylation (Lerouxel et al. 2006). Auxin perception in infected plant cells might therefore be necessary to allow the growth of infection threads (Benjamin et al. 2008).

13 IAA in the Development of Arbuscular Mycorrhizal Symbioses

Arbuscular mycorrhiza (AM), a symbiosis between plants and members of an ancient phylum of fungi, the *Glomeromycota*, improves the supply of water and nutrients, such as phosphate and nitrogen, to the host plant. In return, up to 20 % of plant-fixed carbon is transferred to the fungus. Nutrient transport occurs through symbiotic structures inside plant root cells known as arbuscules. The complex relationship between host roots and AM fungi requires a continuous exchange of signals, which results in the proper development of the symbiosis (Gianinazzi-Pearson 1996; Hause and Fester 2005). Plant hormones are signal molecules known to regulate many developmental processes in plants and are therefore suitable candidates to function in the colonization process and likely during the establishment of an AM symbiosis (Barker and Tagu 2000; Ludwig-Muller and Güther 2007). IAA may facilitate the colonization of a host by increasing the number of lateral roots as preferential colonization sites for the fungi during early growth phases (Kaldorf and Ludwig-Muller 2000). It is suggested that increased IAA levels and subsequent IAA-induced gene expression might contribute to the phenotypical changes during mycorrhizal colonization (Ludwig-Muller and Güther 2007). Although reports on IAA levels during AM in different plant species are contradictory, the contribution of IAA to the establishment of an AM symbiosis might be an important factor especially for the development of lateral roots which are the preferred infection sites for the fungi (Ludwig-Muller and Güther 2007). Recent findings about the role of fungal-produced IAA in different plant–fungus interacting systems open the possibility that fungi may use IAA and related compounds to interact with plants as part of its colonization strategy, leading to plant growth stimulation and modification of basal plant defense mechanisms (Prusty et al. 2004; Contreras-Cornejo et al. 2009). In maize/*Zea mays* and *A. thaliana*, *Trichoderma* inoculation affected root system architecture, which was related to increased yield of plants. Reported developmental effects include increased lateral root formation and root hair growth (Bjorkman et al. 1998; Harman et al. 2004; Contreras-Cornejo et al. 2009). Studies also indicate that the effects of inoculation with IAA-producing fungi in plants under natural conditions may depend on the type and concentration of IAA produced by the fungi. In general, the increased IAA levels lead to the formation of more lateral roots, which constitute preferential penetration sites for the AM hyphae, thus closing the infection cycle. Future research has to provide functional proof for these hypotheses.

14 IAA and Environmental Stresses

Studies have shown that IAA triggers an increased level of protection against external adverse conditions by coordinately enhancing different cellular defense systems (Lindberg et al. 1985; Frankenberger and Arshad 1995; Bianco et al. 2006; Bianco and Defez 2009). These authors investigated the effect of IAA treatment on bacterial cells and demonstrated that the cells were tolerant to a variety of stress conditions. The role of IAA produced by PGPRs in the promotion of plant growth during stress conditions such as salinity or drought has also been demonstrated (Bianco and Defez 2009; Egamberdieva and Kucharova 2009). Since, indigenously produced IAA in plants decreases in salt stress conditions, salt tolerant PGPRs may increase plant growth and lengthen the root by supplying IAA synthesized by them. Spaepen et al. (2007) reported the role of IAA in response to stress as evident from its increased production of IAA in *Azospirillum* sp. during carbon limitation and acidic pH. An increased tolerance of *M. truncatula* against salt stress was also observed in plants inoculated by the IAA-overproducing strain *S. meliloti* DR-64 (Bianco and Defez 2009). Plants inoculated with this mutant accumulated a high amount of proline and showed enhanced levels of the antioxidant enzymes superoxide dismutase, peroxidase, glutathione reductase and ascorbate peroxidase compared with plants inoculated with the parental strain. In general, IAA-producing bacteria may enhance growth of plant in drought conditions by stimulating formation of well-developed root system enough for providing sufficient water from soil.

15 Ethylene

The phytohormone ethylene (C_2H_4), a unique plant growth hormone, is found only in gaseous form and produced endogenously by almost all plants (Babalola 2010). Ethylene can function as an efficient plant growth regulator at very low concentrations as low as 0.05 μL^{-1} (Abeles et al. 1992). This phytohormone is involved in the regulation of numerous physiological processes in plants including modulating the growth and cellular metabolism of plants, disease-resistant biotic/abiotic stress tolerance, plant–microbe partnership and plant nutrient cycle (Ping and Boland 2004; Babalola 2010). However, stress conditions such as flooding, wounding, drought, chilling temperature, exposure to chemicals and pathogen attack may induce the production of ethylene substantially (Gnanamanickam 2006; Babalola 2010). The term stress ethylene is used to describe the acceleration of ethylene biosynthesis associated with environmental and biological stresses (Morgan and Drew 1997). The overproduction of ethylene can cause the inhibition of root elongation, lateral root growth and root hair formation (Mayak et al. 2004; Pierik et al. 2006; Saleem et al. 2007; Belimov et al. 2009).

15.1 Ethylene and the Inhibition of Endophytic Colonization

The increased level of ethylene formed in response to stress conditions can be both the cause of some of the symptoms of stress, and the inducer of defense responses, which help to enhance survival of the plant under adverse conditions. The host plant induces defense mechanisms against pathogens. However, in contrast to the plant response to phytopathogens only few defense responses have been described in plant response to endophytes. These differences can be probably explained by the secretion of different compounds or by the amount of secreted metabolites, which may be very low in the case of endophytes (James et al. 2002). However, it has been reported that plants may show defense reactions controlling endophytic colonization (Iniguez et al. 2005). Some plants are known to use salicylic acid (SA), jasmonic acid (JA) and ethylene as signaling molecules, which control colonization by some endophytes inside the root system (Iniguez et al. 2005; Miché et al. 2006). Ethylene has been known as signal molecule and secondary messenger in the induction of a salicylic acid (SA)-independent plant defense pathway referred to as induced systemic resistance (ISR) in plants, decreasing endophytic colonization (Knoester et al. 1998; Pieterse et al. 1998; Ton et al. 2001, 2002; Wildermuth et al. 2001; Audenaert et al. 2002; Iniguez et al. 2005). In a study,

Iniguez et al. (2005) showed addition of ACC to the growth media significantly reduced endophytic colonization in wild-type *Medicago sativa* by *Klebsiella pneumoniae* 342 and *Salmonella enteric*. These evidences suggest that ethylene can significantly inhibit invasion of bacterial cells into plants.

16 ACC Deaminase-Containing PGPR

PGPRs containing ACC deaminase activity can affect plant growth directly through various ways such as nitrogen fixation, solubilization of phosphorus, and increasing growth by regulating endogenous level of plant hormones or indirectly by increasing the natural resistance of the host against pathogens and other environmental stresses (Glick 2004; Lugtenberg and Kamilova 2009; Spaepen et al. 2009). A particular bacterium may affect plant growth using any one, or more, of these mechanisms. Moreover, a bacterium may provide different benefits at various times during the life cycle of the plant. These bacteria can cleave the plant ethylene precursor ACC, and thereby lower the level of ethylene in a developing or stressed plant (Jacobson et al. 1994; Glick 1995, 1998). Under stress conditions, a sustained high level of ethylene may inhibit root elongation (Jackson 1991). Thus, ACC deaminase-producing PGPRs, when bound to the seed coat of a developing seedling, may act as a mechanism for ensuring that the ethylene level does not become elevated to the point where root growth is impaired. By facilitating the formation of longer roots, these bacteria may enhance the survival of some seedlings, especially during the first few days after the seeds are planted. Similarly, ACC deaminase-containing PGPRs bound to the roots of plants can act as a sink for ACC and protect stressed plants from some of the deleterious effects of stress ethylene (Arshad et al. 2008; Belimov et al. 2009). ACC deaminase has been widely reported in numerous species of PGPRs such as *V. paradoxus*, *Agrobacterium genomovars*, *Azospirillum lipoferum*, *Alcaligenes*, *Bacillus*, *Burkholderia*, *Enterobacter*, *Methylobacterium fujisawaense*, *Pseudomonas*, *R. solanacearum*, *Rhizobium*, *Rhodococcus* and *S. meliloti* (Belimov et al. 2001; Dobbelaere et al. 2003; Blaha et al. 2006; Rasche et al. 2006; Pandey and Maheshwari 2007a; Belimov et al. 2009; Duan et al. 2009; Sharp et al. 2011; Jiang et al. 2012; Chen et al. 2013).

16.1 ACC Deaminase in Promotion of Plant Growth

Stimulation of root elongation and biomass production of different plant species by inoculations with PGPRs having ACC deaminase activity has been repeatedly documented, particularly when the plants were subjected to stressful growth conditions (Hall et al. 1996; Burd et al. 1998; Glick 1998; Belimov et al. 2001, 2005; Madhaiyan et al. 2006; Safronova et al. 2006; Glick et al. 2007; Belimov et al.

2009). *P. putida* UW4 deficient in ACC deaminase activity simultaneously lost the ability to elongate roots in infected canola plants (Li et al. 2000). Inoculation of plants with PGPRs containing ACC deaminase activity may lead to various subsequent physiological changes in plants (Glick et al. 2007; Saleem et al. 2007). Considerable evidences have demonstrated the beneficial role of bacterial ACC deaminase in decreasing stress reactions in plant growth under different stresses, including range of pathogenic agents (Arshad and Frankenberger 2002; Arshad et al. 2007; Saleem et al. 2007), salinity, flooding, drought, toxicity of high concentrations of heavy metals present in pollutant soils (Grichko et al. 2000; Grichko and Glick 2001; Nie et al. 2002; Kausar and Shahzad 2006; Zahir et al. 2007; Gamalero et al. 2009; Nadeem et al. 2009) and the presence of toxic organic compounds (Arshad and Frankenberger 2002; Arshad et al. 2007; Glick et al. 2007). Saleem et al. (2007) reviewed the role of PGPRs containing ACC deaminase activity in stress management in agriculture. Following inoculation of pea with the ACC deaminase containing rhizobacterium *V. paradoxus* 5C-2 obtained from pea increased seed nitrogen concentration in plants grown and enhanced vegetative growth and seed yield in drying soil (Dey et al. 2004; Belimov et al. 2009) that may have been due to enhanced nodulation, since ethylene typically inhibits nodulation (Guinel and Geil 2002), attenuated a drought-induced increase in xylem sap ACC concentration in non-nodulated plants and prevented drought-induced decrease in seed nitrogen content of nodulated plants respectively. In addition, adding the ACC deaminase-containing rhizobacterium *V. paradoxus* 5C-2 to the substrate of well-watered, well-fertilized pea plants increased root and shoot growth by 20 and 15 %, respectively (Jiang et al. 2012). Since bacterial mutants having low ACC deaminase activity (including a transposome mutant of *V. paradoxus* 5C-2) did not stimulate plant growth (Glick et al. 1994; Belimov et al. 2007, 2009) and the growth promotion observed was most probably due to decreased plant production of the growth-inhibitory phytohormone ethylene. In other study, Inoculation of *V. paradoxus* 5C-2 significantly ($P < 0.01$) increased fresh biomass of *A. thaliana* by 34–47 % throughout development (Chen et al. 2013). Furthermore, transposon mutagenesis of microorganisms to downregulate ACC deaminase activity reduced or eliminated their growth-promoting effect, in plant–microbe interactions such as canola–*Enterobacter cloacae* (Li et al. 2000), tomato–*Pseudomonas brassicacearum* (Belimov et al. 2007), pea–*V. paradoxus* (Belimov et al. 2009) and canola–*Trichoderma asperellum* (Viterbo et al. 2010). These findings suggested that ACC deaminase plays a key role in promoting plant growth. In general, inoculation with ACC deaminase-containing bacteria induce longer roots which might be helpful in the uptake of relatively more water from deep soil under drought stress conditions, thus increasing water use efficiency of the plants (Zahir et al. 2007). Many studies showed using ACC deaminase-producing bacteria in association with plants subjected to a wide range of different kinds of biotic and abiotic stresses, in all instances tested, resulted in enhanced plant tolerance to the stresses (Table 2). Thus, use of these microorganisms per se can alleviate stresses in agriculture thus opening a new and emerging application of microorganisms.

Table 2 PGPRs conferring abiotic and biotic stress tolerance in crop plants by ACC deaminase activity

Host plant	Bacterial strain	Type of environmental stress	Bacterial effects on plants	References
Ocimum sanctum	*Achromobacter xylosoxidans* Fd2, *Serratia ureilytica* Bac5, *Herbaspirillum seropedicae* Oci9, *Ochrobactrum rhizosphaerae* Oci13,	Flooding	Increase of foliar nutrient uptake, growth and yield	Barnawal et al. (2012)
Tomato (*Lycopersicon esculentum*)	*Enterobacter cloacae* UW4, *E. cloacae* CAL2, *Pseudomonas putida* ATCC17399	Flooding	Improvement of root and shoot growth	Grichko and Glick (2001)
Pea (*Pisum sativum*)	*Variovorax paradoxus* 5C-2	Drought	Improving growth, yield and water use efficiency and increased nodulation by symbiotic nitrogen-fixing bacteria	Belimov et al. (2009)
Wheat (*Triticum aestivum* L.)	SBW17 and SBW27	Drought	Increased root-shoot length, root-shoot mass and lateral root number	Shakir et al. (2012)
Pea (*P. sativum*)	*V. paradoxus*	Drought	Stimulated root biomass by 20–25 %, whole plant biomass was stimulated also by 25 %	Dodd et al. (2005)
Tomato and Pepper	*A. piechaudii* ARV8	Drought	Enhancing the fresh and dry weights	Mayak et al. (2004)
Tomato	*A. piechaudii*	Salinity	Increasing the fresh and dry weights and water use efficiency (WUE)	Mayak et al. (2004)
Canola (*Brassica napus*)	*P. putida* UW4	Salinity	Improved plant growth	Cheng et al. (2007)
Canola (*B. napus*)	*P. putida* UW4	Salinity	Improved plant growth	Cheng et al. (2012)
Mung bean	*Pseudomonas fluorescens* Mk20, *Rhizobium phaseoli* M6	Salinity	Improving seedling growth and nodulation	Amhad et al. (2011)

(continued)

Table 2 (continued)

Host plant	Bacterial strain	Type of environmental stress	Bacterial effects on plants	References
Cucumber	*P. putida* UW4	Salinity	Promoted plant growth, effects on plant biomass, total root length and total leaf projected area, increased AM fungus *Gigaspora rosea* colonization and arbuscule abundance	Gamalero et al. (2008)
Tomato (*Licopersicon esculentum* Mill)	*Bacillus licheniformis* B2r	Salinity	A significant increase in the germination percentage, germination index, root length and seedling dry weight	Chookietwattana and Maneewan (2012)
Cucumber	*P. putida* UW4	Salinity	Increased plant growth, affected root architecture and improved photosynthetic activity	Gamalero et al. (2010)
Canola (*B. napus* L.)	*P. fluorescens*, *P. putida*	Salinity	Increased seedling growth and the rate of germinating seeds	Jalili et al. (2009)
Catharanthus roseus	*A. xylosoxidans* AUM54	Salinity	Increased germination percentage, vigor index, plant height, root dry weight, increased the antioxidative enzyme content, ascorbate peroxidase (APX) activity, superoxide dismutase (SOD) activity and catalase (CAT)	Karthikeyan et al. (2012)
Maize	*Pseudomonas syringae* S5, *Enterobacter aerogenes* S14, *P. fluorescens* S20	Salinity	Improved the growth and yield, increased plant height, root length, total biomass, cob mass and grain yield	Nadeem et al. (2007)

(continued)

Table 2 (continued)

Host plant	Bacterial strain	Type of environmental stress	Bacterial effects on plants	References
Wheat (*T. aestivum* L.)	*Hallobacillus* sp. SL3, *Bacillus halodenitrificans* PU62	Salinity	Increased root elongation and dry weight	Ramadoss et al. (2013)
Wheat (*T. aestivum* L.)	*P. putida* (W2), *P. fluorescens* (W17)	salinity	Increased plant height, root length, plant biomass and grain yield	Nadeem et al. (2010)
Tomato	*Pseudomonas mendocina*	Salinity	Increasing content of growth	Sadrnia et al. (2011)
Groundnut (*Arachis hypogea*)	*P. fluorescens* TDK1	Salinity	Improving the plant growth parameters, increased yield	Saravanakumar and Samiyappan (2006)
Canola (*B. napus*)	*Brevibacterium epidermidis* RS15, *Micrococcus yunnanensis* RS222, *Bacillus aryabhattai* RS341	Salinity	Increase in root length, dry weight	Siddikee et al. (2010)
Groundnut	*P. fluorescens*	Salinity	Increased yield	Saravanakumar and Samiyappan (2007)
Red pepper	*Brevibacterium iodinum* RS16, *B. licheniformis* RS656, *Zhihengliuela alba* RS111	Salinity	Significantly increase the growth, increase of nutrient uptakes	Siddikee et al. (2011)
Canola (*B. napus*)	*Pseudomonas asplenii* AC	Organics	Significantly increased root and shoot biomass	Reed and Glick (2005)
Mini carnation	*P. fluorescens* YsS6, *Pseudomonas migulae* 8R6, *P. putida* UW4	Flower wilting	Decreased levels of flower senescence	Ali et al. (2012)
Tomato and Castor bean	*P. putida* UW4	Pathogens	Inhibited tumour development on plant	Hao et al. (2007)
Castor bean	*Agrobacterium tumefaciens* D3	Pathogens	Inhibited tumour development on plant, promoted plant root elongation	Hao et al. (2011)

(continued)

Table 2 (continued)

Host plant	Bacterial strain	Type of environmental stress	Bacterial effects on plants	References
Cucumber potato	*P. fluorescens* CHA0	Pathogens	protected cucumber against *Pythium* damping-off and potato tubers against *Erwinia* soft rot	Wang et al. (2000)
Chamaecytisus proliferus	*P. fluorescens*	Pathogens	Controlling the growth of *Fusarium oxysporum* and *Fusarium proliferatum*	Donate-Correa et al. (2005)
Mimosa pudica	*Burkholderia* sp.	Pathogens	Exhibited antagonistic activity against *Rhizoctonia solani* and *Sclerotinia sclerotiorum*	Pandey et al. (2005)
Tomato (*Solanum lycopersicum*)	*P. putida* UW4, *Burkholderia phytofirmans* PsJN, *Azospirillum brasilense* Cd1843	Pathogens	Reduced the development of tumours on plants	Toklikishvili et al. (2010)
Grapevine (*Vitis vinifera* L.)	*B. phytofirmans* PsJN	Low temperature	Increased grapevine growth and physiological activity at a low temperature	Ait Bakra et al. (2006)
Potato	*Burkholderia phytofrmans* PsJN	Heat stress	Maintaining normal growth	Bensalim et al. (1998)
Canola	*P. putida*	Low temperature	Increased yield	Chang et al. (2007)
P. sativum L.	*Pseudomonas brassicacearum*, *Pseudomonas marginalis*, *Pseudomonas oryzihabitans*, *P. putida*, *Pseudomonas* sp., *Alcaligenes xylosoxidans*, *Alcaligenes* sp., *V. paradoxus*, *Bacillus pumilus*, *Rhodococcus* sp.	Heavy metal	The bacteria were tolerant to Cd^{2+} toxicity and stimulated root elongation of rape seedlings in the presence of 300 μM $CdCl_2$ in the nutrient solution	Belimov et al. (2001)

(continued)

Table 2 (continued)

Host plant	Bacterial strain	Type of environmental stress	Bacterial effects on plants	References
Phragmites australis	*P. asplenii* AC[a]	Heavy metal	Inoculation resulted in normal plant growth under high levels of Cu^{2+} and creosote	Reed et al. (2005)
Lycopersicum esculentum Mill	*Kluyvera ascorbata* SUD165 *K. ascorbata* SUD165/26	Heavy metal	Toxic effects of the heavy metals (Ni^{2+}, Pb^{2+} and Zn^{2+}) were not pronounced in inoculated plants	Burd et al. (2000)
Brassica juncea L.	*P. brassicacearum*, *P. marginalis*, *P. oryzihabitans*, *P. putida*, *Pseudomonas* sp., *A. xylosoxidans*, *Alcaligenes* sp., *V. paradoxus*, *B. pumilus*, *Rhodococcus* sp.	Heavy metal	The bacteria were tolerant to Cd^{2+} toxicity and stimulated root elongation of rape seedlings in the presence of 300 μM $CdCl_2$ in the nutrient solution	Belimov et al. (2001)
B. juncea L.	*V. paradoxus*, *Rhodococcus* sp.	Heavy metal	Plant growth was improved in Cd^{2+} supplemented media in response to inoculation.	Belimov et al. (2005)
B. juncea L.	*K. ascorbata* SUD165, *K. ascorbata* SUD165/26	Heavy metal	Toxic effects of heavy metals (Ni^{2+}, Pb^{2+} and Zn^{2+}) were not pronounced in inoculated plants.	Burd et al. (2000)
B. napus	*K. ascorbata* SUD165	Heavy metal	Plant demonstrated normal growth under high levels of Ni^{2+}, Pb^{2+}, Zn^{2+} and CrO_4^{-2}	Burd et al. (1998)
Indian mustard	*V. paradoxus*, *Rhodoccus* sp., *Flavobacterium* sp.	Heavy metal	Increased root length	Belimov et al. (2005)
Canola	*E. cloacae*	Heavy metal	Increased biomass	Nie et al. (2002)
Indian mustard	*A. xylosoxidans*	Heavy metal	Increased root and shoot length and biomass	Ma et al. (2009)

(continued)

Table 2 (continued)

Host plant	Bacterial strain	Type of environmental stress	Bacterial effects on plants	References
P. sativum	*P. brassicacearum* Am3	Heavy metal	Increased Root and shoot biomass	Safronova et al. (2006)
B. juncea	*Pseudomonas* sp. PsA, *Bacillus* sp. Ba32 (RS)	Heavy metal	Plant growth	Rajkumar et al. (2006)
Rape	*P. fluorescens* G10, *Microbacterium* sp. G16	Heavy metal	Increased in biomass production and total Pb uptake, root elongation	Sheng et al. (2008)
Zea mays	*Burkholderia* sp. J62 (RS)	Heavy metal	Increased root and shoot dry weight	Jiang et al. (2008)
Indian mustard, corn, tomato	*Burkholderia* sp.	Heavy metal	Increased biomass	Jiang et al. (2008)

16.2 ACC Deaminase in Endophytic Bacterial Colonization

Successful colonization of the root surface is considered as a key property of prospective inoculants. PGPRs that produce the enzyme ACC deaminase promote plant growth by sequestering and cleaving plant-produced ACC and thereby lowering the level of ethylene in the plant. In experiments, the colonization of root systems with *P. fluorescens*, *P. putida*, *Bacillus pumilus* and *Serratia marcescens* was protected against foliar diseases (Pieterse et al. 2002). Decreased ethylene levels allow the plant to be more resistant to a wide variety of environmental stresses. It has been suggested that the ability to utilize ACC may contribute to the root-colonization ability of bacterial strains. It has been discovered that some PGPRs possess the enzyme ACC deaminase which can cleave ACC, the immediate precursor of ethylene in plants, to α-ketobutyrate and ammonia. The products of this hydrolysis are used by the ACC-degrading PGPRs as nitrogen and carbon sources, and thereby, lower the level of ethylene in a developing seedling or stressed plant. Treatment of plant seeds or roots with ACC deaminase-containing PGPRs typically reduces ACC and ethylene levels about two- to fourfold (Grichko and Glick 2001a; Mayak et al. 2004a). Under stress conditions, ACC deaminase-producing bacteria are able to utilize ACC, thereby increasing the root surface in contact with soil. Since a dynamic equilibrium of ACC concentration exists between root, rhizosphere and bacterium, bacterial uptake of rhizospheric ACC stimulates plant ACC efflux, decreases root ACC concentration and root ethylene evolution, and can increase root growth (Glick 1998). Accordingly, rhizosphere inoculation with ACC deaminase-containing bacteria decreases root ACC levels and ethylene evolution (Belimov et al. 2002; Mayak et al. 2004a). Previous results indicated that ethylene is a key regulator of the colonization of plant tissue by bacteria and that this regulation is most likely mediated by its effect on the plant signaling pathways. In this context, bacterial endophytes with high locally induced ACC deaminase activities might be excellent plant growth promoters, because they ameliorate plant stress by efficiently blocking ethylene production (Cheng et al. 2007). Furthermore, IAA-producing bacteria known to stimulate plant growth might even increase plant ethylene levels (Glick 1995). To avoid the deleterious effects of ethylene, plants might actually select for ACC deaminase-producing bacteria to become endophytic, thereby lessening plant stress caused by excessive ethylene levels. The selection of such beneficial endophytes might take place at an earlier stage (Kucera 2005). Thus, colonization by bacteria with high ACC deaminase activities might reduce the stress imposed by excessive ethylene to the plant originating from biotic and abiotic stresses (Arshad 2007). Hence, IAA and ACC deaminase production are being deployed as tools for identification and screening of endophytes (Khalid et al. 2005; Shaharoona et al. 2006; Etesami et al. 2014a, b). Therefore, trait of ACC utilization ability as a nutrient substance gives ACC deaminase-producing isolates advantages in more colonization and increase of root length of plants (Etesami et al. 2014a, b). For example, presence of PGPRs containing ACC deaminase on the roots of legume could suppress accelerated endogenous synthesis of ethylene during the rhizobial infection and thus may facilitate nodulation. Therefore, co-inoculation of legumes with competitive rhizobia and PGPRs containing ACC deaminase could be an effective and novel approach to achieve successful and dense nodulation in legumes. It is highly expected that inoculation with PGPRs containing ACC deaminase hydrolyzed endogenous ACC instead of ethylene and subsequently legume plant as well as nodulation can be promoted (Garcia Lucas et al. 2004; Remans et al. 2007). ACC deaminase-containing PGPRs can derepress the expression of auxin response genes in the shoots (Glick et al. 2007) and also suppress the expression or functioning of other plant signaling molecules such as jasmonic acid and giberellin (Czarny et al. 2006; Cheng et al. 2010). Therefore, these bacteria may have a competitive edge over other microorganisms in the rhizosphere because of use of ACC (Glick and Bashan 1997) that helps plants to overcome many detrimental effects of biotic and abiotic stresses (Glick et al. 2007; Saleem et al. 2007). In general, a decreased level of ACC results in a lower level of endogenous ethylene, which eliminates the inhibitory effect of high ethylene concentrations (Shaharoona et al. 2006) and contribute to their root colonization (Etesami et al. 2014a).

16.3 IAA and ACC Deaminase-Producing PGPRs in Phytoremediation

Phytoremediation is the direct use of green plants and their associated microorganisms to stabilize or reduce contamination in soils, sludges, sediments, surface water, or ground water. Plant species are selected for use based on factors such as ability to extract or degrade the contaminants of concern, adaptation to local climates, high biomass, depth root structure, compatibility with soils, growth rate, ease of planting and maintenance, and ability to take up large quantities of water through the roots. Since the activity of inoculated microbes is necessary to exhibit beneficial traits for improving the plant growth and overall phytoremediation process in metal contaminated soils, the colonization and survival in metal stress field environment are considered as important factors. Plant-associated bacteria can potentially improve phytoextraction by altering the solubility, availability, and transport of heavy metal and nutrients by reducing soil pH, release of chelators, P solubilization or redox changes (Gadd 2000, 2004). In addition to improving plant's nutrient uptake and growth, the plant-associated microbes alleviate heavy metal toxicity by reducing stress ethylene production. In general, heavy metal stress induces endogenous ethylene production in plants, which can affect the root growth and consequently the growth of the whole plant. Under such conditions, in order to maintain the equilibrium between the rhizosphere and root interior ACC levels, the plants release more ACC through exudation and thus results decrease in the production of stress ethylene (Adams and Yang 1979). Recent studies have revealed that plants inoculated with PGPRs containing ACC were better able to thrive in metal polluted soils (Rodriguez et al. 2008). Madhaiyan et al. (2007) reported that *M. oryzae* strain CBMB20 having ACC deaminase activity increased the growth of tomato seedlings grown in Ni and Cd polluted soils. The bacterium reduced the production of ethylene, which was otherwise stimulated when seedlings were challenged with increasing Ni and Cd. Zhang et al. (2011) have also confirmed that Pb-resistant and ACC deaminase-producing endophytic bacteria conferred metal tolerance onto plants by lowering the synthesis of metal-induced stress ethylene and promoted the growth of rape. Ma et al. (2011b) have also observed similar results in the case of *Allysusm serpyllifolium* and *Brassica juncea* growth under Ni stress in response to inoculation with ACC deaminase-producing endophytic bacteria. We anticipate that manipulating the rhizosphere processes for example increasing rhizosphere microbial population (by IAA-producing bacteria), inoculating the microbial strains with various PGP features as well as co-inoculating ecologically diverse microbes would yield better results for effective phytoremediation. In view of role of bacterial IAA and ACC deaminase activity in stimulation of root elongation and biomass production, increasing root exudates, enhancing plant tolerance to stresses, decreasing stresses and effective colonization, IAA and ACC deaminase-producing PGPRs can be used for effective phytoremediation of contaminated soil environment (Arshad and Frankenberger 2002; Glick et al. 2007; Saleem et al. 2007) (Fig. 4).

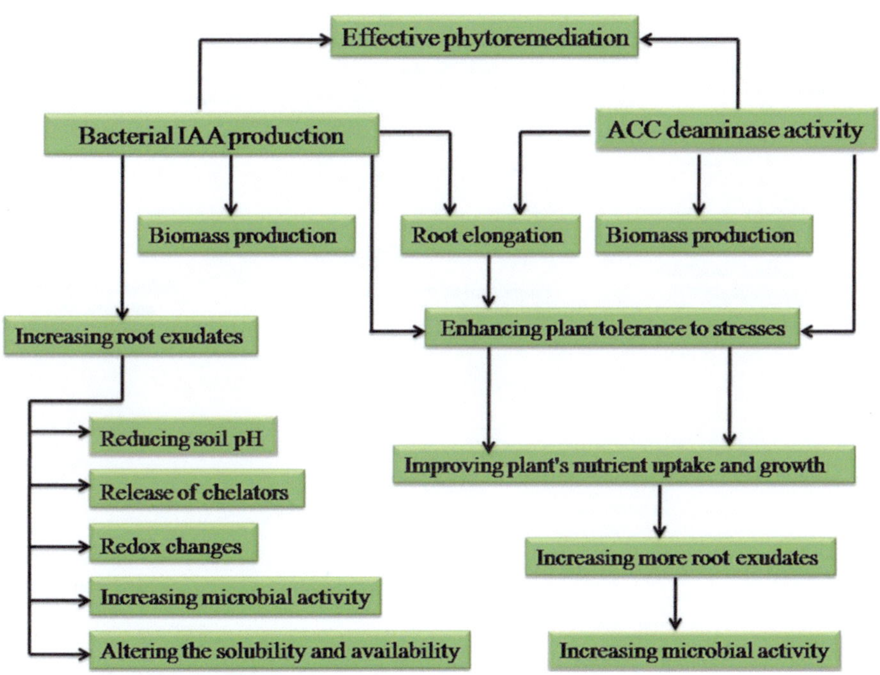

Fig. 4 Acceleration of phytoremediation by IAA and ACC deaminase-producing PGPR. Abbreviations: indole-3-acetic acid (IAA), 1-aminocyclopropane-1-carboxylate (ACC)

17 Mechanism of Action of IAA-Producing Bacteria in Nitrogen Uptake

The mechanism most often invoked to explain the direct effects of PGP bacteria on plants is the production of phytohormones, including IAA (Brown 1974; Patten and Glick 1996, 2002). The IAA containing PGPRs stimulate root proliferation and increase the root surface area or the general root architecture (Biswas et al. 2000; Lucy et al. 2004; Aloni et al. 2006). These bacteria enhance uptake of soil minerals and nutrients by the host plant. The plants growing better in turn release higher amounts of C in root exudates. The release of more C prompts increase in microbial activity, and this process continues in a cycle. The whole process makes more N available from the soil pool, influencing N flux into plant roots, and the plant is able to take up more available N (Adesemoye et al. 2009) (Fig. 5).

For various PGPRs, it has been demonstrated that enhanced root proliferation is related to bacterial IAA biosynthesis. The plant growth promotion observed after inoculation with *A. brasilense* is mainly caused by biosynthesis and secretion of bacterial IAA. In addition to providing the mechanical support and facilitating water and nutrient uptake, plant roots also synthesize, accumulate and secrete a diverse array of compounds (Walker et al. 2003). Because of

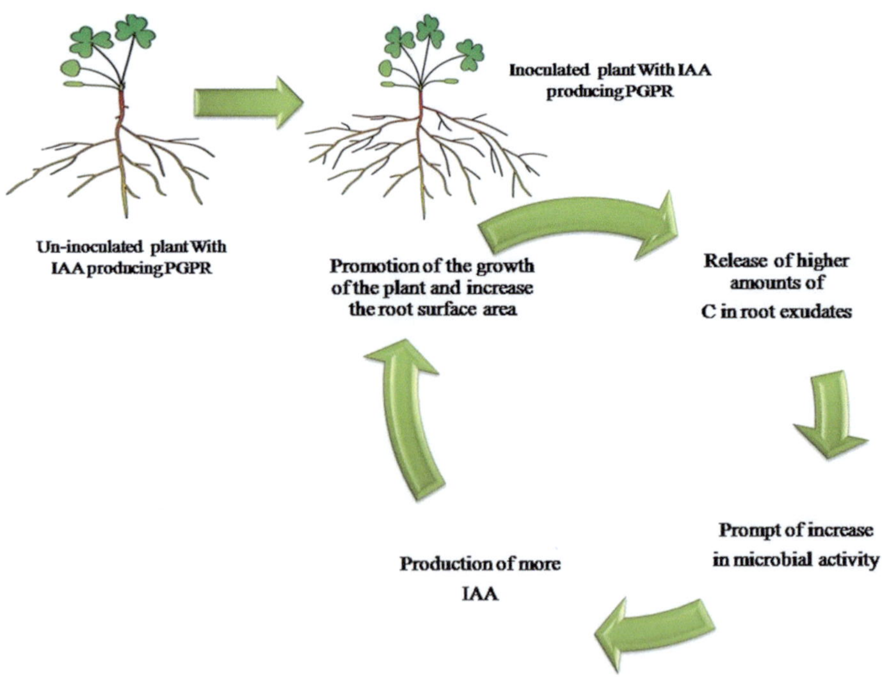

Fig. 5 Action mechanism of IAA-producing bacteria in uptake of nitrogen

the growth and development of the root system by bacterial IAA, an extremely diverse range of organic and inorganic compounds (substantial amounts of C- and N-containing compounds) can be taken up or released by seeds and roots into the soil. Microorganisms are attracted to this nutritious environment and use the root exudates and lysates for growth and multiplication on the surface of root and in the adjacent rhizosphere soil. These compounds secreted by plant roots act as chemical attractants for a vast number of heterogeneous, diverse and actively metabolizing soil microbial communities. Many organic compounds and enzymes are released by plants in root exudates that Faure et al. (2009) have reviewed their functions in the rhizosphere. Through the exudation of a wide variety of compounds, roots may regulate the soil microbial community in their immediate vicinity, cope with herbivores, encourage beneficial symbioses, change the chemical and physical properties of the soil and inhibit the growth of competing plant species (Nardi et al. 2000). Moreover, the microbial community influences the composition of the exudates to its advantage (Paterson et al. 2006; Shaw et al. 2006). The exudation of a wide range of chemical compounds modifies the chemical and physical properties of the soil and thus regulates the structure of soil microbial community in the immediate vicinity of root surface (Dakora and Phillips 2002). A fraction of these plant-derived small organic molecules is further metabolized by microorganisms in the vicinity as carbon and nitrogen sources, and some microbe-oriented molecules are subsequently retaken up by plants for growth and development (Kang et al. 2010). Indeed, carbon fluxes are critical determinants of rhizosphere function. It is reported that approximately 5–21 % of photosynthetically fixed carbon is transported to the rhizosphere through root exudation (Marschner 1995). The higher plant root system significantly contributes to the establishment of the microbial population in the rhizosphere (Dakora and Philips 2002). PGPRs often help increase root surface area to increase nutrient uptake and in turn enhance plant production (Mantelin and Touraine 2004). Application of several genera, such as *B. licheniformis* RC02, *Rhodobacter capsulatus* RC04, *Paenibacillus polymyxa* RC05, *P. putida* RC06, *Bacillus* OSU-142, *B. megaterium* RC01 and *Bacillus* M-13, showed increased root and shoot weight along with nutrient uptake in barley (Cakmacki et al. 1999). Studies with *Azospirillum* mutants altered IAA production support the view that increased rooting is caused by *Azospirillum* IAA synthesis (Dobbelaere et al. 1999). This increased rooting enhances plant mineral uptake and root exudation, which in turn stimulates bacterial colonization and thus amplifies the inoculation effect (Dobbelaere et al. 1999; Lambrecht et al. 2000; Steenhoudt and Vanderleyden 2000). It was demonstrated that *M. fujisawaense* promoted root elongation in canola (Madhaiyan et al. 2006). Ghosh et al. (2003) observed that *Bacillus circulans* DUC1, *B. wrmus* DUC2 and *Bacillus globisporus* DUC3 enhanced root and shoot elongation in *Brassica campestris*. Some compounds identified in root exudates that have been shown to play an important role in root microbe interactions include flavonoids present in the root exudates of legumes that activate *Rhizobium meliloti* genes responsible for the nodulation process (Peters et al. 1986).

17.1 IAA and ACC Deaminase in Reduced Application Rates of Chemical Fertilizers

Some chemical fertilizers have low use efficiency, meaning that only a portion of the applied nutrients are taken up by plants (Gyaneshwar et al. 2002) especially in the case of phosphorous fertilizers. One of the important mechanisms for the beneficial effects of PGPRs is stimulated nutrient availability and increase in nutrient use efficiency. Overall, results suggest that inoculants could be used to allow reductions in the current high rates of fertilizer and the resulting environmental problems (Malakoff 1998; Gyaneshwar et al. 2002; Shaharooma et al. 2008) without compromising plant productivity. In addition, under stress conditions resulting from reduced rates of inorganic fertilizers, ACC deaminase activity might have produced better root growth in the initial stages of crop growth. There has also been much recent interest in using PGPRs inoculants to decrease the application of chemical fertilizers (Adesemoye et al. 2009), either by stimulating root growth (thereby increasing root foraging for nutrients) or by directly stimulating plant nutrient uptake. Some ACC deaminase-containing PGPRs increased shoot and grain nutrient concentrations in specific plant–microbe interactions: pea and *Pseudomonas brassicacearum* Am3, *Pseudomonas marginalis* Dp1, or *Rhodococcus* sp. Fp2 (Safronova et al. 2006); peanut (*Arachis hypogea*) and various *Pseudomonas* spp. isolates (Dey et al. 2004); and wheat (*Triticum aestivum*) and *A. brasilense* Sp245 (Creus et al. 2004). Therefore, the PGPRs enhance the access of plants to the nutrient and more uptake of it by increasing the root growth of plant. For example, applied N can be lost through nitrate leaching (Biswas et al. 2000). However, a plant with a good root growth can uptake more nutrient than the same plant without a good root growth during a given period (Fig. 5). In a study, Adesemoye et al. (2009) showed PGPRs or combinations of PGPRs and Arbuscular mycorrhizal fungi (AMF) can improve the nutrient use efficiency of fertilizers. When the percentage of recommended fertilizer was reduced and inoculants were used, plant growth parameters and nutrient uptake were comparable to those with the full rate of fertilizer without inoculants. After testing different reduced fertilizer rates, under these experimental conditions, 75 % fertilizer was the stable minimum to which fertilizer could be reduced if supplemented with PGPRs to achieve growth equivalent to 100 % fertilizer without PGPRs. Shaharoona et al. (2008) reported that N use efficiency increased in response to inoculation with *P. fluorescens* at all fertilizer levels in wheat, causing 115, 52, 26 and 27 % increase over the noninoculated control at N, P and K application rates of 25, 50, 75 and 100 % recommended doses, respectively. Plants inoculated with the PGPRs together with one-third of the normal rate (33 kg N ha^{-1}) gave the highest storage root dry weight compared to noninoculated control sweet potato plants. Inoculation also increased the concentrations of N, P and K in shoots and storage root (Farzana et al. 2007). Many reports indicated that the enhancement of N uptake by plants inoculated with the PGPRs strains was not via associative N fixation (Malakoff 1998; Gyaneshwar et al. 2002; Shaharooma et al. 2008; Adesemoye et al. 2009) and the resulting enhancement of N uptake has been attributed to alternative bacterial effects. Use of mutant strains (carrying nif D::kan interposan mutation that prevents N fixation entirely) proved the participation of *Gluconacetobacter diazotrophicus* in N fixation. It is an established fact that the growth hormones, auxins (IAA), cytokinins and gibberellins, play a role in enhancing the growth of grasses associated with diazotrophs (Bottini et al. 2004). Apart from N fixation, *G. diazotrophicus* is also reported to benefit sugarcane through production of PGP factors (Fuentes-Ramirez et al. 2001). As previously suggested, the effect of *Azotobacter* and *Azospirillum* species is attributed not only to the amounts of fixed nitrogen but also to the production of plant growth regulators such as IAA, gibberellic acid, cytokinins and vitamins (Rodelas et al. 1999; Arkhipova et al. 2007). Similarly, *Azospirillum* is also known to secrete phytohormones, induce root cell differentiation and increase water uptake (Bashan and Holguin 1997). As stated earlier, Gyaneshwar et al. (2001) also showed inoculation of *S. marcescens* IRBG500 with rice variety IR72 resulted in a significant increase in root length and root dry weight but not in total N content of rice, suggesting that the growth promotion was probably due to mechanisms other than N_2 fixation. Furthermore, *S. marcescens* IRBG500 did not show acetylene reduction activity (ARA) in association with rice.

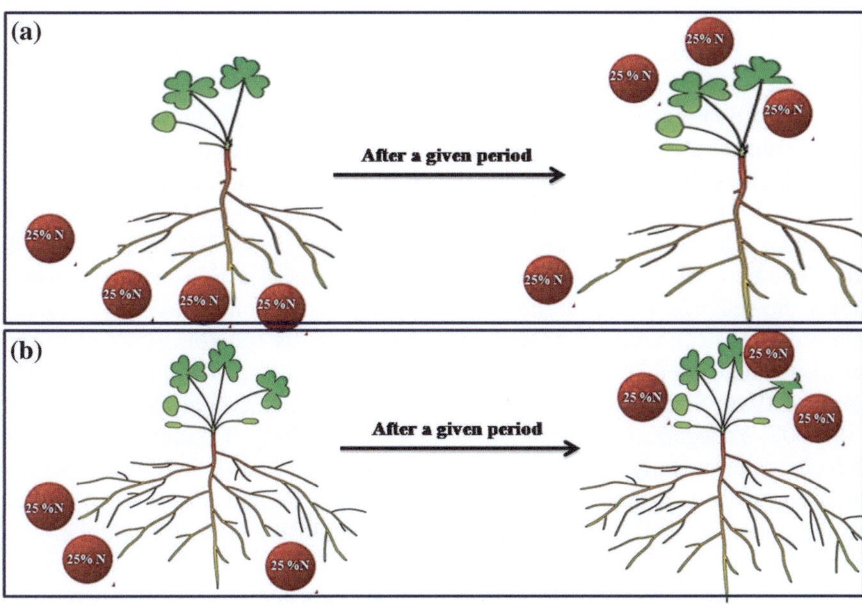

Fig. 6 Improving the nutrient use efficiency of N-fertilizers by IAA-producing PGPR. **a** 100 % recommended fertilizer without IAA-producing PGPR. **b** 75 % recommended fertilizer with IAA-producing PGPR. IAA-producing PGPR by increasing root surface area led to reduction of the percentage of recommended fertilizer (25 % reduction). Nutrient (N) uptake was comparable to those with the full rate of fertilizer without inoculants (75 % N in each plant) during a given period

Enhanced root growth following *V. paradoxus* 5C-2 inoculation probably improved nutrient uptake. These nutritional effects seem partially specific to *V. paradoxus* 5C-2, as other ACC deaminase-containing PGPRs (*P. brassicacearum* Am3, *P. marginalis* Dp1, or *Rhodococcus* sp. Fp2) had positive effects on pea foliar N, Ca, S and Fe concentrations (Safronova et al. 2006). A combination of the activities of plant and inoculants may be proposed as a model for PGPRs-enhanced N uptake in plants (Adesemoye et al. 2009) (Fig. 5). PGPRs promote the growth of the plant and increase the root surface area or the general root architecture (Biswas et al. 2000; Lucy et al. 2004). Plants growing better in turn release higher amounts of C in root exudates. The release of more C prompts increase in microbial activity and this process continues in a cycle. The whole process makes more N available from the soil pool, influencing N flux into plant roots and the plant is able to take up more available N. Figure 6 shows Improving the nutrient use efficiency of N-fertilizers by IAA-producing PGPR.

18 Co-inoculation of Multiple PGPR Strains as Way to Enhance the Performance of PGPR

Because soil is an unpredictable environment, the effect of PGPRs in crop productivity varies under laboratory, greenhouse and field trials. Climatic variations also have a large impact on the effectiveness of PGPRs but sometimes unfavorable growth conditions in the field are to be expected as normal functioning of agriculture (Zaidi et al. 2009a). To overcome the inconsistencies, one way that some previous studies have used to enhance the performance of PGPRs is co-inoculation of multiple PGPRs strains (Belimov et al. 1995; Raupach and Kloepper 2000; Bai et al. 2003; Kloepper et al. 2007; Pandey and Maheshwari 2007b; Elkoca et al. 2008). The best PGPRs may use multiple mechanisms of action on plant growth. Studies showed a promising trend in the field of inoculation technology, which is the use of mixed inoculants or application of consortia (combinations of microorganisms) that interact synergistically are currently being devised (Parmar and Dadarwal 1999; Steenhoudt and Vanderleyden 2000; Kumar et al. 2007; Rokhzadi et al. 2008; Yadegari et al. 2008; Pirlak and Kose 2009). Tittabutr et al. (2008) conducted such a study to evaluate effect of ACC deaminase activity on nodulation and growth of *Leucaena leucocephala*. Further, Remans et al. (2007) examined the potential of ACC deaminase producing PGPRs to enhance nodulation of common bean (*P. vulgaris*). Shaharoona et al. (2006) observed that co-inoculation with *Pseudomonas* and *Bradyrhizobium* species significantly improved root length, total biomass and nodulation in mung bean. Co-inoculation of a variety of PGPRs such as *Azotobacter chroococcum* and *P. putida* with *Rhizobium* sp. (AR-2-2 k) showed increased plant growth, nodulation and improved nitrogenase activity. The association of *Rhizobium* sp. with *P. putida*, *P. fluorescens* and *Bacillus cereus* seem to produce the best agronomical results (Tilak and Ranganayaki 2006). Belimov et al. (1995) reported significantly greater uptake of P in shoot of barley with co-inoculation of *A. lipoferum* 137 and *Arthrobacter mysorens* 7 or *A. lipoferum* 137 and *Agrobacterium radiobacter* 10 than single inoculation of any of the three organisms. Microbial interaction studies performed without plants indicate that some bacterial genera allow each other to interact synergistically providing nutrients, removing inhibitory products and stimulating each other through physical or biochemical activities that may enhance some beneficial aspects of their physiology such as nitrogen fixation (Pandey and Maheshwari 2007; Arora et al. 2008). Plant studies have shown that these beneficial effects of *Azospirillum* on plants can be enhanced by co-inoculation with other microorganisms (Alagawadi and Gaur 1992; Belimov et al. 1995). Co-inoculation frequently increased growth and yield compared to single inoculation, which provided the plants with more balanced nutrition and improved absorption of nitrogen, phosphorus and mineral nutrients (Kumar et al. 2009). There is a great advantage of using phosphate-solubilizing bacteria in co-inoculation with rhizobia. This is because increased P mobilization in soil improves P deficiency. Deficit P severely limits plant growth and productivity particularly with legumes, where both plants and their symbiotic bacteria are affected. Iron availability is one of the limiting factors for poor rhizospheric colonization. The successful performance of rhizobial inoculant strain depends upon their capability to outcompete the indigenous soil bacteria, survive, propagate and enter into effective symbiosis with host plant. Many studies have indicated that efficient utilization of siderophores by rhizobia is a positive fitness factor with respect to its survival in soil (Carson et al. 2000). Further, Joshi et al. (2009) observed increase in nodule occupancy and higher rhizospheric colonization by pigeon pea-nodulating rhizobia expressing engineered siderophore cross-utilizing abilities. Thus, iron availability is one of the major factors determining rhizospheric colonization. This fact is further evidenced by work of Mahmoud and Abd-Alla (2001) where authors showed that co-inoculation of siderophore-producing PGPRs significantly enhanced nodulation and nitrogen fixation in mung bean compared to plants infected with rhizobial strain alone. There are more reports that specific siderophore-producing PGPRs stimulated the nodulation, nitrogen fixation and plant growth of leguminous plants (Grimes and Mount 1987; Omar and Abd-Alla 1994; Shenker et al. 1999). Application of PGPRs could not only produce significant benefits that require minimal or reduced levels of fertilizers but also consequently produce a synergistic effect on root growth and development (Kumar et al. 2009). Figueiredo et al. (2008) reported increased plant growth, N content and nodulation of *P. vulgaris* L. under drought stress due to co-inoculation of *Rhizobium tropici* and *P. polymyxa*. *P. vulgaris* (common bean) plants inoculated with *Rhizobium etli* overexpressing trehalose-6-phosphate synthase gene had more nodules with increased nitrogenase activity and high biomass compared with plants inoculated with wild-type *R. etli*. Three weeks old plants subjected to drought stress fully recovered, whereas plants inoculated with a wild-type *R. etli* did not survive. Decreased ethylene levels allow the plant to be more resistant to a wide variety of environmental stresses. Indeed PGP microorganisms have multifaceted beneficial effects (Avis et al. 2008) that can complement each other due to multifarious phenomenon (Maheshwari et al. 2014).

19 Biological Fertilizers Based on Bacterial Hormones

Chemical fertilizers are essential components of modern agriculture because they provide essential plant nutrients. For example, rice is the most important staple food in several developing countries and chemical fertilizers (especially N) are the most important input required for its cultivation. However, overuse of the fertilizers can cause unanticipated environmental impacts. The search for microorganisms that improve soil fertility and enhance plant nutrition has continued to attract attention due to the increasing cost of fertilizers and some of their negative environmental impacts. One potential way is the use PGPRs in order to make its cultivation sustainable and less dependent on chemical fertilizers. It is important to know how to use PGPRs that can biologically fix nitrogen, solubilize phosphorus and iron and induce some substances like IAA that could contribute to the improvement of plant growth. Nevertheless, PGPRs often fail to confer these beneficial effects when applied in the field, which is often due to insufficient rhizo- and/or endosphere colonization. The major limitation today for use of these organisms is the lack of consistent effects in PGP traits under field conditions. This is likely due to competition with the native microflora and environmental factors that either limit the population size (poor colonization) or activity of the PGPRs. Thus, the ability of a bacterial inoculant to promote plant growth can only be fully evaluated when they are tested in association with all of the components of the rhizosphere (Schroth and Weinhold 1986). Physical and chemical (abiotic soil factors) factors, such as soil texture, pH, nutrient status, high osmotic conditions, moisture, temperature, organic matter content and biological interactions in the rhizosphere are also known to impose stresses on microorganisms that may affect the establishment, survival and activity of certain organisms, whereas other organisms may remain unaffected (van Elsas and van Overbeek 1993; van Veen et al. 1997; Schroth and Weinhold 1986; Glick 1995). Bashan et al. (1995) demonstrated that concentrations of nitrogen, potassium and phosphate in soil are correlated with survival of *A. brasilense*. Despite inconsistency in field performance, PGPRs are considered as an alternative or a supplemental way of reducing chemical fertilizer in agroecosystem. In natural ecosystems, the behavior of introduced bacterial inoculants (e.g. PGPRs) and the subsequent expression of PGP represent a complex set of multiple interactions between introduced bacteria, associated crops and indigenous soil microflora. The expression of a particular trait under soil conditions is governed by the interaction of the inoculant strain with the host plant, other microorganisms in the rhizosphere, environmental factors and its own genetic makeup. In general, root elongation changes qualitatively are based on the IAA level, therefore, the amount of released IAA could have an important role in modulating the plant–microbe interaction. The property of synthesizing IAA and ACC deaminase activity is considered as effective tool for screening beneficial microorganisms suggesting that IAA-producing bacteria have profound effect on plant growth (Wahyudi et al. 2011). In view of role of bacterial IAA together with ACC deaminase activity in root elongation, enhancing root surface area, decreasing environmental stresses and more colonization, it may be suggested the production of biological fertilizer based on bacterial hormones can be effectively used for a sustainable crop management under field conditions. Production of IAA and ACC deaminase by PGPRs, result in increased root length, root surface area and number of root tips, leading to enhanced uptake of nutrients thereby improving plant health under stress conditions (IAA by better root growth and nutrient uptake and ACC deaminase by reducing stress ethylene) (Egamberdieva and Kucharova 2009).

20 Conclusion and Future Prospects

The regulation of growth and functioning of plant root systems has attracted increased scientific attention in studies which aim to increase crop production but decrease negative environmental impacts of agriculture by decreasing water and nutrient inputs (Lynch 2007; Ghanem et al. 2011). This can be achived by using ACC deaminase and IAA-producing bacteria. These PGPRs potentially offer a low cost and flexible method to increase plant growth by regulating the growth and functioning of the root system and can stimulate plant growth directly by producing or metabolizing plant hormones or enhancing plant nutrient uptake (Arshad and Frankenberger 1991; Vessey 2003; Dodd et al. 2010; Dodd and Ruiz-Lozano 2012). It has been documented that the IAA-producing bacteria together with ACC deaminase activity exert stimulatory effects on the growth of plants. The beneficial effects of these PGPRs are mostly related to the changes in IAA concentration. At the same time, modification of phytohormone levels by microbes can lead to characteristic changes in plant growth development such as phytohormones produced by the bacteria, which can increase root area, leading to higher water and other nutrients uptake from soil. These bacteria, therefore, can be effectively used for plant growth improvement.

Further investigations about the mechanisms involved would help to improve the understanding of plant growth promotion by microorganisms. IAA accumulation in the rhizosphere contributes to an increase in the root surface area and to alterations in root exudation. As a result, plant nutrition and growth are improved, new niches for plant colonization by the bacteria are formed, and bacterial IAA production is corrected again. A better understanding of the basic principles of the rhizosphere ecology, including the function and diversity of inhabiting microorganisms is on the way but further knowledge is necessary to optimize soil microbial technology to the benefit of plant growth and health in the natural environment. Therefore, current production methods in agriculture, e.g. the improper use of chemical pesticides and fertilizers creating a long list of environmental and health problems, should be reduced. Our understanding of plant–microbe interactions in rhizosphere must increase before we can presume that utilization of PGPRs as biofertilizers will determine a sustainable promotion of host plants growth. While considerable research has demonstrated their potential utility, the successful application of PGPRs in the field has been limited by a lack of knowledge of ecological factors that determine their survival and activity in the plant rhizosphere. Therefore, the practical application of these techniques should be further evaluated in field experiments.

The finding that IAA is used as a signal for gene regulation in some bacteria, both in IAA producers and nonproducers further supports the idea of IAA being part of genetic networks in some microorganisms. When these microorganisms interact with plants as part of their ecological habitat, it becomes obvious that a reciprocal IAA-mediated signaling process in microbe–plant interactions is likely to occur (Lambrecht et al. 2000). Our further understanding of bacteria–plant interactions be it pathogenic or beneficial, needs detailed studies that examine hormonal dynamics throughout the course of the interaction. Nevertheless, these conditions were removed from real conditions where the inoculum strain has to compete with a wide variety of soil microorganisms. Therefore, experiments under real conditions are necessary to clarify if the strain is able to promote the growth of plants under real soil conditions. However, the application of inocula in agriculture needs further research to better understand the interactions between plants and microorganisms. Not only is it necessary to provide the right microorganisms, but also the correct techniques to check the fate of the inoculum in order to establish the most suitable way to use the microorganisms in agriculture. The lack of such information has been shown to be the main cause of failure in the use of PGPRs. It is also suggested that PGPRs need to be reinoculated every year/season as they will not live forever in the soil. A large body of knowledge suggests that root exudates may act as messengers that communicate and initiate biological and physical interactions between roots and soil organisms. Although root exudation clearly represents a significant carbon cost to the plant, the mechanisms and regulatory processes controlling root secretion are just now beginning to be examined. In conclusion, this review and our studies (Etesami et al. 2014a, b, 2015) also signify that screening of effective bacterial strains under controlled conditions based on IAA and ACC deaminase production and growth promotion may be a useful strategy for the selection of efficient isolates.

Acknowledgments We wish to thank the head of the Soil Science Department, the Vice-Chanceller for research at University of Tehran for providing the necessary facilities for this study.

References

Abeles FB, Morgan PW, Saltveit ME Jr (1992) Ethylene in plant biology, 2nd edn. Academic Press, New York

Abou-Shanab RA, Angle JS, Delorme TA, Chaney RL, van Berkum P, Moawad H, Ghanem K, Ghozlan HA (2003) Rhizobacterial effects on nickel extraction from soil and uptake by *Alyssum murale*. New Phytol 158:219–224

Abou-Shanab RA, Angle JS, Chaney RL (2006) Bacterial inoculants affecting nickel uptake by *Alyssum murale* from low, moderate and high Ni soils. Soil Biol Biochem 38:2882–2889

Adams PD, Kloepper JW (2002) Effect of host genotype on indigenous bacterial endophytes of cotton (*Gossypium hirsutum* L.). Plant Soil 240:181–189

Adams DO, Yang SF (1979) Ethylene biosynthesis: identification of l-aminocyclopropanecarboxylic acid as an intermediate in the conversion of methionine to ethylene. Proc Natl Acad Sci USA 76:170–174

Adesemoye AO, Torbert HA, Kloepper JW (2009) Plant growth promoting rhizobacteria allow reduced application rates of chemical fertilizers. Microbial Ecol 58:921–929

Adhikari T, Joseph C, Yang G, Phillips D, Nelson L (2001) Evaluation of bacteria isolated from rice for plant growth promotion and biological control of seedling disease of rice. Can J Microbiol 47(10):916–924

Adriano-Anaya ML, Salvador-Figueroa M, Ocampo JA, Garcia Romero I (2006) Hydrolytic enzyme activities in maize (*Zea mays*) and sorghum (*Sorghum bicolor*) roots inoculated with *Gluconacetobacter diazotrophicus* and *Glomus* intraradices. Soil Biol Biochem 38:879–886

Ahmad F, Ahmad I, Khan MS (2008) Screening of free-living rhizospheric bacteria for their multiple plant growth promoting activities. Microbiol Res 163:173–181

Ait bakra E, Nowak J, Clement C (2006) Enhancement of chilling resistance of inoculated grapevine plantlets with a plant growth promoting rhizobacterium, *Burkholderia phytofirmans* strain PsJN. Appl Environ Microbiol 72(11):7246–7252

Akhtar MS, Siddiqui ZA (2008) Biocontrol of a root-rot disease complex of chickpea by *Glomus intraradices*, *Rhizobium* sp. and *Pseudomonas straita*. Crop Prot 27:410–417

Alagawadi AR, Gaur AC (1992) Inoculation of *Azospirillum brasilense* and phosphate solubilizing bacteria on yield of sorghum (*Sorghum bicolor* L. Moench) in dry land. Trop Agric 69:347–350

Ali S, Charles TC, Glick BR (2012) Delay of carnation flower senescence by bacterial endophytes expressing ACC deaminase. J Appl Microbiol 113:1139–1144

Aloni R, Aloni E, Langhans M, Ullrich CI (2006) Role of cytokinin and auxin in shaping root architecture: regulating vascular differentiation, lateral root initiation, root apical dominance and root gravitropism. Ann Bot 97:883–893

Altomare C, Norvell WA, Bjorkman T, Harman GE (1999) Solubilization of phosphates and micronutrients by the plant growth-promoting and biocontrol fungus *Trichoderma harzianum Rifai* 1295–22. Appl Environ Microbiol 65:2926–2933

Alvarez R, Nissen SJ, Sutter EG (1989) Relationship between indole-3-acetic acid levels in apple (*Malus pumila* Mill.) rootstocks cultured in vitro and adventitious root formation in the presence of indole-3-butyric acid. Plant Physiol 89:439–443

Amhad M, Zahir ZA, Asghar HN, Asghar M (2011) Inducing salt tolerance in mung bean through coinoculation with rhizobia and plant growth-promoting rhizobacteria containing 1-aminocyclopropane-1-carboxylate deaminase. Can J Microbiol 57:578–589

Amir H, Pineau R (2003) Release of Ni and Co by microbial activity in New Caledonian ultramafic soils. Can J Microbiol 49:288–293

Amir HG, Shamsuddin ZH, Halimi MS, Marziah M, Ramlan MF (2005) Enhancement in nutrient accumulation and growth of oil palm seedlings caused by PGPR under field nursery conditions. Commun Soil Sci Plant Anal 36:2059–2066

Andrews J, Harris R (2000) The ecology and biogeography of microorganisms on plant surface. Ann Rev Phytopathol 38:145–180

Araujo WL, Maccheroni W Jr, Aguilar-Vildoso CI, Barroso PAV, Saridakis HO, Azevedo JL (2001) Variability and interactions between endophytic bacteria and fungi isolated from leaf tissues of citrus rootstocks. Can J Microbiol 47:229–236

Araujo WL, Marcon J, Macceroni Junior W, van Elsas JD, van Vuured JWL, Azevedo JL (2002) Diversity of endophytic bacterial populations and their interaction with *Xylella fastidiosa* in citrus plants. Appl Environ Microbiol 68:4906–4914

Arkhipova TN, Prinsen E, Veselov SU, Martinenko EV, Melentiev AI, Kudoyarova GR (2007) Cytokinin producing bacteria enhance plant growth in drying soil. Plant Soil 292:305–315

Arnold AE, Lutzoni F (2007) Diversity and host range of foliar fungal endophytes: are tropical leaves biodiversity hotspots. Ecol 88:541–549

Arora NK, Kang SC, Maheshwari DK (2001) Isolation of siderophore-producing strains of *Rhizobium meliloti* and their biocontrol potential against *Macrophomina phaseolina* that causes charcoal rot of groundnut. Curr Sci 81:673–677

Arora NK, Khare E, Oh JH, Kang SC, Maheshwari DK (2008) Diverse mechanisms adopted by *fluorescent Pseudomonas* PG during the inhibition of *Rhizoctonia solani* and *Phytophthora capsici*. World J Microbiol Biotechnol 24:581–585

Arshad M (2007) Perspectives of bacterial ACC deaminase in phytoremediation. Trends Biotechnol 25:356–362

Arshad M, Frankenberger WT (1991) Microbial production of plant hormones. Plant Soil 133:1–8

Arshad M, Frankenberger WT (2002) Ethylene: agricultural sources and applications. Kluwer Academic/Plenum, New York

Arshad M, Saleem M, Hussain S (2007) Perspectives of bacterial ACC deaminase in phytoremediation. Trends Biotechnol 25:356–362

Arshad M, Shaharoona B, Mahmood T (2008) Inoculation with plant growth promoting rhizobacteria containing ACC-deaminase partially eliminates the effects of water stress on growth, yield and ripening of pea (*Pisum sativum* L.). Pedosph 18:611–620

Aseri GK, Jain N, Panwar J, Rao AV, Meghwal PR (2008) Biofertilizers improve plant growth, fruit yield, nutrition, metabolism and rhizosphere enzyme activities of pomegranate (*Punica granatum* L.) in Indian Thar Desert. Sci Hortic 117:130–135

Asghar HN, Zahir ZA, Arshad M, Khaliq A (2002) Relationship between in ritro production of auxins by rhizobacteria and their growth-promoting activities in *brassicajunceal*. Biol Fertil Soils 35:231–237

Audenaert K, Pattery T, Cornelis P, Hofte M (2002) Induction of systemic resistance to Botrytis cinerea in tomato by *Pseudomonas aeruginosa* 7NSK2: role of salicylic acid, pyochelin, and pyocyanin. Mol Plant Microbe Interact 15:1147–1156

Avis TJ, Gravel V, Antoun H, Tweddell RJ (2008) Multifaceted beneficial effects of rhizosphere microorganisms on plant health and productivity. Soil Biol Biochem 40:1733–1740

Ayala S, Rao EVSP (2002) Perspective of soil fertility management with a focus on fertilizer use for crop productivity. Curr Sci 82:797–807

Azevedo JL, Maccheroni JJ, Pereira JO, Ara WL (2000) Endophytic microorganisms: a review on insect control and recent advances on tropical plants. Electron J Biotechnol 3:40–65

Babalola OO (2010) Beneficial bacteria of agricultural importance. Biotechnol Lett 32:1559–1570

Babalola OO, Glick BR (2012) Indigenous African agriculture and plant associated microbes: current practice and future transgenic prospects. Sci Res Essays 7(28):2431–2439

Babana AH, Antoun H (2006) Effect of Tilemsi phosphate rock-solubilizing microorganisms on phosphorus uptake and yield of field grown wheat (*Triticum aestivum* L.) in Mali. Plant Soil 28:51–58

Bacilio-Jimenez M (2003) Chemical characterization of root exudates from rice (*Oryza sativa*) and their effects on the chemotactic response of endophytic bacteria. Plant Soil 249:271–277

Badenochjones J, Rolfe BG, Letham DS (1983) Phytohormones, *Rhizobium* mutants, and nodulation in legumes. 3. Auxin metabolism in effective and ineffective pea root nodules. Plant Physiol 73:347–352

Badri DV, Vivanco JM (2009) Regulation and function of root exudates. Plant, Cell Environ 32:666–681

Bai Y, Zhou X, Smith DL (2003) Enhanced soybean plant growth resulting from coinoculation of *Bacillus* strains with *Bradyrhizobium japonicum*. Crop Sci 43:1774–1781

Bais HP, Loyola Vargas VM, Flores HE, Vivanco JM (2001) Root specific metabolism: the biology and biochemistry of underground organs. In Vitro Cell Dev Biol Plant 37:730–741

Bais HP, Park SW, Weir TL, Callaway RM, Vivanco JM (2004) How plants communicate using the underground information superhighway. Trends Plant Sci 9:26–32

Bais HP, Weir TL, Perry LG, Gilroy S, Vivanco JM (2006) The role of root exudates in rhizosphere interactions with plants and other organisms. Annu Rev Plant Biol 57:233–266

Baldani VLD, Dobereiner J (1980) Host-plant specificity in the infection of cereals with *Azospirillum* spp. Soil Biol Biochem 12:439–443

Barea JM, Azcon R, Azcon-Aguilar C (2002) Mycorrhizosphere interactions to improve plant fitness and soil quality. Antonie Van Leeuwenhoek 81:343–351

Barker SJ, Tagu D (2000) The roles of auxins and cytokinins in mycorrhizal symbiosis. J Plant Growth Regul 19:144–154

Barnawal D, Bharti N, Maji D, Chanotiya CS, Kaira A (2012) 1-Aminocyclopropane-1-carboxylic acid (ACC) deaminase-containing rhizobacteria protect *Ocimumsanctum* plants during waterlogging stress via reduced ethylene generation. Plant Physiol Biochem 58:227–235

Barzanti R, Ozino F, Bazzicalupo M, Gabbrielli R, Galardi F, Gonnelli C et al (2007) Isolation and characterization of endophytic bacteria from the nickel hyperaccumulator plant *Alyssum bertolonii*. Microb Ecol 53:306–316

Bashan Y (1986) Inoculation of rhizosphere bacteria *Azospirillum brasilense* and *Pseudomonas fluorescens* towards wheat roots in the soil. J Gen Microbiol 132:3407–3414

Bashan Y, Holguin G (1997) *Azospirillum*-plant relationships: environmental and physiological advances (1990–1996). Can J Microbiol 43:102–121

Bashan Y, Holguin G (1998) Proposal for the division of plant growth-promoting rhizobacteria into two classifications: biocontrol-PGPB (plant growth promoting bacteria) and PGPB. Soil Biol Biochem 30:1225–1228

Bashan Y, Puente ME, Rodriguez-Mendoza MN, Toledo G, Holguin G, Ferrera-Cerrato R, Pedrin S (1995) Survival of *Azospirillum brasilense* in the bulk soil and rhizosphere of 23 soil types. Appl Environ Microbiol 61:1938–1945

Bastián F, Cohen A, Piccoli P, Luna V, Baraldi R, Bottini R (1998) Production of indole-3-acetic acid and gibberellin A1 and A3 *by Acetobacter diazotrophicus* and *Herbaspirillum seropedicae* in chemically-defined culture media. Plant Growth Regul 24:7–11

Beattie GA (2006) Plant-associated bacteria: survey, molecular phylogeny, genomics and recent advances. In: Gnanamanickam SS (ed) Plant-associated bacteria. Springer, The Netherlands, pp 1–56

Begonia MFT, Kremer RJ (1994) Chemotaxis of deleterious rhizobacteria to velvet leaf (*Abutilon theophrasti Medik.*) seeds and seedlings. FEMS Microbiol Ecol 15:227–236

Belimov AA (2012) Interactions between associative bacteria and plants: the role of biotic and abiotic factors. Palmarium Academic Publishers, Moscow

Belimov AA, Kojemiakov AP, Chuvarliyea CV (1995) Interaction between barley and mixed cultures of nitrogen fixing and phosphate solubilizing bacteria. Plant Soil 173:29–37

Belimov AA, Safronova VI, Sergeyeva TA, Egorova TN, Matveyeva VA, Tsyganov VE, Borisov AY, Tikhonovich IA, Kluge C, Preisfeld A, Dietz KJ, Stepanok VV (2001) Characterization of plant growth promoting rhizobacteria isolated from polluted soils and containing 1-aminocyclopropane-1-carboxylate deaminase. Can J Microbiol 47:242–252

Belimov AA, Safronova VI, Mimura T (2002) Response of spring rape to inoculation with plant growth-promoting rhizobacteria containing 1-aminocyclopropane-1-carboxylate deaminase depends on nutrient status of the plant. Can J Microbiol 48:189–199

Belimov AA, Hontzeas N, Safronova VI, Demchinskaya SV, Piluzza G, Bullitta S, Glick BR (2005) Cadmium-tolerant plant growth-promoting bacteria associated with the roots of Indian mustard (*Brassica juncea* L. Czern.). Soil Biol Biochem 37:241–250

Belimov A, Dodd IC, Safronova VI, Hontzeas N, Davies WJ (2007) Pseudomonas brassicacearum strain Am 3 containing 1-aminocyclopropane-1-carboxylate deaminase can show both pathogenic and growth-promoting properties in its interaction with tomato. J Exp Bot 58:1485–1495

Belimov AA, Dodd IC, Hontzeas N, Theobald JC, Safronova VI, Davies WJ (2009a) Rhizosphere bacteria containing 1-aminocyclopropane-1-carboxylate deaminase increase yield of plants grown in drying soil via both local and systemic hormone signaling. New Phytol 181:413–423

Belimov AA, Dodd IC, Safronova VI, Davies WJ (2009b) ACC deaminase-containing rhizobacteria improve vegetative development and yield of potato plants grown under water-limited conditions. Aspects Appl Biol 98:163–169

Bellis P, Ercolani GL (2001) Growth interactions during bacterial colonization of seedling rootlets. Appl Environ Microbiol 67:1945–1948

Bender C, Alarcón-Chaidez F, Gross D (1999) *Pseudomonas syringae* phytotoxins: mode of action, regulation, and biosynthesis by peptide and polyketide synthetases. Microbiol Mol Biol Rev 63:266–292

Benizri E, Courtade A, Picard C, Guckert A (1998) Role of maize root exudates in the production of auxins by *Pseudomonas fluorescens* M.3.1. Soil Biol Biochem 30:1481–1484

Benizri E, Baudoin E, Guckert A (2001) Root colonization by inoculated plant growth promoting rhizobacteria. Biocontrol Sci Technol 11:557–574

Benjamin P, Sergio S, Benoit L, Florence A, Carole S, Patrick D, Laurent LA (2008) Role for auxin during actinorhizal symbioses formation? Plant Signal Behav 3(1):34–35

Bensalim S, Nowak J, Asiedu SK (1998) A plant growth promoting rhizobacterium and temperature effects on performance of 18 clones of potato. Am J Potato Res 75:145–152

Benson DR, Silvester WB (1993) Biology of *Frankia* strains, actinomycete symbionts of actinorhizal plants. Microbiol Rev 57:293–319

Bent E, Tuzun S, Chanway CP, Enebak S (2001) Alterations in plant growth and in root hormone levels of lodgepole pines inoculated with rhizobacteria. Can J Microbiol 47:793–800

Berg G (2009) Plant-microbe interactions promoting plant growth and health: perspectives for controlled use of microorganisms in agriculture. Appl Microbiol Biotechnol 84:11–18

Bhattacharjee RB, Singh A, Mukhopadhyay SN (2008) Use of nitrogen-fixing bacteria as biofertiliser for non-legumes: prospects and challenges. Appl Microbiol Biotechnol 80:199–209

Bhattacharyya PN, Jha DK (2012) Plant growth-promoting rhizobacteria (PGPR): emergence in agriculture. World J Microbiol Biotechnol 28:1327–1350

Bhusan Bal H, Das S, Tushar K, Adhya TK (2013) ACC deaminase and IAA producing growth promoting bacteria from the rhizosphere soil of tropical rice plants. J Basic Microbiol 00:1–13

Bianco C, Defez R (2009) *Medicago truncatula* improves salt tolerance when nodulated by an indole-3-acetic acid overproducing *Sinorhizobium meliloti* strain. J Exp Bot 60:3097–3107

Bianco C, Defez R (2010) Improvement of phosphate solubilization and *Medicago* plant yield by an indole-3-acetic acid-overproducing strain of *Sinorhizobium meliloti*. Appl Env Microbiol 76:4626–4632

Bianco C, Imperlini E, Calogero R, Senatore B, Pucci P, Defez R (2006) Indole-3-acetic acid regulates the central metabolic pathways in *Escherichia coli*. Microbiol-Sgm 152:2421–2431

Biswas JC, Ladha JK, Dazzo FB (2000a) Rhizobial inoculation improves uptake and growth of lowland rice. Soil Sci Soc Am J 64:1644–1650

Biswas JC, Ladha JK, Dazzo FB, Yanni YG, Rolfe BG (2000b) Rhizobial inoculation influences seedling vigor and yield of rice. Agron J 92:880–886

Bjorkman T, Blanchard LM, Harman GE (1998) Growth enhancement of shrunken-2 sweet corn when colonized with *Trichoderma harzianum* 1295-22: effect of environmental stress. J Am Soc Hortic Sci 123:35–40

Blaha D, Prigent-Combaret C, Mirza MS, Moenne-Loccoz Y (2006) Phylogeny of the 1-aminocyclopropane-1-carboxylic acid deaminase-encoding gene *acdS* in phytobeneficial and pathogenic Proteobacteria and relation with strain biogeography. FEMS Microbiol Ecol 56:455–470

Boddey RM, De Olivera OC, Urquaga S, Reis VM, Olivares FL, Baldani VLD, Döbereiner J (1995) Biological nitrogen fixation associated with sugarcane and rice: contributions and prospects for improvement. Plant Soil 174:195–209

Böhm M, Hurek T, Reinhold-Hurek B (2007) Twitching motility is essential for endophytic rice colonization by the N2-fixing endophyte *Azoarcus* sp. strain BH72. Mol Plant Microbe Interact 20:526–533

Boiero L, Perrig D, Masciarelli O, Penna C, Cassan F, Luna V (2007) Phytohormone production by three strains of *Bradyrhizobium japonicum* and possible physiological and technological implications. Appl Microbiol Biotechnol 74:874–880

Bonaterra A, Ruz L, Badosa E, Pinochet J, Montesinos E (2003) Growth promotion of Prunus root stocks by root treatment with specific bacterial strains. Plant Soil 255:555–569

Bottini R, Cassan F, Picolli P (2004) Gibberellin production by bacteria and its involvement in plant growth promotion. Appl Microbiol Biotechnol 65:497–503

Bowen GD, Rovira AD (1991) The rhizosphere, the hidden half of the hidden half. In: Waisel Y, Eshel A, Kafkafi U (eds) The hidden half plant roots. Marcel Dekker, New York, pp 641–649

Brandl MT, Lindow SE (1996) Cloning and characterization of a locus encoding an indolepyruvate decarboxylase involved in indole-3- acetic acid synthesis in *Erwinia herbicola*. Appl Environ Microbiol 62:4121–4128

Brandl MT, Lindow SE (1997) Environmental signals modulate the expression of an indole-3-acetic acid biosynthetic gene in *Erwinia herbicola*. Mol Plant Microbe Interact 10:450–499

Brandl MT, Lindow SE (1998) Contribution of indole- 3-acetic acid production to the epiphytic fitness of *Erwinia herbicola*. Appl Environ Microbiol 64:3256–3263

Brandl M, Clark EM, Lindow SE (1996) Characterization of theindole-3-acetic acid (IAA) biosynthetic pathway in an epiphytic strain of *Erwinia herbicola* and IAA production in vitro. Can J Microbiol 42:586–592

Brandl MT, Quinones B, Lindow SE (2001) Heterogeneous transcription of an indoleacetic acid biosynthetic gene in *Erwinia herbicola* on plant surfaces. Proc Natl Acad Sci USA 98:3454–3459

Bringhurst RM, Cardon ZG, Gage DJ (2001) Galactosides in the rhizosphere: utilization by *Sinorhizobium meliloti* and development of a biosensor. Proc Natl Acad Sci USA 98:4540–4545

Brown ME (1974) Seed and root bacterization. Ann Rev Phytopathol 12:181–197

Buell CR, Joardar V, Lindeberg M et al (2003) The complete genome sequence of the Arabidopsis and tomato pathogen *Pseudomonas syringae* pv. tomato DC3000. Proc Natl Acad Sci USA 100:10181–10186

Bull CT, Weller DM, Thomashow LS (1991) Relationship between root colonization and suppression of *Gaeumannomyces graminis* var. *triciti* by *Pseudomonas fluorescens* strain 2–79. Phytopathology 81:954–959

Burd GI, Dixon DG, Glick BR (1998) A plant growth-promoting bacterium that decreases nickel toxicity in seedlings. Appl Environ Microbiol 64:3663–3668

Burd GI, Dixon DG, Glick BR (2000) Plant growth-promoting bacteria that decrease heavy metal toxicity in plants. Can J Microbiol 46:237–245

Burdman S, Volpin H, Kigel J, Kapulnik Y, Okon Y (1996) Promotion of *nod* gene inducers and nodulation in common bean (roots inoculated with *Azospirillum brasilense* Cd. Appl Environ Microbiol 62:3030–3033

Cakmacki R, Kantar F, Algur OF (1999) Sugar beet and barley yields in relation to *Bacillus polymyxa and Bacillus megaterium* var. phosphaticum inoculation. J Plant Nutr Soil Sci 162:437–442

Camerini S, Senatore B, Lonardo E, Imperlini E, Bianco C, Moschetti G, Rotino GL, Campion B, Defez R (2008) Introduction of a novel pathway for IAA biosynthesis to rhizobia alters vetch root nodule development. Arch Microbiol 190:67–77

Canbolat MY, Bilen S, Cakmakci R, Sahin F, Aydin A (2006) Effect of plant growth-promoting bacteria and soil compaction on barley seedling growth, nutrient uptake, soil properties and rhizosphere microflora. Biol Fertil Soils 42:350–357

Carrillo-Castañeda G, Munoz JJ, Peralta-Videa JR, Gomez E, Gardea-Torresdey JL (2003) Plant growth-promoting bacteria promote copper and iron translocation from root to shoot in alfalfa seedlings. J Plant Nutr 26:1801–1814

Carson KC, Meyer JM, Dilworth MJ (2000) Hydroxamate siderophores of root nodule bacteria. Soil Biol Biochem 32:11–21

Cattelan AJ, Hartela PG, Fuhrmann JJ (1999) Screening for plant growth-promoting rhizobacteria to promote early soybean growth. Soil Sci Soc Am J 63:1670–1680

Chaintreuil C, Giraud E, Prin Y, Lorquin J, Ba A, Gillis M, de Lajudie P, Dreyfus B (2000) Photosynthetic *bradyrhizobia* are natural endophytes of the African wild rice *Oryza breviligulata*. Appl Environ Microbiol 66:5437–5447

Chandra S, Choure K, Dubey RC, Maheshwari DK (2007) Rhizosphere competent *Mesorhizobium loti* MP6 induces root hair curling of Indian mustard (*Brassica compestris*). Braz J Microbiol 38:124–130

Chang WS, van de Mortel M, Nielsen L, de Guzman GN, Li X, Halverson LJ (2007) Alginate production by *Pseudomonas putida* creates a hydrated microenvironment and contributes to biofilm architecture and stress tolerance under water-limiting conditions. J Bacteriol 189:8290–8299

Chanway CP, Holl FB (1993) First year yield performance of spruce seedlings inoculated with plant growth promoting rhizobacteria. Can J Microbiol 39:1084–1088

Chen L, Dodd IC, Theobald JC, Belimov AA, Davies WJ (2013) The rhizobacterium *Variovorax paradoxus* 5C-2, containing ACC deaminase, promotes growth and development of *Arabidopsis thaliana* via an ethylene-dependent pathway. J Exp Bot 64(6):1565–1573

Cheng Z, Park E, Glick BR (2007) 1-aminocyclopropane-1-carboxylate deaminase from *Pseudomonas putida* UW4 facilitates the growth of canola in the presence of salt. Can J Microbiol 53:912–918

Cheng ZY, McConkey BJ, Glick BR (2010) Proteomic studies of plant–bacterial interactions. Soil Biol Biochem. doi:10.1016/j.soilbio.2010.05.033

Cheng Z, Woody OZ, McConkey BJ, Glick BR (2012) Combined effects of the plant growth-promoting bacterium *Pseudomonas putida* UW4 and salinity stress on the Brassica napus proteome. Appl Soil Ecol 61:255–263

Chet I, Zilberstein Y, Henis Y (1973) Chemotaxis of *Pseudomonas lachrymans* to plant extracts and to water droplets collected from the leaf surface of resistant and susceptible plants. Physiol Plant Pathol 3:473–479

Chi F, Shen SH, Cheng HP, Jing YX, Yanni YG, Dazzo FB (2005) Ascending migration of endophytic rhizobia, from roots to leaves, inside rice plants and assessment of benefits to rice growth physiology. Appl Environ Microbiol 71:7271–7278

Chookietwattana K, Maneewan K (2012) Selection of efficient salt-tolerant bacteria containing ACC deaminase for promotion of tomato growth under salinity stress. Soil Environ 31:30–36

Chung H, Park M, Madhaiyan M, Seshadri S, Song J, Cho H, Sa T (2005) Isolation and characterization of phosphate solubilizing bacteria from the rhizosphere of crop plants of Korea. Soil Biol Biochem 37:1970–1974

Compant S, Duffy B, Nowak J, Clement C, Barka A (2005a) Use of plant growth-promoting bacteria for biocontrol of plant diseases: principles, mechanisms of action, and future prospects. Appl Environ Microbiol 71(9):4951–4959

Compant S, Reiter B, Sessitsch A, Nowak J, Clément C, Ait Barka E (2005b) Endophytic colonization of *Vitis vinifera* L. by plant growth-promoting bacterium *Burkholderia* sp. strain PsJN. Appl Environ Microbiol 71:1685–1693

Compant S, Cle´ ment C, Sessitsch A (2010) Plant growth-promoting bacteria in the rhizo- and endosphere of plants: their role, colonization, mechanisms involved and prospects for utilization. Soil Biol Biochem 42:669–678

Compant S, Mitter B, Colli-Mull JG, Gangl H, Sessitsch A (2011) Endophytes of grapevine flowers, berries, and seeds: identification of cultivable bacteria, comparison with other plant parts, and visualization of niches of colonization. Microb Ecol 62:188–197

Contesto C, Desbrosses G, Lefoulon C, Béna G, Borel F, Galland M, Gamet L, Varoquaux F, Touraine B (2008) Effects of rhizobacterial ACC deaminase activity on *Arabidopsis* indicate that ethylene mediates local root responses to plant growth-promoting rhizobacteria. Plant Sci 175:178–189

Contreras-Cornejo HA, Macías-Rodríguez LI, Cortés-Penagos C, López-Bucio J (2009) *Trichoderma virens*, a plant beneficial fungus, enhances biomass production and promotes lateral root growth through an auxin-dependent mechanism in *Arabidopsis*. Plant Physiol 149:1579–1592

Costacurta A, Vanderleyden J (1995) Synthesis of phytohormones by plant-associated bacteria. Crit Rev Microbiol 21:1–18

Creus CM, Sueldo RJ, Barassi CA (2004) Water relations and yield in *Azospirillum* inoculated wheat exposed to drought in the field. Can J Bot 82:273–281

Czarny JC, Grichko VP, Glick BR (2006) Genetic modulation of ethylene biosynthesis and signaling in plants. Biotechnol Adv 24:410–419

Czarny JC, Shah S, Glick BR (2007) Response of canola plants at the transcriptional levelto expression of a bacterial ACC deaminase in the roots. In: Ramina A, Chang C, Giovannoni J, Klee H, Perata P, Woltering E (eds) Advances in plant ethylene research. Springer, Dordrecht, pp 377–382

Dakora FD, Philips DA (2002) Root exudates as mediators of mineral acquisition in low-nutrient environments. Plant Soil 245:35–47

Dangar TK, Basu PS (1987) Studies on plant growth substances, IAA metabolism and nitrogenase activity in root nodules of *Phaseolus aureus Roxb*. var. *mungo*. Biol Plant 29:350–354

Datta M, Banish S, Gupta RK (1982) Studies on the efficacy of a phytohormone producing phosphate solubilizing *Bacillus firmus* in augmenting paddy yield in acid soils of Nagaland. Plant Soil 69:365–373

Davies PJ (2004) The plant hormones: their nature, occurrence, and functions. In: Davies PJ (ed) Plant hormones, biosynthesis, signal transduction, action. Kluwer Academic Publishers, Dordrecht, pp 1–15

de Salamone IEG, Hynes RK, Nelson LM (2005) Role of cytokinins in plant growth promotion by rhizosphere bacteria. In: Siddiqui ZA (ed) PGPR: biocontrol and biofertilization. Springer, The Netherlands, pp 173–195

Dey R, Pal KK, Bhatt DM, Chauhan SM (2004) Growth promotion and yield enhancement of peanut (*Arachis hypogaea* L.) by application of plant growth-promoting rhizobacteria. Microbiol Res 159:371–394

Dobbelaere S, Croonenborghs A, Thys A, Broek AV, Vanderleyden J (1999) Phytostimulatory effect of *Azospirillum brasiliense* wild type and mutant strains altered in IAA production on wheat. Plant Soil 212:155–164

Dobbelaere S, Croonenborghs A, Thys A, Ptacek D, Vanderleyden J, Dutto P, Labandera-Gonzalez C, Caballero-Mellado J, Anguirre JF, Kapulnik Y, Brener S, Burdman S, Kadouri D, Sarig S, Okon Y (2001) Response of agronomically important crops to inoculation with *Azospirillum*. Aust J Plant Physiol 28:871–879

Dobbelaere S, Vanderleyden J, Okon Y (2003) Palnt growth promoting effects of diazotrophs in the rhizosphere. Crit Rev Plant Sci 22:107–149

Dodd IC, Ruiz-Lozano JM (2012) Microbial enhancement of crop resource use efficiency. Currt Opinion Biotechnol 23:236–242

Dodd IC, Belimov AA, Sobeih WY, Safronova VI, Grierson D, Davies WJ (2005) Will modifying plant ethylene status improve plant productivity in water-limited environments? In: 4th international crop science congress

Dodd IC, Zinovkina NY, Safronova VI, Belimov AA (2010) Rhizobacterial mediation of plant hormone status. Ann Appl Biol 157:361–379

Domka J, Lee J, Wood TK (2006) YliH (BssR) and YceP (BssS) regulate *Escherichia coli* K-12 biofilm formation by influencing cell signaling. Appl Environ Microbiol 72:2449–2459

Donate-Correa J, Leon-Barrios M, Perez-Galdona R (2005) Screening for plant growth-promoting rhizobacteria in *Chamaecytisus proliferus* (tagasaste), a forage tree-shrub legume endemic to the Canary Islands. Plant Soil 266:261–272

Dörr J, Hurek T, Reinhold-Hurek B (1998) Type IV pili are involved in plantemicrobe and fungusemicrobe interactions. Mol Microbiol 30:7–17

Duan J, Muller KM, Charles TC, Vesely S, Glick BR (2009) 1-Aminocyclopropane-1-Carboxylate (ACC) deaminase genes in Rhizobia from Southern Saskatchewan. Microb Ecol 57:423–436

Duineveld BM, Rosado AS, Elsas JD, van Veen JA (1998) Analysis of the dynamics of bacterial communities in the rhizosphere of the *Chrysanthemum* via denaturing gradient gel electrophoresis and substrate utilization patterns. Appl Environ Microbiol 64:4950–4957

Egamberdieva D, Kucharova Z (2009) Selection for root colonizing bacteria stimulating wheat growth in saline soils. Biol Fert Soil. doi:10.1007/s00374-009-0366-y

Elbeltagy A, Nishioka K, Suzuki H, Sato T, Sato Y, Morisaki H, Mitsui H, Minamisawa K (2000) Isolation and characterization of endophytic bacteria from wild and traditionally cultivated rice varieties. Soil Sci Plant Nutr 46:617–629

Elkoca E, Kantar F, Sahin F (2008) Influence of nitrogen fixing and phosphorus solubilizing bacteria on the nodulation, plant growth, and yield of chickpea. J Plant Nutr 31:157–171

Elliot LF, Lynch JM (1984) *Pseudomonads* as a factor in the growth of winter wheat (*Triticum aestivum* L.). Soil Biol Biochem 16:69–71

Elvira-Recuenco M, Van Vuurde JW (2000) Natural incidence of endophytic bacteria in pea cultivars under field conditions. Can J Microbiol 46:1036–1041

Etesami H, Mirseyedhosseini H, Alikhani HA (2013) Rapid screening of berseem clover (*Trifolium alexandrinum*) endophytic bacteria for rice plant seedlings growth-promoting agents. ISRN Soil Sci

Etesami H, Mirseyedhosseini H, Alikhani HA (2014a) Bacterial biosynthesis of 1-aminocyclopropane-1-caboxylate (ACC) deaminase, a useful trait to elongation and endophytic colonization of the roots of rice under constant flooded conditions. Physiol Mol Biol Plants 20(4):425–434

Etesami H, Mirsyed Hosseini H, Alikhani HA, Mohammadi L (2014b) Bacterial biosynthesis of 1-aminocyclopropane-1-carboxylate (ACC) deaminase and indole-3-acetic acid (IAA) as endophytic preferential selection traits by rice plant seedlings. J Plant Growth Regul 33:654–670

Etesami H, Alikhani HA, Mirsyed Hosseini H (2015) Indole-3-acetic acid (IAA) production trait, a useful screening to select endophytic and rhizosphere competent bacteria for rice growth promoting agents. Methodsx. doi:10.1016/j.mex.2015.02.008

Evans ML, Ishikawa H, Estelle MA (1994) Responses of arabidopsis roots to auxin studied with high temporal resolution: comparison of wild type and auxin-response mutants. Planta 194:215–222

Fallik E, Okon Y, Fisher M (1988) Growth response of maize roots to *Azospirillum* inoculation: effect of soil organic matter content, number of rhizosphere bacteria and timing of inoculation. Soil Biol Biochem 20:45–49

Farrar J, Hawes M, Jones D, Lindow S (2003) How roots control the flux of carbon to the rhizosphere. Ecology 84:827–837

Farzana Y, Radziah O, Kamaruza-man S, Saad MS (2007) Effect of PGPR inoculation on growth and yield of sweet potato. J Biol Sci 7:421–424

Faure D, Vereecke D, Leveau JHJ (2009) Molecular communication in the rhizosphere. Plant Soil 321:279–303

Fett WF, Osman SG, Dunn MF (1987) Auxin production by plant-pathogenic *pseudomonads* and *xanthomonads*. Appl Environ Microbiol 53:1839–1845

Figueiredo MVB, Burity HA, Martinez CR, Chanway CP (2008) Alleviation of drought stress in common bean (*Phaseolus vulgaris* L.) by co-inoculation with *Paenibacillus polymyxa* and Rhizobium tropici. Appl Soil Ecol 40:182–188

Fokkema NJ, Lorbeer JW (1974) Interactions between *Alternariaporri* and the saprophytic mycoflora of onion leaves. Phytopathology 64:1128–1133

Foster RC, Rovira AD (1978) The ultrastructure of the rhizosphere of *Trifolium subterraneum* L. In: Loutit MW, Miles JAR (eds) Microbial Ecol. Springer, Berlin, pp 278–290

Fouts DE, Abramovitch RB, Alfano JR (2002) Genomewide identification of *Pseudomonas syringae* pv. tomato DC3000 promoters controlled by the HrpL alternative sigma factor. Proc Natl Acad Sci USA 99:2275–2280

Frankenberger WT Jr, Arshad M (1995) Phytohormones in soil: microbial production and function. Dekker, New York

Fromin N, Achouak W, Thiery JM, Heulin T (2001) The genotypic diversity of *Pseudomonas brassicacearum* populations isolated from roots of *Arabidopsis thaliana*: influence of plant genotype. FEMS Microb Ecol 37:21–29

Fry SC (1989) Cellulases, hemicelluloses and auxin-stimulated growth: a possible relationship. Physiol Plant 75:532–536

Fuentes-Ramirez LE, Bustillos-Cristales R, Tapia-Hernandez A, Jimenez-Salgado T, Wang ET, Martinez-Romero E, Caballero-Mellado J (2001) Novel nitrogen fixing acetic acid bacteria, *Gluconacetobacter johannae* sp. nov. and *Gluconacetobacter azotocaptans* sp. nov., associated with coffee plants. Int J Syst Evol Microbiol 51:1305–1314

Fujishige NA, Kapadia NN, De Hoff PL, Hirsch AM (2006) Investigations of *Rhizobium* biofilm formations. FEMS Microbiol Ecol 56:195–206

Fukuhara H, Minakawa Y, Akao S, Minamisawa K (1994) The involvement of indole-3-acetic acid produced by *Bradyrhizobium elkanii* in nodule formation. Plant Cell Physiol 35:1261–1265

Gadd GM (2000) Bioremedial potential of microbial mechanisms of metal mobilization and immobilization. Curr Opin Biotechnol 11:271–279

Gadd GM (2004) Microbial influence on metal mobility and application for bioremediation. Geoderma 122:109–119

Gamalero E, Berta G, Massa N, Glick BR, Lingua G (2008) Synergistic interactions between the ACC deaminase-producing bacterium *Pseudomonas putida* UW4 and the AM fungus *Gigaspora rosea* positively affect cucumber plant growth. FEMS Microbiol Ecol 64:459–467

Gamalero E, Berta G, Glick BR (2009) The use of microorganisms to facilitate the growth of plants in saline soils. In: Khan MS, Zaidi A, Musarrat J (eds) Microbial strategies for crop

Gamalero E, Berta G, Massa N, Glick BR, Lingua G (2010) Interactions between *Pseudomonas putida* UW4 and *Gigaspora rosea* BEG9 and their consequences on the growth of cucumber under salt stress conditions. J Appl Microbiol 108:236–245

García de Salamone IE, Hynes RK, Nelson LM (2001) Cytokinin production by plant growthpromoting rhizobacteria and selected mutants. Can J Microbiol 47(5):404–411

Garcia Lucas JA, Probanza A, Ramos B, Colon Flores JJ, Gutierrez Manero FJ (2004) Effects of plant growth promoting rhizobacteria (PGPRs) on the biological nitrogen fixation, nodulation and growth of Lupinus albus L. cv. Multolupa. Eng Life Sci 4(1):71–77

Ghanem ME, Hichri I, Smigocki AC, Albacete A, Fauconnier ML, Diatloff E, Martinez-Andujar C, Lutts S, Dodd IC, Perez-Alfocea F (2011) Root-targeted biotechnology to mediate hormonal signaling and improve crop stress tolerance. Plant Cell Rep 30:807–823

Ghosh S, Basu PS (2006) Production and metabolism of indole acetic acid in roots and root nodules of *Phaseolus mungo*. Microbiol Res 161:362–366

Ghosh S, Penterman JN, Little RD, Chavez R, Glick BR (2003) Three newly isolated plant growthpromoting bacilli facilitate the seedling growth of canola Brassica campestris. Plant Physiol Biochem 41:277–281

Ghosh S, Sengupta C, Maiti TK, Basu PS (2008) Production of 3-indolylacetic acid in root nodules and culture by a *Rhizobium* species isolated from root nodules of the *leguminous* pulse *Phaseolus mungo*. Folia Microbiol 53:351–355

Gianinazzi-Pearson V (1996) Plant cell responses to arbuscular mycorrhizal fungi: getting to the roots of the symbiosis. Plant Cell 8:1871–1883

Glass ADM (1989) Plant nutrition: an introduction to current concepts. Jones and Bartlett Publishers, Boston, p 234

Glick BR (1995) The enhancement of plant growth by free-living bacteria. Can J Microbiol 4:109–117

Glick BR (1998) A model for the lowering of plant ethylene concentrations by plant growth-promoting bacteria. J Theor Biol 190:63–68

Glick BR (2004) Bacterial ACC deaminase and the alleviation of plant stress. Adv Appl Microbiol 56:291–312

Glick BR (2005) Modulation of plant ethylene levels by the bacterial enzyme ACC deaminase. FEMS Microbiol Lett 251:1–7

Glick BR (2012) Plant growth-promoting bacteria: mechanisms and applications. Scientifica 2012:1–15

Glick BR (2013) Bacteria with ACC deaminase can promote plant growthand help to feed the world. Microbiol Res 169(1):30–39

Glick BR, Bashan Y (1997) Genetic manipulation of plant growth-promoting bacteria to enhance biocontrol of phytopathogens. Biotechnol Adv 15:353–378

Glick BR, Jacobson CB, Schwarze MMK, Pasternak JJ (1994) 1-Aminocyclopropane-1-carboxylic acid deaminase mutants of the plant-growth promoting rhizobacterium *Pseudomonas putida* GR12-2 do not stimulate canola root elongation. Can J Microbiol 40:911–915

Glick BR, Karaturovic DM, Newell PC (1995) A novel procedure for rapid isolation of plant growth promoting Pseudomonads. Can J Microbiol 41:533–536

Glick BR, Cheng Z, Czarny J, Duan J (2007a) Promotion of plant growth by ACC deaminase-producing soil bacteria. Eur J Plant Pathol 119:329–339

Glick BR, Todorovic B, Czarny J, Cheng Z, Duan J, McConkey B (2007b) Promotion of plant growth by bacterial ACC deaminase. Crit Rev Plant Sci 26:227–242

Glickmann E, Gardan L, Jacquet S, Hussain S, Elasri M, Petit A, Dessaux Y (1998) Auxin production is a common feature of most pathovars of *Pseudomonas syringae*. Mol Plant Microbe Interact 11:156–162

Gnanamanickam SS (2006) Plant-associated bacteria. Springer wwwspringercom, pp 195–218

Goldberg R (1980) Cell wall polysaccharidase activities and growth processes: a possible relationship. Physiol Plant 50:261–264

Grichko VP, Glick BR (2001a) Amelioration of flooding stress by ACC-deaminase containing plant growth promoting bacteria. Plant Physiol Bioch 39:11–17

Grichko VP, Glick BR (2001b) Flooding tolerance of transgenic tomato plants expressing the bacterial enzyme ACC deaminase controlled by the 35 S, rolD or PRB-1b promoter. Plant Physiol Biochem 39:19–25

Grichko VP, Filby B, Glick BR (2000) Increased ability of transgenic plants expressing the bacterial enzyme ACC deaminase to accumulate Cd Co, Cu, Ni, Pb, and Zn. J Biotechnol 81:45–53

Grimes HD, Mount MS (1987) Influence of *Pseudomonas putida* on nodulation of Phaseolus vulgaris. Soil Biol Biochem 6:27–30

Guan LL, Kanoh K, Kamino K (2001) Effect of exogenous siderophores on iron uptake activity of marine bacteria under ironlimited conditions. Appl Environ Microbiol 67:1710–1717

Guinel FC, Geil RD (2002) A model for the development of the rhizobial and arbuscular mycorrhizal symbioses in legumes and its use to understand the roles of ethylene in the establishment of these two symbioses. Can J Bot 80:695–720

Gulati A, Sharma N, Vyas P, Sood S, Rahi P, Pathania V, Prasad R (2010) Organic acid production and plant growth promotion as a function of phosphate solubilization by *Acinetobacter rhizosphaerae* strain BIHB 723 isolated from the cold deserts of the trans-Himalayas. Arch Microbiol 192:975–983

Gutierrez-Manero FJ, Ramos-Solano B, Probanza A, Mehouachi J, Tadeo Francisco R, Talon M FR (2001) The plant growth-promoting rhizobacteria *Bacillus pumilus* and *Bacillis licheniformis* produce high amounts of physiologically active gibberellins. Physiol Plant 111:206–211

Gyaneshwar P, James EK, Mathan N, Reddy PM, Reinhold-Hurek Ladha JK (2001) Endophytic colonization of rice by a diazotrophic strain of *Serratia marcescens*. J Bacteriol 183:2634–2645

Gyaneshwar P, Kumar GN, Parekh LJ, Poole PS (2002) Role of soil microorganisms in improving P nutrition of plants. Plant Soil 245:83–93

Halda-Alija L (2003) Identification of indole-3- acetic acid producing freshwater wetland rhizosphere bacteria associated with *Juncus effusus* L. Can J Microbiol 49:781–787

Hall JA, Peirson D, Ghosh S, Glick BR (1996) Root elongation in various agronomic crops by the plant growth promoting rhizobacterium *Pseudomonas putida* GR 12–2. Isr J Plant Sci 44:37–42

Hallmann J (1997) Bacterial endophytes in agricultural crops bacterial endophytes in agricultural crops. Can J Microbiol 43:895–914

Hammad Y, Nalin R, Marechal J, Fiasson K, Pepin R, Berry AM, Normand P, Domenach AM (2003) A possible role for phenyl acetic acid (PAA) on *Alnus glutinosa* nodulation by *Frankia*. Plant Soil 254:193–205

Hao Y, Charles TC, Glick BR (2007) ACC deaminase from plant growth promoting bacteriaaffects crown gall development. Can J Microbiol 53:1291–1299

Hao Y, Charles TC, Glick BR (2011) An ACC deaminase containing A. tumefaciens strain D3 shows biocontrol activity to crown gall disease. Can J Microbiol 57:278–286

Hardoim PR, van Overbeek LS, Elsas JD (2008) Properties of bacterial endophytes and their proposed role in plant growth. Trends Microbiol 16:463–471

Hariprasad P, Niranjana SR (2009) Isolation and characterization of phosphate solubilizing rhizobacteria to improve plant health of tomato. Plant Soil 316:13–24

Harman GE, Petzoldt R, Comis A, Chen J (2004) Interaction between *Trichoderma harzianum* strain T-22 and maize inbred line Mo17 and effects of these interactions on disease caused by *Phytium ultimum* and *Colletotrichum graminicola*. Phytopathology 94:147–153

Hartman A, Schmid M, van Tuinen D, Berg G (2009) Plant-driven selection of microbes. Plant Soil 321:235–257

Hause B, Fester T (2005) Molecular and cell biology of arbuscular mycorrhizal symbiosis. Planta 221:184–196

He SY, Nomura K, Whittam TS (2004) Type III protein secretion mechanism in mammalian and plant pathogens. BBA-Mol Cell Res 1694:181–206

Hider RC, Kong X (2010) Chemistry and biology of siderophores. Nat Prod Rep 27:637–657

Himanen K, Boucheron E, Vanneste S, de Almeida Engler J, Inzé D, Beeckman T (2002) Auxin-mediated cell cycle activation during early lateral root initiation. Plant Cell 14:2339–2351

Howie WJ, Cook RJ, Weller DM (1987) Effects of soil matrix potential and cell motility on wheat root colonization by fluorescent pseudomonads suppressive to take-all. Phytopathology 77:286–292

Hunter WJ (1989) Indole-3-acetic acid production by bacteroids from soybean root nodules. Physiol Plant 76:31–36

Idris EES, Iglesias DJ, Talon M, Borriss R (2007) Tryptophan-dependent production of indole-3-acetic acid (IAA) affects level of plant growth promotion by *Bacillus amyloliquefaciens* FZB42. Mol Plant Microbe Interact 20:619–626

Iniguez AL, Dong Y, Carter HD, Ahmer BMM, Stone JM, Triplett EW (2005) Regulation of enteric endophytic bacterial colonization by plant defenses. Mol Plant Microbe Interact 18:169–178

Jackson MB (1991) Ethylene in root growth and development. In: Matoo AK, Suttle JC (eds) The plant hormone ethylene. CRC Press, Boca Raton, pp 159–181

Jacobson CB, Pasternak JJ, Glick BR (1994) Partial purification and characterization of 1-aminocyclopropane-1-carboxylate deaminase from PGPR, *Pseudomonas putida* GR12-2. Can J Microbiol 40:1019–1025

Jalili F, Khavazi K, Pazira E, Nejati A, Rahmani HA, Sadaghiani HR et al (2009) Isolation and characterization of ACC deaminase-producing fluorescent pseudomonads, to alleviate salinity stress on canola (*Brassica napus* L.) growth. J Plant Physiol 166:667–674

James EK, Gyaneshwar P, Manthan N, Barraquio WL, Reddy PM, Ianetta PPM, Olivares FL, Ladha JK (2002) Infection and colonization of rice seedlings by the plant growth-promoting bacterium *Herbaspirillum seropedicae* Z67. Mol Plant Microbe Interact 15:894–906

Jameson PE (2000) Cytokinins and auxins in plant-pathogen interactions—an overview. Plant Growth Reg 32:369–380

Jiang CY, Sheng XF, Qian M, Wang QY (2008) Isolation and characterization of a heavy metal resistant *Burkholderia* sp. from heavy metal-contaminated paddy field soil and its potential in promoting plant growth and heavy metal accumulation in metal polluted soil. Chemosphere 72:157–164

Jiang F, Chen L, Belimov AA, Shaposhnikov AI et al (2012) Multiple impacts of the plant growth-promoting rhizobacterium *Variovorax paradoxus* 5C-2 on nutrient and ABA relations of *Pisum sativum*. J Exp Bot 63(18):6421–6430

Jin QL, Thilmony R, Zwiesler-Vollick J, He SY (2003) Type III protein secretion in *Pseudomonas syringae*. Microbes Infect 5:301–310

Jones DL (1998) Organic acids in the rhizosphere e a critical review. Plant Soil 205:25–44

Jones DL, Dennis PG, Owen AG, van Hees PAW (2003) Organic acid behavior in soils—misconceptions and knowledge gaps. Plant Soil 248:31–41

Joshi FR, Dhwani KD, Archana G, Desai AJ (2009) Enhanced survival of and nodule occupancy of Pigeon pea nodulating *Rhizobium* sp ST1 expressing *feg*A gene of *Bradyrhizobium japonicum* 61A152. On line J Boil Sci 9(2):40–51

Jouanneau JP, Lapous D, Guern J (1991) In plant protoplasts, the spontaneous expression of defense reactions and the responsiveness to exogenous elicitors are under auxin control. Plant Physiol 96:459–466

Jungk A (2001) Root hair and acquisition of plant nutrients from soils. J Plant Nutr Soil Sci 164:121–129

Kaldorf M, Ludwig-Muller J (2000) AM fungi might affect the root morphology of maize by increasing indole-3-butyric acid biosynthesis. Physiol Plant 109:58–67

Kamilova F, Validov S, Azarova T, Mulders I, Lugtenberg B (2005) Enrichment for enhanced competitive plant root tip colonizers selects for a new class of biocontrol bacteria. Environ Microbiol 7(11):1809–1817

Kaneshiro T, Kwolek WF (1985) Stimulated nodulation of soybeans by *Rhizobium japonicum* mutant (B-14075) that catabolizes the conversion of tryptophan to indol-3yl-acetic acid. Plant Sci 42:141–146

Kang BG, Kim WT, Yun HS, Chang SC (2010) Use of plant growth-promoting rhizobacteria to control stress responses of plant roots. Plant Biotechnol Rep 4:179–183

Karthikeyan B, Joe MM, Islam MR, Sa T (2012) ACC deaminase containing diazotrophicendophytic bacteria ameliorate salt stress in *Catharanthus roseus* through reduced ethylene levels and induction of antioxidative defense systems. Symbiosis 56:77–86

Kausar R, Shahzad SM (2006) Effect of ACC-deaminase containing rhizobacteria on growth promotion of maize under salinity stress. J Agri Social Sciences 2:216–218

Kennedy IR, Pereg-Gerk LL, Wood C, Deaker R, Gilchrist K, Katupitiya S (1997) Biological nitrogen fixation in non-leguminous field crops: facilitating the evolution of an effective association between *Azospirillum* and wheat. Plant Soil 194:65–79

Khalid A, Arshad M, Zahir ZA (2004) Screening plant growthpromoting rhizobacteria for improving growth and yield of wheat. J Appl Microbiol 96:473–480

Khalid A, Arshad M, Zahir Z (2005) Screening plant growth promoting rhizobacteria for improving growth and yield of wheat. J Appl Microbiol 96:473–480

Khan MS, Zaidi A, Wani PA (2006) Role of phosphate solubilizing microorganisms in sustainable agriculture-a review. In: Lichtfouse E, Navarrete M, Debaeke P, Veronique S, Alberola C (eds) Sustainable agriculture, vol 5. Springer, The Netherlands, pp 551–570

Khan MS, Zaidi A, Wani PA, Oves M (2009) Role of plant growth promoting rhizobacteria in the remediation of metal contaminated soils. Environ Chem Lett 7:1–19

Kim J, Rees DC (1994) Nitrogenase and biological nitrogen fixation. Biochemistry 33:389–397

Kinkel LL, Wilson M, Lindow SE (2000) Plant species and plant incubation conditions influence variability in epiphytic bacterial population size. Microb Ecol 39:1–11

Kloepper JW, Beauchamp CJ (1992) A review of issues related to measuring colonization of plant roots by bacteria. Can J Microbiol 38:1219–1232

Kloepper JW, Lifshitz R, Zablotowich RK (1989) Free living bacterial inocula for enhancing crop productivity. Trends Biotechnol 7:39–43

Kloepper JW, Gutierrez-Estrada A, McInroy JA (2007) Photoperiod regulates elicitation of growth promotion but not induced resistance by plant growth-promoting rhizobacteria. Can J Microbiol 53:159–167

Knoester M, van Loon LC, van den Heuvel J, Hennig J, Bol JF, Linthorst HJM (1998) Ethylene-insensitive tobacco lacks nonhost resistance against soil-borne fungi. Proc. Natl. Acad. Sci. U. S. A. 95:1933–1937

Kraffczyk I, Trolldeiner G, Beringer H (1984) Soluble root exudates of maize: influence of potassium supply and rhizosphere microorganisms. Soil Biol Biochem 16:315–322

Kravchenko LV, Leonava EI, Tikhonovich IA (1994) Effect of root exudates of non-legame plants on the response of auxin production by associated diazotrophs. Microb Release 2:267–271

Kucera B (2005) Plant hormone interactions during seed dormancy release and germination. Seed Sci Res 15:281–307

Kuklinsky-Sobral J, Araújo WL, Mendes R, Geraldi IO, Pizzirani- Kleiner AA, Azevedo JL (2004) Isolation and characterization of soybean-associated bacteria and their potential for plant growth promotion. Environ Microbiol 6:1244–1251

Kumar B, Trivedi P, Pandey A (2007a) *Pseudomonas corrugata*: a suitable bioinoculant for maize grown under rainfed conditions of Himalayan region. Soil Biol Biochem 39:3093–3100

Kumar R, Bhatia R, Kukreja K, Behl RK, Dudeja SS, Narula N (2007b) Establishment of *Azotobacter* on plant roots: chemotactic response, development and analysis of root exudates of cotton (*Gossypium hirsutum* L.) and wheat (*Triticum aestivum* L.). J Basic Microbiol 47:436–439

Kumar S, Pandey P, Maheshwari DK (2009) Reduction in dose of chemical fertilizers and growth enhancement of sesame (*Sesamum indicum* L.) with application of rhizospheric competent *Pseudomonas aeruginosa* LES4. Eur J Soil Biol 45:334–340

Lacava PT, Li WB, Araujo WL, Azevedo JL, Hartung JS (2007) The endophyte *Curtobacterium flaccumfaciens* reduces symptoms caused by *Xylella fastidiosa* in *Catharanthus roseus*. J Microbiol 45:388–393

Lalonde M, Knowles R (1975) Ultrastructure, composition, and biogenesis of the encapsulation material surrounding the endophyte in Alnus crispa var. mollis root nodules. Can J Bot 53:1951–1971

Lambrecht M, Vande Broek A, Dosselaere F, Vanderleyden J (1999) The *ipd*C promoter auxin-responsive element of *Azospirillum brasilense*, a prokaryotic ancestral form of the plant AuxRE. Mol Microbiol 32:889–890

Lambrecht M, Okon Y, Vande Broek A, Vanderleyden J (2000) Indole-3-acetic acid: a reciprocal signalling molecule in bacteria–plant interactions. Trends Microbiol 8:298–300

Landa BB, Mavrodi OV, Raaijmakers JM, McSpadden-Gardener BB, Thomashow LS, Weller DM (2002) Differential ability of genotypes of 2,4-diacetylphloroglucinol producing *Pseudomonas fluorescens* to colonize the roots of pea. Appl Environ Microbiol 68:3226–3237

Landa BB, Mavrodi DM, Thomashow LS, Weller DM (2003) Interactions between strains of 2, 4-diacetylphloroglucinol-producing *Pseudomonas fluorescens* in the rhizosphere of wheat. Phytopathol 93:982–994

Lee S, Flores-Encarnacion M, Contreras-Zentella M, Garcia-Flores L, Escamilla J, Kennedy C (2004) Indole-3-acetic acid biosynthesis is deficient in *Gluconacetobacter diazotrophicus* strains with mutations in cytochrome C biogenesis genes. J Bacteriol 186:5384–5391

Lerouxel O, Cavalier DM, Liepman AH, Keegstra K (2006) Biosynthesis of plant cell wall polysaccharides—a complex process. Curr Opin Plant Biol 9:621–630

Li J, Glick BR (2001) Transcriptional regulation of the *Enterobacter cloacae* UW41-aminocyclopropane-1-carboxylate (ACC) deaminase gene (acdS). Can J Microbiol 47:359–367

Li JP, Daniel H, Charles TC, Glick BR (2000) An ACC deaminase minus mutant of *Enterobacter cloacae* UW4 no longer promotes root elongation. Curr Microbiol 41:101–105

Li W, Ye Z, Wong M (2009) Metal mobilization and production of short-chain organic acids by rhizosphere bacteria associated with a Cd/Zn hyperaccumulating plant, Sedum alfredii. Plant Soil 326:453–467

Lim HS, Kim SD (1995) The role and characterization of β-1, 3-glucanase in biocontrol of *Fusarium solani* by *Pseudomonas stutzeri* YLP-1. Curr Microbiol 33(4):295–301

Lindberg T, Granhall U, Tomenius H (1985) Infectivity and acetylene reduction of diazotrophic rhizosphere bacteria in wheat (*Triticum aestivum*) seedlings under gnotobiotic conditions. Biol Fert Soils 1:123–129

Lindow SE, Brandl MT (2003) Microbiology of the phyllosphere. Appl Environ Microbiol 69:1875–1883

Lindow SE, Desurmont C, Elkins R, McGourty G, Clark E, Brandl MT (1998) Occurrence of indole-3-acetic acid-producing bacteria on pear trees and their association with fruit russet. Phytopathology 88:1149–1157

Liu P, Nester EW (2006) Indoleacetic acid, a product of transferred DNA, inhibits vir gene expression and growth of *Agrobacterium tumefaciens* C58. Proc Natl Acad Sci USA 103:4658–4662

Lodewyckx C, Vangronsveld J, Porteous F, Moore ERB, Taghavi S, Mezgeay M, van der Lelie D (2002) Endophytic bacteria and their potential applications. Crit Rev Plant Sci 21:583–606

Loper JE, Schroth MN (1986) Influence of bacterial sources of indole-3-acetic acid biosynthetic on root elongation of sugar beet. Phytopathology 76:386–389

Lucy M, Reed E, Glick BR (2004) Application of free living plant growth-promoting rhizobacteria. Antonie Van Leeuwenhoek 86:1–25

Ludwig-Müller J (2004) From auxin homeostasis to understanding plant pathogen and plant symbiont interaction: editor's research interests. J Plant Growth Regul 23:1–8

Ludwig-Muller J, Güther M (2007) Auxins as signals in arbuscular mycorrhiza formation. Plant Signal Behav 3:194–196

Lugtenberg BJJ, Bloemberg GB (2004) In: Ramos JL (ed) Pseudomonas, vol 1. Kluwer, Academic Publishers, New York, pp 403–430

Lugtenberg BJJ, Dekkers LC (1999) What makes *Pseudomonas* bacteria rhizosphere competent? Environ Microbiol 1:9–13

Lugtenberg B, Kamilova F (2009) Plant-growth-promoting rhizobacteria. Annu Rev Microbiol 63:541–556

Lugtenberg BJJ, Kravchenko LV, Simons M (1999) Tomato seed and root exudate sugars: composition, utilization by *Pseudomonas* biocontrol strains and role in rhizosphere colonization. Environ Microbiol 1:439–446

Lugtenberg BJJ, Dekkers LC, Bloemberg GV (2001) Molecular determinants of rhizosphere colonization by Pseudomonas. Annu Rev Phytopathol 39:461–490

Lugtenberg B, Chin-A-Woeng T, Bloemberg G (2002) Microbe–plant interactions: principles and mechanisms. Antonie Van Leeuwenhoek 81:373–383

Lynch JM (1990) The rhizosphere. Wiley-Interscience, Chichester

Lynch JP (2007) Roots of the second green revolution. Aust J Bot 55:493–512

Lynch JM, Whipps JM (1990) Substrate flow in the rhizosphere. Plant Soil 129:1–10

Ma Y, Rajkumar M, Freitas H (2009) Improvement of plant growth and nickel uptake by nickel resistant-plant growth promoting bacteria. J Hazard Mater 166:1154–1161

Ma Y, Rajkumar M, Vicente J, Freitas H (2010) Inoculation of Ni-resistant plant growth promoting bacterium *Psychrobacter* sp. strain SRS8 for the improvement of nickel phytoextraction by energy crops. Int J Phytoremediation 13:126–139

Ma Y, Prasad MN, Rajkumar M, Freitas H (2011a) Plant growth promoting rhizobacteria and endophytes accelerate phytoremediation of metalliferous soils. Biotechnol Adv 29:248–258

Ma Y, Rajkumar M, Luo Y, Freitas H (2011b) Inoculation of endophytic bacteria on host and non-host plants—effects on plant growth and Ni uptake. J Hazard Mater 196:230–237

Madhaiyan M, Poonguzhali S, Ryu J, Sa T (2006) Regulation of ethylene levels in canola (Brassica campestris) by 1-aminocyclopropane-1-carboxylate deaminase containing *Methylobacterium fujisawaense*. Planta 224:268–278

Madhaiyan M, Poonguzhali S, Sa T (2007) Metal tolerating methylotrophic bacteria reduces nickel and cadmium toxicity and promotes plant growth of tomato (*Lycopersicon esculentum* L.). Chemosphere 69:220–228

Maheshwari DK (2010) Plant growth and health promoting bacteria. Springer-Berlin Heidelberg, Germany, p 448

Maheshwari DK, Aeron A, Dubey RC, Agarwal M, Dheeman S, Shukla S (2014) Multifaceted beneficial associations with *Pseudomonas* and *Rhizobia* on growth promotion of *Mucuna pruriens* L. J Pure Appl Microbiol 8(6):4657–4667

Mahmoud ALE, Abd-Alla MH (2001) Siderophore production by some microorganisms and their effect on B*radyrhizobium*-Mung Bean symbiosis. Int J Agric Biol 03(2):157–162

Malakoff D (1998) Coastal ecology: death by suffocation in the Gulf of Mexico. Science 281:190–192

Malboobi M, Owlia P, Behbahani M, Sarokhani E, Moradi S, Yakhchali B, Deljou A, Heravi KM (2009) Solubilization of organic and inorganic phosphates by three highly efficient soil bacterial isolates. World J Microbiol Biotechnol 25:1471–1477

Mantelin S, Touraine B (2004) Plant growth-promoting bacteria and nitrate availability: impacts on root development and nitrate uptake. J Exp Bot 55:27–34

Manulis S, Haviv-Chesner A, Brandl MT, Lindow SE, Barash I (1998) Differential involvement of indole-3-acetic acid biosynthetic pathways in pathogenicity and epiphytic fitness of *Erwinia herbicola* pv. *gypsophilae*. Mol Plant Microbe Interact 11:634–642

Marschner H (1995) Mineral nutrition of higher plants, 2nd edn. Academic Press, London

Marschner P, Yang CH, Lieberei R, Crowley DE (2001) Soil and plant specific effects on bacterial community composition in the rhizosphere. Soil Biol Biochem 33:1437–1445

Martínez-Morales LJ, Soto-Urzúa L, Baca BE, Sánchez-Ahédo JA (2003) Indole-3-butyric acid (IBA) production in culture medium by wild strain *Azospirillum brasilense*. FEMS Microbiol Lett 228:167–173

Martinez-Viveros O, Jorquera M, Crowley DE, Gajardo G, Mora ML (2010) Mechanisms and practical considerations involved in plant growth promotion by rhizobacteria. J Soil Sci Plant Nutr 10:293–319

Mathesius U, Schlaman HR, Spaink HP, Of Sautter C, Rolfe BG, Djordjevic MA (1998) Auxin transport inhibition precedes root nodule formation in white clover roots and is regulated by flavonoids and derivatives of chitin oligosaccharides. Plant J 14:23–34

Mattos KA, Vania LM, Pádua, Alexandre R, Hallack LF, Bianca C et al (2008) Endophytic colonization of rice (*Oryza sativa* L.) by the diazotrophic bacterium *Burkholderia kururiensis* and its ability to enhance plant growth. Anais da Academia Brasileira de Ciências 80(3):477–493

Mayak S, Tirosh T, Glick BR (1999) Effect of wild-type and mutant plant growth-promoting rhizobacteria on the rooting of mung bean cuttings. J Plant Growth Regul 18:49–53

Mayak S, Tirosh T, Glick BR (2004a) Plant growth promoting bacteria confer resistance in tomato plants to salt stress. Plant Physiol Biochem 42:565–572

Mayak S, Tirosh T, Glick BR (2004b) Plant growth-promoting bacteria that confer resistance to water stress in tomato and pepper. Plant Sci 166:525–530

Mazzola M, White FF (1994) A mutation in the indole-3-aceticacid biosynthesis pathway of *Pseudomonas syringae* pv *syringae* affects growth in *Phaseolus vulgaris* and syringomycin production. J Bacteriol 176:1374–1382

McCully ME (2001) Niches for bacterial endophytes in crop plants: a plant biologist's view. Aust J Plant Physiol 28:983–990

Mendes R, Pizzirani-Kleiner AA, Araujo WL, Raaijmakers JM (2007) Diversity of cultivated endophytic bacteria from sugarcane: genetic and biochemical characterization of *Burkholderia cepacia* complex isolates. Appl Environ Microbiol 73(22):7259–7267

Meuwley P, Pilet PE (1991) Local treatment with indole-3-acetic acid induces differential growth responses in Zea mays L. roots. Planta 185:58–64

Miché L, Battistoni F, Gemmer S, Belghazi M, Reinhold-Hurek B (2006) Upregulation of jasmonate-inducible defense proteins and differential colonization of roots of *Oryza sativa* cultivars with the endophyte *Azoarcus* sp. Mol Plant Microbe Interact 19:502–511

Mills HA, Jones JB (1996) Plant analysis handbook II: a practical sampling, preparation, analysis, and interpretation guide. Micro-macro Publishing, Athens, pp 6–18, 69, 81

Mocali S, Bertelli E, Di Cello F, Mengoni A, Sfalanga A, Viliani F, Caciotti A, Tegli S, Surico G, Fani R (2003) Fluctuation of bacteria isolated fromelm tissues during different seasons and from different plant organs. Res Microbiol 154:105–114

Mohnen D, Shinshi H, Felix G, Meins F (1985) Hormonal regulation of β-1,3-glucanase messenger RNA levels in cultured tobacco tissues. EMBO J 4:1631–1635

Mole BM, Baltrus DA, Dangl JL, Grant SR (2007) Global virulence regulation networks in phytopathogenic bacteria. Trends Microbiol 15:363–371

Molina-Favero C, Creus CM, Simontacchi M, Puntarulo S, Lamattina L (2008) Aerobic nitric oxide production by *Azospirillum brasilense* Sp245 and its influence on root architecture in tomato. Mol Plant Microbe Interact 21:1001–1009

Morgan PW, Drew CD (1997) Ethylene and plant responses to stress. Physiol Plant 100:620–630

Morris RO (1995) Genes specifying auxin and cytokinin biosynthesis in prokaryotes. In: Davies PJ (ed) Plant hormones: physiology, biochemistry and molecular biology. Kluwer Academic Publishers, pp 318–339

Müller M, Deigele C, Ziegler H (1989) Hormonal interactions in the rhizosphere of maize (*Zea mays* L.) and their effects on plant development. Z Pflanzenernähr Bodenk 152:247–254

Nadeem SM, Zahir ZA, Naveed M, Arshad M (2007) Preliminary investigations on inducing salt tolerance in maize through inoculation with rhizobacteria containing ACC deaminase activity. Can J Microbiol 53:1141–1149

Nadeem SM, Zahir ZA, Naveed M, Arshad M (2009) Rhizobacteria containing ACC-deaminase confer salt tolerance in maize grown on salt-affected fields. Can J Microbiol 55:1302–1309

Nadeem SM, Zahair ZA, Naveed M, Asghar HN, Asghar M (2010) Rhizobacteria capable of producing ACC-deaminase may mitigate salt stress in wheat. Soil Sci Soc Am J 74:533–542

Nardi S, Concheri G, Pizzeghello D, Sturaro A, Rella R, Parvoli G (2000) Soil organic matter mobilization by root exudates. Chemosphere 5:653–658

Navarro L, Dunoyer P, Jay F, Arnold B, Dharmasiri N, Estelle M et al (2006) A plant miRNA contributes to antibacterial resistance by repressing auxin signaling. Science 312:436–439

Nie L, Shah S, Burd GI, Dixon DG, Glick BR (2002) Phytoremediation of arsenate contaminated soil by transgenic canola and the plant growth-promoting bacterium *Enterobacter cloacae* CAL2. Plant Physiol Biochem 40:355–361

Notz R, Maurhofer M, Schnider-Keel U, Duffy B, Haas D, De´fago G (2001) Biotic factors affecting expression of the 2,4-diacetylphloroglucinol biosynthesis gene *phl*A in *Pseudomonas fluorescens* biocontrol strain CHA0 in the rhizosphere. Phytopathology 91:873–881

Obertello M, Sy MO, Laplaze L, Santi C, Svistoonoff S, Auguy F, Bogusz D, Franche C (2003) Actinorhizal nitrogen fixing nodules: infection process, molecular biology and genomics. Afr J Biotechnol 2:528–538

Okon Y, Itzigsohn R (1995) The development of *Azospirillum* as a commercial inoculant for improving crop yields. Biotechnol Adv 13:415–424

Okon Y, Vanderleyden J (1997) Root-associated *Azospirillum* species can stimulate plants. Am Soc Microbiol News 63:366–370

Omar SA, Abd-Alla MH (1994) Enhancement of faba bean nodulation, nitrogen fixation, and growth by different microorganisms. Biol Plant 36:295–300

Ortiz-Castro R, Contreras-Cornejo HA, Macias-Rodriguez L, Lopez-Bucio J (2009) The role of microbial signals in plant growth and development. Plant Signal Behav 4:701–712

Pandey P, Maheshwari DK (2007a) Bioformulation of *Burkholderia* sp. MSSP multispecies consortium growth promotion Cajanus cajan. Can J Microbiol 53:213–222

Pandey P, Maheshwari DK (2007b) Two-species microbial consortium for growth promotion of *Cajanus cajan*. Curr Sci 92:1137–1142

Pandey P, Kang SC, Maheshwari DK (2005) Isolation of endophytic plant growth promoting *Burkholderia* sp. MSSP from root nodules of *Mimosa pudica*. Curr Sci 89:170–180

Parke JL (1991) Root colonization by indigenous and introduced microorganisms. In: Keister DL, Gregan PB (eds) The rhizosphere and plant growth. Kluwer Academic Publishers, Dordrecht, pp 33–42

Parmar N, Dadarwal KR (1999) Stimulation of nitrogen fixation and induction of flavonoids like compounds by rhizobacteria. J Appl Microbiol 86:36–44

Parmar N, Dadarwal KR (2000) Stimulation of plant growth of chickpea by inoculation of fluorescent Pseudomonads. J Appl Microbiol 86:36–44

Paterson E, Sim A (2000) Effect of nitrogen supply and defoliation on loss of organic compounds from roots of *Festuca rubra*. J Exp Bot 51:1449–1457

Paterson E, Gebbing T, Abel C et al (2006) Rhizodeposition shapes rhizosphere microbial community structure in organic soil. New Phytol 173:600–610

Patten C, Glick B (1996) Bacterial biosynthesis of indole-3-acetic acid. Can J Microbiol 42:207–220

Patten C, Glick B (2002) Role of *Pseudomonas putida* indoleacetic acid in development of the host plant root system. Appl Environ Microb 68:3795–3801

Pawlowski K, Bisseling T (1996) Rhizobial and actinorhizal symbioses: what are the shared features. Plant Cell 8:1899–1913

Peck SC, Kende H (1995) Sequential induction of the ethylene biosynthetic enzymes by indole-3-acetic acid in etiolated peas. Plant Mol Biol 28:293–301

Pedrosa FO, Monteiro RA, Wassem R (2011) Genome of *Herbaspirillum seropedicae* strain SmR1, a specialized diazotrophic endophyte of tropical grasses. PLoS Genet 7:e1002064

Penrose DM, Glick BR (2003) Methods for isolating and characterizing ACC deaminase-containing plant growthpromoting rhizobacteria. Physiol Plant 118:10–15

Persello-Cartieaux F, Nussaume L, Robaglia C (2003) Tales from the underground: molecular plant-rhizobacterial interactions. Plant, Cell Environ 26:189–199

Peters NK, Frost JW, Long SR (1986) A plant flavone, luteolin, induces expression of *Rhizobium meliloti* nodulation genes. Science 233:977–980

Petrini LE, Petrini O, Laflamme G (1989) Recovery of endophytes of *Abiens balsamea* from needles and galls of *Paradiplosis tumifex*. Phytoprotection 70:97–103

Phi QT, Park YM, Ryu CM, Park SH, Ghim SY (2008) Functional identification 1 and expression of indole-3-pyruvate decarboxylase from *Paenibacillus polymyxa* E681. J Microbiol Biotechnol 18:1235–1244

Pielach CA, Roberts DP, Kobayashi DY (2008) Metabolic behavior of bacterial biological control agents in soil and plant rhizospheres. In: Laskin AI, Sariaslani S, Gadd GM (eds) Applied microbiology, vol 65. Academic, New York, pp 199–215

Pierik R, Tholen D, Poorter H, Visser EJ, Voesenek LA (2006) The Janus face of ethylene: growth inhibition and stimulation. Trends Plant Sci 11:176–183

Pieterse CMJ, van Wees SCM, van Pelt JA, Knoester M, Laan R, Gerrits N, Weisbeek PJ, van Loon LC (1998) A novel signaling pathway controlling induced systemic resistance in *Arabidopsis*. Plant Cell 10:1571–1580

Pieterse CM, Van Der Ent S, Van Pelt JA, Van Loon LC (2002) The role of ethylene in rhizobacteria- induced systemic resistance (ISR). In: Ramina A, Chang C, Giovannoni J, Klee H, Perata P, Woltering E (eds) Advances in ethylene research. Springer, The Netherlands, pp 325–331

Pilet PE, Saugy M (1987) Effect on root growth of endogenous and applied IAA and ABA. Plant Physiol 83:33–38

Ping L, Boland W (2004) Signals from the underground: bacterial volatiles promote growth in *Arabidopsis*. Trends Plant Sci 9:263–266

Pinton R, Varanini Z, Nannipieri P (2007) The rhizosphere. In: Biochemistry and organic substances at the soil plant interface, 2nd edn. CRC, Taylor and Francis, Boca Raton

Pirlak M, Kose M (2009) Effects of plant growth promoting rhizobacteria on yield and some fruit properties of strawberry. J Plant Nutr 32:1173–1184

Prayitno J, Rolfe BG, Mathesius U (2006) The ethylene-insensitive sickle mutant of Medicago truncatula shows altered auxin transport regulation during nodulation. Plant Physiol 142:168–180

Prieto P, Schilirò E, Maldonado-González MM, Valderrama R, Barroso-Albarracín JB, Mercado-Blanco J (2011) Root hairs play a key role in the endophytic colonization of olive roots by *Pseudomonas* spp. with biocontrol activity. Microb Ecol 62:435–445

Prinsen E, Chauvaux N, Schmidt J, John M, Wieneke U, Degreef J, Schell J, Vanonckelen H (1991) Stimulation of indole-3- acetic acid production in *Rhizobium* by flavonoids. FEBS Lett 282:53–55

Prusty R, Grisafi P, Fink GR (2004) The plant hormone indoleacetic acid induces invasive growth in *Saccharomyces cerevisiae*. Proc Natl Acad Sci USA 101:4153–4157

Quadt-Hallman A, Benhamou N, Kloepper JW (1997) Bacterial endophytes in cotton: mechanisms of entering the plant. Can J Microbiol 43:577–582

Raja P, Uma S, Gopal H, Govindarajan K (2006) Impact of bioinoculants on rice root exudates, biological nitrogen fixation and plant growth. J Biol Sci 6(5):815–823

Rajkumar M, Nagendran R, Kui JL, Wang HL, Sung ZK (2006) Influence of plant growth promoting bacteria and Cr (VI) on the growth of Indian mustard. Chemosphere 62:741–748

Rajkumar M, Ae N, Prasad MNV, Freitas H (2010) Potential of siderophore-producing bacteria for improving heavy metal phytoextraction. Trends Biotechnol 28:142–149

Ramadoss D, Lakkineni VK, Bose P, Ali S, Annapurna K (2013) Mitigation of salt stress in wheat seedlings by halotolerant bacteria isolated from saline habitats. Springer-Plus 2:6

Randy OC, Angel Contreras-Cornejo H, Macías-Rodríguez L, López-Bucio J (2009) The role of microbial signals in plant growth and development. Plant Signal Behav 4(8):701–712

Rasche F, Marco-Noales E, Velvis H, Overbeek LS, Lo'pez MM, Elsas JD, Sessitsch A (2006) Structural characteristics and plant-beneficial effects of bacteria colonizing the shoots of field grown conventional and genetically modified T4-lysozyme producing potatoes. Plant Soil 298:123–140

Raupach GS, Kloepper JW (2000) Biocontrol of cucumber diseases in the field by plant growth-promoting rhizobacteria with and without methyl bromide fumigation. Plant Dis 84:1073–1075

Reed MLE, Glick BR (2005) Growth of canola (*Brassica napus*) in the presence of plant growth-promoting bacteria and either copper or polycyclic aromatic hydrocarbons. Can J Microbiol 51:1061–1069

Reed MLE, Warner BG, Glick BR (2005) Plant growth-promoting bacteria facilitate the growth of the common reed *Phragmites australis* in the presence of copper or polycyclic aromatic hydrocarbons. Curr Microbiol 51:425–429

Reinhold-Hurek B, Hurek T (1998) Life in grasses: diazotrophic endophytes. Trends Microbiol 6:139–144

Reinhold-Hurek B, Hurek T (2011) Living inside plants: 1 bacterial endophytes. Curr Opin Plant Biol 14:435–443

Remans R, Spaepen S, Vanderleyden J (2006) Auxin signaling in plant defense. Science 313:171
Remans R, Croonenborghs A, Gutierrez RT, Michiels J, Vanderleyden J (2007) Effects of plant growth-promoting rhizobacteria on nodulation of *Phaseolus vulgaris* L. are dependent on plant P nutrition. Eur J Plant Pathol 119:341–351
Remans R, Beebe S, Blair M, Manrique G, Tovar E, Rao I, Croonenborghs A, Torres-Gutierrez R, El-Howeity M, Michiels J, Vanderleyden J (2008) Physiological and genetic analysis of root responsiveness to auxin-producing plant growth-promoting bacteria in common bean (*Phaseolus vulgaris* L.). Plant Soil 302:149–161
Rengel Z, Marschner P (2005) Nutrient availability and management in the rhizosphere: exploiting genotypic differences. New Phytol 168:305–312
Rezzonico E, Flury MJ, Beffa R (1998) Transcriptional down-regulation by abscisic acid of pathogenesis-related β-1,3-glucanase genes in tobacco cell cultures. Plant Physiolo 117:585–592
Richardson AE (1994) Soil microorganisms and phosphorus availability. In: Pankhurst CE (ed) Soil biota. Management in sustainable farming systems. CSIRO, Melbourne, pp 50–62
Riov J, Yang SF (1989) Ethylene and auxin-ethylene interaction in adventitious root formation in mung bean (*Vigna radiata*) cuttings. J Plant Growth Regul 8:131–141
Rodelas B, Gonza'lez-Lo'pez J, Martı'nez-Toledo MV, Pozo C, Salmero'n V (1999) Influence of *Rhizobium/Azotobacter* and *Rhizobium/Azospirillum* combined inoculation on mineral composition of faba bean (Vicia faba L). Biol Fertil Soils 29(2):165–169
Rodnguez H, Fraga R (1999) Phosphate solubilizing bacteria and their role in plant growth promotion. Biotechnol Adv 17:319–339
Rodriguez H, Gonzalez T, Goire I, Bashan Y (2004) Gluconic acid production and phosphate solubilization by the plant growth-promoting bacterium *Azospirillum* spp. Naturwissenschaften 91:552–555
Rodriguez H, Vessely S, Shah S, Glick BR (2008) Effect of a nickel-tolerant ACC deaminase-producing *Pseudomonas* strain on growth of nontransformed and transgenic canola plants. Curr Microbiol 57:170–174
Rodriguez-Navarro DN, Dardanelli MS, Ruiz-Sainz JE (2007) Attachment of bacteria to the roots of higher plants. FEMS Microbiol Lett 272:127–136
Rokhzadi A, Asgazadeh A, Darvish F, Nour-Mohammed G, Majidi E (2008) Influence of plant growth promoting rhizobacteria on dry matter accumulation and yield of chick pea (*Cicer arietinium* L.) under filed conditions. American-Eurasian J Agric Environ Sci 3:253–257
Rosenblueth M, Martínez-Romero E (2006) Bacterial endophytes and their interaction with hosts. Mol Plant Microbe Interact 19:827–837
Rothballer M, Schmid M, Fekete A, Hartmann A (2005) Comparative in situ analysis of ipdC-gfpmut3 promoter fusions of *Azospirillum brasilense* strains Sp7 and Sp245. Environ Microbiol l7:1839–1846
Ryan RP, Germaine K, Franks A, Ryan DJ, Dowling DN (2008) Bacterial endopHytes: recent developments and applications. FEMS Microbiol Lett 278:1–9
Ryan RP, Monchy S, Cardinale M, Taghavi S, Crossman L, Avison MB, Berg G, van der Lelie D, Maxwell Dow J (2009) The versatility and adaptation of bacteria from the genus StenotropHomonas. Nature Rev Microbiol 7:514–525
Sadrnia M, Maksimava N, Khromsova E, Stanislavich S, Arjomandzadegan M (2011) Study the effect of bacterial 1-aminocyclopropane-1-carboxylate deaminase (ACC deaminase) on resistance to salt stress in tomato plant. Anal Univ Oradea 18:120–123
Safronova VI, Stepanok VV, Engqvist GL, Alekseyev YV, Belimov AA (2006) Root-associated bacteria containing 1-aminocyclopropane-1-carboxylate deaminase improve growth and nutrient uptake by pea genotypes cultivated in cadmium supplemented soil. Biol Fertil Soils 42:267–272
Saleem M, Arshad M, Hussain S, Bhatti AS (2007) Perspective of plant growth promoting rhizobacteria (PGPR) containing ACC deaminase in stress agriculture. J Ind Microbiol Biotechnol 34:635–648

Saravanakumar D, Samiyappan R (2006) ACC deaminase from *Pseudomonas fluorescens* mediated saline resistance in groundnut (*Arachis hypogea*) plants. J Appl Microbiol 102:1283–1292
Saravanakumar D, Samiyappan R (2007) Effects of 1-aminocyclopropane- 1-carboxylic acid (ACC) deaminase from *Pseudomonas fluorescence* against saline stress under in vitro and field conditions in groundnut (*Arachis hypogea* L.) plants. J Appl Microbiol 102:1283–1292
Saravanan VS, Madhaiyan M, Thangaraju M (2007) Solubilization of zinc compounds by the diazotrophic, plant growth promoting bacterium *Gluconacetobacter diazotrophicus*. Chemosphere 66:1794–1798
Saubidet MI, Fatta N, Barneix AJ (2002) The effect of inoculation with *Azospirillum brasilense* on growth and nitrogen utilization by wheat plants. Plant Soil 245:215–222
Schippers B, Baker AW, Bakker PAHM (1987) Interactions of deleterious and beneficial rhizosphere microorganisms and the effect of cropping practices. Annu Rev Phytopathol 25:339–358
Schroth MN, Weinhold AR (1986) Root-colonizing bacteria and plant health. Horts 21:1295–1298
Schulz B, Boyle C, Sieber N (2006) Microbial root endophytes. Springer, Berrlin
Senthilkumar M, Anandham R, Madhaiyan M, Venkateswaran V (2011) Endophytic bacteria: perspectives and applications in agricultural crop production. In: Maheshwari DK (ed) Bacteria in agrobiology: crop ecosystems. Springer, Berlin Heidelberg, pp 61–96
Sevilla M, Burris RH, Gunapala N, Kennedy C (2001) Comparison of benefit to sugarcane plant growth and $^{15}N_2$ incorporation following inoculation of sterile plants with *Acetobacter diazotrophicus* wild-type and nif mutant strains. Mol Plant Microbe Interact 14:358–366
Shah S, Li J, Moffatt BA, Glick BR (1998) Isolation and characterization of ACC deaminase genes from two different plant growth-promoting rhizobacteria. Can J Microbiol 44:833–843
Shaharooma B, Naveed M, Arshad M, Zahir ZA (2008) Fertilizer-dependent efficiency of Pseudomonads for improving growth, yield, and nutrient use efficiency of wheat (*Triticum aestivum* L.). Appl Microbiol Biotechnol 79:147–155
Shaharoona B, Arshad M, Zahir Z (2006) Effect of plant growth promoting rhizobacteria containing ACC-deaminase on maize (*Zea mays* L.) growth under axenic conditions and on nodulation in mung bean (Vigna radiata L.). Lett Appl Microbiol 42:155–159
Shakir MA, Asghari B, Arshad M (2012) Rhizosphere bacteria containing ACC deaminase conferred drought tolerance in wheat grown under semi-arid climate. Soil Environ 31:108–112
Sharp RG, Chen L, Davies WJ (2011) Inoculation of growing media with the rhizobacterium *Variovorax paradoxus* 5C-2 reduces unwanted stress responses in hardy ornamental species. Sci Hortic 129(4):804–811
Shaw LJ, Morris P, Hooker JE (2006) Perception and modification of plant flavonoid signals by rhizosphere microorganisms. Environ Microbiol 8:1867–1880
Sheng X, He L, Wang Q, Ye H, Jiang C (2008) Effects of inoculation of biosurfactant-producing *Bacillus* sp. J119 on plant growth and cadmium uptake in a cadmium-amended soil. J Hazard Mater 155:17–22
Shenker M, Hadar Y, Chen Y (1999) Kinetics of iron complexing and metal exchange in solution by rhioferrin, a fungal siderophore. Soil Sci Soc Am J 63:1681–1687
Shinshi H, Mohnen D, Meins F Jr (1987) Regulation of a plant pathogenesis-related enzyme: Inhibition of chitinase and chitinase mRNA accumulation in cultured tobacco tissues by auxin and cytokinin. Proc Natl Acad Sci Am 84:89–93
Shoebitz M, Ribaudo CM, Pardo MA, Cantore ML, Ciampi L, Curá JA (2009) Plant growth promoting properties of a strain of *Enterobacter ludwigii* isolated from Lolium perenne rhizosphere. Soil Biol Biochem 41(9):1768–1774
Siciliano SD, Goldie H, Germida JJ (1998) Enzymatic activity in root exudates of dahurian wild rye (*Elymus dauricus*) that degrades 2-chlorobenzoic acid. J Agric Food Chem 46:5–7

Siddikee MA, Chauhan PS, Anandham R, Han GH, Sa T (2010) Isolation, characterization, and use for plant growth promotion under salt stress, of ACC deaminase-producing halotolerant bacteria derived from coastal soil. J Microbiol Biotechnol 20:1577–1584

Siddikee MA, Glick BR, Chauhan PS, Yim WJ, Sa T (2011) Enhancement of growth and salt tolerance of red pepper seedlings (*Capsicum annuum* L.) by regulating stress eth-ylene synthesis with halotolerant bacteria containing ACC deaminase activity. Plant Physiol Biochem 49:427–434

Silva HSA, Silva RDS, Mounteer A (2003) Development of a root colonization bioassay for rapid screening of rhizobacteria for potential biocontrol agents. J Phytopathol 151:42–46

Singh B, Satyanarayana T (2011) Microbial phytases in phosphorus acquisition and plant growth promotion. Physiol Mol Biol Plants 17:93–103

Smidt M, Kosuge T (1978) The role of indole-3-acetic acid accumulation by alpha methyl tryptophan-resistant mutants of *Pseudomonas savastanoi* in gall formation on oleanders. Physiol Plant Pathol 13:203–214

Smith KP, Goodman RM (1999) Host variation for interactions with beneficial plant associated microbes. Annu Rev Phytopathol 37:473–491

Somers E, Vanderleyden J, Srinivasan M (2004) Rhizosphere bacterial signalling: a love parade beneath our feet. Crit Rev Microbiol 30:205–240

Soto MJ, Sanjua'n J, Olivares J (2006) Rhizobia and plant-pathogenic bacteria: common infection weapons. Microbiology 152:3167–3174

Spaepen S, Vanderleyden J, Remans R (2007) Indole- 3-acetic acid in microbial and microorganism-plant signaling. FEMS Microbiol Rev 31:425–448

Spaepen S, Vanderleyden J, Okon Y (2009) Plant growth-promoting actions of rhizobacteria. In: van Loon LC (ed) Plant innate immunity book series: advances in botanical research, vol 51. Academic, San Diego, pp 283–320

Stearns JC, Woody OZ, McConkey BJ, Glick BR (2012) Effects of bacterial ACC deaminase on *Brassica napus* gene expression measured with an Arabidopsis thaliana microarray. Mol Plant Microb Interact 25:668–676

Steenhoudt O, Vanderleyden J (2000) *Azospirillum* a free-living nitrogen-fixing bacterium closely associated with grasses: genetic, biochemical and ecological aspects. FEMS Microbiol Rev 24:487–506

Sturz AV, Nowak J (2000) Endophytic communities of rhizobacteria and the strategies required to create yield enhancing associations with crops. Appl Soil Ecol 15:183–190

Sturz AV, Cristie BR, Nowak J (2000) Bacterial endophytes: potential role in developing sustainable systems of crop production. Crit Rev Plant Sci 19:1–30

Suzuki S, He YX, Oyaizu H (2003) Indole-3-acetic acid production in *Pseudomonas fluorescens* HP72 and its association with suppression of creeping bentgrass brown patch. Curr Microbiol 47:138–143

Swain M, Naskar S, Ray R (2007) Indole-3-acetic acid production and effect on sprouting of yam (*Dioscorea rotundata* L.) minisets by *Bacillus subtilis* isolated from culturable cow dung microflora. Pol. J Microbiol 56(2):103–110

Taiz L, Zeiger E (1998) Plant physiology. Sinauer Associates, Sunderland

Tang WH (1994) Yield increasing bacteria (YIB) and biological control of sheath blight of rice. In: Ryder MH, Stephens PM, Bowen GD (eds) Improving plant productivity with rhizosphere bacteria. Commonwealth Scientific and Industrial Research Organization, Adelaide, pp 267–278

Tanimoto E (2005) Regulation of root growth by plant hormones: roles for auxin and gibberellin. Crit Rev Plant Sci 24:249–265

Taurian T, Anzuay MS, Angelini JG, Tonelli ML, Ludueña L, Pena D et al (2010) Phosphate-solubilizing peanut associated bacteria: screening for plant growth-promoting activities. Plant Soil 329:421–431

Teaumroong N, Teamtaisong K, Sooksa-ngun T, Boonkerd N (2001) The dizotrophic endophytic bacteria in Thai rice. In: Suriyaphan O, Hansakdi E, Jongruaysup S, Simons R (eds) Proceeding of the fifth ESAFS international conference on rice environments and rice products. Krabi, Thailand, pp 147–160

Theunis M (2005) IAA biosynthesis in rhizobia and its potential role in symbiosis. In: Ph.D. thesis, Universiteit Antwerpen

Thomas FC, Woeng CA, Lugtenberg BJJ (2008) Root colonisation following seed inoculation. Springer,

Tien TM, Gaskins MH, Hubbell DH (1979) Plant growth substances produced by *Azospirillum brasilense* and their effect on the growth of pearl millet (*Pennisetum americanum* L.). Appl Environ Microbiol 37:1016–1024

Tilak KV, Ranganayaki N (2006) Synergistic effects of plant-growth promoting rhizobacteria and *Rhizobium* on nodulation and nitrogen fixation by pigeon pea (*Cajanus cajan*). Eur J Soil Sci 57:67–71

Tittabutr P, Awaya JD, Li QX, Borthakur D (2008) The cloned 1-aminocyclopropane-1-carboxylate (ACC) deaminase gene from *Sinorhizobium* sp. strain BL3 in *Rhizobium* sp. strain TAL1145 promotes nodulation and growth of *Leucaena leucocephala*. Syst Appl Microbiol 31:141–150

Toklikishvili N, Dandurishvili N, Tediashvili M, Giorgobiani N, Szegedi E, Glick BR et al (2010) Inhibitory effect of ACC deaminase-producing bacteria on crown gall for-mation in tomato plants infected by *Agrobacterium tumefaciens* or A. vitis. Plant Pathol 59:1023–1030

Ton J, Davison S, Van Wees SCM, Van Loon LC, Pieterse CMJ (2001) The *Arabidopsis* isr1 locus controlling rhizobacteria-mediated induced systemic resistance is involved in ethylene signaling. Plant Physiol 125:652–661

Ton J, De Vos M, Robben C, Buchala A, Metraux JP, Van Loon LC, Pieterse CMJ (2002) Characterization of *Arabidopsis* enhanced disease susceptibility mutants that are affected in systemically induced resistance. Plant J 29:11–21

Tsakelova EA, Klimova SY, Cherdyntseva TA, Netrusov AI (2006) Microbial producers of plant growth stimulators and their practical use: a review. Appl Biochem Microbiol 42:117–126

Tsavkelova EA, Cherdyntseva TA, Botina SG, Netrusov AI (2007) Bacteria associated with orchid roots and microbial production of auxin. Microbiol Res 162:69–76

van der Heijden MGA, Bardgett RD, van Straalen NM (2008) The unseen majority: soil microbes as drivers of plant diversity and productivity in terrestrial ecosystems. Ecol Lett 11:296–310

van Elsas JD, van Overbeek LS (1993) Bacterial responses to soil stimuli. In: Kjelleberg S (ed) Starvation in bacteria. Plenum, New York, pp 55–79

van Loon LC (2007) Plant responses to plant-growth promoting bacteria. Eur J Plant Pathol 119:243–254

van Noorden GE, Ross JJ, Reid JB, Rolfe BG, Mathesius U (2006) Defective long distance auxin transport regulation in the *Medicago truncatula* super numerary nodules mutant. Plant Physiol 140:1494–1506

Van Noorden GE, Kerim T, Goffard N, Wiblin R, Pellerone FI, Rolfe BG, Mathesius U (2007) Overlap of proteome changes in *Medicago truncatula* in response to auxin and *Sinorhizobium meliloti*. Plant Physiol 144:1115–1131

Van Puyvelde S, Cloots L, Engelen K, Das F, Marchal K, Vanderleyden J, Spaepen S (2011) Transcriptome analysis of the rhizosphere bacterium *Azospirillum brasilense* reveals an extensive auxin response. Microb Ecol 61:723–728

van Veen JA, van Overbeek LS, van Elsas JD (1997) Fate and activity of microorganisms introduced into soil. Microbiol Mol Biol Rev 61:121–135

Vande Broek A, Lambrecht M, Eggermont K, Vanderleyden J (1999) Auxins upregulate expression of the indole-3-pyruvate decarboxylase gene in *Azospirillum brasilense*. J Bacteriol 181:1338–1342

Vandeputte O, Oden S, Mol A, Vereecke D, Goethals K, El Jaziri M, Prinsen E (2005) Biosynthesis of auxin by the Gram positive phytopathogen *Rhodococcans fascians* is controlled by compounds specific to infect plant tissues. Appl Environ Microbiol 71:1169–1170

Vanderhoff LN, Dute RR (1981) Auxin-regulated wall loosening and sustained growth in elongation. Plant Physiol 67:146–149

Varvaro L, Martella L (1993) Virulent and avirulent isolates of *Pseudomonas syringae* subsp. savastanoi as colonizers of olive leaves: evaluation of possible biological control of the olive knot pathogen. EPPO Bull 23:423–427

Verma SC, Ladha JK, Tripathi AK (2001) Evaluation of plant growth promoting and colonization ability of endophytic diazotrophs from deep water rice. J Biotechnol 91:127–141

Vessey JK (2003) Plant growth promoting rhizobacteria as biofertilizers. Plant Soil 255:571–586

Viterbo A, Landau U, Kim S, Chernin L, Chet I (2010) Characterization of ACC deaminase from the biocontrol and plant growth-promoting agent *Trichoderma asperellum* T203. FEMS Microbiol Lett 305:42–48

Vivas A, Barea JM, Azcón R (2005) Brevibacillus brevis isolated from cadmium- or zinc-contaminated soils improves in vitro spore germination and growth of Glomus mosseae under high Cd or Zn Concentrations. Microb Ecol 49:416–442

Wahyudi AT, Astuti RP, Widyawati A, Meryandini A, Nawangsih AA (2011) Characterization of *Bacillus* sp. strains isolated from rhizosphere of soybean plants for their use as potential plant growth for promoting Rhizobacteria. J Microbiol Antimicrobial 3:34–40

Walker R, Rossall S, Asher MJ (2002) Colonization of the developing rhizosphere of sugar beet seedlings by potential biocontrol agents applied as seed treatments. J Appl Microbiol 92:228

Walker TS, Bais HP, Grotewold E, Vivanco JM (2003) Root exudation and rhizosphere biology. Plant Physiol 132:44–51

Wang C, Knill E, Glick BR, Defago G (2000) Effect of transferring 1-amino-cyclopropane-1-carboxylic acid (ACC) deaminase genes into *Pseudomonas fluorescens strain* CHA0 and its gacA derivative CHA96 on their growth-promoting and disease-suppressive capacities. Can J Microbiol 46:898–907

Wang C, Ramette A, Punjasamamwong P, Zala M, Natsch A, Moenne-Loccoz Y, Defago G (2001) Cosmopolitan distribution of phlD-containing dicotyledonous crop-associated *pseudomonads* of worldwide origin. FEMS Microbiol Ecol 37:105–116

Weller DM (1988) Biological control of soilborne pathogens in the rhizosphere with bacteria. Annu Rev Phytopathol 26:379–407

Wen F, VanEtten HD, Tsaprailis G, Hawes MC (2007) Extracellular proteins in pea root tip and border cell exudates. Plant Physiol 143:773–783

Weyens N, van der Lelie D, Taghavi S, Vangronsveld J (2009) Phytoremediation: planteendophyte partnerships take the challenge. Curr Opin Biotechnol 20:248–254

Wheeler CT, Henson IE, MacLaughlin ME (1979) Hormones in plants bearing actinomycete nodules. Bot Gaz 140:52–57

Whipps JM (2001) Microbial interactions and biocontrol in the rhizosphere. J Exp Bot 52:487–512

Wiehe W, Flich HG (1995) Establishment of plant growth promoting bacteria in the rhizosphere of subsequent plants after harvest of the inoculated precrops. Microbiol Res 150:331–336

Wildermuth MC, Dewdney J, Wu G, Ausubel FM (2001) Isochorismate synthase is required to synthesize salicylic acid for plant defence. Nature 414:562–565

Wu SC, Cao ZH, Li ZG, Cheung KC, Wong MH (2005) Effects of biofertilizer containing N-fixer, P and K solubilizers and AM fungi on maize growth: a greenhouse trial. Geoderma 125:155–166

Xie H, Pasternak JJ, Glick BR (1996) Isolation and characterization of mutants of the plant growth-promoting rhizobacterium *Pseudomonas putida* GR12-2 that overproduce indoleacetic acid. Curr Microbiol 32:67–71

Yadegari M, Rahmani HA, Noormohammadi G, Ayneband A (2008) Evaluation of bean (*Phaseolus vulgaris*) seeds inoculation with *Rhizobium phaseoli* and plant growth promoting rhizobacteria on yield and yield components. Pak J Biol Sci 11:1935–1939

Yamada T (1993) The role of auxin in plant-disease development. Annu Rev Phytopathol 31:253–273

Yang SH, Zhang Q, Guo JH, Charkowski AO, Glick BR, Ibekwe AM (2007) Global effect of indole-acetic acid biosynthesis on multiple virulence factors of *Erwinia chrysanthemi* 3937. Appl Environ Microbiol 73:1079–1088

Yanni YG, Rizk RY, Corich V, Squartini A, Ninke K, Philip-Holingsworth S, Orgambide G, de Bruijn FJ, Stoltzfus J, Buckley D, Schmidt TM, Mateos PF, Ladha JK, Dazzo FB (1997) Natural endophytic association between *Rhizobium leguminosarum* bv. *trifolii* and rice roots and assessment of its potential to promote rice growth. Plant Soil 194:99–114

You CB, Lin M, Fang XJ, Song W (1995) Attachment of Alcaligenes to rice roots. Soil Biol Biochem 7:463–466

Yuan Z, Haudecoeur E, Faure D, Kerr K, Nester E (2008) Comparative transcriptome 1 analysis of *Agrobacterium tumefaciens* in response to plant signal salicylic acid, indole-3-acetic acid and gamma-amino butyric acid reveals signaling cross talk and *Agrobacterium*-plant co-evolution. Cell Microbiol 10:2339–2354

Zahir AZ, Arshad M, Frankenberger WT (2004) Plant growth promoting rhizobacteria: applications and perspectives in agriculture. Adv Agron 81:97–168

Zahir ZA, Munir A, Asghar HN, Shaharoona B, Arshad M (2007) Effectiveness of rhizobacteria containing ACC-deaminase for growth promotion of pea (*Pisum sativum*) under drought conditions. J Microbiol Biotechnol 18:958–963

Zaidi S, Usmani S, Singh BR, Musarrat J (2006) Significance of *Bacillus subtilis* strain SJ-101 as a bioinoculant for concurrent plant growth promotion and nickel accumulation in *Brassica juncea*. Chemosphere 64:991–997

Zaidi A, Khan MS, Ahemad M, Oves M (2009a) Plant growth promotion by phosphate solubilizing bacteria. Acta Microbiol Immunol Hung 56:263–284

Zaidi A, Khan MS, Ahemad M, Oves M, Wani PA (2009b) Recent advances in plant growth promotion by phosphate-solubilizing microbes. In: Khan MS, Zaidi A, Musarrat J (eds) Microbial strategies for crop improvement. Springer, Berlin-Heidelberg, pp 23–51

Zhang YF, He LY, Chen ZJ, Zhang WH, Wang QY, Qian M et al (2011) Characterization of lead-resistant and ACC deaminase-producing endophytic bacteria and their potential in promoting lead accumulation of rape. J Hazard Mater 186:720–725

Zhender GW, Yao C, Murphy JF, Sikora ER, Kloepper JW, Schuster DJ, Polston JE (1999) Microbeinduced resistance against pathogens and herbivores: evidence of effectiveness in agriculture. In: Agarwal AA, Tuzun S, Bent E (eds) Induced plant defenses against pathogens and herbivores: biochemistry, ecology and agriculture. APS Press, St Paul, p 33

Role of Abscisic Acid Producing PGPR in Sustainable Agriculture

Ana Carmen Cohen, Ruben Bottini and Patricia Piccoli

Abstract The global climate is predicted to change the environment drastically over the next century. Increase in CO_2 and temperature and decrease in soil water content leading to enhance drought in several areas of the world are expected. In the last few years, it has been increased the interest in environmental friendly, sustainable, and organic cultural practices that warrant high yield and quality in agricultural crops. Plant growth-promoting rhizobacteria (PGPR) have an important role in the growth and metabolism of plants. The beneficial effects of PGPRs have been demonstrated for many agricultural crop species. Numerous studies indicated that PGPR allow plants survive to biotic and abiotic stresses. Production of phytohormones is one of the main mechanisms to explain the beneficial effects that modified plant growth and development. In this review we are focusing on drought tolerance through ABA regulation and we showed that PGPR act as important agent for influencing the beneficial response of plants to climate change.

Keywords Abscisic acid · Phytohormones · Drought stress · PGPR · Agriculture

1 Introduction

According to the United Nations estimates, the global human population to meet out their requirement projected to reach nine billion in 2050, and as a consequence roughly calculation of 50 % more food to be produced (Tomlinson 2013). Green

A.C. Cohen (✉) · R. Bottini · P. Piccoli
Laboratorio de Bioquímica Vegetal, Instituto de Biología Agrícola de Mendoza,
Consejo Nacional de Investigaciones Científicas y Técnicas-Universidad
Nacional de Cuyo, Almirante Brown 500, M5528AHB Chacras de Coria, Argentina
e-mail: acohen@fca.uncu.edu.ar

D.K. Maheshwari (ed.), *Bacterial Metabolites in Sustainable Agroecosystem*,
Sustainable Development and Biodiversity 12, DOI 10.1007/978-3-319-24654-3_9

revolution, together with intensification of agriculture triggered a dramatic use of chemical fertilizer, pesticide and insecticides. In order to increase world food production and protect crops, the farmers have been applying high amounts of fertilizers which are very costly and make the environment hazardous especially when are used indiscriminately. The projections based on the crop intensification will need change to a more sustainable agriculture so as to preserve the different ecosystem in the world. Because of this, sustainable agriculture should be able to: (1) produce enough food for an increasing world population, (2) protect the environment (micro and macro-environment), (3) maintain and improve human health, and (4) should be economically beneficial to producers and consumers.

Moreover, drought and salinity are two of the main abiotic stress factors that decrease agricultural productivity most dramatically (Boyer 1982; Bray et al. 2000). Global warming and decreases in annual precipitations produce more droughts areas with negative effects on agricultural productivity (Pandey et al. 2007; Barrios et al. 2008; Ziska 2011). Climate change models indicate that warmer temperatures and increases in the frequency and duration of drought during the twenty-first century could adversely affect agriculture (St Clair and Lynch 2010). Drought reduces the availability of CO_2 for photosynthesis, leading to the formation of reactive oxygen species. Abscisic acid (ABA), a sesquiterpenic plant growth regulator (PGR), is the signal that induces different adaptive responses, mainly the closure of stomata to avoid water loss (Zhang and Outlaw 2001; Sansberro et al. 2004). Under water stress, plants increase ABA biosynthesis and/or decrease its catabolism (Bray 2002; Seki et al. 2002; Christmann et al. 2007; Zeller et al. 2009). Further, ABA plays a role in mediating root branching, thereby improving the plant water uptake capacity (De Smet et al. 2006; Cohen et al. 2015). Tardieu et al. (2010) proposed that ABA induces leaf growth by augmenting water movement in the plant because of increased tissue hydraulic conductivity by other way.

Furthermore, global changes, as dry weather and relatively warm conditions, offer the opportunity for a wide variety of living organisms, typical of more arid habitats. According to the soil conditions, the diversity of microorganism vary, i.e. it could be replaced by others lesser or more beneficial micoorganisms for plants. So “native bacteria” need to be redefined in more specific concept to a spatial and physical environment. These soil bacteria, indispensable part of rhizosphere biota, produce positive effects on plant growth and development. When PGPR interact with plants can either directly or indirectly promote rooting, help host plants to establish and improve growth. Some PGPR antagonize the deleterious effects of phytopathogens by inducing systemic resistance (ISR, indirect mechanism) or by metabolites production such as antibiotics (that prevent the growth of the pathogens), siderophores (so, pathogenic microorganisms with nutritional deficit cannot attack the crop) or by synthesis of volatile metabolites such as hydrogen cyanide (Bano and Musarrat 2003; Bashan and de-Bashan 2005). Among PGPR, *Bacillus*, *Streptomyces*, *Pseudomonas*, *Burkholderia* and *Agrobacterium* are predominant genera marketed as the biological control agents.

PGPR provide plants with diverse mechanism like via production of plant growth regulators (PGR), (Arshad and Frankenberger 1993; Costacurta and

Vanderleyden 1995; Glick 1995; Bastián et al. 1998; Bloemberg and Lugtenberg 2001; Bottini et al. 2004; Cohen et al. 2008; Piccoli et al. 2011), biological nitrogen fixation (FBN), (Boddey and Dobereiner 1995), or nutrient mobilization from the surrounding environment (Glick 1995; De Freitas et al. 1997; Rodriguez and Fraga 1999; Richardson 2001; Chen et al. 2006; Rodriguez and Fraga 2006; Hu et al. 2009; Sharma et al. 2011). Additionally, the use of PGPR increases plant's resistance to adverse environmental stresses, drought, salinity, nutrient deficiency and heavy metal contamination. PGPR could alleviate the stress in plant growth caused by drought (summarized in Table 1), salt (Egamberdieva 2008; Mayak et al. 2004a; Kaymak et al. 2009; Ahmad et al. 2011) and some other unfavorable environmental conditions. However, fewer reports have been published on PGPR related with mechanism of tolerance to abiotic stresses (Yang et al. 2009; Mahehswari 2012). Thus, PGPR are used to remediate and rehabilitate nonfertile and contaminated soils into fertile ones (Glick 2010).

In the recent years exist a pronounced interest in eco-friendly and sustainable agriculture, proved potential for the use of PGPR as inoculants for biofertilization,

Table 1 Drought stress alleviation by PGPR in different crops

PGPR	Plants	Effect of inoculation	References
Azospirillum sp.	Corn, wheat, sorghum and other grasses	Improves plant–water relationships and grain yield	Okon (1985)
Azospirillum brasilense	Sorghum	Improvements of water status and yield of field-grown grain sorghum	Sarig et al. (1988)
A. brasilense Sp 245	Wheat	Increases shoot growth, water status and rate of coleoptile growth	Creus et al. (1997)
A. brasilense Sp 245	Wheat	Stimulates water status, cell wall elasticity and (or) apoplastic water	Creus et al. (1998)
Paenibacillus polymyxa	*Arabidopsis*	Induces changes in gene expression associated with drought	Timmusk and Wagner (1999)
Bacillus sp.	Lettuce	Increases arbuscular mycorrhizal fungus colonization in roots and enhances photosynthesis rate	Vivas et al. (2003)
A. brasilense Sp 245	Wheat	Improves water status, induces elastic adjustment, grain yield and mineral quality	Creus et al. (2004)

(continued)

Table 1 (continued)

PGPR	Plants	Effect of inoculation	References
Streptomyces padanus AOK-30	Laurel	Enhancements of osmotic pressure in leaf cells and induces modifications in cell wall	Hasegawa et al. (2004)
Achromobacter piechaudii ARV8	Tomato and pepper	Increases fresh and dry weight	Mayak et al. (2004a)
Variovorax paradoxus 5C-2	Pea	Stimulates plant growth	Dodd et al. (2005)
Pseudomonas sp. ACP *Pseudomonas putida* GR12-2 *Hansenula saturnus* *Penicillium citrinum* *P. putida* UW4	*Arabidopsis*	Modulate plant ethylene levels by the bacterial enzyme ACC deaminase	Glick (2005)
S. padanus AOK-30	Laurel	Accelerates callose accumulation and lignification	Hasegawa et al. (2005)
Bacillus subtilis GB03 *B. amyloliquefaciens* IN937a *B. pumilus* SE-34 *B. pumilus* T4 *B. pasteurii* C9 *P. polymyxa* E681 *Pseudomonas fluorescens* 89B-61 *Serratia marcescens* 90-166	*Arabidopsis*	Increase foliar and fresh weight	Ryu et al. (2005)
Bacillus thuringiensis	Broom	Increases plant water uptake	Marulanda et al. (2006)
Bradyrhizobium *P. putida* biotype A *P. fluorescens*	Maize and bean	Increase root and shoot growth and enhance nodulation in legume	Shaharoona et al. (2006a)
Achromobacter xiloxidans *Bacillus pumilus*	LB media	Increase ABA and JA	Forchetti et al. (2007)
P. fluorescens	Periwinkle	Enhances biomass yield and ajmalicine production	Jaleel et al. (2007)
Pseudomonas sp.	Pea	Increases plant growth, yield and ripening	Arshad et al. (2008)

(continued)

Table 1 (continued)

PGPR	Plants	Effect of inoculation	References
A. brasilense Sp 245	*Arabidopsis*	Increases growth and ABA levels	Cohen et al. (2008)
Acinetobacter *Alcaligenes faecalis* *Bacillus cereus* *Enterobacter hormaechei* *Pantoae* *P. aeruginosa*	Wheat	Improve plant growth and nutrition	Egamberdieva (2008)
Paenibacillus polymyma *Rhizobium tropici*	Mung bean	Alter hormonal balance and stomatal conductance. Improve plant growth, nitrogen content, nodule number	Figueiredo et al. (2008)
Pseudomona mendonica	Lettuce	Increases phosphatase activity in roots and proline accumulation in leaves	Kohler et al. (2008)
P. fluorescens biotype G ACC-5 *P. fluorescens* ACC-14 *P. putida biotype* A Q-7	Pea	Improve fresh and dry weight, root length, shoot length, number of leaves per plant and water-use efficiency	Zahir et al. (2008)
V. paradoxus 5C-2	Pea	Improves growth, yield and water-use efficiency. Increases nodulation by *Rhizobium leguminosarum*	Belimov et al. (2009)
A. lipoferum USA 59b	Maize	Reverses effect of fluridone (F) and prohexadione-Ca (P, inhibitors of ABA and GA synthesis, respectively), increases relative water content and enhances growth in P, F, or P + F and enhances GAs and ABA levels in plants	Cohen et al. (2009)

(continued)

Table 1 (continued)

PGPR	Plants	Effect of inoculation	References
Agrobacterium rubi A 16 *Burkholderia gladii* BA 7 *P. putida* BA 8 *B. subtilis* BA 142 *B. megaterium* M 3	Radish	Improve the percentage of seed germination	Kaymak et al. (2009)
A. brasilense mutant	Maize	Increases biomass and trehalose	Rodríguez-Salazar et al. (2009)
Pseudomonas BA-8 *Bacillus* OSU-142 *Bacillus* M-3	Strawberry	Increase total soluble solids, total sugar and reduce sugar	Pirlak and Kose (2009)
P. putida *Pseudomonas* sp. *Bacillus megaterium*	Clover	Increase IAA production, shoot and root biomass and water content	Marulanda et al. (2009)
P. entomophila BV-P13 *P. stutzeri* GRFHAP-P14 *P. putida* GAP-P45 *P. syringae* GRFHYTP52 *P. monteilli* WAPP53	Maize	Improve plant biomass, relative water content, leaf water potential, root adhering soil/root tissue ratio, aggregate stability and mean weight diameter, levels of proline, sugars and free amino acids. Decrease electrolyte leakage, leaf water loss and activities of antioxidant enzymes	Sandhya et al. (2010)
Pseudomonas sp. *Bacillus lentus*	Basil	Improve growth, photosynthesis, mineral content and antioxidant enzymes	Golpayegani and Tilebeni (2011)
Pseudomonas sp.	Basil	Improves plant growth, auxin and protein contents	Heidari et al. (2011)
Burkholdera cepacia SE4 *Promicromonospora* sp. SE188 *Acinetobacter calcoaceticus* E370	Cucumber	Increase biomass and chlorophyll contents, water potential and decreased electrolytic leakage. Reduce sodium ion concentration and activities of catalase, peroxidase, polyphenol oxidase and total polyphenol and ABA. Increase salicylic acid and gibberellin	Kang et al. (2014)

(continued)

Table 1 (continued)

PGPR	Plants	Effect of inoculation	References
Proteus penneri Pp1 *Pseudomonas aeruginosa* Pa2 *A. faecalis* AF3	Maize	Increase exopolysaccharide, relative water content, protein, sugar, proline and decrease in the activities of antioxidant enzymes	Naseem and Bano (2014)
Bacillus licheniformis Rt4M10 *P. fluorescens* Rt6M10	Grapevine	Decrease plant water loss rate in correlation with increments of ABA. Increase monoterpenes and sesquiterpenes	Salomon et al. (2014)
P. aeruginosa GGRJ21	Mung bean	Accelerates the accumulation of levels of antioxidant enzymes, cell osmolytes and consistently expediting the upregulation of stress-responsive genes	Sarma and Saikia (2014)
A. brasilense Sp 245	*Arabidopsis*	Augments plant biomass, alters root architecture, stimulates photosynthetic and photoprotective pigments and retards water loss in correlation with incremented ABA levels. Improves plants seed yield, plants survival, proline levels and relative leaf water content; it also decreases stomatal conductance, malondialdehyde and relative soil water	Cohen et al. (2015)

phytostimulation and biocontrol (Lugtenberg and Kamilova 2009; Babalola 2010; Maheshwari 2010). These bioinoculants contribute to the development of sustainable agriculture under stressed conditions (Glick et al. 2007a; Dodd and Perez-Alfocea 2012; Berg et al. 2013). Also, with the rise of organic agriculture, the demand of PGPR biofertilizers has been increasing. These are promising solution for sustainable, environmentally friendly agriculture (Tsavkelova et al. 2006). Biofertilizer and biopesticide containing efficient PGPR may improve crop production, reduce agrochemical use, and support eco-friendly sustainable food production.

2 Plant Growth Regulators

One of the most important mechanisms in plant growth promotion is the production of phytohormones (Bottini et al. 2004; Spaepen et al. 2007; Bashan and de-Bashan 2010). PGPR synthesize different group of phytohormones, such as gibberellins (GAs), cytokinins, auxins (IAA), ABA, ethylene and nitric oxide (NO), which regulate plant growth and development throughout their life cycle.

Auxins are a group of compounds in which the most common and active auxin recognized in plants is indole-3-acetic acid (IAA). Different bacteria from plant rhizosphere possess the ability to produce auxin (Tien et al. 1979; Atzorn et al. 1988; Costacurta and Vanderleyden 1995; Patten and Glick 1996; Bastián et al. 1998; Barazani and Friedman 1999; Patten and Glick 2002; Salomon et al. 2014). Auxin is responsible to control processes such as expansion growth, vascular tissue differentiation, root initiation and development, gravitropism, phototropism, apical dominance and stimulation of cell division among others (Bartel 1997; Ross and O'Neill 2001; Zhao 2010). In bacteria, the main precursor for the synthesis of IAA is tryptophan, however, *Gluconoactebacter diazotrophicus*, *Herbaspirillum seropedicae* and few others produce IAA without tryptophan in the culture medium (Bastián et al. 1998). In PGPR at least five different pathways for biosynthesis of IAA have been described (Hartmann et al. 1983; Spaepen et al. 2007). Moreover, several recent reports indicate that IAA can also act as signaling molecule in bacteria and therefore, has a direct effect on bacterial physiology (Spaepen et al. 2009). *Azospirillum*, *Burkholderia*, *Marinomonas*, *Pseudomonas*, *Rhodococcus* and *Sphingomonas* genera can metabolize IAA and the genes and enzymes involved in these reactions have been described (Leveau and Gerards 2008).

Cytokinins are adenine derivatives regulating cell division and differentiation process in plant. At least 90 % of rhizobacteria produce these compounds in culture medium (Barea et al. 1976; Tien et al. 1979). Arshad and Frankenberger (1993) and later Arkhipova et al. (2007) observed strains of *Pseudomonas* and *Bacillus* in the vicinity of wheat and lettuce roots able to produce cytokinins, so helping plant in growth and development. Also, many rhizobacteria are capable of synthesizing kinetins, zeatins, isopentenyladenines and other cytokinin derivatives (Tsavkelova et al. 2006).

PGPR also produce GAs, a group of ent-kaurene-derived diterpenoid phytohormones. More than 130 GAs have been identified, in plants, fungi and bacteria (Bottini et al. 1989; Hedden and Phillips 2001; Yamaguchi 2008). GAs are primarily responsible for stem elongation (Crozier et al. 2001; Davies 2005). They are essential in many other developmental processes in plants, including seed germination, leaf expansion, trichome development, pollen maturation and the induction of flowering (Achard and Genschik 2009). In bacteria, the ability of *Azospirillum lipoferum* to produce GA_1 and GA_3 was confirmed by Bottini et al. (1989) using GC-MS analysis. Different strains like *A. brasilense*, *G. diazotrophicus*, *H. seropedicae*, *Bacillus pumilus*, *B. licheniformis*, *Pseudomonas fluorescens,* etc. are able to produce GAs in vitro (Janzen et al. 1992; Bastián et al. 1998;

Gutiérrez-Mañero et al. 2001; Probanza et al. 2002; Bottini et al. 2004; Salomon et al. 2014). Moreover, rhizobacteria increase the endogenous levels of GAs in plant thus stimulating plant growth (Fulchieri et al. 1993; Lucangeli and Bottini 1997; Gutiérrez-Mañero et al. 2001; Cassán et al. 2001a, b). Also, *Azospirillum* species have capacity to metabolize GAs (Piccoli and Bottini 1994a, b; Piccoli et al. 1996, 1997; Bottini et al. 2004). It has been observed that inoculation with *G. diazotrophicus* and applications of GA_3 enhance fructose and glucose levels in shoots of sorghum (Bastián et al. 1999). *A. lipoferum* USA 5b and *A. brasilense* Cd have the ability to reverse dwarfism in d*1* maize and in d*x* rice mutants similar to that of exogenous application of GA_3 (Lucangeli and Bottini 1997). In addition, *B. pumilus*, *B. licheniformis* and *A. lipoferum* USA 5b reverse the effect of GAs inhibitors in *Alnus* and maize (Gutiérrez-Mañero et al. 2001; Cohen et al. 2009)

Ethylene is other PGR synthesized by majority of bacterial species (Primrose and Dilworth 1976). However, if ethylene concentration remains high after germination, either root elongation, or symbiotic N_2 fixation in leguminous plants are inhibited (Dobbelaere et al. 2003; Lohar et al. 2009). Under stress conditions, the endogenous synthesis of ethylene is accelerated and affects root growth and consequently the entire growth of plant. Certain PGPR produce enzymes like 1-aminocyclopropane-1-carboxylate (ACC) deaminase, which hydrolyze ACC (an immediate precursor of ethylene biosynthesis) and lowers the level of ethylene in crop rhizosphere (Glick 2005; Glick et al. 2007a, b). The products of this hydrolysis, ammonia and α-ketobutyrate, can be used by the bacterium as source of nitrogen and carbon for growth. In this way the bacterium acts as a sink for ACC and thus diminished the ethylene level in plants, preventing some of the potentially deleterious consequences of high ethylene concentrations (Glick et al. 1998). PGPR with ACC deaminase characteristics improve crop growth and yield (Glick et al. 1998; Grichko and Glick 2001; Shaharoona et al. 2006a, b; Glick et al. 2007a; Zahir et al. 2008; Belimov et al. 2009; Dodd et al. 2010).

ABA is related mainly with responses to biotic and abiotic stresses (Davies 1995). Increases in ABA levels have been reported under salt, cold, drought and wounding conditions (Zeevaart and Creelman 1988; Peña-Cortés et al. 1989; Shinozaki and Yamaguchi-Shinozaki 2000; Nambara and Marion-Poll 2005). ABA is formed by cleavage of C40 carotenoids and can be produced ubiquitously by higher plants, algae, fungi and bacteria (Zeevaart 1999; Cutler and KrochKo 1999). More recently, it has been identified as an endogenous proinflammatory cytokine in human's granulocytes (Bruzzone et al. 2007; Bassaganya-Riera et al. 2011). Reports on ABA production by PGPR were detected by radioimmunoassay (Kolb and Martin 1985; Belimov et al. 2001): ABA has been characterized with more accuracy by full scan mass spectrometry in chemically defined growth cultures of *A. brasilense* Sp 245, *Arthrobacter koereensis* and *B. licheniformis*, to avoid contamination of PGR (Cohen et al. 2008; Perrig et al. 2007; Sgroy et al. 2009; Piccoli et al. 2011). Cohen et al. (2008) observed that *A. brasilense* Sp 245 increased ABA production in culture medium supplement with NaCl. Moreover, *Corynebacterium* sp. converts ABA to dehydrovomifoliol in vitro and possessed vomifoliol dehydrogenase activity (Hasegawa et al. 1984). *Rhodococcus* sp. P1Y and *Novosphingobium* sp. P6W metabolize ABA in vitro as a sole carbon and energy source using ABA-supplemented medium (Belimov et al. 2014). Information about bacterial ABA metabolism is very limited. It was proposed that bacterial-synthesized ABA is a product of carotenoid metabolism (Marasco and Schmidt-Dannert 2008) although there are some evidences that the gene *CtrZ*, responsible of the synthesis of xanthoxine in plants, is expressed in *P. fluorescens* Rt6M10 cultures (Domínguez et al. 2011).

Some PGPR such as *A. brasilense* Sp 245 in *Arabidopsis* enhance the plant's ABA levels (Cohen et al. 2008, 2015), for example, *A. lipoferum* USDA 59b in maize (Cohen et al. 2009); *B. licheniformis* Rt4M10 and *P. fluorescens* Rt6M10 in grape plants (Salomon et al. 2014); and *B. subtilis* in shoots of lettuce (Arkhipova et al. 2005). By contrary, *Variovorax paradoxus* 5C-2 does not stimulate ABA signaling in maize in either well-watered or drying soil (Dodd et al. 2006); *Raoultella planticola* Rs-2 and *Promicromonospora* sp. SE188 inoculated in cotton and tomato, respectively, also decreased ABA concentrations (Wu et al. 2012; Kang et al. 2012).

3 Other PGR Compounds

Nitric Oxide (NO) is a volatile, lipophilic free radical that participates in metabolic, signaling, defense and developmental pathways in plants (Lamattina and Polacco 2007). Many bacterial physiological processes involve NO participation both as an intermediate in metabolic pathways and as a regulatory signal molecule. NO modulates biofilm production by *A. brasilense* in NFB modified medium with the addition of NaCl (Arruebarrena Di Palma et al. 2013). NO plays a major role in the IAA signaling pathway and its participation leads to lateral and adventitious root formation (Creus et al. 2005).

Jasmonates are derived from oxygenated fatty acids via the octadecanoid pathway and are characterized by a pentacyclic ring structure. These have diverse functions ranging from the initiation of biotic and abiotic stress responses to the regulation of plant growth and development. Some rhizobacteria such as *B. pumilus* and *Achromobacter* sp. (Forchetti et al. 2007) and *Arthrobacter koreensis* (Piccoli et al. 2011) produce jasmonic acid.

4 Role of ABA Produced by PGPR in Plant–PGPR Interactions in Drought-Stressed Soils

Drought stress is one of the main adverse environmental conditions that limit crop growth and productivity worldwide, especially in arid and semiarid regions (Boyer 1982). Plants must avoid or tolerate cellular dehydration to survive drought (Seki et al. 2007), particularly, morphological adaptation and responses at biochemical and genetic levels. Water balance regulations play an important role in plant adaptation to these environmental conditions. Water deficits increase ABA biosynthesis and/or ABA deactivation (Bray 2002; Ren et al. 2007; Huang et al. 2008), preparing the plant to resist water loss. Physiological responses to drought include stomatal closure (Zhang and Outlaw 2001), decrease in photosynthetic activity, modification in cell wall elasticity and generation of toxic metabolites causing plant death (Ahuja et al. 2010).

ABA acts as an endogenous messenger that modules several physiological processes controlling plant response to biotic and abiotic stresses. Different abiotic stress-inducible genes are controlled by ABA, however, others are ABA-independent (Yamaguchi-Shinozaki and Shinozaki 2005). In addition, ABA regulates a variety of plant processes as it was mentioned, thereby improving the plant water uptake capacity (De Smet et al. 2006). Sansberro et al. (2004) observed that ABA sprayed on leaves promotes growth in *Ilex paraguariensis* by alleviating diurnal water stress and more recently Tardieu et al. (2010), observed that ABA induces leaf growth in maize by augmenting water movement in the plant because of increased tissue hydraulic conductivity. In tomato, ABA overproduction enhanced transpiration efficiency and root hydraulic conductivity, thereby affecting leaf expansion through improvements in water status (Thompson et al. 2007). Also, ABA increases leaf carotenoid content and allocation of carbohydrate in wheat and soybean grains (Travaglia et al. 2007, 2009), and augments yields in field-grown wheat with a moderate water restriction (Travaglia et al. 2010). In grape, ABA enhances fruit yield (Quiroga et al. 2009) and increases sugar transport and promotes carbon allocation toward sink organs involved in plant survival (roots and fruits; Moreno et al. 2011).

In general, PGPR play an essential role in improving crop growth especially under stress conditions. Maize plants inoculated with *A. lipoferum* increase ABA levels and reverse the effects of inhibitors of ABA and GA synthesis (fluridone and prohexadione-Ca, respectively). Fluridone (inhibitor) application decreases the ABA levels and it affects growth in well-watered plants to a level found in drought-stressed. Fluridone plants are as short as those submitted to a period of water stress and they did not control water loss efficiently, which in turn reduce cell turgidity, decrease growth and as consequence reduce shoot and root dry weight. In fluridone plants, *A. lipoferum* reverse growth parameters at the level of control unstressed (well-watered) suggesting that it might supply the plant with ABA so as to cover the deficit produced by fluridone. This inhibitor also affects the relative water content (RWC) in both, well-watered and drought-stressed plants, and *Azospirillum* reverses this effect (Cohen et al. 2009). These results are further evidenced using the *Arabidopsis* mutant *aba*2-1, defective in ABA biosynthesis and wild type Col-0 (Cohen et al. 2015).

A. brasilense Sp 245 colonizes the roots and rosettes of Col-0 and *aba*2-1 plants and increases main root length, lateral roots number and fresh weight. The roots of *aba*2-1 plants have a larger number of lateral roots that grew longer than those of Col-0, something previously observed by Deak and Malamy (2005). In order to observe the effect of inhibitor similar effect is also founded applying fluridone in Col-0 plants where *Azospirillum* inoculation increases the length and number of lateral roots (Cohen et al. 2007). Although the bacterium-produced ABA might reduce the amount of lateral roots, such effect is counteracted by IAA and GAs produced by *Azospirillum*. There are many reports that supporting *Azospirillum* sp. produces both IAA (Crozier et al. 1988; Bashan and de-Bashan 2010) and GAs (Bottini et al. 1989, 2004). Different PGPR produce small amounts of IAA increasing considerably the development of roots, plant growth and its crop productivity (maize, rice, sorghum, potato, canola).

Azospirillum increase the root system and leaf area of pot-grown Col-0 plants in respect to that of the control. In this case, inoculated plants improve root proliferation, suggesting an increased ability to uptake water from the water stress, as confirmed by soil RWC determinations (Cohen et al. 2015). *Azospirillum* strains improve plant–water relationships and cell wall elasticity with higher seed yield in sorghum and wheat (Sarig et al. 1988; Creus et al. 1997, 1998, 2004). It was also observed in *Retama sphaerocarpa* due to *B. thuringiensis* (Marulanda et al. 2006). Such ability may be related to the presence of aquaporins, as was observed in case of *Azospirillum*-inoculated barley seedlings whereas higher root expression of aquaporin gene was detected (Zawoznik et al. 2011). On the other hand, Dardanelli et al. (2008) reported that *A. brasilense* promotes root branching in bean seedling and increased secretion of flavonoids and lipochitooligosaccharides. Also, *Proteus penneri* Pp1, *P. aeruginosa* Pa2 and *Alcaligenes faecalis* AF3 increase exopolysaccharide (EPS) in maize plants. The EPS produced by bacteria protects the microbes against inhospitable conditions and enables their survival.

Azospirillum affects the whole life cycle of a plant, accelerating its growth rate and shortening its vegetative period, both effects relevant for most crops. Further, *Azospirillum* sp. increases both vegetative (rosettes size and dry weight as consequence of root branching that improves the area active in water and nutrient uptake) and reproductive parameters (number of inflorescences and flowers, inflorescences dry weight and seed production). Some of these effects have also been observed in *Arabidopsis* inoculated with *B. subtilis* GB03 that increase foliar and fresh weight (Ryu et al. 2005). On the other hand, *Burkholderia phytofirmans* PsJN produces bigger rosette areas and early flowering times (Poupin et al. 2013). The emission of volatile compounds by *B. subtilis* GB03 increases growth and photosynthesis through modulation of ABA signaling in *Arabidopsis* (Zhang et al. 2008) and delayed flowering (Xie et al. 2009; Bresson et al. 2013), thereby flowering time may depend more on the bacterium strain itself rather than on the assembly PGPR–plant variety, probably because of differential modulation in the plant hormonal homeostasis.

Many reports have described the essential role of ABA in plant responses to drought during plant–bacteria interaction, it was observed that the mutant *aba*2-1 has only 37 % of the total ABA measured in the wild-type Col-0, but when *aba*2-1 is inoculated, it produced higher ABA levels than that of Col-0. Thus, confirming the ability of *A. brasilense* Sp 245 to produce ABA per se increases the plant biosynthesis of ABA in both Col-0 and aba2-1. This suggests that *A. brasilense* Sp 245 has the enzyme/s involved in this reaction. The plant–bacteria association generates higher ABA levels than the sum of plant plus bacteria alone shown previously (Cohen et al. 2008). Salomon et al. (2014) founded that *B. licheniformis* increased ABA content by 70-fold and *P. fluorescens* to that of 40-fold in grape leaf tissues. This was correlated with water loss rate assay, where the plants bacterized with *P. fluorescens* and *B. licheniformis* lost 4 and 10 % less water, respectively, than controls. During drought, ABA induces stomatal closure to minimize water loss through transpiration. Also, *A. brasilense* Sp 245 inoculation delayed water losses after cutting rosettes by controlling stomatal closure through increasing ABA levels (Cohen et al. 2015). This augment in ABA levels may provide the plant to cope better with unfavorable environmental conditions. The highest leaf RWC found in Col-0+Sp 245 plants confirms once again how inoculated plants were able to control water loss. Stomatal conductance (*gs*) value is a crucial characteristic that determines plant water status. Although inoculated plants had greater leaf area; the *gs* diminishes in these plants and reaches the wilting point later than the Col-0 probably because they have more ABA than the non-inoculated. Drought caused an accentuated increase in ABA levels when compared to those watered; however, *Azospirillum* increased the ABA levels under both watered and drought conditions (Cohen et al. 2015). Similarly, *Paenibacillus polymyma* and *Rhizobium tropici* inoculated in bean plants altered hormonal balance and stomatal conductance (Figueiredo et al. 2008). Nevertheless, tomato *flacca* and *notabilis* mutants deficient in ABA inoculated with *Rhodococcus* sp. P1Y and *Novosphingobium* sp. P6W decreased root and/or leaf ABA concentrations (Belimov et al. 2014).

Plants inoculated with *A. brasilense* Sp 245 are greener than non-inoculated plants, with increase in both photosynthetic and photoprotective pigments (Cohen et al. 2015). The increment in chlorophyll and enhanced photosynthesis are well-known responses of plants to inoculation with several PGPR (Deka and Dileep 2002; Bashan et al. 2006; Zhang et al. 2008). Wheat and grapevine plants treated with ABA present higher content of carotenoids (Travaglia et al. 2007, 2010; Berli et al. 2010) indicating that inoculation with *A. brasilense* Sp 245 may be involved in this process. Also, total phenolic compound and anthocyanins were strongly augmented in inoculated *A. brasilense* Sp 245 under drought treatment because these compounds are related with stress conditions in grapevine with Solar UV-B and ABA treatments (Berli et al. 2010, 2011). The photoprotective role of anthocyanins can be due to radiation filtering and/or to ROS quenching through the powerful antioxidative capacity (Sperdouli and Moustakas 2012). Additionally, phenolic compounds may also enhance protection against oxidative stress, as they possess chemical structures capable of scavenging free radicals (Blokhina et al. 2002; Berli et al. 2010). Drought stress induced changes in lipid peroxidation, can be quantified by malondialdehyde (MDA) levels. We observed an increased MDA in drought-Col-0 plants, whereas Col-0 plants inoculated with *A. brasilense* Sp 245 (Col-0+Sp 245) or Col-0+ABA showed lesser damages, indicating that these plants are protected against the adverse effects of oxidative stress and demonstrating the efficiency of both *A. brasilense* Sp 245 and ABA to induce antioxidative defense mechanisms. Proline levels were increased in Col-0+Sp 245 since irrigation was suspended, and the highest value was recorded at day 7 since water was withheld before entering wilt. Afterward, Col-0+Sp 245 survive better a second cycle of drought and also increase seed yield (Fig. 1) (Cohen et al. 2015). This osmolyte contributes to osmotic adjustment during stress allowing the plant to obtain water even with very low soil water potentials, and it protects the structure of macromolecules and membranes during extreme dehydration (Meloni et al. 2001). Sandhya et al. (2010) observed that *Pseudomonas* sp. inoculated in maize increased solutes and modified antioxidants status in drought conditions. Earlier, Timmusk and Wagner (1999) reported that inoculation with *P. polymyxa* enhanced the drought tolerance of *A. thaliana*. Using RNA display, they concluded that mRNA transcriptions of a drought-response gene, Early Response to Dehydration 15 (ERD15), were augmented in inoculated plants compared to uninoculated controls. *P. aeruginosa* GGRJ21 strain elicits water stress tolerance in mung bean plants by accelerating the accumulation of antioxidant enzymes, cell osmolytes and consistently expediting the upregulation of stress-responsive genes: dehydration responsive element binding protein (DREB2A), catalase and dehydrin in PGPR-treated plants (Sarma and Saikia 2014). Also, under drought stress, wheat plants inoculated with *Azospirillum* showed an enhanced osmotic adjustment that maintains cell turgor, thus, preventing degenerative processes (Creus et al. 2004). Proline synthesis in stressed plants was reported in other PGPR such as *Burkholderia* sp., *Arthrobacter* sp. and *Bacillus* sp. (Dodd and Pérez-Alfocea 2012). Also, the inoculation of basil with *Pseudomonas* sp. and *Bacillus lentus* alleviated the salinity effects on growth, photosynthesis, mineral content and antioxidant enzymes (Golpayegani and Tilebeni 2011). *Ocimum basilicum* inoculated with *Pseudomonas* sp. increases plant growth, as well as auxin and protein

Fig. 1 Representative images of the *Arabidopsis thaliana* Col-0 pot-grown plants under drought conditions treated with 100 μM ABA (Col-0+ABA), control, (Col-0) and inoculated with *Azospirillum brasilense* Sp 245 (Col-0+Sp 245)

contents under drought stress conditions (Heidari et al. 2011). Similarly, *P. mendocina* inoculated in lettuce plants increase phosphatase activity in roots and proline accumulation in leaves (Kohler et al. 2008).

These differences in the physiologic response of inoculated plants to drought are explained by a better control of stomatal closure and opening mediated by ABA as has been shown in *Arabidopsis*. Inoculation of *Arabidopsis* with *A. brasilense* Sp 245 enhances plant biomass, root surface, accelerate different stages of *Arabidopsis* growth, augment photosynthetic, photoprotective pigments and proline levels. Augmented ABA levels corresponded to that of decrease in stomatal conductance and retard in water losses.

5 Conclusion

The productivity of important agricultural crops is drastically reduced due to both biotic and abiotic stresses. Climate model projections indicate large increases in drought frequency, duration and extent. Taking into account the research material related to the capacity of PGPR to help agricultural yield to increase their tolerance and adaptation to drought conditions as well as pathogens attack, it is relevant to consider that PGPR alone or with ABA applications act as useful tools for increasing crop yield in an efficient and ecological way.

Acknowledgments This paper was supported by Fondo para la Investigación Científica y Tecnológica (FONCYT, PICT 2008 1666 to R. Bottini and PICT 2007 02190 to P. Piccoli), Consejo Nacional de Investigaciones Científicas y Tecnológicas (CONICET, PIP 2008 to P. Piccoli) and Secretaría de Ciencia y Técnica-Universidad Nacional de Cuyo to A. Cohen, R. Bottini and P. Piccoli. A. Cohen, R. Bottini and P. Piccoli are fellows of CONICET.

References

Achard P, Genschik P (2009) Releasing the brakes of plant growth: how GAs shutdown DELLA proteins. J Exp Bot 60(4):1085–1092

Ahmad M, Zahir ZA, Asghar N, Asghar M (2011) Inducing salt tolerance in mung bean through coinoculation with rhizobia and plant-growth promoting rhizobacteria containing 1-amino-cyclopropane-1-carboxylate deaminase. Can J Microbiol 57:578–589

Ahuja I, de Vos RC, Bones AM, Hall RD (2010) Plant molecular stress responses face climate change. Trends Plant Sci 15(12):664–674

Arkhipova TN, Veselov SU, Melentiev AI, Martynenko EV, Kudoyarova GR (2005) Ability of bacterium *Bacillus subtilis* to produce cytokinins and to influence the growth and endogenous hormone content of lettuce plants. Plant Soil 272:201–209

Arkhipova TN, Veselov SY, Melentiev AI, Martynenko EV, Kudoyarova GR (2006) Comparison of effects of bacterial strains differing in their ability to synthesize cytokinins on growth and cytokinin content in wheat plants. Russ J Plant Physiol 53:507–513

Arkhipova TN, Prinsen E, Veselov SU, Martinenko EV, Melentiev AI, Kudoyarova GR (2007) Cytokinin producing bacteria enhances plant growth in drying soil. Plant Soil 292:305–315

Arruebarrena Di Palma A, Pereyra CM, Moreno Ramirez L, Xiqui Vázquez ML, Baca BE, Pereyra MA, Lamattina L, Creus CM (2013) Denitrification-derived nitric oxide modulates biofilm formation in *Azospirillum brasilense*. FEMS Microbiol Lett 338(1):77–85

Arshad M, Frankenberger WT Jr (1993) Microbial production of plant growth regulators. In: Blaine F, Metting Jr. (eds) Soil microbial ecology: Marcel and Dekker, Inc., New York, pp 307–347

Arshad M, Sharoona B, Mahmood T (2008) Inoculation with *Pseudomonas* spp. containing ACC deaminase partially eliminate the effects of drought stress on growth, yield and ripening of pea (*Pisum sativum* L.). Pedosphere 18:611–620

Attia M, Ahmed MA, El-Sonbaty MR (2009) Use of biotechnologies to increase growth, productivity and fruit quality of *Maghrabi Banana* under different rates of phosphorous. World J Agric Sci 5(2):211–220

Atzorn R, Crozier A, Wheeler C, Sandberg G (1988) Production of gibberellins and Indole-3-acetic acid by *Rhizobium phaseoli* in relation to nodulation of *Phaseolus vulgaris* roots. Planta 175:532–538

Babalola OO (2010) Beneficial bacteria of agricultural importance. Biotechnol Lett 32:1559–1570

Bano N, Musarrat J (2003) Characterization of a new *Pseudomonas aeruginosa* strain NJ-15 as a potential biocontrol agent. Curr Microb 46(5):324–328

Bano A, Fatima M (2009) Salt tolerance in *Zea mays* (L) following inoculation with *Rhizobium* and *Pseudomonas*. Biol Fertil Soils 45:405–413

Barazani OZ, Friedman J (1999) Is IAA the major root growth factor secreted from plant-growth-mediating bacteria? J Chem Ecol 25:2397–2406

Barea J, Navarro M, Montoya E (1976) Production of plant growth regulators by rhizosphere phosphate-solubilizing bacteria. J Appl Bacteriol 40:129–134

Barrios S, Ouattara B, Strobl E (2008) The impact of climatic change on agricultural production: is it different for Africa? Food Policy 33:287–298

Bartel B (1997) Auxin biosynthesis. Annu Rev Plant Physiol Plant Mol Biol 48:51–66

Bashan Y, de-Bashan LE (2005) Fresh-weigth measurements of roots provide inaccurate estimates of the effects of plant growth-promoting bacteria on root growth: a critical examination. Soil Biol Biochem 37:1795–1804

Bashan Y, Bustillos JJ, Leyva LA, Hernandez J-P, Bacilio M (2006) Increase in auxiliary photoprotective photosynthetic pigments in wheat seedlings induced by *Azospirillum brasilense*. Biol Fertil Soils 42:279–285

Bashan Y, de-Bashan L (2010) How the plant growth-promoting bacterium *Azospirillum* promotes plant growth. A critical assessment. Adv Agron 108:77–136

Bassaganya-Riera J, Guri AJ, Lu P, Climent M, Carbo A, Sobral BW, Horne WT, Lewis SN, Bevan DR, Hontecillas R (2011) Abscisic acid regulates inflammation via ligand-binding domain-independent activation of peroxisome proliferator-activated receptor gamma. J Biol Chem 286(4):2504–2516

Bastián F, Cohen AC, Piccoli P, Luna V, Baraldi R, Bottini R (1998) Production of indole-3-acetic acid and gibberellins A_1 and A_3 by *Acetobacter diazotrophicus* and *Herbaspirillum seropedicae* in chemically-defined culture media. Plant Growth Regul 24:7–11

Bastián F, Rapparini F, Baraldi R, Piccoli P, Bottini R (1999) Inoculation with *Acetobacter diazotrophicus* increases glucose and fructose content in shoots of *Sorghum bicolor* (L.) Moench. Symbiosis 27:147–156

Belimov AA, Dodd IC, Hontzeas N, Theobald JC, Safronova VI, Davies WJ (2009) Rhizosphere bacteria containing 1-aminocyclopropane-1-carboxylate deaminase increase yield of plants grown in drying soil via both local and systemic hormone signalling. New Phytol 181(2):413–423

Belimov AA, Safronova VI, Sergeyeva TA, Egorova TN, Matveyeva VA, Tsyganov VE, Borisov AY, Tikhonovich IA, Kluge C, Preisfeld A, Dietz KJ, Stepanok VV (2001) Characterization of plant growth promoting rhizobacteria isolated from polluted soils and containing 1-aminocyclopropane-1-carboxylate deaminase. Can J Microbiol 47:242–252

Belimov AA, Dodd IC, Safronova VI, Dumova VA, Shaposhnikov AI, Ladatko AG, Davies WJ (2014) Abscisic acid metabolizing rhizobacteria decrease ABA concentrations in planta and alter plant growth. Plant Physiol Biochem 74:84–91

Berg G, Alavi M, Schmidt CS, Zachow C, Egamberdieva D, Kamilova F, Lugtenberg B (2013) Biocontrol and osmoprotection for plants under salinated conditions. In: de Bruijn FJ (ed) Molecular microbial ecology of the rhizosphere. Wiley, Hoboken, pp 561–573

Berli FJ, Moreno D, Piccoli P, Hespanhol-Viana L, Silva MF, Bressan-Smith R, Cavagnaro JB, Bottini R (2010) Abscisic acid is involved in the response of grape (*Vitis vinifera* L.) cv. Malbec leaf tissues to ultraviolet-B radiation by enhancing ultraviolet-absorbing compounds, antioxidant enzymes and membrane sterols. Plant Cell Environ 33:1–10

Berli FJ, Fanzone M, Piccoli P, Bottini R (2011) Solar UV-B and ABA are involved in phenol metabolism of *Vitis vinifera* L. increasing biosynthesis of berry skin polyphenols. J Agric Food Chem 59:4874–4884

Bloemberg GV, Lugtenberg BJ (2001) Molecular basis of plant growth promotion and biocontrol by rhizobacteria review. Curr Opin Plant Biol 4(4):343–350

Blokhina O, Virolainen E, Fagerstedt KV (2002) Antioxidants, oxidative damage and oxygen deprivation stress: a review. Ann Bot 91:179–194

Boddey RM, Dobereiner J (1995) Nitrogen fixation associated with grasses and cereals: recent progress and perspectives for the future. Fert Res 42:241–250

Bottini R, Fulchieri M, Pearce D, Pharis RP (1989) Identification of gibberellins A_1, A_3 and iso-A_3 in cultures of *Azospirillum lipoferum*. Plant Physiol 90:45–47

Bottini R, Cassán F, Piccoli P (2004) Gibberellin production by bacteria and its involvement in plant growth promotion and yield increase. Appl Microbiol Biotechnol 65:497–503

Boyer JS (1982) Plant productivity and environment. Science 218:443–448

Bray EA, Bailey-Serres J, Weretilnyk E (2000) Responses to abiotic stresses. In: Gruissem W, Buchannan B, Jones R (eds) Biochemistry and molecular biology of plants: American society of plant physiologists, pp 1158–1203

Bray EA (2002) Abscisic acid regulation of gene expression during water deficit stress in the era of the Arabidopsis genome. Plant Cell Environ 25:153–161

Bresson J, Varoquaux F, Bontpart T, Touraine B, Vile D (2013) The PGPR strain *Phyllobacterium brassicacearum* STM196 induces a reproductive delay and physiological changes that result in improved drought tolerance in *Arabidopsis*. New Phytol 200:558–569

Bruzzone S, Moreschi I, Usai C, Guida L, Damonte G, Salis A, Scarfì S, Millo E, De Flora A, Zocchi E (2007) Abscisic acid is an endogenous cytokine in human granulocytes with cyclic ADP-ribose as second messenger. Proc Natl Acad Sci U. S. A. 104(14):5759–5764

Cassán F, Bottini R, Schneider G, Piccoli P (2001a) *Azospirillum brasilense* and *Azospirillum lipoferum* hydrolyze conjugates of GA_{20} and metabolize the resultant aglycones to GA_1 in seedlings of rice dwarf mutants. Plant Physiol 125:2053–2058

Cassán F, Lucangeli C, Bottini R, Piccoli P (2001b) *Azospirillum* spp. metabolize [17,17-$^2H_2$] Gibberellin A_{20} to [17,17-$^2H_2$]Gibberellin A_1 in vivo in dry rice mutant seedlings. Plant Cell Physiol 42:763–767

Chen YP, Rekha PD, Arun AB, Shen FT, Lai WA, Young CC (2006) Phosphate solubilizing bacteria from subtropical soil and their tricalcium phosphate solubilizing abilities. Appl Soil Ecol 34(1):33–41

Christmann A, Weiler EW, Steudle E, Grill E (2007) A hydraulic signal in root-to-shoot signalling of water shortage. Plant J 52:167–174

Cohen AC, Pontin M, Bottini R, Piccoli P (2007) *Azospirillum brasilense* and ABA improve growth in *Arabidopsis thaliana*. In: 19th international plant growth substances association meeting, Puerto Vallarta, México

Cohen AC, Bottini R, Piccoli P (2008) *Azospirillum brasilense* Sp 245 produces ABA in chemically-defined culture medium and increases ABA content in Arabidopsis plants. Plant Growth Regul 54:97–103

Cohen AC, Travaglia C, Bottini R, Piccoli P (2009) Participation of abscisic acid and gibberellins produced by entophytic *Azospirillum* in the alleviation of drought effects in maize. Botany 87:455–462

Cohen AC, Bottini R, Pontin M, Berli FJ, Moreno D, Boccanlandro H, Travaglia CN, Piccoli PN (2015) *Azospirillum brasilense* ameliorates the response of *Arabidopsis thaliana* to drought mainly via enhancement of ABA levels. Physiol Plant 153(1):79–90

Costacurta A, Vanderleyden J (1995) Synthesis of phytohormones by plant-associated bacteria. Crit Rev Microbiol 21(1):1–18

Creus C, Sueldo R, Barassi C (1997) Shoot growth and water status in *Azospirillum*-inoculated wheat seedlings grown under osmotic and salt stresses. Plant Physiol Biochem 35:939–944

Creus CM, Sueldo RJ, Barassi CA (1998) Water relations in Azospirillum-inoculated wheat seedlings under osmotic stress. Can J Bot 76:238–244

Creus CM, Sueldo RJ, Barassi CA (2004) Water relations and yield in *Azospirillum* inoculated wheat exposed to drought in the field. Can J Bot 82:273–281

Creus CM, Graziano M, Casanovas EM, Pereyra MA, Simontacchi M, Puntarulo S, Barassi CA, Lamattina L (2005) Nitric oxide is involved in the *Azospirillum brasilense*-induced lateral root formation in tomato. Planta 221:297–303

Crozier A, Arruda P, Jasmim JM, Monteiro AM, Sandberg G (1988) Analysis of indole-3-acetic acid and related indoles in culture medium from *Azospirillum lipoferum* and *Azospirillum brasilense*. Appl Environ Microbiol 54:2833–2837

Crozier A, Kamiya Y, Bishop G, Yokota T (2001) Biosynthesis of hormones and elicitors molecules. In: Buchanan BB, Grussem W, Jones RL (eds) Biochemistry and molecular biology of plants, pp 850–900

Cutler AJ, Krochko JE (1999) Formation and break-down of ABA. Trends Plant Sci 4:472–478

Dardanelli MS, Fernandez de Cordoba FJ, Espuny MR, Rodriguez Carvajal MA, Soria Diaz ME, Gil Serrano A, Okon Y, Megias M (2008) Effect of *Azospirillum brasilense* coinoculated with *Rhizobiumon on Phaseolus* vulgaris flavonoids and nod factor production under salt stress. Soil Biol Biochem 40:2713–2721

Davies P (1995) Plant hormones. Physiology, biochemistry and molecular biology. Kluwer Acad Press, Dordrecht, p 833

Davies P (2005) The plant hormones; their nature occurrence and function. In: Davies P (ed) Plant hormones. biosynthesis, signal transduction, action. Kluwer Academic Press, Dordrecht, pp 1–15

Deka BHP, Dileep KBS (2002) Plant disease suppression and growth promotion by a *fluorescent Pseudomonas* strain. Folia Microbiol 47:137–143

De Freitas JR, Banerjee MR, Germida JJ (1997) Phosphate solubilizing rhizobacteria enhance the growth and yield but not phosphorus uptake of canola (*Brassica napus* L.). Biol Fertil Soil 24:358–364

De Smet I, Zhang H, Inze D, Beeckman T (2006) A novel role for abscisic acid emerges from underground. Trends Plant Sci 11:434–439

Deak KI, Malamy J (2005) Osmotic regulation of root system architecture. Plant J 43:17–28

Dobbelaere S, Vanderleyden J, Okon Y (2003) Plant growth promoting effects of diazotrophs in the rhizosphere. Crit Rev Plant Sci 22:107–149

Dodd IC, Belimov AA, Sobeih WY, Safronova VI, Grierson D, Davies WJ (2005) Will modifying plant ethylene status improve plant productivity in water-limited environments? In: 4th international crop science congress

Dodd IC, Theobald JC, Bacon MA, Davies WJ (2006) Alternation of wet and dry sides during partial root zone drying irrigation alters root-to-shoot signalling of abscisic acid. Funct Plant Biol 33(12):1081–1089

Dodd IC, Zinovkina NY, Safronova VI, Belimov A (2010) Rhizobacterial mediation of plant hormone status. Ann Appl Biol 157:361–379

Dodd IC, Pérez-Alfocea F (2012) Microbial amelioration of crop salinity stress. J Exp Bot 63:3415–3428

Domínguez JE, García Lampasona S, Cohen AC, Salomon MV (2011) Piccoli. Estudio del gen *CrtZ*, intermediario clave en la vía del Ácido Abscísico en *Pseudomonas fluorescens*, aisladas de raíces de plantas de vid. Tucumán, Argentina: XVII Congr SAMIGE
Egamberdieva D (2008) Plant growth promoting properties of rhizobacteria isolated from wheat and pea grown in loamy sand soil. Turk J Biol 32(1):9–15
Figueiredo MVB, Burity HA, Martinez CR, Chanway CP (2008) Alleviation of drought stress in common bean (*Phaseolus vulgaris* L.) by co-inoculation with *Paenibacillus polymyxa* and *Rhizobium tropici*. Appl Soil Ecol 40:182–188
Forchetti G, Masciarelli O, Alemano S, Alvarez D, Abdala G (2007) Endophytic bacteria in sunflower (*Helianthus annuus* L.) isolation, characterization, and production of jasmonates and abscisic acid in culture medium. Appl Microbiol Biotechnol 76:1145–1152
Fulchieri M, Lucangeli C, Bottini R (1993) Inoculation with *Azospirillum lipoferum* affects growth and gibberellin status of corn seedling roots. Plant Cell Physiol 34:1305–1309
Glick BR (1995) The enhancement of plant growth by free-living bacteria. Can J Microbiol 41:109–117
Glick BR, Penrose DM, Li J (1998) A model for the lowering of plant ethylene concentrations by plant growth promoting bacteria. J Theor Biol 190:63–68
Glick BR (2005) Modulation of plant ethylene levels by the bacterial enzyme ACC deaminase. FEMS Microbiol Lett 251:1–7
Glick BR, Cheng Z, Czarny J, Duan J (2007a) Promotion of plant growth by ACC deaminase-producing soil bacteria (review). Euro J Plant Pathol 119(3):329–339
Glick BR, Todorovic B, Czarny J, Cheng ZY, Duan J, McConkey B (2007b) Promotion of plant growth by bacterial ACC deaminase. Crit Rev Plant Sci 26:227–242
Glick BR (2010) Using soil bacteria to facilitate phytoremediation. Biotech Adv 28:367–374
Golpayegani A, Tilebeni HG (2011) Effect of biological fertilizers on biochemical and physiological parameters of Basil (*Ociumum basilicm* L.) medicine plant. Am-Eur J Agric Environ Sci 11(3):411–416
Grichko VP, Glick BR (2001) Amelioration of flooding stress by ACC deaminase-containing plant growth promoting bacteria. Plant Physiol Biochem 39:11–17
Gutiérrez-Mañero FJ, Ramos-Solano B, Probanza A, Mehouachi J, Tadeo FR, Talon M (2001) The plant-growth-promoting rhizobacteria *Bacillus pumilus* and *Bacillus licheniformis* produce high amounts of physiologically active gibberellins. Physiol Plant 111:206–211
Hartmann A, Singh M, Klingmüller W (1983) Isolation and characterization of *Azospirillum* mutants excreting high amounts of indole acetic acid. Can J Microbiol 29:916–923
Hasegawa S, Poling SM, Maier VP, Bennett RD (1984) Metabolism of abscisic acid: bacterial conversion to dehydrovomifoliol and vomifoliol dehydrogenase activity. Phytochemistry 23:2769–2771
Hasegawa S, Meguro A, Nishimura T, Kunoh H (2004) Drought tolerance of tissue-cultured seedlings of mountain laurel (*Kalmia latifolia* L.) induced by an endophytic actinomycete. I. enhancement of osmotic pressure in leaf cells. Actinomycetologica 18:43–47
Hasegawa S, Meguro A, Toyoda K, Nishimura T, Kunoh H (2005) Drought tolerance of tissue-cultured seedlings of mountain laurel (*Kalmia latifolia* L.) induced by an endophytic actinomycete. II. Acceleration of callose accumulation and lignification. Actinomycetologica 19:13–17
Hedden P, Phillips A (2000) Gibberellin metabolism: new insights revealed genes. Trends Plant Sci 5:523–530
Hedden P, Phillips AL, Rojas MC, Carrera E, Tudzynski B (2001) Gibberellin biosynthesis in plants and fungi: a case of convergent evolution? J Plant Growth Regul 20:319–331
Heidari M, Mousavinik SM, Golpayegani A (2011) Plant growth promoting rhizobacteria (PGPR) effect on physiological parameters and mineral uptake in basil (*Ociumum basilicm* L.) under water stress. ARPN J Agric Biol Sci 6(5):6–11
Hu J, Lin X, Wang J, Chu H, Yin R, Zhang J (2009) Population size and specific potential of P-mineralizing and P-solubilizing bacteria under long-term P-deficiency fertilization in a sandy loam soil. Pedobiologia 53:49–58

Huang D, Wu W, Abrams SR, Cutler AJ (2008) The relationship of drought-related gene expression in *Arabidopsis thaliana* to hormonal and environmental factors. J Exp Bot 59:2991–3007
Jaleel CA, Manivannan P, Sankar B, Kishorekumar A, Gopi R, Somasundaram R, Panneerselvam R (2007) *Pseudomonas fluorescens* enhances biomass yield and ajmalicine production in *Catharanthus roseus* under water deficit stress. Colloids Surf B 60:7–11
Janzen R, Rood S, Dormar J, McGill W (1992) *Azospirillum brasilense* produces gibberellins in pure culture and chemically-medium and in co-culture on straw. Soil Biol Biochem 24:1061–1064
Kang S-M, Khana AL, Waqas M, You Y-H, Kim J-H, Kim J-G, Hamayun M, Lee I-J (2014) Plant growth-promoting rhizobacteria reduce adverse effects of salinity and osmotic stress by regulating phytohormones and antioxidants in *Cucumis sativus*. J Plant Interact 9(1):673–682
Kang S-M, Khan AL, Hamayun M, Shinwari ZK, Kim Y-H, Joo G-J, Lee I-J (2012) *Acinetobacter calcoaceticus* ameliorated plant growth and influenced gibberellins and functional biochemical. Pak J Bot 44(1):365–372
Kaymak HC, Guvenc I, Yarali F, Donmez MF (2009) The effects of bio-priming with PGPR on germination of radish (*Raphanus sativus* L.) seeds under saline conditions. Turk J Agric For 33(2):173–179
Kloepper JW, Schroth MN (1978) Plant growth-promoting rhizobacteria on radishes. Proc. 4:879–882
Kloepper J, Lifshitz R, Schroth M (1989) *Pseudomonas* inoculants to benefit plant production. ISI Atlas Sci Anim Plant Sci 8:6064
Kohler J, Hernández JA, Caravaca F, Roldán A (2008) Plant-growth promoting rhizobacteria and arbuscular mycorrhizal fungi modify alleviation biochemical mechanisms in water-stressed plants. Funct Plant Biol 35:141–151
Kolb W, Martin P (1985) Response of plant roots to inoculation with *Azospirillum brasilense* and to application of indole acetic acid. In: Klingmüller W (ed) Azospirillum III: genetics, physiology, ecology. Springer, Berlin, pp 215–221
Lamattina L, Polacco J (2007) Nitric oxide in plant growth development and stress physiology. Springer, Berlin, p 283
Leveau JHJ, Gerards S (2008) Discovery of a bacterial gene cluster for catabolism of the plant hormone indole-3-acetic acid. FEMS Microbiol Ecol 65:238–250
Lohar D, Stiller J, Kam J, Stacey G, Gresshoff PM (2009) Ethylene insensitivity conferred by a mutated *Arabidopsis* ethylene receptor gene alters nodulation in transgenic *Lotus japonicas*. Ann Bot-london 1–9
Lucangeli C, Bottini R (1997) Effects of *Azospirillum* spp. on endogenous gibberellin content and growth of maize (*Zea mays* L.) treated with uniconazole. Symbiosis 23:63–72
Lugtenberg B, Kamilova F (2009) Plant-growth-promoting rhizobacteria. Annu Rev Microbiol 63:54–56
Maheshwari DK (2010) Plant growth and health promoting rhizobacteria. In: Microbiology monograph, Springer, Heidelberg Dordrecht London New York, vol 18, p 453
Maheshwari DK (2012) Bacteria in agrobiology: stress management. Springer, Germany 341
Marasco EK, Schmidt-Dannert C (2008) Exploring and accessing plant natural product biosynthesis in engineered microbial hosts. Medicinal plant biotechnology: from basic research to industrial applications, pp 287–317
Marulanda A, Barea JM, Azcón R (2006) An indigenous drought-tolerant strain of *Glomus intraradices* associated with a native bacterium improves water transport and root development in *Retama sphaerocarpa*. Microb Ecol 52:670–678
Marulanda A, Barea J-M, Azcón R (2009) Stimulation of plant growth and drought tolerance by native microorganisms (AM fungi and bacteria) from dry environments: Mechanisms related to bacterial effectiveness. J Plant Growth Regul 28(2):115–124
Mayak S, Tirosh T, Glick BR (1999) Effect of wild-type and mutant plant growth promoting rhizobacteria on the rooting of mung bean cuttings. J Plant Growth Regul 18:49–53

Mayak S, Tirosh T, Glick BR (2004a) Plant growth promoting bacteria confers resistance in tomato plants to salt stress. Plant Physiol Biochem 42:565–572

Mayak S, Tirosh T, Glick BR (2004b) Plant growth-promoting bacteria that confer resistance to water stress in tomatoes and pepper. Plant Sci 166:525–530

Meloni DA, Oliva MA, Ruiz HA, Martinez CA (2001) Contribution of proline and inorganic solutes to osmotic adjustment in cotton under salt stress. J Plant Nutr 24:599–612

Mishra G, Zhang W, Deng F, Zhao J, Wang X (2006) A bifurcating pathway directs abscisic acid effects on stomatal closure and opening in Arabidopsis. Science 312:264–266

Moreno D, Berli FJ, Piccoli P, Bottini R (2011) Gibberellins and abscisic acid promote carbon allocation in roots and berries of grape plants. J Plant Growth Regul 30:220–228

Nambara E, Marion-Poll A (2005) Abscisic acid biosynthesis and catabolism. Annu Rev Plant Biol 56:165–185

Naseema H, Bano A (2014) Role of plant growth-promoting rhizobacteria and their exopolysaccharide in drought tolerance of maize. J Plant Inter 9(1):689–701

Okon Y (1985) *Azospirillum* as a potential inoculant for agriculture. Trends Biotechnol 3:223–228

Pandey S, Bhandari H, Ding PP, Sharan R, Naik D, Taunk SK, Sastri A (2007) Coping with drought in rice farming in Asia: insights from a cross-country comparative study. Agric Econ 37:213–224

Patten CL, Glick BR (1996) Bacterial biosynthesis of indole-3-acetic acid. Can J Microbiol 42:207–220

Patten CL, Glick BR (2002) The role of bacterial indole acetic acid in the development of the host plant root system. Appl Environ Microbiol 68:3795–3801

Peña-Cortés H, Sánchez-Serrano JJ, Mertenst R, Willmitzer L, Prat S (1989) Abscisic acid is involved in the wound-induced expression of the proteinase inhibitor II gene in potato and tomato. Proc Natl Acad Sci U.S.A. 86:9851–9855

Perrig D, Boiero L, Masciarelli O, Penna C, Cassa´n F, Luna V (2007) Plant growth promoting compounds produced by two agronomically important strains of *Azospirillum brasilense*, and their implications for inoculant formulation. Appl Microbiol Biotechnol 75:1143–1150

Piccoli P, Bottini R (1994a) Metabolism of 17,17-($^2H_2$)gibberellin A_{20} to 17,17-($^2H_2$)gibberellin A1 by *Azospirillum lipoferum* cultures. Agri-Scientia 11:13–15

Piccoli P, Bottini R (1994b) Effects of C/N relationships, N content, pH, and time of culture on growth and gibberellin production of *Azospirillum lipoferum* cultures. Symbiosis 17:229–236

Piccoli P, Masciarelli O, Bottini R (1996) Metabolism of 17,17($^2H_2$)-Gibberellins A_4, A_9, and A_{20} by *Azospirillum lipoferum* in chemically-defined culture medium. Symbiosis 21:263–274

Piccoli P, Lucangeli D, Schneider G, Bottini R (1997) Hydrolysis of (17,17-$^2H_2$)gibberellin A_{20}-glucoside and (17,17-$^2H_2$)gibberellin A_{20}-glucosyl ester by *Azospirillum lipoferum* cultured in a nitrogen free biotin-based chemically-defined medium. Plant Growth Regul 23:179–182

Piccoli P, Travaglia C, Cohen A, Sosa CP, Masuelli R, Bottini R (2011) An endophytic bacterium isolated from roots of the halophyte *Prosopis strombulifera* produces ABA, IAA, gibberellins A_1 and A_3 and jasmonic acid in chemically-defined culture medium. Plant Growth Regul 64:207–210

Pirlak M, Kose M (2009) Effects of plant growth promoting rhizobacteria on yield and some fruit properties of strawberry. J Plant Nutr 32:1173–1184

Poupin MJ, Timmermann T, Vega A, Zuñiga A, González B (2013) Effects of the plant growth-promoting bacterium *Burkholderia phytofirmans* PsJN throughout the life cycle of *Arabidopsis thaliana*. PLoS ONE 8:e69435

Primrose S, Dilworth M (1976) Ethylene production by bacteria. J Gen Microbiol 93:177–181

Probanza A, García JAL, Palomino MR, Ramos B, Manero FJG (2002) *Pinus pinea* L. seedling growth and bacterial rhizosphere structure after inoculation with PGPR *Bacillus (B. licheniformis* CECT 5106 and *B. pumilus* CECT 5105). Appl Soil Ecol 20:75–84

Quiroga AM, Berli F, Moreno D, Cavagnaro JB, Bottini R (2009) Abscisic acid sprays significantly increase yield per plant in vine yard grown wine grape (*Vitis vinifera* L.) cv. *Cabernet sauvignon* through increased berry set with no negative effects on anthocyanin content and total polyphenol index of both juice and wine. J Plant Growth Regul 28:28–35

Ren H, Gao Z, Chen L, Wei K, Liu J, Fan Y, Davies WJ, Jia W, Zhang J (2007) Dynamic analysis of ABA accumulation in relation to the rate of ABA catabolism in maize tissues under water deficit. J Exp Bot 58:211–219

Richardson AE (2001) Prospects for using soil microorganisms to improve the acquisition of phosphorus by plants. Aust J Plant Physiol 28:897–906

Rodríguez-Salazar J, Suárez R, Caballero-Mellado J, Iturriaga G (2009) Trehalose accumulation in *Azospirillum brasilense* improves drought tolerance and biomass in maize plants. FEMS Microbiol Lett 296:52–59

Rodriguez H, Fraga R (1999) Phosphate solubilizing bacteria and their role in plant growth promotion. Biotechnol Adv 17:319–339

Rodriguez H, Fraga R, Gonzalez T, Bashan Y (2006) Genetics of phosphate solubilization and its potential applications for improving plant growth-promoting bacteria. Plant Soil 287(Suppl 1–2):15–21

Ross J, O'Neill D (2001) New interactions between classical plant hormones. Trends Plant Sci 6:2–4

Ryu CM, Hu CH, Locy RD, Kloepper JW (2005) Study of mechanisms for plant growth promotion elicited by rhizobacteria in *Arabidopsis thaliana*. Plant Soil 268(1):285–292

Salisbury FB, Ross CW (1992) Plant physiology. Wadsworth Publishing Company, Belmont, pp 329–407

Salomon MV, Bottini R, de Souza Filho GA, Cohen AC, Moreno D, Gil M, Piccoli P (2014) Bacteria isolated from roots and rhizosphere of *Vitis vinifera* retard water losses, induce abscisic acid accumulation and synthesis of defense-related terpenes in in vitro cultured grapevine. Physiol Plant 151(4):359–374

Sandhya V, Ali SKZ, Grover M, Reddy G, Venkateswarlu B (2010) Effect of plant growth promoting *Pseudomonas* spp. on compatible solutes, antioxidant status and plant growth of maize under drought stress. Plant Growth Regul 62(1):21–30

Sansberro P, Mroginski L, Bottini R (2004) Abscisic Acid promotes growth of *Ilex paraguariensis* plants by alleviating diurnal water stress. Plant Growth Regul 42:105–111

Sarig S, Blum A, Okon Y (1988) Improvement of water status and yield of field-grown grain sorghum (*Sorghum bicolor*) by inoculation with *Azospirillum brasilense*. J Agr Sci 110:271–277

Sarma RK, Saikia R (2014) Alleviation of drought stress in mung bean by strain *Pseudomonas aeruginosa* GGRJ21. Plant Soil 377:111–126

Seki M, Narusaka M, Ishida J, Nanjo T, Fujita M, Oono Y, Kamiya A, Nakajima M, Enju A, Sakurai T, Satou M, Akiyama K, Taji T, Yamaguchi-Shinozaki K, Carninci P, Kawai J, Hayashizaki Y, Shinozaki K (2002) Monitoring the expression pattern of around 7000 Arabidopsis genes under ABA treatments using a full-length cDNA microarray. Funct Integr Genomics 2:282–291

Seki MT, Umezawa KU, Shinozaki K (2007) Regulatory metabolic networks in drought stress responses. Curr Opin Plant Biol 10:296–302

Sgroy V, Cassán F, Masciarelli O, Del Papa M, Lagares A, Luna V (2009) Isolation and characterization of endophytic plant growth-promoting (PGPB) or stress homeostasis regulating (PSHB) bacteria associated to the halophyte *Prosopis strombulifera*. Appl Microbiol Biotechnol 85:371–381

Shaharoona B, Arshad M, Zahir ZA, Khalid A (2006a) Performance of *Pseudomonas* spp. containing ACC-deaminase for improving growth and yield of maize (*Zea mays* L.) in the presence of nitrogenous fertilizer. Soil Biol Biochem 38:2971–2975

Shaharoona B, Arshad M, Zahir ZA (2006b) Effect of plant growth promoting rhizobacteria containing ACC-deaminase on maize (*Zea mays* L.) growth under axenic conditions and on nodulation in mung bean (*Vigna radiata* L.). Lett Appl Microb 42:155–159

Sharma S, Kumar V, Tripathi RB (2011) Isolation of Phosphate Solubilizing Microorganism (PSMs) From Soil. J Microbiol Biotechnol Res 1:90–95

Shinozaki K, Yamaguchi-Shinozaki K (2000) Molecular responses to dehydration and low temperature: differences and cross-talk between two stress signaling pathways. Curr Opin Plant Biol 3:217–223

Spaepen S, Vanderleyden J, Remans R (2007) Indole-3-acetic acid in microbial and microorganism-plant signalling. FEMS Microbiol Rev 31(4):425–448

Spaepen S, Vanderleyden J, Okon Y (2009) Chapter 7 plant growth-promoting actions of rhizobacteria (review). Adv Bot Res 51:283–320

Sperdouli I, Moustakas M (2012) Interaction of proline, sugars, and anthocyanins during photosynthetic acclimation of *Arabidopsis thaliana* to drought stress. J Plant Physiol 169:577–585

St Clair SB, Lynch J (2010) The opening of pandora's box: climate change impacts on soil fertility and crop nutrition in developing countries. Plant Soil 335:101–115

Tardieu F, Parent B, Simonneau T (2010) Control of leaf growth by abscisic acid: hydraulic or non-hydraulic processes? Plant, Cell Environ 33:636–647

Thompson AJ, Andrews J, Mulholland BJ, McKee JM, Hilton HW, Horridge JS, Farquhar GD, Smeeton RC, Smillie IR, Black CR, Taylor IB (2007) Overproduction of abscisic acid in tomato increases transpiration efficiency and root hydraulic conductivity and influences leaf expansion. Plant Physiol 143:1905–1917

Tien TM, Gaskin MH, Hubbel DH (1979) Plant growth substances produced by *A. brasilense* and their effect on the growth of pearl millet (*Pennisetum americanum* L.). Appl Environ Microbiol 37:1016–1024

Timmusk S, Wagner GH (1999) The plant-growth-promoting rhizobacterium *Paenibacillus polymyxa* induces changes in *Arabidopsis thaliana* gene expression: a possible connection between biotic and abiotic stress responses. Mol Plant Microbe Interact 12:951–959

Tomlinson I (2013) Doubling food production to feed the 9 billion: a critical perspective on a key discourse of food security in the UK. J Rural Stud 21:81–90

Travaglia C, Cohen AC, Reinoso H, Castillo C, Bottini R (2007) Exogenous abscisic acid increases carbohydrate accumulation and redistribution to the grains in wheat grown under field conditions of soil water restriction. J Plant Growth Regul 26:285–289

Travaglia C, Reinoso H, Bottini R (2009) Application of abscisic acid promotes yield in field-cultured soybean by enhancing production of carbohydrates and their allocation in seed. Crop Pasture Sci 60:1131–1136

Travaglia C, Reinoso H, Cohen AC, Luna C, Castillo C, Bottini R (2010) Exogenous ABA increases yield in field-grown wheat with a moderate water restriction. J Plant Growth Regul 29:366–374

Tsavkelova EA, Klimova SY, Cherdyntseva TA, Netrusov AI (2006) Microbial producers of plant growth stimulators and their practical use: a review. Appl Biochem Microb 42:117–126

Vivas A, Marulanda A, Ruiz-Lozano JM, Barea JM, Azcon R (2003) Influence of a *Bacillus* sp. on physiological activities of two arbuscular mycorrhizal fungi and on plant responses to PEG-induced drought stress. Mycorrhiza 13:249–256

Wu Z, Yue H, Lu J, Li C (2012) Characterization of rhizobacterial strain Rs-2 with ACC deaminase activity and its performance in promoting cotton growth under salinity stress. World J Microbiol Biotechnol 28:2383–2393

Xie X, Zhang H, Paré PW (2009) Sustained growth promotion in *Arabidopsis* with long-term exposure to the beneficial soil bacterium *Bacillus subtilis* (GB03). Plant Signal Behav 4:948–953

Yamaguchi-Shinozaki K, Shinozaki K (2005) Organization of cis-acting regulatory elements in osmotic- and cold-stress-responsive promoters. Trends Plant Sci 10:88–94

Yamaguchi S (2008) Gibberellin metabolism and its regulation. Annu Rev Plant Biol 59:225–251

Yang J, Kloepper JW, Ryu CM (2009) Rhizosphere bacteria help plants tolerate abiotic stress. Trends Plant Sci 14:1–4

Zahir ZA, Munir A, Asghar HN, Shaharoona B, Arshad M (2008) Effectiveness of rhizobacteria containing ACC deaminase for growth promotion of peas (*Pisum sativum*) under drought conditions. J Microbiol Biotechnol 18(5):958–963

Zhao Y (2010) Auxin biosynthesis and its role in plant development. Annu Rev Plant Biol 61:49–64

Zawoznik MS, Ameneiros M, Benavides MP, Vázquez S, Groppa MD (2011) Response to saline stress and aquaporin expression in *Azospirillum*-inoculated barley seedlings. Appl Microbiol Biotechnol 90:1389–1397

Zeevaart JAD, Creelman RA (1988) Metabolism and physiology of abscisic acid. Annu Rev Plant Physiol Plant Mol Biol 39:439–473

Zeevaart J (1999) Abscisic acid metabolism and its regulation. In: Hooykaas P, Hall M, Libbenga K (eds) Biochemistry and molecular biology of plant hormones. Elsevier, Amsterdam, pp 189–207

Zeller G, Henz SR, Widmer CK, Sachsenberg T, Rätsch G, Weigel D, Laubinger S (2009) Stress-induced changes in the *Arabidopsis thaliana* transcriptome analyzed using whole-genome tiling arrays. Plant J 58:1068–1082

Zhang H, Xie X, Kim MS, Kornyeyev DA, Holaday S, Paré PW (2008) Soil bacteria augment Arabidopsis photosynthesis by decreasing glucose sensing and abscisic acid levels in planta. Plant J 56:264–273

Zhang SQ, Outlaw WH Jr (2001) Abscisic acid introduced into the transpiration stream accumulates in the guard-cell apoplast and causes stomatal closure. Plant, Cell Environ 24:1045–1054

Ziska LH (2011) Climate change, carbon dioxide and global crop production: food security and uncertainty. In: Dinar A, Mendelsohn R (eds) Handbook on climate change and agriculture. Edward Elgar Publishing Ltd., ltenham and Camberley, UK, pp 9–31

Microbial Phytohormones Have a Key Role in Mitigating the Salt-Induced Damages in Plants

Dilfuza Egamberdieva, Abeer Hashem and Abdulaziz A. Alqarawi

Abstract Salinity is among the most challenging and devastating environmental problems which cause drastic decline in normal growth and developmental processes in crop plants. Plants have evolved several tolerance strategies to avert the damaging effects of high salinity. During the past few years most of the research is focused on increasing the salt tolerance of major food crops through the application of phytohormone producing beneficial microorganisms. During stress microbial phytohormones are having critical roles in modulating the physiology and biochemistry of plants so as to elicit a tolerance response to avoid stress. Induced plant growth and development of various plants by inoculation with PGPR having phytohormone, such as indole-3-acetic acid (IAA), cytokinins (CK), gibberelic acid (GA), salicylic acid (SA) and abscisic acid (ABA) producing ability, has been repeatedly documented. Present review discusses the role of phytohormones in ameliorating the salt stress-induced changes in plants and provides valuable insight into microbes evolved interactions with plant under hostile environmental conditions.

Keywords Salt stress · Phytohormones · Plant growth promoting rhizobacteria

D. Egamberdieva (✉)
Institute for Landscape Biogeochemistry, Leibniz Centre for Agricultural Landscape Research (ZALF), 15374 Müncheberg, Germany
e-mail: egamberdieva@yahoo.com

A. Hashem
Department of Botany and Microbiology, College of Science, King Saud University, 2460, Riyadh 11451, Saudi Arabia

A.A. Alqarawi
Department of Plant Production, College of Food and Agricultural Sciences, King Saud University, 2460, Riyadh 11451, Saudi Arabia

D.K. Maheshwari (ed.), *Bacterial Metabolites in Sustainable Agroecosystem*, Sustainable Development and Biodiversity 12, DOI 10.1007/978-3-319-24654-3_10

1 Introduction

Plants are often exposed to several environmental stresses, such as water, salinity, high temperature, freezing, radiation, heavy metals, pathogens, etc. These stresses alter almost every metabolic and physiological pathway leading to altered growth and development ultimately leading to the reduced crop yield (Alqarawi et al. 2014). Among these stresses, salinity is the most important and common environmental constraint restricting plant growth. Soil salinization is rapidly increasing reduced land area available for crop cultivation resulting in declining yields of major important crops (Ashraf et al. 2002). Salt stress-induced reduction in growth and development of plants is due to the collective effect of toxic salts on effect ion homeostasis, water imbalance and several other important metabolic pathways like photosynthesis, respiration as well as enzyme activity (Khan et al. 2014). Exposure to higher levels of salt causes oxidative stress through enhanced production of toxic reactive oxygen species (ROS) which can disrupt functions of chloroplast (Apel and Hirt 2004). Toxic effects of salt stress on plant growth and metabolism are also related to the changes in endogenous levels of phytohormones (Khan et al. 2014). Plant hormones, such as auxins, abscisic acid (ABA), cytokinins (CK), ethylene, gibberellic acid (GA), salicylic acid (SA), play an important role in plant physiology and are involved in the alleviation of salt stress (Teale et al. 2006). The inhibition of endogenous levels of phytohormones, such as auxins, gibberellins, ABA, jasmonic acid and SA by salinity stress decrease root growth and development (Debez et al. 2001; Egamberdieva 2009). The decrease of hormone levels in root system of plants resulted in disturbance of nutrient uptake by plant roots from soil (Egamberdieva 2012; Egamberdieva et al. 2013).

Different salinity tolerance mechanisms are employed by plants for ameliorating salinity-induced deleterious effects (Ahmad 2010; Jamil et al. 2007). Under salt stress enhanced activities of antioxidant enzymes mediate scavenging of toxic free radicals reducing effects of oxidative stress. In addition efficient and selective uptake as well as compartmentation of ions contributes to enhance salt tolerance (Ahanger et al. 2014; Khan et al. 2014). Moreover, under salt stress synthesis and accumulation of compatible organic osmolytes is enhanced. Compatible osmolytes include proline, glycine betaine, free sugars, etc. which help to maintain water content under stressful conditions (Ahmad et al. 2014). The exogenous application of phytohormones, such as GB (Afzal et al. 2005), auxins such as indole-3-acetic acid (IAA) (Egamberdieva 2009), and CK (Maggio et al. 2010) mitigate salt stress and stimulate root and shoot growth under stress condition. In salt stressed citrus plants, exogenously applied ABA reduces release of ethylene, and hence, reducing leaf abscission and possible reason for this is believed to be the restricted accumulation of toxic Cl^- in leaves (Gómez-Cadenas et al. 2002). Buran et al. (2012) have reported that exogenous application of ABA resulted in enhanced antioxidant potential in high bush blueberries (*Vaccinium darrowii*). Use of plant growth regulators have been suggested as an effective strategy for enhancing growth of crop plants under stressful conditions (Egamberdieva 2008, 2011; Ahanger et al. 2014).

Root associated bacteria isolated from various plant species has been found to produce phytohormones, including IAA, GA, CK and ABA (Spaepen et al. 2007; Egamberdieva et al. 2014). Since they have potential root colonizing abilities, continuous slow release of phytohormones produced by microbes taken up by roots positively affect on plant growth and development (Ali et al. 2009). Phytohormones were detected in the culture medium of diversified PGPR, such as *Pseudomonas putida, P. extremorientalis, P. chlororaphis* (Egamberdieva and Kucharova 2009), *Bacillus subtilis, B, megaterium* (Egamberdieva 2009), CK were detected in *Halomonas desiderata*, *B. megaterium*, *B. cereus*, *B. subtilis* and *Escherichia coli* (Karadeniz et al. 2006), and gibberellins in *Acetobacter* sp., *Bacillus* sp., *Azospirillium* sp. (Bottini et al. 2004). It has been observed that the content of phytohormones in plants may affect by microbes colonized in root system (Turan et al. 2014). For example, soybean inoculated with PGPR strains exhibited higher amounts of GA and IAA and stimulated root shoot growth under water stress condition (Bano et al. 2010). Several other bacterial strains, such as *P. alcaligenes P. aurantiaca, P. aureofaciens* and *P. chlororaphis* isolated from saline arid soil produce phytohormones and promote growth of cotton, wheat, maize and pea under hostile condition (Egamberdiyeva 2005; Egamberdiyeva and Hoflich 2003; Egamberdieva et al. 2008, 2011). The inoculation of plants with CK producing PGPR strains stimulated plant growth and mitigated adverse effect of drought stress on plant (Arkipova et al. 2007). The ability of PGPR strains to produce IAA at higher saline conditions could balance the decrease in the IAA levels of the roots and thus alleviate salt stress in plants.

2 Abscisic Acid

Several reports are available showing positive role of ABA in mediating normal plant growth. Besides having several key metabolic functions like seed germination, maturity and dormancy, it has been proposed to mediate plant responses to a range of environmental stresses like salinity and water stress (Baumann 2010). ABA serves as major internal signal providing plants with the potential to escape from the adverse environmental conditions (Keskin et al. 2010). Salinity causes increase in endogenous levels of ABA which is often correlated with the change in water potential of leaf and hence suggesting that increase in ABA due to salinity stress is as the result of reduced water potentials as well as toxicity of specific ions (Zhang et al. 2006). Increased endogenous levels of ABA often lead to the growth inhibition as has been reported in several plants like *Zea mays* (Kramer et al. 2002) and *Phaseolus vulgaris* (Cabot et al. 2009). Kang et al. (2005) demonstrated that salinity stressed rice plants maintained higher levels of ABA. However, role of ABA in mediating signal from the root zone during stress is still not fully understood. Nevertheless significant evidences are available pointing ABA as important root-to-shoot stress signal. Fricke et al. (2004) demonstrated contribution of ABA for increasing water potential of xylem sap through its role in mediating water uptake under salinity stress. In root tissues ABA has been reported to stimulate accumulation of compatible ions and mediate compartmentation of toxic ones into vacuoles, an essential tolerance strategy to avoid salinity-induced toxic effects (Jeschke et al. 1997). The germination, plant growth and nitrogen fixation were improved under salt stress condition after treatment of seeds with ABA (Khadri et al. 2006). In *Agrostis stolonifera* and kentucky bluegrass (*Poa pratensis*), exogenous application of ABA mitigated the effect of salinity and drought stress by enhancing the activities of antioxidant enzymes and reducing electrolyte leakage and lipid peroxidation (Yang and Yu 2012). ABA is also synthesized in roots of plants by root colonizing bacteria, which often demonstrate plant growth stimulation. For example, bacterial strains *Proteus mirabilis*, *P. vulgaris, Klebsiella pneumoniae, Bacillus megaterium* and *B. cereus* found to synthesize ABA (Karadeniz et al. 2006). The plant growth promoting rhizobacteria (PGPR) strains *Bacillus licheniformis* Rt4M10 and *P. fluorescens* Rt6M10 isolated from the rhizosphere of *Vitis vinifera* produced ABA and was able to stimulate plant growth under water stress and induce ABA synthesis in plants. The level of ABA levels in 45 days-old in vitro plants were increased 76-fold by *B. licheniformis* and 40-fold by *P. fluorescens* as compared to control plants and strains reduced plant water loss rate in correlation with increments of ABA (Salomon et al. 2014). In other study, halophyte (*Prosopis strombulifera*) associated bacteria *P. putida* produced 4.27 μg ml^{-1}, *A. xylosoxidans* 0.127 μg ml^{-1}, *L. fusiformis* 0.15 μg ml^{-1}, *B. licheniformis* 0.32 μg ml^{-1}, *B. pumilus* 0.06 μg ml^{-1}, *B. subtilis* 1.79 μg ml^{-1} and *B. halotolerans* 0.18 μg ml^{-1}. PGPR strains with improved capacity of ABA synthesis could help plants tolerate abiotic stress.

3 Salicylic Acid

In plants, SA shows active participation in mediating and regulating several essential physiological and biochemical processes like growth, photosynthesis, nitrogen metabolism, production of ethylene and flowering (Hayat et al. 2010). In addition of its diverse physiological roles SA provides protection against various environmental stresses including water stress (Senaratna et al. 2000), freezing (Tasgin et al. 2003), salinity (Azooz et al. 2011), heavy metals (Ahmad et al. 2011), etc. On exposure to stress exogenously applied SA acts as a signal which is involved in activation of specific response mechanisms in plants. Exogenous application of SA have been reported to protect plants from deleterious impact of stress factors by promoting several processes that contribute to enhanced stress tolerance (Azooz et al. 2011; Khan et al. 2014). The role of SA in defense mechanism to alleviate salt stress in plants has been extensively studied (Afzal et al. 2006). SA-induced amelioration of salinity stress has been observed in several crops like faba bean (Azooz et al. 2011), maize (Gunes et al. 2007), *Vigna radiata* L. (Khan et al. 2014) and wheat (Shakirova et al. 2003). Direct addition of SA can be also an effective strategy to avoid salinity stress-induced damage to crop plants. In salinity stressed maize, Gunes et al. (2007)

reported that addition of SA to soil mitigates the salinity-induced negative impact by reducing the uptake of toxic ions like Na and hence reducing their accumulation within the sensitive plant parts. SA alleviates salinity stress-induced oxidative damage due to accumulation of sufficient levels of hydrogen peroxide (Wahid et al. 2007). In salt stressed *Vigna radiata* reduction in endogenous levels of ethylene observed due to SA application (Khan et al. 2014). The microbes which able to colonize plant root able to produce SA proved significant as important component in the induction of plant mediated defense enzymes. For example, the production of SA found in bacterial strains *B. licheniformis* MML2501 (18 μg/ml) (Shanmugam and Narayanasamy 2008), *Serratia* sp. PSGB13 (10.0 μg/ml), *Acinetobacter* sp. PRGB16 (7.2 μg/ml) and *Pseudomonas* sp. PRGB06 (6.8 μg/ml) (Indiragandhi et al. 2008). Those PGPR strains elicit-induced systemic tolerance to salt and drought stress. Forchetti et al. (2010) reported production of SA in endophytic bacterial strains isolated from sunflower grown under drought condition *Achromobacter xylosoxidans* and *Bacillus pumilus*. The strains enhanced the root and shoot growth of sunflower seedlings under water stress condition. Recently, Lavania and Nautiyal (2013) observed production of SA in salt-tolerant *Serratia marcescens* NBRI1213 strains. The strain colonized the rhizosphere of maize, stimulated root, shoot growth and N, P, K uptake by plants and increased salt stress tolerance of plants.

4 Indole Acetic Acid

Indole acetic acid (IAA) is a naturally occurring auxin which is having major role in plant growth regulation. It is involved in controlling the vascular tissue development, cell elongation, apical dominance, etc. (Wang et al. 2001). Scanty information is available regarding the relationship and impact of auxin with the salt tolerance and amelioration of salt stress. However, stress-induced alterations in levels of IAA is somehow similar to that of ABA. Therefore, applying growth hormone exogenously gives an attractive vision and potent approach for counteracting stress-induced changes. Most of the research work has quoted that IAA levels decrease under salinity, for example, a reduction in IAA level has been reported by Prakash and Prathapasenan (1990) in NaCl stressed rice leaves. They further reported that applying GA3 under such conditions mitigates the effect of salinity on IAA levels showing that salinity affects hormonal balance by influencing plant growth and development. Significant reduction in levels of IAA has been reported in salt stressed tomato (Dunlap and Binze 1996). Nevertheless several researchers have also reported IAA mitigates salinity-induced damage in plants, e.g. in wheat IAA has been reported to alleviate the inhibitory effects of salt stress (Afzal et al. 2005). Gulnaz et al. (1999) reported that salinity reduced the growth of wheat considerably and exogenous application of IAA mitigated the adverse effects to considerable extent. Akbari et al. (2007) also demonstrated that exogenous application of auxin enhanced length of hypocotyls, fresh biomass as well as dry biomass of wheat cultivars under saline conditions.

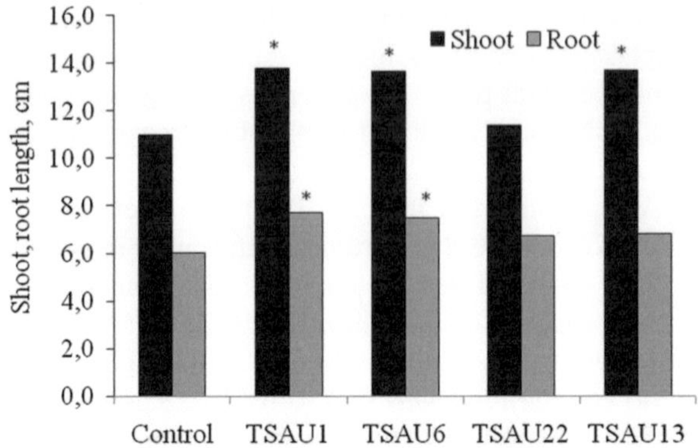

Fig. 1 The effect of IAA producing root associated bacteria (*P. putida* TSAU1, *P. extremorientalis* TSAU6, *P. chlororaphis* TSAU13 and *P. aureantiaca* TSAU22) on shoot and root growth of cucumber under saline soil condition

The root associated PGPR strains able to stimulate plant growth and stress tolerance of plants through supplying additional phytohormones into root system (Khan et al. 2011; Berg et al. 2013; Egamberdieva et al. 2013). In our previous studies, we observed that goat's rue, wheat, cotton and chickpea growth improved by IAA producing PGPR strains *P. extremorientalis* TSAU6, *P. putida* TSAU1, *P. aureantiaca* TSAU22 and *P. chlororaphis* TSAU13 due to increase nutrient absorbing surfaces (Egamberdieva and Kucharova 2009; Egamberdieva et al. 2011, 2013). In other studies, we observed that IAA producing, root associated bacteria were able to relieve salt stress and promote the growth of cucumber seedlings in salinated soil (Fig. 1). Moreover, shoot and root length of cucumber increased by 25 and 27 %, respectively.

The salt-tolerant strains *Serratia plymuthica* RR-2-5-10, *Stenotrophomonas rhizophila* e-p10, *P. fluorescens* SPB2145, *P. extremorientalis* TSAU20 and *P. fluorescens* PCL1751 also able to produce IAA under higher NaCl condition significantly increased dry weight of whole cucumber plants up to 62 % in comparison to the non-bacterized control. The strains also increased fruit yield in greenhouse varying from 9 to 32 % under saline soil condition (Egamberdieva et al. 2011). Similar observations reported by Figueiredo et al. (2008) where co-inoculation of bean (*Phaseolus vulgaris* L.) with *Rhizobium tropici* and two strains of *P. polymyxa* resulted in higher uptake of nutrient from soil and nodule formation. IAA producing PGPR strain alleviated abiotic stress in sunflower (Fassler et al. 2010) and groundnut (Kishore et al. 2005). Bianco and Defez (2009) explained this process due to enhanced cellular defence systems by IAA which protect plants from external adverse conditions. In our previous work we observed that IAA producing strains were able to improve symbiotic performance of legumes under salt stress condition. The plant growth and nodule number of soybean inoculated with *Bradyrhizobium japonicum* USDA110 reduced at 75 mM NaCl. In that condition the co-inoculation of salt stressed soybean with *B. japonicum* USDA110 and IAA producing *P. putida* TSAU1 improved plant growth, and nodulation compared that to plants inoculated

Table 1 The effect of IAA producing *Pseudomonas putida* TSAU1 combined with *Bradyrhizobium japonicum* USDA110 on root, shoot length and dry weight of soybean under salt stress condition

Treatments	Root length (cm)	Shoot length (cm)	Dry weight (g/plant)	Nodule number
0 mM NaCl				
USDA110	11.7	20.6	0.086	6.3
USDA110 + TSAU1	13.4*	23.4	0.1	8
50 mM NaCl				
USDA110	10.2	10.6	0.067	4.2
USDA110 + TSAU1	12.4*	16.0*	0.088*	4.6
75 mM NaCl				
USDA110	9	8.2	0.053	3
USDA110 + TSAU1	10.2	12.2*	0.084*	4

* Significantly different from the plant inoculated with rhizobia alone at $P < 0.05$; plants were grown under hydroponic sand system for 14 days

with USDA110 alone. The co-inoculation of soybean gave a significantly higher proportion of nodules compared to the numbers induced by USDA110 alone (Table 1). The production of IAA by bacterial inoculants might be responsible for the enlarged root system and number of infection sites prior to nodulation (Tilak et al. 2006).

5 Cytokinins

Cytokinins (CKs) are in regulation of various plant growth attributes and developmental events. They are involved in apical dominance, cell division, increasing biogenesis of chloroplast and synthesis of anthocyanin, tissue differentiation (both vascular and shoot), mobilization and assimilation of nutrients, photomorphogenesis and senescence (Kang et al. 2012; Kunikowska et al. 2013). In addition to this CK can also promote plant growth under stress conditions and impart salt tolerance to plants (Kang et al. 2012; Lubovska et al. 2014). CK can lead to increased salt tolerance through its vital interactions with the other phytohormones like auxins and ABA (Frebort et al. 2011). During stress, reduced supply of CK from root tissues bring alterations in expression of genes in shoot so as to elicit a response leading to amelioration of the encountering stress (Haare et al. 1997). CK have the potential to break stress-induced seed dormancy of several plants, whereas kinetin increases plant growth and development under salt stress (Boucaud and Ungar 1976). In barley, application of CK reduces growth of salt sensitive variety while enhances growth, shoot/root ratio and endogenous CK levels in salt-tolerant variety (Kuiper et al. 1990).

Several PGPR species, such as *Arthrobacter, Bacillus, Azospirillum* and *Pseudomonas* synthesize CK which positively affects root development (Naz et al. 2009). Salmone et al. (2001) reported that plant inoculation with CK producing

PGPR strains increased the level of CK in root tissues, through which strains has impact on root growth and development. In other study, CK producing bacteria *Micrococcus luteus* chp37 isolated from desert of Pakistan increased shoot, root length, dry weight of maize (up to 54 %) and photosynthetic pigments, such as chlorophyll a, chlorophyll b and total carotenoids as compared to non-inoculated control plants under drought condition (Raza and Faizal 2013). Naz et al. (2009) observed CK production by root associated bacteria isolated from weed grown under salt affected soil. The strains stimulated root and shoot growth of soybean and proline content under 20 dS/m NaCl condition. More recently, Liu et al. (2013) studied the effect of CK producing *Bacillus subtilis* strain on the growth of *Platycladus orientalis* (oriental thuja) seedlings under drought conditions. They observed that plant inoculation with *B. subtilis* stimulated the shoot dry weight by 19.2 %, as well as the root by 13.9 %, and alleviated drought stress. PGPR can affect on root system architecture through producing phytohormones including auxins and CK (Dodd et al. 2010).

6 Gibberellic Acid

Gibberellins are another group of plant growth regulators having important role in growth and development. As stated earlier, gibberellins are involved in seed germination, leaf expansion and stem elongation as well as flowering (Iqbal et al. 2014; Manjili et al. 2012). It is well accepted that cross-talking between GA and several growth hormones during signaling in response to environmental stress mediated controlled and regulate plant growth and development.

Both developmental and environmental factors contribute to synthesis of gibberellins in plants (Olszewski et al. 2002). Accumulation of GA occurs at higher rates when plants are exposed to environmental extremes. Plant scientists are widely studying the use of exogenous application of phytohormones for improving growth and yield of important crop plants. For instance, Ahmad (2010) observed increased growth in salinity stressed *Brassica* spp. due to exogenous application of GA. Moreover, an increase in content of osmotic constituents was reported in salinity stressed plants which were further increased by application of GA leading to better osmotic adjustment in GA treated plants even under salinity stress conditions (Ahmad 2010). Maintained tissue water content and GA-induced mitigation of salinity effects on water content has been reported in wheat (Manjili et al. 2012) and maize (Tuna et al. 2008). Exogenous application of GA enhanced activity of antioxidant enzyme activity promoting better and quick removal of toxic free radicals under salt stress. Several workers have reported efficiency of GA to ameliorate salinity-induced deleterious changes and result in maintained growth of salt stressed wheat and rice (Prakash and Prathapasenan 1990).

It has been also found that the content of phytohormones in plants may affect by root associated microorganisms (Turan et al. 2014). In earlier studies, Fulchieri et al. (1993) and Lucangeli and Bottini (1997) reported that corn seedlings

Fig. 2 The effect of gibberellic acid producing *Bacillus subtilus* on the growth of *Ephedra aphylla* under salt stressed soil condition. **a** Uninoculated control plants. **b** Plants inoculated with *B. subtilis*)

inoculated with PGPR strains exhibited relatively higher amounts of IAA, and gibberellin GA_3, as compared to non-inoculated controls. PGPR inoculations *P. agglomerans* RK-92, and *B. subtilis* TV-17C increased GA, and SA concentrations in cabbage. In other study, the inoculation of chickpea with *Bradyrhizobium japonicum* caused a marked increase in GA and IAA content (Bano et al. 2010). *Azospirillum* strains increased levels of GA_3 in the roots after inoculation of maize seedling (Fulchieri et al. 1993). Endophytic fungi *Aspergillus fumigatus* isolated from soybean roots grown under saline soil condition produced gibberellins (GAs), such as GA_4 (24.8 ng/ml), GA_9 (1.2 ng/ml) and GA_{12} (9.8 ng/ml) (Khan et al. 2011). The strain significantly increased shoot length, shoot fresh and dry biomass, leaf area, chlorophyll contents and photosynthetic rate under salt stress (70 and 140 mM) as compared to non-inoculated plants (Khan et al. 2011). Creus et al. (2004) observed mitigation of drought stress in wheat by *Azospirillum lipoferum,* which synthesized GA. While, Rodríguez et al. (2006) reported improved salt tolerance of lupine by GA_3 producing *Cyanobacteria* strain. GA producing *Bacillus subtilis* stimulated root and shoot growth of *Ephedra aphylla* under salt and drought stress condition (Fig. 2).

7 Conclusion

It is widely accepted that salinity is having damaging effects on growth and physiology of plants. Restrictions in uptake of water and essential mineral elements caused due to salinity results in perturbed growth and hence yield reductions. Plants have developed several tolerance strategies which help mitigate the deleterious impact of salinity. However, salinity-induced changes can be more precisely avoided and combated by exploiting potential of the growth regulators. Among the several approaches suggested and adapted worldwide exogenous application of plant growth regulators and phytohormone producing microbes (PGPR) are now a day extensively used. PGPR increase levels of auxin, gibberellins, CK in roots and may affect the metabolism of endogenous phytohormones in the plant and plays a vital role in mitigation of salt stress in plants. Thus, applying genomic and proteomic approaches to study the changes induced by interactions of phytohormone producing microbes in plants may prove quite significant for future studies.

Acknowledgments The research activity of Dilfuza Egamberdieva was supported by a Georg Forster Research Fellowship for experienced Researchers (HERMES), Alexander Von Humboldt Foundation.

References

Afzal I, Basara SMA, Faooq M, Nawaz A (2006) Alleviation of salinity stress in spring wheat by hormonal priming with ABA, salicylic acid and ascorbic acid. Int J Agric Biol 8:23–28

Afzal I, Basra S, Iqbal A (2005) The effect of seed soaking with plant growth regulators on seedling vigor of wheat under salinity stress. J Stress Physiol Biochem 1:6–14

Ahanger MA, Tyagi SR, Wani MR, Ahmad P (2014) Drought tolerance: roles of organic osmolytes, growth regulators and mineral nutrients. In: Ahmad P, Wani MR (eds) Physiological mechanisms and adaptation strategies in plants under changing environment. Springer, New York, pp 25–56

Ahmad P (2010) Growth and antioxidant responses in mustard (*Brassica juncea* L.) plants subjected to combined effect of gibberellic acid and salinity. Arch Agron Soil Sci 56(5):575–588

Ahmad P, Nabi G, Ashraf M (2011) Cadmium-induced oxidative damage in mustard (Brassica juncea (L.) Czern. & Coss.) plants can be alleviated by salicylic acid. S Afr J Bot 77:36–44

Ahmad P, Ashraf M, Azooz MM, Rasool S, Akram NA (2014) Potassium starvation induced oxidative stress and antioxidant defense responses in *Brassica juncea*. J Plant Int 9(1):1–9

Akbari G, Sanavy SA, Yousefzadeh S (2007) Effect of auxin and salt stress (NaCl) on seed germination of wheat cultivars (*Triticum aestivum* L.). Pak J Biol Sci 10:2557–2561

Ali B, Sabri AN, Ljung K, Hasnain S (2009) Auxin production by plant associated bacteria: impact on endogenous IAA content and growth of *Triticum aestivum* L. Lett Appl Microbiol 48:542–547

Alqarawi AA, Hashem A, Abd-Allah EF, Alshahrani TS, Huqail AA (2014) Effect of salinity on moisture content, pigment system, and lipid composition in *Ephedra alata* decne. Acta Biol Hung 65(1):61–71

Apel K, Hirt H (2004) Reactive oxygen species: metabolism, oxidative stress, and signal transduction. Annu Rev Plant Biol 55:373–399

Arkhipova TN, Prinsen E, Veselov SU, Martineko EV, Melentiev AI, Kudoyarova GR (2007) Cytokinin producing bacteria enhances plant growth in drying soil. Plant Soil 292:305–315

Ashraf MY, Sarwar G, Ashraf M, Afaf R, Sattar A (2002) Salinity induced changes in α-amylase activity during germination and early cotton seedling growth. Biol Plant 45:589–591

Azooz MM, Youssef AM, Ahmad P (2011) Evaluation of salicylic acid (SA) application on growth, osmotic solutes and antioxidant enzyme activities on broad bean seedlings grown under diluted seawater. Int J Plant Physiol Biochem 3(14):253–264

Bano A, Batool R, Dazzo F (2010) Adaptation of chickpea to desiccation stress is enhanced by symbiotic rhizobia. Symbiosis 50:129–133

Baumann K (2010) Signalling: ABA's greatest hits. Nat Rev Mol Cell Biol 11(1):2

Berg G, Alavi M, Schmidt CS, Zachow C, Egamberdieva D, Kamilova F, Lugtenberg B (2013) Biocontrol and osmoprotection for plants under saline conditions. In: de Bruijn Frans J (ed) Molecular microbial ecology of the rhizosphere. Wiley, USA

Boucaud J, Ungar IA (1976) Hormonal control of germination under saline conditions of three halophyte taxa in genus Suaeda. Physiol Plant 36:197–200

Bianco C, Defez R (2009) *Medicago truncatula* improves salt tolerance when nodulated by an indole-3-acetic acid-overproducing *Sinorhizobium meliloti* strain. J Exp Bot 60:3097–3107

Bottini R, Cassán F, Piccoli P (2004) Gibberellin production by bacteria and its involvement in plant growth promotion and yield increase. Appl Microbiol Biotechnol 65:497–503

Buran TJ, Sandhu AK, Azeredo AM, Bent AH, Williamson JG, Gu L (2012) Effects of exogenous abscisic acid on fruit quality, antioxidant capacities, and phytochemical contents of southern high bush blueberries. Food Chem 132:1375–1381

Cabot C, Sibole JV, Barcelo J, Poschenrieder C (2009) Abscisic acid decreases leaf Na + exclusion in salt-treated *Phaseolus vulgaris* L. J Plant Growth Regul 28:187–192

Cramer GR, Quarrie SA (2002) Abscsic acid is correlated with the leaf growth inhibition of four genotypes of maize differing in their response to salinity. Funct Plant Biol 29:111–115

Creus DM, Sueldo RJ, Rolando J, Barassi CA (2004) Water relations and yield in Azospirillum inoculated wheat exposed to drought in the field. Can J Bot 82:273–281

Debez A, Chaibi W, Bouzid S (2001) Effect du NaCl et de regulatoeurs de croissance sur la germination d' *Atriplex halimus L.* Cahiers Agric 10:135–138

Dodd IC, Zinovkina NY, Safronova VI, Belimov AA (2010) Rhizobacterial mediation of plant hormone status. Ann Appl Biol 157:361–379

Dunlap JR, Binzel ML (1996) NaCl reduces indole-3-acetic acid levels in the roots of tomato plants independent of stress induced abscisic acid. Plant Physiol 112:379–384

Egamberdieva D, Botir H, Hashem A, Abd_Allah EF (2014) Characterization of salt tolerant *Enterobacter hormaechei* strain associated with tomato root grown in arid saline soil. Pure Appl Microbiol 8(5):4231–4239

Egamberdieva D, Berg G, Lindström K, Räsänen LA (2013) Alleviation of salt stress of symbiotic *Galega officinalis* L. (goat's rue) by co-inoculation of *Rhizobium* with root colonising *Pseudomonas*. Plant Soil 369(1):453–465

Egamberdieva D (2012) *Pseudomonas chlororaphis*: a salt tolerant bacterial inoculant for plant growth stimulation under saline soil conditions. Acta Physiol Plant 34:751–756

Egamberdieva D, Kucharova Z, Davranov K, Berg G, Makarova N, Azarova T, Chebotar V, Tikhonovich I, Kamilova F, Validov S, Lugtenberg B (2011) Bacteria able to control foot and root rot and to promote growth of cucumber in salinatedsoils. Biol Fertil Soils 47:197–205

Egamberdieva D (2011) Survival of *Pseudomonas extremorientalis* TSAU20 and *P. chlororaphis* TSAU13 in the rhizosphere of common bean (*Phaseolus vulgaris*) under saline conditions. Plant Soil Environ 57(3):122–127

Egamberdieva D, Kucharova Z (2009) Selection for root colonising bacteria stimulating wheat growth in saline soils. Biol Fert Soils 45:561–573

Egamberdieva D (2009) Alleviation of salt stress by plant growth regulators and IAA producing bacteria in wheat. Acta Physiol Plant 31:861–864

Egamberdieva D (2008) Alleviation of salinity stress in radishes with phytohormone producing rhizobacteria. Biotechnol J 136S:262

Egamberdieva D, Kamilova F, Validov S, Gafurova L, Kucharova Z, Lugtenberg B (2008) High incidence of plant growth-stimulating bacteria associated with the rhizosphere of wheat grown in salinated soil in Uzbekistan. Environ Microbiol 19:1–19

Egamberdiyeva D (2005) Characterization of *Pseudomonas* species isolated from the rhizosphere of plants grown in serozem soil, semi arid region of Uzbekistan. Sci World J 5:501–509

Egamberdiyeva D, Höflich G (2003) The effect of associative bacteria from different climates on plant growth of pea at different soils and temperatures. Arch Agr Soil Sci 49(2):203–213

Fassler E, Evangeloua MW, Robinson BH, Schulin R (2010) Effects of indole-3-acetic acid (IAA) on sunflower growth and heavy metal uptake in combination with ethylene diamine disuccinic acid (EDDS). Chemosphere 80:901–907

Figueiredo MVB, Burity HA, Martınez CR, Chanway CP (2008) Alleviation of drought stress in the common bean (*Phaseolus vulgaris* L.) by co-inoculation with *Paenibacillus polymyxa* and *Rhizobium tropici*. Appl Soil Ecol 4:182–188

Forchetti G, Masciarelli O, Izaguirre MJ, Alemano S, Alvarez D, Abdala G (2010) Endophytic bacteria improve seedling growth of sunflower under water stress, produce salicylic acid, and inhibit growth of pathogenic fungi. Curr Microbiol 61(6):485–493

Frebort I, Kowalsk M, Hluska T, Frebortova J, Galuszka P (2011) Evolution of cytokinin biosynthesis and degradation. J Exp Bot 62(8):2431–2452

Fricke W, Akhiyarova G, Veselov D, Kudoyarova G (2004) Rapid and tissue-specific changes in ABA and in growth rate in response to salinity in barley leaves. J Exp Bot 55:1115–1123

Fulchieri M, Lucangeli C, Bottini R (1993) Inoculation with *Azospirillum lipoferum* affects growth and gibberellin status of corn seedling roots. Plant Cell Physiol 34:1305–1309

Gómez-Cadenas A, Arbona V, Jacas J, Primo-Millo E, Talon M (2002) Abscisic acid reduces leaf abscission and increases salt tolerance in citrus plants. J Plant Growth Reg 21(3):234–240

Gulnaz AJ, Iqbal J, Azam F (1999) Seed treatment with growth regulators and crop productivity. II. Response of critical growth stages of wheat (*Triticum aestivum* L.) under salinity stress. Cereal Res 27:419–426

Gunes A, Inal A, Alpaslam M, Erslan F, Bagsi EG, Cicek N (2007) Salicylic acid induced changes on some physiological parameters symptomatic for oxidative stress and mineral nutrition in maiz (*Zea mays* L.) grown under salinity. J Plant Physiol 164:728–736

Hare PD, Cress WA, van Staden J (1997) The involvement of cytokinins in plant responses to environmental stress. Plant Growth Regul 23:79–103

Hayat Q, Hayat S, Irfan M, Ahmad A (2010) Effect of exogenous salicylic acid under changing environment: a review. Environ Exp Bot 68:14–25

Indiragandhi P, Anandham R, Madhaiyan M, Sa TM (2008) Characterization of plant growth promoting traits of bacteria isolated from larval guts of diamondback moth Plutella xylostella (Lepidoptera: Plutellidae). Current Microbiology 56: 327–333

Iqbal N, Umar S, Khan NA, Khan MIR (2014) A new perspective of phytohormones in salinity tolerance: regulation of proline metabolism. Environ Exp Bot 100:34–42

Jamil M, Lee KB, Jung KY, Lee DB, Han MS, Rha ES (2007) Salt stress inhibits germination and early seedling growth in cabbage (*Brassica oleracea capitata* L.). Pak J Biol Sci 10:910–914

Jeschke WD, Peuke AD, Pate JS, Hartung W (1997) Transport, synthesis and catabolism of abscisic acid (ABA) in intact plants of castor bean (*Ricinus communis* L.) under phosphate deficiency and moderate salinity. J Exp Bot 48:1737–1747

Kang DJ, Seo YJ, Lee JD, Ishii R, Kim KU, Shin DH, Park SK, Jang SW, Lee IJ (2005) Jasmonic acid differentially affects growth, ion uptake and abscisic acid concentration in salt-tolerant and salt-sensitive rice cultivars. J Agron Crop Sci 191:273–282

Kang NY, Cho C, Kim NY, Kim J (2012) Cytokinin receptor-dependent and receptor-independent pathways in the dehydration response of *Arabidopsis thaliana*. J Plant Physiol 169:1382–1391

Karadeniz A, Topcuoğlu SF, İnan S (2006) Auxin, gibberellin, cytokinin and abscisic acid production in some bacteria. World J Microbiol Biotechnol 22(10):1061–1064

Keskin BC, Sarikaya AT, Yuksel B, Memon AR (2010) Abscisic acid regulated gene expression in bread wheat. Aust J Crop Sci 4:617–625

Khan MIR, Asgher M, Khan NA (2014) Alleviation of salt-induced photosynthesis and growth inhibition by salicylic acid involves glycine betaine and ethylene in mungbean (*Vigna radiata* L.). Plant Physiol Biochem 80:67–74

Khan AL, Hamayun M, Kim Yoon-Ha, Kang SM, Lee JH, Lee IN (2011) Gibberellins producing endophytic *Aspergillus fumigatus* sp. LH02 influenced endogenous phytohormonal levels, iso flavonoids production and plant growth in salinity stress. Process Biochem 46:440–447

Khadri M, Tejera NA, Lluch K (2006) Alleviation of salt stress in common bean (*Phaseolus vulgaris*) by exogenous abscisic acid supply. J Plant Growth Regul 25:110–119

Kishore GK, Pande S, Podile AR (2005) Phylloplane bacteria increase seedling emergence, growth and yield of field-grown groundnut (*Arachis hypogaea* L.). Lett Appl Microbiol 40:260–268

Kuiper D, Schuit J, Kuiper PJC (1990) Actual cytokinin concentrations in plant tissue as an indicator for salt resistance in cereals. Plant Soil 123:243–250

Kunikowska A, Byczkowska A, Doniak M, Kazmierczak A (2013) Cytokinins resume: their signaling and role in programmed cell death in plants. Plant Cell Rep 32:771–780

Lavania M, Nautiyal CS (2013) Solubilization of tricalcium phosphate by temperature and salt tolerant *Serratia marcescens* NBRI1213 isolated from alkaline soils. Afr J Microbiol Res 7(34):4403–4413

Liu F, Xing S, Ma H, Du Z, Ma B (2013) Cytokinin-producing, plant growth-promoting rhizobacteria that confer resistance to drought stress in *Platycladus orientalis* container seedlings. Microbiol Biotechnol 97(20):9155–9164

Lubovska Z, Dobra J, Storchova H, Wilhelmova N, Vankova R (2014) Cytokinin oxidase/dehydrogenase over expression modifies antioxidant defense against heat, drought and their combination in *Nicotiana tabacum* plants. J Plant Physiol 171:1625–1633

Lucangeli C, Bottini R (1997) Effects of *Azospirillum* spp. on endogenous gibberellin content and growth maize (*Zea mays* L.) treated with uniconazole. Symbiosis 23:63–71

Maggio A, Barbieri G, Raimondi G, De Pascale S (2010) Contrasting effects of GA3 treatments on tomato plants exposed to increasing salinity. J Plant Growth Regul 29:63–72

Manjili FA, Sedghi M, Pessarakli M (2012) Effects of phytohormones on proline content and antioxidant enzymes of various wheat cultivars under salinity stress. J Plant Nut 35:1098–1111

Naz I, Bano A, Ul-Hassan T (2009) Isolation of phytohormones producing plant growth promoting rhizobacteria from weeds growing in Khewra salt range, Pakistan and their implication in providing salt tolerance to *Glycine max* L. Afr J Biotech 8(21):5762–5766

Olszewski N, Sun TP, Gubler F (2002) Gibberellin signaling, biosynthesis, catabolism, and response pathways. Plant Cell 14:561–580

Prakash L, Prathapasenan G (1990) NaCl and gibberellic acid induced changes in the content of auxin, the activity of cellulose and pectin lyase during leaf growth in rice (*Oryza sativa*). Ann Bot 365:251–257

Raza A, Faisal M (2013) Growth promotion of maize by desiccation tolerant Micrococcus luteus-chp37 isolated from Cholistan desert, Pakistan. Austr J Crop Sci 7(11):1693–1698

Ribaut JM, Pilet PE (1994) Water stress and indole-3ylacetic acid content of maize roots. Planta 193:502–507

Rodriguez H, Fraga R, Gonzalez T, Bashan Y (2006) Genetics of phosphate solubilization and its potential applications for improving plant growth-promoting bacteria. Plant Soil 287:15–21

Salamone DG, Hynes RK, Nelson LM (2001) Cytokinin production by plant growth promoting rhizobacteria and selected mutants. Can J Microbiol 47:404–411

Senaratna T, Touchell D, Bunns E, Dixon K (2000) Acetyl salicylic acid (aspirin) and salicylic acid induce multiple stress tolerance in bean and tomato plants. Plant Growth Reg 30:157–161

Shakirova FM, Sakhabutdinova AR, Bezrukova MV, Fatkhutdinova RA, Fatkhutdinova DR (2003) Changes in the hormonal status of wheat seedlings induced by salicylic acid and salinity. Plant Sci 164:317–322

Salomon MV, Bottini R, de Souza Filho GA, Cohen AC, Moreno D, Gil M, Piccoli P (2014) Bacteria isolated from roots and rhizosphere of *Vitis vinifera* retard water losses, induce abscisic acid accumulation and synthesis of defense-related terpenes in vitro cultured grapevine. Physiol Plant 51(4):359–374

Shanmugam P, Narayanasamy M (2008) Optimization and production of salicylic acid by rhizobacterial strain *Bacillus licheniformis* MML2501. Internet J Microbiol 6:1

Sgroy V, Cassán F, Masciarelli O, Florencia M, Papa D, Lagares A, Luna V (2009) Isolation and characterization of endophytic plant growth-promoting (PGPB) or stress homeostasis-regulating (PSHB) bacteria associated to the halophyte *Prosopis strombulifera*. Appl Microbiol Biotechnol 85(2):371–381

Spaepen S, Vanderleyden J, Remans R (2007) Indole-3-acetic acid in microbial and microorganism-plant signalling. FEMS Microbiol Rev 31:425–448

Tasgin E, Attici O, Nalbantogly B (2003) Effect of salicylic acid and cold on freezing tolerance in winter wheat leaves. Plant Growth Reg 41:231–236

Teale WD, Paponov IA, Palme K (2006) Auxin in action: signalling, transport and the control of plant growth and development. Nat Rev Mol Cell Biol 7:847–859

Tilak KVB, Ranganayaki N, Manoharachari C (2006) Synergistic effects of plant-growth promoting rhizobacteria and Rhizobium on nodulation and nitrogen fixation by pigeonpea (Cajanus cajan). Eur J Soil Sci 57:67–71

Tuna AL, Kaya C, Dikilitas M, Higgs D (2008) The combined effects of gibberellic acid and salinity on some antioxidant enzyme activities, plant growth parameters and nutritional status in maize plants. Environ Expt Bot 63:1–9

Turan M, Ekinci M, Yildirim E, Gneş A, Karagz K, Kotan R, Dursun A (2014) Plant growth-promoting rhizobacteria improved growth, nutrient, and hormone content of cabbage (*Brassica oleracea*) seedlings. Turk J Agric For 38:327–333

Wahid A, Perveen M, Gelani S, Basra SMA (2007) Pretreatment of seed with H_2O_2 improves salt tolerance of wheat seedlings by alleviation of oxidative damage and expression of stress proteins. J Plant Physiol 164:283–294

Wang Y, Mopper S, Hasentein KH (2001) Effects of salinity on endogenous ABA, IAA, JA, and SA in *Iris hexagona*. J Chem Ecol 27:327–342

Yang Z, Yu J (2012) Differential effects of abscisic acid and glycine betaine on physiological responses to drought and salinity stress for two perennial grass species. J Am Soc Hortic Sci 137(2):96–106

Zhang J, Jia W, Yang J, Ismail AM (2006) Role of ABA in integrating plant responses to drought and salt stresses. Field Crop Res 97:111–119

Exploitation of Phytohormone-Producing PGPR in Development of Multispecies Bioinoculant Formulation

Sandeep Kumar, Mohit Agarwal, Shrivardhan Dheeman and Dinesh Kumar Maheshwari

Abstract Bacterial communities constitute an excellent plant-specific micro-ecosystem in the rhizosphere. Microbes establish a sophisticated relationship in the rhizosphere with plants to enhance productivity through the production of an array of metabolites, especially phytohormones. Plant growth and development is substantially influenced by plant hormones; it is well known that plant hormones regulate the growth and development of plants. PGPR bioinoculant not only exerts a positive effect on growth and yield but also triggers biocontrol against a broad spectrum of pathogens. It is important to use selective PGPR as consortia, which are individually able to produce certain phytohormone in dexterity against mono-species bioinoculant with multifarious activity. Understanding the application of bioinoculant having biocoenotic consortia of bacteria capable of producing phyto-hormone will serve as the basis for future research to elucidate the role of bacterial communities in crop productivity and sustainable agriculture.

Keywords Bioinoculant · Bioformulation · PGPR · Phytohormone · Bacterial metabolite · Consortia · IAA · Cytokinin · Ethylene · Gibberellin · Abscisic acid

S. Kumar
Regional Experiment Station, Rubber Research Institute of India, Rubber Board, Nagrakata, Jalpaiguri, West Bengal, India

M. Agarwal · S. Dheeman · D.K. Maheshwari (✉)
Department of Botany and Microbiology, Gurukul Kangri University, Haridwar, Uttarakhand, India
e-mail: maheshwaridk@gmail.com

D.K. Maheshwari (ed.), *Bacterial Metabolites in Sustainable Agroecosystem*, Sustainable Development and Biodiversity 12, DOI 10.1007/978-3-319-24654-3_11

1 Introduction

The role of plant growth-promoting rhizobacteria (PGPRs) in plant growth promotion and biological control of soil-borne phytopathogens has been intensively investigated. PGPRs influence plant growth directly due to the production of plant growth hormones, auxin, cytokinin, ethylene, gibberellin and abscisic acid (Glick 2015). Such phytohormone-producing PGPRs facilitate the availability and uptake of certain nutrients from the root environment (Barea et al. 2004). Some bacterial strains directly regulate plant physiological mimicking due to their synthesis of hormones, whereas others increase mineral and nitrogen availability in the soil as a way to augment growth. Apart from this, the release of volatile organic compounds (VOCs) blends by PGPRs establishes an additional function for VOCs as signaling molecules mediating plant–microbe interaction. The plant responds to these external stimuli (VOC's) and triggers growth promotion (Ryu et al. 2003).

PGPRs as microbial inoculants proved as an effective tool for boosting agricultural productivity, thereby helping to feed the growing world population to ensure food security. Recently, there has been a shift in the approach of workers, as, instead of using a single strain of PGPR as inoculant, co-inoculation of two or multiple PGPR is the need of the hour to transform green revolution to evergreen revolution. The microbial consortium is ecofriendly and easy to apply, resulting in improved soil health-related problems, viz. decrease in soil pathogenic load, increase in soil micro and macronutrient status, etc. The specifically designed carrier-based poly-microbial formulations further offer nutrient management, its availability, beating deleterious phytopathogens besides producing nitrogen economy, thus lowering the requirements of chemical fertilizers and pesticides. In fact, the substitution of chemical fertilizers with bacterial inoculants, especially PGPRs that show multifarious activity including production of plant hormones are a promising approach to sustain plant growth and health. Various bacterial genera of the PGPR group are a potential source of plant growth regulators. The concept of combined use of PGPRs is an effort to shift microbiological equilibrium in favor of increased plant growth production, nutrient uptake and protection (Khalid et al. 2004).

Selection of efficient indigenous strains is important so as to obtain desirable crop yields that bring sustainability to the crop ecosystem. The indigenous microorganisms proved more efficient than that of non-indigenous strains in the improvement of crop yield (Aeron et al. 2010). Ecofriendly bioformulations of multi-microbial consortia significantly support and induce both vegetative and reproductive parameters of plant growth and development. In this review, we emphasized to rediscover the knowledge for two- or multi-strain bacterial consortium over mono-inoculant excel to resist environmental stress and having long shelf life with low expenditure on production and storage. Secondly, the exploration of agricultural importance of microbial formulations comprising phytohormone in particular is covered in this chapter. In the recent scenario, different modes of inoculation of these bioinoculants, such as via soil drench, seed bacterization, seedling deep treatment or in the form of foliar spray, have lacunae within laboratory to farmer's field. Hence, a good quality bioinoculant with its suitable mode of application is necessary to investigate into crop growth. Further, to explore the applications of bioformulations for crop agroecosystem, this review focuses on consortial bioinoculant technology development for scientific as well as commercial up-gradation.

2 PGPRs-Involvement of Phytohormones

Microbial populations in the rhizosphere provide benefits to plants by a variety of ways, such as availability of mineral nutrients, synthesis of phytohormones, antagonism against plant pathogens through competition or due to the presence of antimicrobial metabolites. In fact, the growth hormones are effective to ameliorate environmental stress susceptible to that of plant growth. Bacterial synthesis of phytohormones, auxins and gibberellin, increases the rate of seed germination, seedling emergence and root system proliferation which further aid in plant growth promotion (Bottini et al. 2004; Hayat et al. 2010).

One of the phytohormones produced by soil microorganisms is indole acetic acid (IAA) (Spaepen et al. 2007). The capacity to synthesize IAA is widespread among soil- and plant-associated bacteria. It has been estimated that 80 % of rhizospheric bacteria produce plant-growth regulator IAA (Patten and Glick 1996). IAA production has been reported in PGPR strains of genera *Enterobacter, Pseudomonas, Azospirillum, Gluconacetobacter, Pseudomonas, Bacillus* and *Rhizobium* spp. (Lucy et al. 2004; Prasad et al. 2015). Various strategies of phytohormone-producing PGPR thus used for plant growth promotion in consortium formulation are summarized in Fig. 1.

Our current knowledge on the role of quorum sensing (QS) acts as a sophisticated co-operative behavior mediated by extracellular signal molecules in nature. QS molecules of the N-acylhomoserine lactone (AHLs) type play a key role in the effective bacterial colonization of plant hosts. On the other hand, AHLs are used by symbiotic, pathogenic and biological control agents to regulate wide array of factors such as virulence, rhizosphere competence, conjugation, secretion of hydrolytic enzymes and production of antimicrobial compounds. Further, AHLs have a key role to play in bacterial-plant cross-signaling, as some plants are able to reprogram gene expression in the presence of these bacterial molecules, whereas, others may interfere with AHL-QS systems by producing small molecules. This role of QS in special reference to exogenous phytohormone signaling is yet to be explored (Valverde et al. 2014).

Most of PGPR genera are able to produce a capricious quantity of phytohormones that enhance inculcation of both abiotic and biotic stresses in nature. Sufficient or excessive production of phytohormone is affected due to availability of suitable metabolic precursors that manage phytohormone production by providing suitable concentration of a specific substrate. It has been observed that the exogenous application of a precursor increased the magnitude of phytohormone

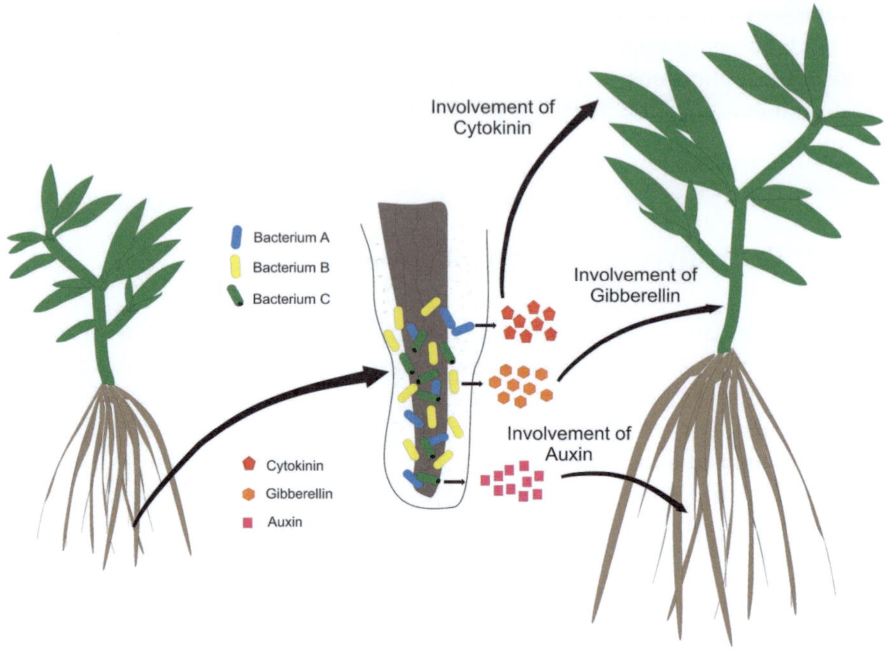

Fig. 1 Plant growth and health promotion using bacterial consortia producing different phytohormone

production many fold in culture as well as in soil. This precursor–inoculum interaction thus is useful for promoting plant growth (Asghar et al. 2002). Several studies put forward the concept of precursor–inoculum interaction that enhanced plant growth (Sarwar and Kremer 1995; Zahir et al. 1997; Arshad and Frankenberger 1997). Zahir et al. (2005) conducted a study to evaluate the effectiveness of precursor–inoculum interaction for enhancing maize growth. In fact, the precursor–inoculum interaction is not only effective under normal conditions but proves useful for promoting plant growth under stress conditions (Ahmad et al. 2012). As stated, PGPRs enhance plant growth by phytohormone production, and this attitude of production of phytohormones depends upon precursor substrate. The application of this technology is suggested to be exploited as per expectations, where sole bacterial inoculation is ineffective and cannot perform efficiently. Advocacies from Zahir et al. (2003) revealed the influence of plant growth regulators (PGRs) in growth and development. Further, Davies (2010) suggested the role of phytohormones in phyto-physiological processes that enable plant growth enhancement by upholding hormonal balance.

The use of phytohormone-producing PGPRs as bioinoculant has certain limitations as well. The application of a particular hormone-producing bacteria (in case of few) is found to be impractical. For example, ethylene is a gaseous hormone that stimulates various physiological processes in plants but, the application of ethylene due to its gaseous nature is difficult. However, its precursor L-methionine

Table 1 List of phytohormone-producing PGPR as biofertilizer and biostimulators

Crop	Bacteria	Associative phytohormone	Effect on yield	References
Ornamental species	*Azospirillum* isolates, Azo 195, Azo 249, Azo 274	Auxin	Inoculation increased shoot height and root dry mass by 15 and 75 % in cv	Moutia et al. (2010)
			M 1176/77 when subjected to drought stress, whereas cv. R 570 responded negatively particularly in the absence of drought stress	
Trifolium repens	*Pseudomonas* sp., *P. putida, B. megaterium*	IAA	Inoculation increased shoot and root biomass and water content under drought conditions	Marulanda et al. (2009)
Catharanthus roseus	*P. fluorescens*	IAA/gibberellin	*P. fluorescens* enhanced the growth parameters and partially ameliorated the drought-induced growth inhibition by increasing the fresh and dry weights significantly	Abdul Jaleel et al. (2007)
Lactuca sativa L.	*A. brasilense* Sp245	IAA	Inoculation increased leaf area, chlorophyll content and dry weight up to 63, 24 and 102 %, respectively, at higher salinity. At 40 mol m^{-3} NaCl, 60 and 73 % of plants remained alive in noninoculated and *Azospirillum* -inoculated plants, respectively	Fasciglione et al. (2012)
Oryza sativa L./ *Abelmoschus esculentus* L.	*Agrobacterium* sp. SUND BDU1, *Bacillus* sp. strains SUND LM2, Can4, Can6	IAA	The *Bacillus* sp. increased yield and N uptake	Barua et al. (2012)
Gossypium hirsutum L.	*Raoultella planticola* Rs-2	ACC-deaminase, IAA	Inoculation increased seed germination, plant height and dry biomass	Wu et al. (2012)
	P. putida Rs-198	IAA	Inoculation increased the seed germination rate, plant height and dry weight	Yao et al. (2010)
Zea mays L.	*Azotobacter* sp. C5, C7, C8 and C9	IAA	*Azotobacter* sp. C9 increased shoot biomass and polyphenol content	Rojas-Tapias et al. (2012)

(continued)

Table 1 (continued)

Crop	Bacteria	Associative phytohormone	Effect on yield	References
Triticum aestivum	*Streptomyces* isolates C	IAA, siderophores	Inoculation increased germination and biomass in saline soil	Sadeghi et al. (2012)
	B. pumilus, Pseudomonas mendocina, Arthrobacter sp., *Halomonas* sp., *Nitrinicola lacisaponensis*	IAA, siderophores	Inoculated plants showed increase in shoot and root length and their biomass	Tiwari et al. (2010)
Helianthus annuus	*P. fluorescens* biotype F and *P. fluorescens* CECT 378^T	IAA, siderophores	Increment in leaves, stems and roots occured	Shilev et al. (2012)
Cicer arietinum	*P. putida* MSC1, *P. pseudoalcaligenes* MSC4	IAA, siderophores	Inoculated plants with bioinoculants showed an increase in the plant growth	Patel et al. (2012)
T. aestivum L.	*Pseudomonas* sp.	IAA, siderophores	Bacterization significantly improved the root/shoot length, biomass and the level of cellular metabolites	Mishra et al. (2011)
	Pseudomonas lurida strain M2RH3	IAA, siderophores	Seed bacterization positively influenced the growth and nutrient uptake	Selvakumar et al. (2011)
	Exiguobacterium acetylicum strain 1P (MTCC 8707)	IAA, siderophores	Seed bacterization influenced the growth and nutrient uptake	Selvakumar et al. (2010)
Cucurbita pepo	*S. marcescens* strain SRM (MTCC 8708)	IAA, siderophores	Seed bacterization significantly enhanced plant biomass and nutrient uptake in cold temperature	Selvakumar et al. (2007)

(continued)

Table 1 (continued)

Crop	Bacteria	Associative phytohormone	Effect on yield	References
Z. mays / H. annuus	*Pseudomonas* sp. DGS6	ACC-deaminase, IAA	Inoculation increased the root–shoot dry weight of maize and sunflower in metal contaminated soil	Yang et al. (2013)
	Pseudomonas strains 3–3.5-1, TLC 6–6.5-1, TLC 6–6.5	IAA	*Pseudomonas* sp. resulted in a significant increase in copper accumulation in maize and sunflower	Li and Ramakrishna (2011)
T. aestivum L.	*Staphylococcus arlettae* Strain Cr11	ACC-deaminase, IAA	Bacterial inoculation showed significant increase in percent seed germination, root and shoot length	Sagar et al. (2012)
S. bicolor L.	*Bacillus* sp. SLS18	IAA, ACC-deaminase, siderophores	Inoculation increased the dry weights of aerial part and root in metal	Luo et al. (2012)
Brassica juncea	*A. xylosoxidans* Ax10	ACC-deaminase, IAA	*A . xylosoxidans* increased the root length, shoot length, fresh weight and dry weight	Ma et al. (2009)
Sedum alfredii	*Burkholderia* sp.D54	IAA, ACC-deaminase, siderophores	Bacterial inoculation significantly enhanced biomass production and increased both shoot and root	Guo et al. (2011)
Lycopersicum esculentum	*Bacillus amyloliquefa-cien* CM-2, T-5	IAA, siderophore	Strains showed strong biocontrol and growth promotion effects	Tan et al. (2013)
T. aestivum	*Pseudomonas, Bacillus*	IAA	Maximum increase in spike length, number of tillers and weight of seeds recorded	Hussain and Hassain (2011)
Cucumis sativus	*Ochrobactrum haematophilum*	IAA	Increased growth of cucumber leaf and root length	Zhao et al. (2012)

(continued)

Table 1 (continued)

Crop	Bacteria	Associative phytohormone	Effect on yield	References
Brassica napus	*Achromobacter* sp., *Klebsiella* sp., *Pseudomonas* sp., *Pantoea* sp., *Chrysobacterium* sp.	IAA and ACC-deaminase	The inoculation increased dry matter	Farina et al. (2012)
Lens culinaris	PGPR strain LCA-1, LCA-2, LCA-3, LCA-4 and LCA-5	IAA	PGPR significantly increased shoot weight, root weight, root length and dry weight	Zafar et al. (2012)
Fragaria virginiana	*Psenibacillus polymyxa* RC05, *Bacillus* sp. RC23	IAA	Root inoculation increased yield, average fruit weight and quality fruit	Erturk et al. (2012)
Azadirachta indica	*Streptomyces* strain AzR-010, 049, 051	IAA	Bacterization improved seed germination	Verma et al. (2011)
Piper nigrum	*Bacillus fequilansis*	IAA, ACC-deaminase	Black pepper cuttings showed more root and shoot length	Dastager et al. (2011)
Beta vulgaris	*Acienobacter johnsonii*	IAA	Inoculation increased plant dry weight and yield	Shi et al. (2011)
Cucumis melo	*B. subtilis* Y-IVI	IAA	Inoculation increased shoot dry weight and length	Zhao et al. (2011)
Nicotiana tabacum	*Pantoea agglomerans* strain PVM	IAA	In-vitro root induction in was observed	Apine and Jadhav (2011)
Triticum oryzae	*Enterobacter cloacae Bacillus* sp. SVPR30, *P. Polymyxa* ATCC-10343	IAA	Bacterization significantly improved the fresh weight, root length, shoot length and nitrogen content	Shankar et al. (2011)
		IAA	Inoculation attained increased in plant dry weight	Beneduzia et al. (2008)

(continued)

Table 1 (continued)

Crop	Bacteria	Associative phytohormone	Effect on yield	References
Zea mays	*Aceinobacter* CR 1.8 *Klebsiella* SN 1.1	IAA	Inoculation of seeds showed non-significant response	Chaiharn and Lumyong (2011)
Sorghum	*Azotobacter brassilense* SM	IAA	Seed bacterization with *A. brassilense* improved root shoot length and seedling dry weight	Malhotra and Srivastava (2009)
Vigna radiate	*Acienobacter* CR 1.8 *Klebsiella* SN 1.1	IAA	The inoculation of bean significantly increased the adventitious root length	Chaiharn and Lumyong (2011)
	Pseudomonas, *Escherichia* *Micrococcus* *Staphylococcus* sp.	IAA	Bacterization significantly enhanced shoot length and biomass	Ali et al. (2010)
Malus domestica	*Bacillus* OSU-142, *Bacillus* M-3 *Burkholderia* OSU-7 *Pseudomonas* BA-8	IAA, Cytokinin	Bacterial inoculation increased average shoot length and fruit yield	Aslantas et al. (2007)

(L-MET) can release ethylene in the presence of soil microflora, ultimately, stimulating plant growth (Arshad and Frankenberger 1997; Arshad et al. 2008).

The mechanism of plant growth-promotion by non-pathogenic, plant-associated bacteria has not been completely elucidated, but the direct plant growth-promoting (PGP) mechanism involving production of plant growth regulators (hormones) has been studied by a number of workers whereas, indirect plant growth-promotion exhibits when PGPRs promote plant growth by improving growth-restricting conditions (Glick 1995). Although, commercially available phytohormones are used for promoting plant growth, microbially synthesized phytohormones are more efficiently and reasonably good and better substitutes to that of chemical inocula. Considerable work on IAA-producing PGPR as biofertilizers and phytostimulators has been demonstrated (Table 1).

3 Bacterial Hormones, Synthesis and Regulations

Phytohormones have a wide range of biological activities that can affect plant growth and development in different ways including promoting root initiation, inhibiting root elongation, promoting fruit ripening, lower wilting, stimulate seed germination, promoting leaf abscission, activating the synthesis of other plant hormones as observed in case of ethylene (Ishibashi et al. 2012). IAA secreted by microorganisms has direct involvement in the enhancement of root proliferation and the increase in nutrient uptake (Reetha et al. 2014). Biosynthesis of IAA in bacteria is usually classified based on the intermittent compound of metabolic pathway (Table 2). Bacterial genera *Agrobacterium tumefaciens, A. rhizogenes,*

Table 2 Mechanisms adopted by microbes for IAA biosynthesis

Pathways	Microorganisms	Mechanisms	References
IAM pathway	*Agribacterium, Pseudomonas, Pantoae, Rhizobium, Bradyrhizobium*	Tryptophan → IAM	Prinsen et al. (1993)
		IAM → IAA	
IPyA pathway	*Rhizobium, Bradyrhizobium*	Tryptophan → IPyA	Hartmann et al. (1983)
		IPyA → IAAld	
		IAAld → IAA	
TAM pathway	*Bacillus*	TAM → IAAld	Perley and Stowe (1966)
		IAAld → IAA	
TSO pathway	*Pseudomonas*	Tryptophan → IAAld	Oberhänsli et al. (1991)
		IAAld → IAA	
Indole-3-acetanitrile pathway	*Rhizobium*	Tryptophan → Indole-3-acetanitrile	Patten and Glick (1996)
		Indole-3-acetanitrile → IAA	

IAM Indole-3-acetamide; *IPyA* Indole-3-pyruvate acetic acid; *TAM* Tryptamine; *IAAld* indole-3-acetaldehyde; *TSO* Tryptophan side chain oxidase

Erwinia herbicola and *Pseudomonas sauastanoi* possess IAM pathway that seems to be adopted for expression in different environment. IAM pathway is involved in gall size, whereas the IPyA pathway determines epiphytic fitness (Lambrecht et al. 2000). Ali et al. (2009) reported auxin (IAA)-producing rhizobacteria exert positive effects on the growth and development of *Triticum aestivum* L. The auxin responses during microbe-induced de novo organ formation seem to be dynamic, suggesting that plant-associated microbes can actively modify their host's auxin transport (Grunewald et al. 2009). Lenin and Jayanthi (2012) reported that *Azospirillum lipoferum, Azotobacter chroococcum, P. fluorescens* and *Bacillus megaterium* from the rhizosphere region of *Catharanthus roseus* produce IAA, GA and siderophore. Recently, Morrone et al. (2009) demonstrated that *Bradyrhizobium japonicum* encodes separate diterpenoid from plant and fungi, *ent*-copalyldiphosphate and *ent*-kaurene synthesis pathways for biosynthesis of gibberellin.

On the other hand, Cytokinins are one of the five (auxin, cytokinin, gibberellin, ethylene, abscisic acid) major groups of PGRs. Cytokinin production by PGPR is an innovative sustainable approach to improve the yield and quality of agricultural crops (de Garcia Salamone et al. 2006). Cytokinin produced by PGPR (Arkhipova et al. 2007; Melnykova et al. 2013) exhibited cell division, regulating apical dominance, branching, leaf senescence, chloroplast and nodule development (Werner et al. 2001) beside inducing growth and control developmental process. Recently, Liu et al. (2013) reported that cytokinin-producing PGPR inoculation alleviates the drought stress and interferes with the suppression of shoot growth, showing a real potential to perform as a drought stress inhibitor in arid environments. However, phytopathogens synthesize cytokinins and can affect plant growth and development (Frugier et al. 2008). Earlier, Arkhipova et al. (2007) studied that cytokinin-producing bacteria are able to enhance plant growth parameter (vegetative and reproductive) in dry-aired soil.

Ethylene is a key phytohormone which inhibits root elongation, nodulation and auxin transport and promotes seed germination, senescence, fruit ripening and abscission of various organs (Bleecker and Kende 2000). Ethylene is used in systemic acquired resistance (SAR) during associative and symbiotic plant–bacterium interactions and, if high concentrations are present, is involved in plant defense pathways against pathogens (Broekaert et al. 2006; Glick et al. 2007). Ethylene level is regulated by 1-aminocyclopropane-1-carboxylic acid (ACC) catalyzed by ACC oxidase. Currently, bacterial strains exhibiting ACC-deaminase activity have been identified in a wide range of growth-promoting bacterial genera such as *Acinetobacter, Agrobacterium, Achromobacter, Alcaligenes, Azospirillum, Bacillus, Burkholderia, Enterobacter, Pseudomonas, Ralstonia, Serratia* and *Rhizobium* etc. (Gupta et al. 2015).

Apart from lowering ethylene levels, production of gibberellins, known as juvenile phytohormones, considered helpful in the postharvest environment (Glick et al. 1998; Lucy et al. 2004). Very less is accounted about work on bio-gibberellin synthesis in rhizospheric bacteria. Some bacterial genera *A. brasilense, A. lipoferum, Bacillus sp., Bradyrhizobium japonicum* and *Rhizobium phaseoli* are able to secrete gibberellin (Frankenberger and Arshad 1995; Gutiérrez-Mañero et al. 2001). Fulchieri et al. (1993) speculate that gibberellins increase root hair density in root zones involved in nutrient and water uptake. In another study, Gutiérrez-Mañero et al. (2001) isolated *B. pumilus* and *B. licheniformis* from the rhizosphere of *Alnus glutinosa* shown to produce physiologically active gibberellins which had strong growth-promoting activity on alder.

4 Consortial Bioformulations

The bacterial consortium is a group of different species of single bacteria or sometimes multi-strain diverse genera, which act together as a community. These include bacterial species which are not only imparting resistant to environmental stress but contain natural biomolecule which have beneficial activity and long shelf life. In certain environmental conditions, where single-strain inoculum is unable to perform better, the development of multi-strain inoculum proved quite effective and significant in nature (Roy et al. 2015). The combined application of *Sinorhizobium fredii* KCC5 and *Pseudomonas fluorescens* LPK2 significantly reduced the wilt disease (Kumar et al. 2010). Similarly, multi-strain bacterial consortia proved useful for enhancing plant growth and development particularly in conditions where single inoculation was not so effective (Maheshwari et al. 2015). Such multi-strain inocula are more effective for enhancing plant growth and development due to the presence of more beneficial traits which might not be possible in single individual strain (Bashan et al. 2014). Broadly, two-species and multi-species microbial consortia in bioformulation have found wide applicability in the growth promotion of several crops (Pandey and Maheshwari 2007a, b; Thakkar and Saraf 2014). These are applied through different modes such as via soil drench, seed bacterization, seedling deep treatment, or in the form of foliar spray. The nature of the formulation and display on the level about their mode of application to the particular crop plants is prerequsite. In the present scenario, individual microbe at one time approach has been replaced with that of a mixture/consortium of microorganisms bearing multifarious characteristics. But prior studies to determine balancing or synergistic behaviors during microbe–microbe interaction provide better and consistent effects. PGPR are known to enhance plant growth by a variety of mechanisms in particular due to phytohormone production although, role of other beneficial traits such as nitrogen fixation, siderophore production, solubilization of minerals, antibiotic production, induced systemic resistance (ISR) and SAR is another added advantage (Podile and Kishore 2006; Maheshwari 2011). PGPR inoculation containing single inoculant (strain) has been proved useful for enhancing plant growth and development even in salt-stress condition but inconsistency in their outcome lies due to various factors. This drawback might be due to adverse effects of biotic and abiotic factors and low-quality inoculum (Moënne-Loccoz et al. 1999). Under the circumstances more effort is to be made on multi-strain inoculum so as to gain agricultural sustainability (Maheshwari 2012).

5 Bioinoculants Characteristics

The bioinoculant products following quality according to the norms of the country have certain key effects to influence good agricultural practices. A variety of factors combine to build "quality" of a bioinoculant. In microbial consortia, individually each strain can compete effectively with the indigenous rhizosphere population and also enhance plant growth with its co-strain. The two-strain consortium showed not only successful competition for rhizospheric colonization but also promoted plant growth (Shenoy and Kalagudi 2003). The evaluation of growth and physiology of strains in consortium is an important aspect, which should not be overlooked. The growth rate affects the stability of artificial microbial ecosystem in the process of establishment of consortium. Estimation of variations in growth rate results revealed out-numbered fast-growing strain, whereas slow-growing strain becomes outcompete. This imbalance may affect the colonizing abilities and the potential of consortia bioinoculant. In one of our studies (author's lab), it was observed that growth of *Burkholderia* sp. MSSP was similar in monospecies and mixed species cultures with *Sinorhizobium meliloti* PP3. However, a 25 % increase in mean growth rate was recorded for *S. meliloti* PP3 when grown in a mixed species of two-species culture with respect to monoculture. Although, the cumulative effect of inoculants cannot be ignored. In the mixed culture, these produced certain PGP effects much earlier than that of the individual. It was observed that *Burkholderia* sp. MSSP and *S. meliloti* PP3 in consortia increased IAA production up to 50 % more in respect to monospecies culture, which further extends the finding of growth dependence between two species. There was no appreciable increment in maximum soluble P level but maximum P was released much earlier in two-species consortium. Further we observed that association with *Burkholderia* sp. favors *S. meliloti* as an adaptation of high rate of reproduction—a well-known evolved strategy that enables bacteria to survive successfully and maintain per se communities. In that phenomenon, bacteria attain homeostasis in which the population of indigenous bacterial population remained on the same order of magnitude throughout (Pandey et al. 2005).

Actually, formulation is one of the crucial steps that determines the success or failure of consortium formulation. The reason of failure of microbial formulation lies mainly in poor quality. In microbial formulation, the maintenance of metabolic and physiological states of bacterium is an important aspect for gaining maximum advantage.

To produce a bio-inoculant, potential microorganism should be introduced into a carrier. Substantial proportion of the inoculant produced using non-strile carrier is unsatisfactory for routine use in farming, either because of low population of effective strain or a large number of contaminants. But, the formulations made using sterile carriers are more expensive than those using non-sterile carriers. Carrier materials are generally intended to provide ecological niche protective in nature to microbial inoculants. Therefore, bioformulations should be composed of a superior carrier material with properties of high water-holding capacity, high water retention capacity, no heat production from wetting, nearly sterile, physically uniform, nontoxic in nature, easily biodegradable, nonpolluting, nearly neutral pH (or easily adjustable pH), that could support bacterial growth and survival over and above carrier should be such, that it could act as buffering agent.

From a microbiological point of view, inoculant formulations are expected to standardize to overcome few problems such as loss of viability during short storage in the grower's warehouse, long shelf life and stability over the range of 20–30 °C in the marketing tenure. Naturally, an inoculant should contain a level of bacteria sufficient to impart effect on plant growth and produce economic gain. However, several common methods of inoculant application are available, yet farmers are not always known to practice them. Hence, ease of application and layman procedure to use should imprint over packing. The new bacterial inoculants must meet these standards if they are to compete with chemicals on the farmer's list.

Regarding the commercialization of inoculant products, R&D laboratories should screen the efficient strains, optimized formulations, cost-effective production and good and practical inoculation techniques to launch a new product in the market. Microbial inoculants have long been incorporated into field practices worldwide, with satisfactory results. Prior to release in the market field, trial in two different climatic zones on two different crops for 2 subsequent years should be carried out. Deshwal et al. (2006) demonstrated long-term effects of *Pseudomonas aeruginosa* GRC1 on a yield of subsequent crops of paddy after mustard. In this study, *Rhizobium* and *Pseudomonas* survive in the rhizosphere for several years and cast its growth-promoting effects on subsequent crops. Research and limited field trials of PGPRs over the last decade have opened up new horizons for the inoculation industry.

For the development of successful bioformulation technology, progress must be made to meet numerous scientific challenges: (1) screening of potential strains having PGP properties, survival during seed coating/pelleting, soil drench application, seedling drop treatment, folier spray and storage at variable temperatures. (2) study of environmental stresses that negatively affect plant growth and development besides soil pH, nutritional deficiencies, salinity, high temperature and presence of toxic elements, (3) efficacy of microbial inoculants varies somewhat from site to site and year to year and this has to be considered and studied elaborately and, (4) understanding of interactions of strain with plant rhizosphere and rhizospheric microbal community (5) study of highly complex and dynamic rhizosphere environments to overcome practical problems such as the inconsistency in field performance (Arora et al. 2010).

Development of PGPR formulation with improved technology is a challenging task for popularization of their beneficial effects particularly in tropical countries where the cold supply chain does not exist. As stated, their maintenance in a metabolically and physiologically competent state remains the primary concern to derive the maximum advantage of a formulation.

The application of microbial inoculants in the form of granular or liquid form is also attaining much attention now a days due to easy application of liquid formulation has also achieved much popularity (Xavier et al. 2004). In bioformulations, bio-inoculants are generally prepared with either carriers (solid/liquid) or primed over seed surface using additives and fixers e.g. sugar syrup, carboxy-methyl-cellulose (CMC) and gum-arabic etc. On the other hand, some amendments indigenous in origin that improve chemical, biological, or physical properties of formulated biomass to prepare bio-formulations for routine farming on commercial level should also be explored.

6 Applications

The use of the consortium probably ensured a broader use of plant growth-promoting trait and also for efficient biocontrol mechanisms under the unpredictable field conditions (Pierson and Weller 1994). Role of bacterial consortium in the development of agriculture has its effectiveness and challenges.

Microbial studies executed that some mixtures of bacteria bear synergistic interactions which provision for nutrient availability, removal of inhibitory products and pathogens, and stimulate growth of each other through physical and biochemical activities that may have cumulative benefits for them and plant physiology. Rajasekar and Elango (2011) observed that PGPR consortia significantly increased plant height, root length and alkaloid content in *Withamia somnifera* when compared to the uninoculated control and single inoculation. Jha and Saraf (2012) observed that growth of Jatropha (*Jatropha curcas*) plant improved maximally when three strains were applied together. Concurrently, Ibiene et al. (2012) demonstrated that the use of microbial consortia containing diverse bacterial genera is an excellent inoculant for growth performance of plants and supported the fruitful application for agricultural benefits. Earlier, Annapurna et al. (2011) studied the effectiveness of PGPRs separately and in combination for reducing the impact of salinity on wheat growth. In that, inoculation with single and dual strains of PGPR strains showed variable effects to increase crop tolerance to salt concentrations. The consortium formulation was found more robust for inducing salinity tolerance in wheat to improve plant performance and development under stress environment. Thus, the inoculation of phytohormone-mediated PGPR is one of the essential and key elements (Bowen and Rovira 1999). An organism with properties like phytohormone production or any other beneficial factor with multifarious functions is thought to be an ideal bioinoculant (Catroux et al. 2001). On the other hand, such bacterial inoculants should possess the ability of multiplication and bear a broad spectrum of phytohormone production in order to substantiate overall growth promotion of plant.

7 Future Prospects and Strategies

The research has to be focused on the current concept of rhizo-engineering based on favorably partitioning of the biomolecules, which create a unique setting for the plant and microbes interactions. The rhizosphere biology will rely on the development of molecular and biotechnological approaches to strengthen our knowledge of rhizosphere biology and to achieve an integrated management of soil microbial populations. Fresh alternatives should be explored for the use of bioinoculants for other high-value crops such as vegetables, fruits and flowers. The application of multi-strain bacterial consortium over single inoculation could be an effective approach for reducing the adverse effect on plant growth.

8 Conclusion

From a truthful perspective, one must receive that, in the predictable future, only a gradual and modest increase in the use of bacterial inoculants is yet to be expected. Nevertheless, special attention should be paid to the need-based bioinoculants which are easy-to-use and inexpensive. For the future perspectives, more research should be emphasized on the development of improved and extra-economical viable, natural bio-inoculant, that could sustain production for their judicious application in crop ecosystems.

Acknowledgements We express our thanks to UCOST, Dehradun, India and University Grant Commission (UGC), New Delhi, India for providing financial support in the form of research project to DKM.

References

Abdul Jaleel C, Manivannan P, Sankar B, Kishorekumar A, Gopi R, Somasundaram R, Panneerselvam R (2007) *Pseudomonas fluorescens* enhances biomass yield and ajmalicine production in *Catharanthus roseus* under water defi cit stress. Colloids Surf B Biointerfaces 60:7–11

Aeron A, Pandey P, Maheshwari DK (2010) Differential response of sesame under influence of indigenous and non-indigenous rhizospheric competent fluorescent Pseudomonads. Curr Sci 99(2):166–168

Ahmad M, Zahir ZA, Asghar HN, Arshad M (2012) The combined application of rhizobial strains and plant growth promoting rhizobacteria improves growth and productivity of mung bean (*Vigna radiata* L.) under salt-stressed conditions. Ann Microbiol 62(3):1321–1330

Ali B, Sabri AN, Hasnain S (2010) Rhizobacterial potential to alter auxin content and growth of *Vigna radiata* (L.). World J Microbiol Biotechnol 26:1379–1384

Ali B, Sabri AN, Ljung K, Hasnain S (2009) Auxin production by plant associated bacteria: impact on endogenous IAA content and growth of *Triticum aestivum* L. Lett Appl Microbiol 48(5):542–547

Annapurna K, Ramadoss D, Vithal L, Bose P, Reddy MS, Wang Q, Yellareddygari SKR (2011) PGPR bioinoculants for ameliorating biotic and abiotic stresses in crop production. In: Proceedings of the 2nd Asian PGPR conference Plant growth-promoting rhizobacteria (PGPR) for sustainable agriculture. Beijing, China, 21–24 August, 2011 Asian PGPR Society, pp 67–72

Apine OA, Jadhav JP (2011) Optimization of medium for indole-3-acetic acid production using *Pantoea agglomerans* strain PVM. J Appl Microbiol 110:1235–1244

Arkhipova TN, Prinsen E, Veselov SU, Martinenko EV, Melentiev AI, Kudoyarova GR (2007) Cytokinin producing bacteria enhance plant growth in drying soil. Plant Soil 292(1–2):305–315

Arora NK, Khare E, Maheshwari DK (2010) Plant growth promoting rhizobacteia: constraints in bioformulation, commercialization and future strategy. In: Maheshwari DK (ed) Plant Growth and Health Promoting Bacteria. Microbiology Monographs, Heidelberg, Germany, pp 99–116

Arshad M, Frankenberger WT (1997) Plant growth-regulating substances in the rhizosphere: microbial production and functions. Adv Agro 62:45–151

Arshad M, Shaharoona B, Mahmood T (2008) Inoculation with *Pseudomonas* spp. containing ACC-deaminase partially eliminates the effects of drought stress on growth, yield, and ripening of pea (*Pisum sativum* L.). Pedosphere 18(5):611–620

Asghar H, Zahir Z, Arshad M, Khaliq A (2002) Relationship between in vitro production of auxins by rhizobacteria and their growth-promoting activities in *Brassica juncea* L. Biol Fertil Soil 35(4):231–237

Aslantas R, Cakmakci R, Sahin F (2007) Effect of plant growth promoting rhizobacteria on young apple tree growth and fruit yield under orchard conditions. Sci Hortic 111:371–377

Barea JM, Azcón R, Azcón-Aguilar C (2004) Mycorrhizal fungi and plant growth promoting rhizobacteria. In: Varma A et al. (eds.) Plant surface microbiology, Springer, Heidelberg, 351–371

Barua S, Tripathi S, Chakraborty A, Ghosh S, Chakrabarti K (2012) Characterization and crop production efficiency of diazotrophic bacterial isolates from coastal saline soils. Microbiol Res 167:95–102

Bashan Y, de-Bashan LE, Prabhu SR, Hernandez JP (2014) Advances in plant growth-promoting bacterial inoculant technology: formulations and practical perspectives (1998–2013). Plant Soil 378(1–2):1–33

Beneduzia A, Peresa D, Vargasb LK, Bodanese-Zanettinia MH, Passaglia LMP (2008) Evaluation of genetic diversity and plant growth promoting activities of nitrogen-fixing bacilli isolated from rice fields in South Brazil. Appl Soil Ecol 39:311–320

Bleecker AB, Kende H (2000) Ethylene: a gaseous signal molecule in plants. Ann Rev Cell Dev Biol 16(1):1–18

Bottini R, Cassán F, Piccoli P (2004) Gibberellin production by bacteria and its involvement in plant growth promotion and yield increase. Appl Microbiol Biotechnol 65(5):497–503

Bowen GD, Pierson AD (1999) The rhizosphere and its management to improve plant growth. Adv Agron 66:1–102

Broekaert WF, Delauré SL, De Bolle MF, Cammue BP (2006) The role of ethylene in host-pathogen interactions. Annu Rev Phytopathol 44:393–416

Catroux G, Hartmann A, Revellin C (2001) Trends in rhizobial inoculant production and use. Plant Soil 230(1):21–30

Chaiharn M, Lumyong S (2011) Screening and optimization of indole-3-acetic acid production and phosphate solubilization from rhizobacteria aimed at improving plant growth. Curr Microbiol 62:173–181

Dastager SG, Deepa CK, Pandey A (2011) Growth enhancement of black pepper (*Piper nigrum)* by a newly isolated *Bacillus tequilensis* NII-0943. Biologia 66:801–806

Davies PJ (2010) The plant hormones: their nature, occurrence, and functions. In: Davis PJ (ed) plant hormones. Springer, Netherlands, pp 1–15

de Garcia Salamone IE, Hynes RK, Nelson LM (2006) Role of cytokinins in plant growth promotion by rhizosphere bacteria. PGPR: biocontrol and biofertilization. Springer, Netherlands, pp 173–195

Deshwal VK, Kumar T, Dubey RC, Maheshwari DK (2006) Long term effect of *Pseudomonas aeruginosa* GRC_1 on yield of subsequent crops of paddy after mustard seed bacterization. Curr Sci 91:423–424

Erturk Y, Ercisli S, Cakmakci R (2012) Yield and growth response of strawberry to plant growth-promoting rhizobacteria inoculation. J Plant Nutr 35:817–826

Farina R, Beneduzi A, Ambrosini A, de Campos SB, Lisboa BB, Wendisch V, Vargas LK, Passaglia LMP (2012) Diversity of plant growth-promoting rhizobacteria communities associated with the stages of canola growth. Appl Soil Ecol 55:44–52

Fasciglione G, Casanovas EM, Yommi A, Sueldo RJ, Barassi CA (2012) *Azospirillum* improves lettuce growth and transplant under saline conditions. J Sci Food Agric 92:2518–2523
Frankenberger Jr WT, Arshad M (1995) Phytohormones in soils: microbial production and function. Marcel Dekker Inc
Frugier F, Kosuta S, Murray JD, Crespi M, Szczyglowski K (2008) Cytokinin: secret agent of symbiosis. Trends Plant Sci 13(3):115–120
Fulchieri M, Lucangeli C, Bottini R (1993) Inoculation with *Azospirillum lipoferum* affects growth and gibberellin status of corn seedling roots. Plant Cell Physiol 34(8):1305–1309
Glick BR (1995) The enhancement of plant growth by free-living bacteria. Can J Microbiol 41(2):109–117
Glick BR (2015) Beneficial Plant-bacterial Interactions. Springer
Glick BR, Cheng Z, Czarny J, Duan J (2007) Promotion of plant growth by ACC deaminase-producing soil bacteria. In: Bakker et al. (eds.) New perspectives and approaches in plant growth-promoting rhizobacteria research, Springer Netherlands, pp 329–339
Glick BR, Penrose DM, Li J (1998) A model for the lowering of plant ethylene concentrations by plant growth-promoting bacteria. J Theoret Biol 190(1):63–68
Grunewald W, Van Noorden G, Van Isterdael G, Beeckman T, Gheysen G, Mathesius U (2009) Manipulation of auxin transport in plant roots during Rhizobium symbiosis and nematode parasitism. Plant Cell 21(9):2553–2562
Guo J, Tang S, Ju X, Ding Y, Liao S, Song N (2011) Effects of inoculation of a plant growth promoting rhizobacterium *Burkholderia* sp. D54 on plant growth and metal uptake by a hyperaccumulator *Sedum alfredii* Hance grown on multiple metal contaminated soil. World J Microbiol Biotechnol 27:2835–2844
Gupta G, Parihar SS, Ahirwar NK, Snehi SK, Singh V (2015) Plant growth promoting rhizobacteria (PGPR): current and future prospects for development of sustainable agriculture. J Microb Biochem Technol 7:096–102
Gutiérrez-Mañero FJ, Ramos-Solano B, Probanza A, Mehouachi JR, Tadeo F, Talon M (2001) The plant-growth-promoting rhizobacteria Bacillus pumilus and Bacillus licheniformis produce high amounts of physiologically active gibberellins. Physiol Plant 111(2):206–211
Hartmann A, Singh M, Klingmüller W (1983) Isolation and characterization of *Azospirillum* mutants excreting high amounts of indoleacetic acid. Canadian J Microbiol 29(8):916–923
Hayat R, Ali S, Amara U, Khalid R, Ahmed I (2010) Soil beneficial bacteria and their role in plant growth promotion: a review. Ann Microbiol 60(4):579–598
Hussain A, Hassain S (2011) Interactions of bacterial cytokinins and IAA in the rhizosphere may alter phytostimulatory efficiency of rhizobacteria. World J Microbiol Biotechnol 27:2645–2654
Ibiene AA, Agogbua JU, Okonko IO, Nwachi GN (2012) Plant growth promoting rhizobacteria (PGPR) as biofertilizer: Effect on growth of *Lycopersicum esculentus*. J Am Sci 8(2):318–324
Ishibashi Y, Koda Y, Zheng SH, Yuasa T, Iwaya-Inoue M (2012) Regulation of soybean seed germination through ethylene production in response to reactive oxygen species. Ann Bot mcs 240
Jha CK, Saraf M (2012) Evaluation of multispecies plant-growth-promoting consortia for the growth promotion of *Jatropha curcas* L. J Plant Growth Regul 31(4):588–598
Khalid A, Arshad M, Zahir ZA (2004) Screening plant growth-promoting rhizobacteria for improving growth and yield of wheat. J Appl Microbiol 96(3):473–480
Kumar H, Bajpai VK, Dubey RC, Maheshwari DK, Kang SC (2010) Wilt disease management and enhancement of growth and yield of *Cajanus cajan* (L) var. Manak by bacterial combinations amended with chemical fertilizer. Crop Protec 29(6):591–598
Lambrecht M, Okon Y, Broek AV, Vanderleyden J (2000) Indole-3-acetic acid: a reciprocal signalling molecule in bacteria–plant interactions. Trends Microbiol 8(7):298–300
Lenin G, Jayanthi M (2012) Efficiency of plant growth promoting rhizobacteria (PGPR) on enhancement of growth, yield and nutrient content of *Catharanthus roseus*. Int J Res Pure Appl Microbiol 2(4):37–42

Li K, Ramakrishna W (2011) Effect of multiple metal resistant bacteria from contaminated lake sediments on metal accumulation and plant growth. J Hazard Mater 189:531–539

Liu F, Xing S, Ma H, Du Z, Ma B (2013) Cytokinin-producing, plant growth-promoting rhizobacteria that confer resistance to drought stress in *Platycladus orientalis* container seedlings. Appl Microbial Biotechnol 97(20):9155–9164

Lucy M, Reed E, Glick BR (2004) Applications of free living plant growth-promoting rhizobacteria. Antonie Van Leeuwenhoek 86(1):1–25

Luo S, Xu T, Chen L, Chen J, Rao C, Xiao X, Wan Y, Zeng G, Long F, Liu C, Liu Y (2012) Endophyteassisted promotion of biomass production and metal-uptake of energy crop sweet sorghum by plantgrowth-promoting endophyte *Bacillus* sp. SLS18. Appl Microbiol Biotechnol 93:1745–1753

Ma Y, Rajkumar M, Freitas H (2009) Inoculation of plant growth promoting bacterium *Achromobacter xylosoxidans* strain Ax10 for the improvement of copper phytoextraction by *Brassica juncea*. J Environ Manage 90:831–837

Maheshwari DK (2011) Bacteria in Agrobiology: Plant Growth Responses. Springer, Heidelberg, Germany, p 377

Maheshwari DK (2012) Bacteria in Agrobiology: Plant Probiotics. Springer, Heidelberg, Germany, p 375

Maheshwari DK, Dubey RC, Agarwal M, Dheeman S, Aeron A, Bajpai VK (2015) Carrier based formulations of biocoenotic consortia of disease suppressive *Pseudomonas aeruginosa* KRP1 and *Bacillus licheniformis* KRB1. Ecol Eng 81:272–277

Malhotra M, Srivastava S (2009) Stress-responsive indole-3-acetic acid biosynthesis by *Azospirillum brasilense* SM and its ability to modulate plant growth. Eur J Soil Biol 45:73–80

Marulanda A, Barea JM, Azcon R (2009) Stimulation of plant growth and drought tolerance by native microorganisms (am fungi and bacteria) from dry environments: mechanisms related to bacterial effectiveness. J Plant Growth Regul 28:115–124

Melnykova N, Gryshchuk O, Mykhalkiv L, Mamenko P, Sergii KOTS (2013) Plant growth promoting properties of bacteria isolated from the rhizosphere of soybean and pea. Natura Montenegrina 12(3–4):915–923

Mishra PK, Bisht SC, Ruwari P, Selvakumar G, Joshi GK, Bisht JK, Bhatt JC, Gupta HS (2011) Alleviation of cold stress in inoculated wheat (*Triticum aestivum* L.) seedlings with psychrotolerant Pseudomonads from NW Himalayas. Arch Microbiol 193:497–513

Moënne-Loccoz Y, Naughton M, Higgins P, Powell J, O'Connor B, O'Gara F (1999) Effect of inoculum preparation and formulation on survival and biocontrol efficacy of *Pseudomonas fluorescens* F113. J Applied Microbiol 86(1):108–116

Morrone D, Chambers J, Lowry L, Kim G, Anterola A, Bender K, Peters RJ (2009) Gibberellin biosynthesis in bacteria: separate ent-copalyl diphosphate and ent-kaurene synthases in *Bradyrhizobium japonicum*. FEBS Lett 583(2):475–480

Moutia JY, Spaepen SSS, Vanderleyden J (2010) Plant growth promotion by *Azospirillum* sp. in sugarcane is infl uenced by genotype and drought stress. Plant Soil 337:233–242

Oberhänsli T, Défago G, Haas D (1991) Indole-3-acetic acid (IAA) synthesis in the biocontrol strain CHA0 of *Pseudomonas fluorescens*: role of tryptophan side chain oxidase. J Gen Microbiol 137(10):2273–2279

Pandey P, Kang SC, Maheshwari DK (2005) Isolation of endophytic plant growth-promoting *Burkholderia* sp. MSSP from root nodules of *Mimosa pudica*. Curr Sci 89:(1)177–180

Pandey P, Maheshwari DK (2007a) Bioformulation of *Burkholderia* sp. MSSP with a multi-species consortium for growth promotion of *Cajanus cajan*. Can J Microbiol 53:213–222

Pandey P, Maheshwari DK (2007b) Two-species microbial consortium for growth promotion of *Cajanus cajan*. Curr Sci 92:1137–1142

Patel D, Jha CK, Tank N, Saraf M (2012) Growth enhancement of chickpea in saline soils using plant growth-promoting rhizobacteria. J Plant Growth Regul 31:53–62

Patten CL, Glick BR (1996) Bacterial biosynthesis of indole-3-acetic acid. Can J Microbiol 42(3):207–220

Perley JE, Stowe BB (1966) The production of tryptamine from tryptophan by *Bacillus cereus* (KVT). Biochem J 100:169–174

Pierson EA, Weller DM (1994) Use of mixtures of fluorescent pseudomonads to suppress take-all and improve the growth of wheat. Phytopathol 84:940–947

Podile AR, Kishore GK (2006) Plant growth-promoting rhizobacteria. In: Gnanamanickam SS (ed) Plant-associated bacteria. Springer, Netherlands, pp 195–230

Prasad R, Kumar M, Varma A (2015) Role of PGPR in soil fertility and plant health. In: Egamberdieva et al. (eds.) Plant-Growth-Promoting Rhizobacteria (PGPR) and Medicinal Plants, Springer International Publishing, pp 247–260

Prinsen E, Costacurta A, Michiels K, Vanderleyden J, Van Onckelen H (1993) *Azospirillum brasilense* indole-3-acetic acid biosynthesis: evidence for a non-tryptophan dependent pathway. MPMI 6:609

Rajasekar S, Elango R (2011) Effect of microbial consortium on plant growth and improvement of alkaloid content in *Withania somnifera* (Ashwagandha). Curr Bot 2(8)

Reetha S, Bhuvaneswari G, Thamizhiniyan P, Mycin TR (2014) Isolation of indole acetic acid (IAA) producing rhizobacteria of *Pseudomonas fluorescens* and *Bacillus subtilis* and enhance growth of onion (*Allim cepa.* L). Int J Curr Microbiol Appl Sci 3(2):568–574

Rojas-Tapias D, Moreno-Galván A, Pardo-Díaz S, Obando M, Rivera D, Bonilla R (2012) Effect of inoculation with plant growth-promoting bacteria (PGPB) on amelioration of saline stress in maize (*Zea mays*). Appl Soil Ecol 61:264–272

Roy M, Saha S, Das J, Srivastava RC (2015) Technologies of microbial inoculation in rice-A Review. Agric Rev 36(2):125–132

Ryu CM, Farag MA, Hu CH, Reddy MS, Wei HX, Paré PW, Kloepper JW (2003) Bacterial volatiles promote growth in Arabidopsis. PNAS 100(8):4927–4932

Sadeghi A, Karimi E, Dahaji PA, Javid MG, Dalvand Y, Askari H (2012) Plant growth promoting activity of an auxin and siderophore producing isolate of *Streptomyces* under saline soil conditions. World J Microbiol Biotechnol 28:1503–1509

Sagar S, Dwivedi A, Yadav S, Tripathi M, Kaistha SD (2012) Hexavalent chromium reduction and plant growth promotion by *Staphylococcus arlettae* strain Cr11. Chemosphere 86:847–852

Sarwar M, Kremer RJ (1995) Enhanced suppression of plant growth through production of L-tryptophan-derived compounds by deleterious rhizobacteria. Plant Soil 172(2):261–269

Selvakumar G, Joshi P, Suyal P, Mishra PK, Joshi GK, Bisht JK, Bhatt JC, Gupta HS (2011) *Pseudomonas lurida* M2RH3 (MTCC 9245), a psychrotolerant bacterium from the Uttarakhand Himalayas, solubilizes phosphate and promotes wheat seedling growth. World J Microbiol Biotechnol 27:1129–1135

Selvakumar G, Kundu S, Joshi P, Nazim S, Gupta AD, Gupta HS (2010) Growth promotion of wheat seedlings by *Exiguobacterium acetylicum* 1P (MTCC 8707) a cold tolerant bacterial strain from the Uttarakhand Himalayas. Ind J Microbiol 50:50–56

Selvakumar G, Mohan M, Kundu S, Gupta AD, Joshi P, Nazim S, Gupta HS (2007) Cold tolerance and plant growth promotion potential of *Serratia marcescens* strain SRM (MTCC 8708) isolated from flowers of summer squash (*Cucurbita pepo*). Lett Appl Microbiol 46:171–175

Shankar M, Ponraj P, Ilakkiam D, Gunasekaran P (2011) Root colonization of a rice growth promoting strain of *Enterobacter cloacae*. J Basic Microbiol 51:523–530

Shenoy VV, Kalagudi GM (2003) Meta-bug and near-isogenic strain consortia concepts for plant growth promoting rhizobacteria. In: 6th International PGPR workshop, India, section VII—mechanism of biocontrol. p 108

Shi Y, Lou K, Li C (2011) Growth promotion effects of the endophyte *Acinetobacter johnsonii* strain 3–1 on sugar beet. Symbiosis 54:159–166

Shilev S, Sancho ED, Benlloch-González M (2012) Rhizospheric bacteria alleviate salt-produced stress in sunfl ower. J Environ Manage 95:S37–S41

Spaepen S, Vanderleyden J, Remans R (2007) Indole-3-acetic acid in microbial and microorganism-plant signaling. FEMS Microbiol Rev 31(4):425–448

Tan S, Jiang Y, Song S, Huang J, Ling N, Xu Y, Shen Q (2013) Two *Bacillus amyloliquefaciens* strains isolated using the competitive tomato root enrichment method and their effects on suppressing *Ralstonia solanacearum* and promoting tomato plant growth. Crop Prot 43:134–140

Thakkar A, Saraf M (2014) Development of microbial consortia as a biocontrol agent for effective management of fungal diseases in *Glycine max* L. Arch Phytopathol Plant Protec 1–16

Tiwari S, Singh P, Tiwari R, Meena KK, Yandigeri M, Singh DP, Arora DK (2010) Salt-tolerant rhizobacteria-mediated induced tolerance in wheat (*Triticum aestivum)* and chemical diversity in rhizosphere enhance plant growth. Biol Fertil Soils 47:907–916

Valverde C, Ramírez C, Kloepper JW, Cassán F (2014) Current research on plant-growth promoting rhizobacteria in Latin America: Meeting report from the 2nd Latin American PGPR Workshop. J Plant Growth Regul 34(1):215–219

Verma VC, Singh SK, Prakash S (2011) Bio-control and plant growth promotion potential of siderophore producing endophytic *Streptomyces* from *Azadirachta indica* A. Juss. J Basic Microbiol 51:550–556

Werner T, Motyka V, Strnad M, Schmülling T (2001) Regulation of plant growth by cytokinin. PNAS 98(18):10487–10492

Wu Z, Yue H, Lu J, Li C (2012) Characterization of rhizobacterial strain Rs-2 with ACC deaminase activity and its performance in promoting cotton growth under salinity stress. World J Microbiol Biotechnol 28:2383–2393

Xavier IJ, Holloway G, Leggett M (2004) Development of rhizobial inoculant formulations. Crop Manag 3(1) doi:10.1094/CM-2004-0301-06-RV

Yang R, Luo C, Chen Y, Wang G, Xu Y, Shen Z (2013) Copper-resistant bacteria enhance plant growth and copper phytoextraction. Int J Phytoremediation 15:573–584

Yao L, Wu Z, Zheng Y, Kaleem I, Li C (2010) Growth promotion and protection against salt stress by *Pseudomonas putida* Rs-198 on cotton. Eur J Soil Biol 46:49–54

Zafar M, Abbasi MK, Khan MA, Khaliq A, Sultan T, Aslam M (2012) Effect of plant growth promoting rhizobacteria on growth, nodulation and nutrient accumulation of lentil under controlled conditions. Pedosphere 22:848–859

Zahir ZA, Arshad M, Azam M, Hussain A (1997) Effect of an auxin precursor tryptophan and Azotobacter inoculation on yield and chemical composition of potato under fertilized conditions. J Plant Nut 20(6):745–752

Zahir ZA, Arshad M, Frankenberger WT (2003) Plant growth promoting rhizobacteria: applications and perspectives in agriculture. Adv Agro 81:97–168

Zahir ZA, Asghar HN, Akhtar MJ, Arshad M (2005) Precursor (L-tryptophan)-inoculum (Azotobacter) interaction for improving yields and nitrogen uptake of maize. J Plant Nut 28(5):805–817

Zhao Q, Shen Q, Ran W, Xiao T, Xu D, Xu Y (2011) Inoculation of soil by *Bacillus subtilis* Y-IVI improves plant growth and colonization of the rhizosphere and interior tissues of muskmelon (*Cucumis melo* L.). Biol Fertil Soils 47:507–514

Zhao L, Teng S, Liu Y (2012) Characterization of a versatile rhizospheric organism from cucumber identifi ed as *Ochrobactrum haematophilum.* J Basic Microbiol 52:232–244

Significance of Biosurfactants as Antibiofilm Agents in Eradicating Phytopathogens

Alwar Ramanujam Padmavathi, Dhamodharan Bakkiyaraj and Shunmugiah Karutha Pandian

Abstract Biofilms are complex aggregation of microbial cells which aid the indwelling cells to survive and flourish well in the hostile environments. Besides, biofilms act as multicellular entities and provide resistance toward antibiotics and other bactericidal agents which makes their eradication cumbersome. Biofilm-related infections are tough to treat in healthcare and they are equally important in agriculture as they afflict the crop survival and productivity. It is indeed important to develop a biofilm control strategy to combat biofilm-related infections in agriculture. In recent years, biosurfactants have been exploited as potential antibiofilm candidates to languish the vigor of biofilm formers by selectively eradicating the biofilms. Biosurfactants are the surface active metabolites produced by microbes and are proven to have multifarious role in many fields right from bioremediation to biomedical applications. Biosurfactants due to their surface modifying property, modulate the biofilm forming ability of pathogens which directly prevents microbial colonization and biofilm formation. This chapter summarizes the importance of antibiofilm agents and the role of biosurfactants in eradicating biofilms formed by disease causing pathogens.

Keywords Phytopathogens · Antibiofilm · Biosurfactant · Colonization · PGPR

1 Introduction

Microorganisms are ubiquitous in nature, and they play decisive roles in any functional ecosystem. Microbes are broadly classified into two distinct categories as beneficial and harmful, based on their association with plants, animals and other

A.R. Padmavathi · D. Bakkiyaraj · S.K. Pandian (✉)
Department of Biotechnology, Alagappa University, Science Campus,
Karaikudi 630004, TN, India
e-mail: sk_pandian@rediffmail.com

D.K. Maheshwari (ed.), *Bacterial Metabolites in Sustainable Agroecosystem*,
Sustainable Development and Biodiversity 12, DOI 10.1007/978-3-319-24654-3_12

organisms. Understanding the relationship between plant and their associated microbial community is indispensable, as they have explicit impact on plant productivity. Pathogenic microorganisms are gaining importance worldwide than beneficial microbes due to their direct influence in loss of plant productivity. Bacteria and fungi occupy prime positions among other pathogens because of their ability to infect and invade diverse hosts and their ability to thrive under hostile environments. Microbicidal agents which can kill these pathogens by targeting their essential cellular components and machineries were employed over years to control these detrimental organisms (Mukhtar and Wright 2005; Drlica et al. 2008; Goo and Sim 2010; Maheshwari 2013). In due course, the disease causing pathogens developed resistance against these antimicrobials by modifying their cell membrane permeability and target receptors, by producing inactivating and modifying enzymes, by activating the efflux pumps, or by forming biofilms (Ozanne et al. 1969; Tsang et al. 1975; Leclercq and Courvalin 1991; Gotoh 2001; Kumar and Worobec 2002; Davies 2003). Among the mechanisms enlisted above, biofilm formation is of great concern as it contributes to nonspecific and broad spectrum resistance toward antibiotics and also against metal ions, biocidal agents, protozoan grazers, desiccation and other hostile environmental conditions (Flemming and Wingender 2010).

Biofilm is defined as a complex microbial aggregate embedded in a hydrated extracellular polymeric substance (EPS) matrix (Stewart and Costerton 2001). Growth of microbial aggregates attached to any solid substratum is the characteristic nature of biofilms, though free flowing microbial aggregates are also reported recently, which usually forms biofilm on air-liquid interfaces. Biofilms render resistance to the harboring microbial cells and they are several times resistant to antimicrobials than their free living planktonic counterparts (Mah 2012). Biofilm formation causes severe complications in healthcare, agriculture and industries which results in recurrent infections, plant diseases and biocorrosion respectively (Busalmen et al. 2002; Von Bodman et al. 2003; Aparna and Yadav 2008). Antibiotics and other bactericidal agents became ineffective in controlling the biofilm formers and the subsequent damages caused by them. Biofilm acts as a protective barrier that prevents the penetration of drug molecules, and also it secretes various inactivating and modifying enzymes which nullify the effects exerted by antimicrobials (Flemming and Wingender 2010). To combat biofilm-related problems in any industry including agriculture, antibiofilm agents are the wisest choice.

Many natural, synthetic and semi-synthetic compounds were explored in the quest of antibiofilm agents in which biosurfactants gain importance as potential antibiofilm agents due to their amenability in extreme environments. Biosurfactants are amphipathic surface active molecules produced by living cells, capable of reducing the surface tensions of liquid. These have profound industrial, environmental, and biomedical applications. Anti-adhesive property of biosurfactants has been explored for antibiofilm applications and found to be successful. The components and complexity of biofilms, application of biosurfactants as antibiofilm agents will be discussed in detail in this chapter.

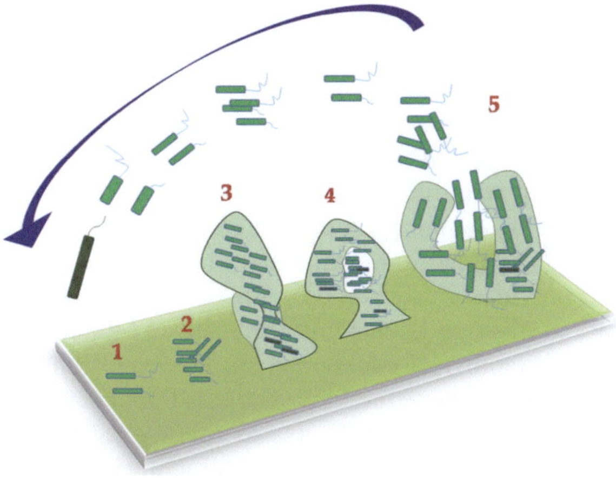

Fig. 1 Developmental stages of biofilm formation. *1* Initial adhesion, *2* irreversible attachment, *3* microcolony formation, *4* maturation, *5* dispersion

2 Biofilm—Intrinsic Guardian of Disease Causing Microbes

Biofilms are complex aggregation of microbial cells encased in EPS matrix which provide protection to the microbial inhabitants from unfavorable environmental conditions. Biofilm formers are differentiated from their free living planktonic counterparts by the secretion of EPS (Donlan 2002) and their collective behavior as a multicellular living system rather than single cell entities. Extensive research on biofilms has exposed five essential steps for biofilm formation: (i) Reversible adsorption, (ii) Irreversible attachment, (iii) Microcolony formation, (iv) Maturation and (v) Dispersion (Fig. 1) (Sauer et al. 2002; Stoodley et al. 2002). Initial adhesion of microbial cells to biotic/abiotic surfaces is the critical prerequisite for the successful formation of a sessile biofilm community. Attached cells secrete EPS to stabilize the irreversible attachment which in turn develop to microcolonies. Biofilms formed on surfaces have a characteristic structure of microcolonies enclosed with EPS.

3 Understanding the Biofilm Architecture

Biofilms have distinct,complex three-dimensional (3D) architecture which selectively prevent the penetration of microbicidal agents and have specific channels for the inflow and exchange of nutrients and water (Donlan 2002; Wilking et al. 2013). The biotic/abiotic solid surface acts as substratum for biofilm formation (Fletcher and Loeb 1979) and the EPS provides mechanical stability, 3D architecture, and physical integrity to biofilms thereby transforming the biofilm formers the most successful forms of life on earth (Flemming and Wingender 2010). The chemical and physical properties of EPS may vary from species to species, but invariably EPS is composed of polysaccharides where some are neutral and some are polyanionic, apart from proteins, lipids and nucleic acids (DNA) are present in trace levels (Whitchurch et al. 2002; Branda et al. 2005; Lasa and Penadés 2006). The presence of uronic acids (D-glucuronic, D-galactouronic and mannuronic acids) or ketal-linked pyruvate confers the anionic properties to the polysaccharides. This anionic nature helps in the association of divalent cations such as calcium and magnesium, which have been shown to interconnect the polymer strands and provides greater binding force in a developed biofilm (O'Toole et al. 2000). EPS production is known to be affected by nutrient status of the growth medium and availability of excess carbon; however, limitation of nitrogen, potassium and phosphate promote the EPS synthesis (Donlan and Costerton 2002).

Biofilms can be of diverse morphological forms from smooth to rough, fluffy or filamentous, flat to raised 3D structures and fruiting bodies (Sutherland 2001; Flemming and Wingender 2010). Alginate in *Pseudomonas aeruginosa* EPS provides notable biofilm architecture and contributes to mucoid phenotype (Franklin and Ohman 1993; Wozniak et al. 2003). Acetyl group substituent of EPS increases the adhesive and cohesive properties which contribute to altered biofilm architecture (Tielen et al. 2005). Studies with mutant strains showed that various factors like hydrodynamic environment, concentration of nutrients, bacterial motility and cell to cell communication between the species, exopolysaccharides and proteins grossly influence the biofilm architecture. Individual EPS components also play major role in biofilm architecture, lack of which leads to impaired biofilm phenotype (Watnick and Kolter 1999; Danese et al. 2000). Role of other abiotic and biotic factors on biofilm formation cannot be ruled out.

Pili, fimbriae and other flagellar structures also help in stabilizing the structure and integrity of biofilms (Zogaj et al. 2001). The term biofilm is often referred to a heterogeneous structure which is formed in response to environmental cues. It contains cells of different physiological states with different genotypes and phenotypes. Oxygen, chemical, nutrient and temperature gradients exist in biofilm and the conditions vary inside this microenvironment (Huang et al. 1995; Schramm et al. 1996). A group of specialized cells known as persisters are present within this biofilm microenvironment which are metabolically inactive, slow growing and extremely resistant to antibiotics which regain their natural growth once the conditions turn favorable (Barth et al. 2014; Butt et al. 2014; Cho et al. 2014). These persisters are the actual reason for the recurrent infections and therapeutic failures (Jubair et al. 2012; Knudsen et al. 2013; Conlon 2014).

4 Molecular Insights on Biofilm Formation

Although the formation of biofilms includes five direct stages, the mechanism behind this is complex and controlled by more than one pathway in most instances. Despite the fact that the roles of individual genes associated with

Table 1 Genes involved in biofilm formation and regulation

Pathogen	Gene(s) involved	References
Bacillus subtilis	*yoaW, sipW* and *kinD* (sensor histidine kinase)	Hamon et al. (2004), Chen et al. (2012a)
Bordetella pertussis	*bpsABCD* operon	Sloan et al. (2007)
Burkholderia cepacia	*cep* quorum sensing system	Huber et al. (2001)
Campylobacter jejuni	*fliS (flagellar protein coding gene), flaA* and *flaB* (genes coding for flagellins), *fliD* (gene coding the filament cap), *flgG* and *flgG2* (genes coding the basal body) and *cheA* (gene coding chemotactic protein)	Joshua et al. (2006), Kalmokoff et al. (2006)
Candida albicans	*bcr1, efg1, ndt80, rob1, tec1, brg1, flo8, gal4* and *rfx2*	Fox et al. (2015)
Escherichia coli	*cyaA* and *crp* (genes for flagella), *lpcA, rfaD* and *rfaE* (genes involved in lipopolysaccharide biosynthesis), *fim* genes (Type 1 fimbriae biosynthesis), *bola* (stress response protein)	Niba et al. (2007), Dressaire et al. (2015)
Klebsiella pneumoniae	*treC* (gene involved in capsular polysaccharide)	Wu et al. (2011)
P. aeruginosa	*pslA-pslO genes* (Polysaccharide synthesis locus), *pelA-pelF* (biosynthesis of Pel polysaccharide), *pilA, pilB, pilD, pilQ, pilT* (Type 4 pili biosynthesis gene), *las* and *rhl* quorum sensing systems	Farinha et al. (1994), Heydorn et al. (2002), Chiang and Burrows (2003), Wei and Ma (2013)
Pantoea stewartii	*esaI/R quorum sensing system*	Koutsoudis et al. (2006)
S. aureus, S. epidermidis	*icaADBC* operon, *bap*	Cramton et al. (1999), Mack et al. (2000), Valle et al. (2012)
S. marcescens	*fimA, fimC, flhD* and *bsmA* (genes involved in biosynthesis of fimbriae)	Padmavathi et al. (2014)
Streptococcus mutans	*gtfBCD, ftf* and *gbpB* (genes involved in EPS biosynthesis), *vicRKX* (two-component regulatory system)	Shemesh et al. (2007)
V. parahaemolyticus	*ScrABC* system	Srivastava and Waters (2012)
V. cholerae	*vps* (genes coding for Vibrio polysaccharides) and *rbm* (genes coding for the protein components of biofilm) operon	Ayala et al. (2015)
Xanthomonas campestris	*vemR, rsmA, gumB-gumM*	Crossman and Dow (2004), Tao and He (2010), Lu et al. (2012)
Yersinia pseudotuberculosis, Y. pestis	*ypsR/I, ytbR/I, hmsHFRS* operon	Jarrett et al. (2004), Atkinson et al. (2006)

biofilm are shown by many studies, complete knowledge about the molecular mechanism behind biofilm formation is available as of date for few bacteria like *Staphylococcus aureus*, *P. aeruginosa* and *Serratia marcescens*. Studies have even shown the interlink between the quorum sensing regulatory circuit and biofilm formation, where the biofilms are modulated by the population density in pathogenic bacteria like *P. aeruginosa, S. marcescens, Vibrio harveyi, V. parahaemolyticus* and *S. aureus* (Stickler et al. 1998; Hardie and Heurlier 2008; Nadell et al. 2008). Brief description of genes involved in biofilm formation in different pathogenic microbes is given in the Table 1.

5 Pathogenic Biofilms in Agriculture

Plant systems possess vast area which can be readily colonized by disease causing bacteria or fungi. Most of these pathogens occupy the plant surfaces in the form of biofilms. Plants provide many microenvironments where most pathogens colonize heterogeneously and form biofilms. Biofilm mode of growth favors these pathogens by rendering resistance toward antibiotics, biocides, toxins, metals, heat, UV, acids and enzymes (Singh et al. 2000; Leid et al. 2002; Hall-Stoodley et al. 2004; Leid et al. 2005; Kolter and Greenberg 2006; Spoering and Gilmore 2006). Most phytopathogens, chiefly those from the genera *Erwinia, Pectobacterium, Pantoea, Agrobacterium, Pseudomonas, Ralstonia, Burkholderia, Acidovorax, Xanthomonas, Clavibacter, Streptomyces, Xylella, Spiroplasma* and *Phytoplasma* form biofilms on the surface, vasculature, root and root hair surfaces in turn causing varied diseases in fruits, vegetables and other parts of crop plants (Carmichael et al. 1998; Von Bodman et al. 2003; Walker et al. 2004; Andersen et al. 2007; Danhorn and Fuqua 2007; Galiana et al. 2008; Zaini et al. 2009; Martinez-Gil et al. 2010). Many pathogens employ biofilm as a tool for the dissemination of infection. *Xanthomonas fragariae* causing angular leaf spot of strawberry forms biofilms on the surfaces and undergo wind-driven dispersal to infect new plants (Allan-Wojtas et al. 2010). *Clavibacter michiganensis* subsp. *michiganensis* causing bacterial wilt and canker of tomato, extensively colonizes the lumen of xylem vessels and reaches the apical region through acropetal movement and infects the whole plant (Chalupowicz et al. 2012). *Acidovorax citrulli* and *Xylella fastidiosa* form biofilms to move upstream and promulgate plant disease. Table 2 summarizes the list of biofilm forming phytopathogens.

Apart from these phytopathogens, some of the enteric pathogens colonize and form biofilms on plants. *Escherichia coli, Salmonella enterica* and *Enterococcus faecalis* were shown to form persistent biofilms on leafy vegetables such as spinach and other plants which led to many outbreaks due to the consumption of improperly processed vegetables (Ronconi et al. 2002; Brandl 2006; Kroupitski et al. 2009; Meric et al. 2012; Patel et al. 2013; Yaron and Romling 2014; Liu et al. 2015).

Table 2 Biofilm forming phytopathogens

Pathogen	Disease	References
Acidovorax citrulli	Blotch in cucurbits	Bahar et al. (2011), Shrestha et al. (2013)
Agrobacterium tumefaciens	Crown gall disease	Morton and Fuqua (2012), Heckel et al. (2014)
B. cepacia	Sour skin of onion	Jacobs et al. (2008), Subramoni et al. (2011)
Clavibacter michiganensis subsp. *michiganensis*	Bacterial wilt and canker of tomato	Gartemann et al. (2003), Chalupowicz et al. (2012)
Erwinia amylovora	Fire blight	McNally et al. (2011), Edmunds et al. (2013)
E. chrysanthemi	Soft rot disease	Toth et al. (2003), Jahn et al. (2011)
Pseudomonas syringae pv *syringae*	Necrotic diseases of fruits and brown spot disease on bean	Rich and Willis (1997), Bais et al. (2004), Penaloza-Vazquez et al. (2010), Keisa et al. (2011)
Pantoea stewartii subsp. *stewartii*	Stewart's wilt disease in maize	Herrera et al. (2008), Roper (2011), Burbank et al. (2014), Ramachandran et al. (2014)
Pectobacterium carotovorum	Soft rot of fruits and vegetables	Kubheka et al. (2013), Lee et al. (2013)
P. corrugate	Pith necrosis of tomato	Ude et al. (2006), Licciardello et al. (2007)
Ralstonia solanacearum	Lethal wilt	Yao and Allen (2007), Chen et al. (2012b)
Xanthomonas axonopodis pv. *Citri*	Citrus canker disease	Golmohammadi et al. (2007), Rigano et al. (2007), Malamud et al. (2010)
X. campestris pv *campestris*	Black rot of crucifers, angular leaf spot of cotton, bacterial spot of pepper and tomato	Torres et al. (2007), McCarthy et al. (2008), Tao et al. (2010)
X. campestris pv. *Vesicatoria*	Bacterial leaf spot on peppers and tomatoes	Jones et al. (1998), Kim et al. (2014), Park et al. (2014)
X. fragariae	Angular leaf spot of strawberry	Allan-Wojtas et al. (2010)
X. oryzae pv. *oryzicola*	Blight of rice	Yang and Bogdanove (2012), Qian et al. (2013)
X. fastidiosa	Pierce's disease in grapes and citrus variegated chlorosis	Guilhabert and Kirkpatrick (2005), Souza et al. (2006), Purcino et al. (2007), Muranaka et al. (2013)

Table 3 Biosurfactants as antibiofilm agents

Biosurfactant	Pathogen(s)	References
Glycolipid from *Brevibacterium casei*	Mixed biofilm bacteria	Kiran et al. (2010)
Glycolipid from Coral associated bacteria	*P. aeruginosa*	Padmavathi and Pandian (2014)
Lipopeptide from *Bacillus circulans*	*E. coli*, *Mycobacterium flavus*, *Proteus vulgaris*, *Serratia marcescens*, *Citrobacter freundii*, *Klebsiella aerogenes*, *Alcaligens faecalis*, *S. typhimurium*	Das et al. (2009)
Rufisan from *C. lipolytica* UCP 0988	*Lactobacillus casei, L. reuteri, Streptococcus mutans, S. oralis, S. sanguis, Rothia dentocariosa, S. salivarius, E. coli, P. aeruginosa, S. aureus, S. epidermidis, S. agalactiae, S. pyogenes*	Rufino et al. (2011)
Lipopeptide from *B. subtilis* AR2	*C. albicans*	Rautela et al. (2014)
Lipopeptide from *B. cereus* NK1	*P. aeruginosa, S. epidermidis*	Sriram et al. (2011)
Biosurfactant from *L. jensenii, L. rhamnosus*	*Acinetobacter baumannii, E. coli*, MRSA	Sambanthamoorthy et al. (2014)
Di-rhamnolipid from *P. aeruginosa* DSVP20	*C. albicans*	Singh et al. (2013)
Rhamnolipid	*P. aeruginosa* PAO1, *Bordetella bronchiseptica*	Irie et al. (2005), Raya et al. (2010)
L. fermentum derived biosurfactant	*S. mutans*	Tahmourespour et al. (2011b)
Lipopeptide from *B. tequilensis* CH	*E. coli, S. mutans*	Pradhan et al. (2013a)
Lipopeptide from *Paenibacillus polymyxa*	Mixed species biofilm	Quinn et al. (2012)
Lunasan from *C. sphaerica* UCP 0995	*P. aeruginosa, S. agalactiae, S. Sanguis*	Luna et al. (2011)
Glycolipid from *Lysinibacillus fusiformis* S9	*E. coli, S. mutans*	Pradhan et al.(2013b)
Glycolipid from *Nocardiopsis* sp. MSA13A	*Vibrio alginolyticus*	Kiran et al. (2014)
Putisolvin I and II from *P. putida*	*Pseudomonas* sp.	Kuiper et al. (2004)
Rhamnolipid	*Yarrowia lipolytica*	Dusane et al. (2012)
Biosurfactant from *S. thermophilus*	*Candida* spp.	Busscher et al. (1997)
Surfactin	*S. typhimurium*	Mireles et al. (2001)
Lipopeptide biosurfactant with antibiotics	*E. coli* CFT073	Rivardo et al. (2011)
Glycolipid- type biosurfactant from *Trichosporon montevideense* CLOA72	*C. albicans*	Monteiro et al. (2011)
Lipopeptide from *B. subtilis* and *B. licheniformis*	*E. coli, S. aureus*	Rivardo et al. (2009)
Pseudofactin II from *P. fluorescens* BD5	*E. coli, E. faecalis, E. hirae, S. epidermidis, P. mirabilis, C. albicans*	Janek et al. (2012)

6 Biosurfactants as Antibiofilm Agents

Biosurfactants are low molecular weight, amphipathic, surface active compounds produced by living cells capable of reducing the surface and interfacial tension of liquids at the surface and interface, respectively (Padmavathi and Pandian 2014). Biosurfactants are chemically diverse group of molecules which comprise glycolipids, lipopeptides, lipoproteins, fatty acids, phospholipids, neutral lipids, polymeric and particulate biosurfactants. Though biosurfactants have versatile applications, their antibiofilm activity has received enormous attention in recent years due to their potential in healthcare, agriculture and industrial applications (Sachdev and Cameotra 2013; Banat et al. 2014; Diaz et al. 2015; Kiran et al. 2015). Biosurfactants due to their surface modifying property, effectively affect the microbial colonization and subsequent biofilm formation. They selectively reduce the hydrophobicity of bacterial cell wall. Hydrophobicity is directly corresponded to the pathogen's biofilm forming ability and reduction of it, will have direct implications on biofilm formation. Apart from efficiently controlling the biofilm formation, biosurfactants proficiently disrupt the preformed biofilms (Irie et al. 2005; Kiran et al. 2010; Dusane et al. 2012; Singh et al. 2013). Certain biosurfactants have pronounced antibiofilm effect which are capable of down-regulating the biofilm and virulence genes in biofilm formers in addition to the phenotypic suppression (Tahmourespour et al. 2011a; Salehi et al. 2014; Savabi et al. 2014). Table 3 summarizes the antibiofilm efficacy of biosurfactants.

Apart from acting as antibiofilm agents, they eradicate the zoospores of pathogenic fungus *Phytophthora capsici* and promote plant growth (Kruijt et al. 2009) and also they act as microbial-Induced Systemic Resistance (ISR) elicitors (Ongena et al. 2007; Tran et al. 2007). Plant-derived biosurfactant has been shown to possess superior antibiofilm activity that effectively reduced the biomass of complex pathogenic biofilms (Quinn et al. 2013).

7 Conclusion

Antibiofilm potential of biosurfactants has been intensely explored in recent years with an intention to develop safer antibiofilm technology in near future. Though several antibiofilm agents have been reported so far, biosurfactants are preferred over their counterparts due to their environmental feasibility, reduced toxicity, better biodegradability and their stability over extreme environmental conditions. In addition to antibiofilm activity, they have multifarious role in agriculture like improving the soil quality and enhancement of plant productivity. Due to these preferable properties, biosurfactants have received enormous attention and hold extensive application in the field of agriculture.

References

Allan-Wojtas P, Hildebrand PD, Braun PG, Smith-King HL, Carbyn S, Renderos WE (2010) Low temperature and anhydrous electron microscopy techniques to observe the infection process of the bacterial pathogen *Xanthomonas fragariae* on strawberry leaves. J Microscop 239:249–258

Andersen PC, Brodbeck BV, Oden S, Shriner A, Leite B (2007) Influence of xylem fluid chemistry on planktonic growth, biofilm formation and aggregation of *Xylella fastidiosa*. FEMS Microbiol Lett 274:210–217

Aparna MS, Yadav S (2008) Biofilms: microbes and disease. Braz J Infect Dis 12:526–530

Atkinson S, Sockett RE, Camara M, Williams P (2006) Quorum sensing and the lifestyle of Yersinia. Curr Issues Mol Biol 8:1–10

Ayala JC, Wang H, Silva AJ, Benitez JA (2015) Repression by H-NS of genes required for the biosynthesis of the *Vibrio cholerae* biofilm matrix is modulated by the second messenger cyclic diguanylic acid. Mol Microbiol 97:630–645. doi:10.1111/mmi.13058

Bahar O, Levi N, Burdman S (2011) The cucurbit pathogenic bacterium *Acidovorax citrulli* requires a polar flagellum for full virulence before and after host-tissue penetration. Mol Plant Microbe Interact 24:1040–1050

Bais HP, Fall R, Vivanco JM (2004) Biocontrol of *Bacillus subtilis* against infection of Arabidopsis roots by *Pseudomonas syringae* is facilitated by biofilm formation and surfactin production. Plant Physiol 134:307–319

Banat IM, De Rienzo MA, Quinn GA (2014) Microbial biofilms: biosurfactants as antibiofilm agents. Appl Microbiol Biotechnol 98:9915–9929

Barth VC Jr, Rodrigues BA, Bonatto GD, Gallo SW, Pagnussatti VE, Ferreira CA, de Oliveira SD (2014) Heterogeneous persister cells formation in *Acinetobacter baumannii*. PLoS ONE 8:e84361. doi:10.1371/journal.pone.0084361

Branda SS, Vik A, Friedman L, Kolter R (2005) Biofilms: the matrix revisited. Trend Microbiol 13:20–26

Brandl MT (2006) Fitness of human enteric pathogens on plants and implications for food safety. Annu Rev Phytopathol 44:367–392

Burbank L, Mohammadi M, Roper MC (2014) Siderophore-mediated iron acquisition influences motility and is required for full virulence of the xylem-dwelling bacterial phytopathogen *Pantoea stewartii* subsp. *stewartii*. Appl Environ Microbiol 81:139–148

Busalmen JP, Vazquez M, de Sanchez SR (2002) New evidences on the catalase mechanism of microbial corrosion. Electrochim Acta 47:1857–1865

Busscher HJ, van Hoogmoed CG, Geertsema-Doornbusch GI, van der Kuijl-Booij M, van der Mei HC (1997) *Streptococcus thermophilus* and its biosurfactants inhibit adhesion by *Candida* spp. on silicone rubber. Appl Environ Microbiol 63:3810–3817

Butt A, Higman VA, Williams C, Crump MP, Hemsley CM, Harmer N, Titball RW (2014) The HicA toxin from *Burkholderia pseudomallei* has a role in persister cell formation. Biochem J 459:333–344

Carmichael I, Harper IS, Coventry MJ, Taylor PW, Wan J, Hickey MW (1998) Bacterial colonization and biofilm development on minimally processed vegetables. J Appl Microbiol 85(Suppl 1):45S–51S

Chalupowicz L, Zellermann EM, Fluegel M, Dror O, Eichenlaub R, Gartemann KH, Savidor A, Sessa G, Iraki N, Barash I, Manulis-Sasson S (2012) Colonization and movement of GFP-labeled *Clavibacter michiganensis* subsp. *michiganensis* during tomato infection. Phytopathology 102:23–31

Chen Y, Cao S, Chai Y, Clardy J, Kolter R, Guo JH, Losick R (2012a) A *Bacillus subtilis* sensor kinase involved in triggering biofilm formation on the roots of tomato plants. Mol Microbiol 85:418–430

Chen Y, Yan F, Chai Y, Liu H, Kolter R, Losick R, Guo JH (2012b) Biocontrol of tomato wilt disease by *Bacillus subtilis* isolates from natural environments depends on conserved genes mediating biofilm formation. Environ Microbiol 15:848–864

Chiang P, Burrows LL (2003) Biofilm formation by hyperpiliated mutants of *Pseudomonas aeruginosa*. J Bacteriol 185:2374–2378

Cho J, Rogers J, Kearns M, Leslie M, Hartson SD, Wilson KS (2014) *Escherichia coli* persister cells suppress translation by selectively disassembling and degrading their ribosomes. Mol Microbiol 95:352–364

Conlon BP (2014) *Staphylococcus aureus* chronic and relapsing infections: evidence of a role for persister cells: an investigation of persister cells, their formation and their role in *S. aureus* disease. BioEssays 36:991–996

Cramton SE, Gerke C, Schnell NF, Nichols WW, Gotz F (1999) The intercellular adhesion (*ica*) locus is present in *Staphylococcus aureus* and is required for biofilm formation. Infect Immun 67:5427–5433

Crossman L, Dow JM (2004) Biofilm formation and dispersal in *Xanthomonas campestris*. Microbes Infect 6:623–629

Danese PN, Pratt LA, Kolter R (2000) Exopolysaccharide production is required for development of *Escherichia coli* K-12 biofilm architecture. J Bacteriol 182:3593–3596

Danhorn T, Fuqua C (2007) Biofilm formation by plant-associated bacteria. Annu Rev Microbiol 61:401–422

Das P, Mukherjee S, Sen R (2009) Anti-adhesive action of a marine microbial surfactant. Colloids Surf B Biointerfaces 71:183–186

Davies D (2003) Understanding biofilm resistance to antibacterial agents. Nat Rev Drug Discov 2:114–122

Diaz DRMA, Banat IM, Dolman B, Winterburn J, Martin PJ (2015) Sophorolipid biosurfactants: possible uses as antibacterial and antibiofilm agent. N Biotechnol pii S1871–6784(15):00034–00035. doi:10.1016/j.nbt.2015.02.009

Donlan RM, Costerton JW (2002) Biofilms: survival mechanisms of clinically relevant microorganisms. Clin Microbiol Rev 15:167–193

Donlan RM (2002) Biofilms: microbial life on surfaces. Emerg Infect Dis 8:881–890

Dressaire C, Moreira RN, Barahona S, Alves de Matos AP, Arraiano CM (2015) BolA is a transcriptional switch that turns off motility and turns on biofilm development. Mol Biol 6:e02352-02314

Drlica K, Malik M, Kerns RJ, Zhao X (2008) Quinolone-mediated bacterial death. Antimicrob Agents Chemother 52:385–392

Dusane DH, Dam S, Nancharaiah YV, Kumar AR, Venugopalan VP, Zinjarde SS (2012) Disruption of *Yarrowia lipolytica* biofilms by rhamnolipid biosurfactant. Aquat Biosyst 8:17. doi:10.1186/2046-9063-8-17

Edmunds AC, Castiblanco LF, Sundin GW, Waters CM (2013) Cyclic Di-GMP modulates the disease progression of *Erwinia amylovora*. J Bacteriol 195:2155–2165

Farinha MA, Conway BD, Glasier LM, Ellert NW, Irvin RT, Sherburne R, Paranchych W (1994) Alteration of the pilin adhesin of *Pseudomonas aeruginosa* PAO results in normal pilus biogenesis but a loss of adherence to human pneumocyte cells and decreased virulence in mice. Infect Immun 62:4118–4123

Flemming HC, Wingender J (2010) The biofilm matrix. Nat Rev Microbiol 8:623–633

Fletcher M, Loeb GI (1979) Influence of substratum characteristics on the attachment of a marine pseudomonad to solid surfaces. Appl Environ Microbiol 37:67–72

Fox EP, Bui CK, Nett JE, Hartooni N, Mui MC, Andes DR, Nobile CJ, Johnson AD (2015) An expanded regulatory network temporally controls *Candida albicans* biofilm formation. Mol Microbiol 96:1226–1239. doi:10.1111/mmi.13002

Franklin MJ, Ohman DE (1993) Identification of *alg*F in the alginate biosynthetic gene cluster of *Pseudomonas aeruginosa* which is required for alginate acetylation. J Bacteriol 175:5057–5065

Galiana E, Fourre S, Engler G (2008) *Phytophthora parasitica* biofilm formation: installation and organization of microcolonies on the surface of a host plant. Environ Microbiol 10:2164–2171

Gartemann KH, Kirchner O, Engemann J, Grafen I, Eichenlaub R, Burger A (2003) *Clavibacter michiganensis* subsp. *michiganensis*: first steps in the understanding of virulence of a Gram-positive phytopathogenic bacterium. J Biotechnol 106:179–191

Golmohammadi M, Cubero J, Penalver J, Quesada JM, Lopez MM, Llop P (2007) Diagnosis of *Xanthomonas axonopodis* pv. *citri*, causal agent of citrus canker, in commercial fruits by isolation and PCR-based methods. J Appl Microbiol 103:2309–2315

Goo KS, Sim TS (2010) Designing new beta-lactams: implications from their targets, resistance factors and synthesizing enzymes. Curr Comput Aided Drug Des 7:53–80

Gotoh N (2001) Antibiotic resistance caused by membrane impermeability and multidrug efflux systems. Nihon Rinsho 59:712–718

Guilhabert MR, Kirkpatrick BC (2005) Identification of *Xylella fastidiosa* antivirulence genes: hemagglutinin adhesins contribute a biofilm maturation to *X. fastidios* and colonization and attenuate virulence. Mol Plant Microbe Interact 18:856–868

Hall-Stoodley L, Costerton JW, Stoodley P (2004) Bacterial biofilms: from the natural environment to infectious diseases. Nat Rev Microbiol 2:95–108

Hamon MA, Stanley NR, Britton RA, Grossman AD, Lazazzera BA (2004) Identification of AbrB-regulated genes involved in biofilm formation by *Bacillus subtilis*. Mol Microbiol 52:847–860

Hardie KR, Heurlier K (2008) Establishing bacterial communities by 'word of mouth': *Lux*S and autoinducer 2 in biofilm development. Nat Rev Microbiol 6:635–643

Heckel BC, Tomlinson AD, Morton ER, Choi JH, Fuqua C (2014) *Agrobacterium tumefaciens exo*R controls acid response genes and impacts exopolysaccharide synthesis, horizontal gene transfer, and virulence gene expression. J Bacteriol 196:3221–3233

Herrera CM, Koutsoudis MD, Wang X, von Bodman SB (2008) *Pantoea stewartii* subsp. *stewartii* exhibits surface motility, which is a critical aspect of Stewart's wilt disease development on maize. Mol Plant Microbe Interact 21:1359–1370

Heydorn A, Ersboll B, Kato J, Hentzer M, Parsek MR, Tolker-Nielsen T, Givskov M, Molin S (2002) Statistical analysis of *Pseudomonas aeruginosa* biofilm development: impact of mutations in genes involved in twitching motility, cell-to-cell signaling, and stationary-phase sigma factor expression. Appl Environ Microbiol 68:2008–2017

Huang CT, Yu FP, McFeters GA, Stewart PS (1995) Nonuniform spatial patterns of respiratory activity within biofilms during disinfection. Appl Environ Microbiol 61:2252–2256

Huber B, Riedel K, Hentzer M, Heydorn A, Gotschlich A, Givskov M, Molin S, Eberl L (2001) The cep quorum-sensing system of *Burkholderia cepacia* H111 controls biofilm formation and swarming motility. Microbiology 147:2517–2528

Irie Y, O'Toole GA, Yuk MH (2005) *Pseudomonas aeruginosa* rhamnolipids disperse *Bordetella bronchiseptica* biofilms. FEMS Microbiol Lett 250:237–243

Jacobs JL, Fasi AC, Ramette A, Smith JJ, Hammerschmidt R, Sundin GW (2008) Identification and onion pathogenicity of *Burkholderia cepacia* complex isolates from the onion rhizosphere and onion field soil. Appl Environ Microbiol 74:3121–3129

Jahn CE, Selimi DA, Barak JD, Charkowski AO (2011) *The Dickeya dadantii* biofilm matrix consists of cellulose nanofibres, and is an emergent property dependent upon the type III secretion system and the cellulose synthesis operon. Microbiology 157:2733–2744

Janek T, Lukaszewicz M, Krasowska A (2012) Antiadhesive activity of the biosurfactant pseudofactin II secreted by the Arctic bacterium *Pseudomonas fluorescens* BD5. BMC Microbiol 12:24. doi:10.1186/1471-2180-12-24

Jarrett CO, Deak E, Isherwood KE, Oyston PC, Fischer ER, Whitney AR, Kobayashi SD, DeLeo FR, Hinnebusch BJ (2004) Transmission of *Yersinia pestis* from an infectious biofilm in the flea vector. J Infect Dis 190:783–792

Jones JB, Bouzar H, Somodi GC, Stall RE, Pernezny K, El-Morsy G, Scott JW (1998) Evidence for the preemptive nature of tomato race 3 of *Xanthomonas campestris* pv. *vesicatoria* in florida. Phytopathology 88:33–38

Joshua GW, Guthrie-Irons C, Karlyshev AV, Wren BW (2006) Biofilm formation in *Campylobacter jejuni*. Microbiology 152:387–396

Jubair M, Morris JG Jr, Ali A (2012) Survival of *Vibrio cholerae* in nutrient-poor environments is associated with a novel "persister" phenotype. PLoS ONE 7:e45187. doi:10.1371/journal.pone.0045187

Kalmokoff M, Lanthier P, Tremblay TL, Foss M, Lau PC, Sanders G, Austin J, Kelly J, Szymanski CM (2006) Proteomic analysis of *Campylobacter jejuni* 11168 biofilms reveals a role for the motility complex in biofilm formation. J Bacteriol 188:4312–4320

Keisa A, Kanberga-Silina K, Nakurte I, Kunga L, Rostoks N (2011) Differential disease resistance response in the barley necrotic mutant nec1. BMC Plant Biol 11:66. doi:10.1186/1471-2229-11-66

Kim JH, Kang WS, Yun SC (2014) Development of a model to predict the primary infection date of bacterial spot (*Xanthomonas campestris* pv. *vesicatoria*) on hot pepper. Plant Pathol J 30:125–135

Kiran GS, Ninawe AS, Lipton AN, Pandian V, Selvin J (2015) Rhamnolipid biosurfactants: evolutionary implications, applications and future prospects from untapped marine resource. Crit Rev Biotechnol 2:1–17

Kiran GS, Nishanth LA, Priyadharshini S, Anitha K, Selvin J (2014) Effect of Fe nanoparticle on growth and glycolipid biosurfactant production under solid state culture by marine *Nocardiopsis* sp. MSA13A. BMC Biotechnol 14:48. doi:10.1186/1472-6750-14-48

Kiran GS, Sabarathnam B, Selvin J (2010) Biofilm disruption potential of a glycolipid biosurfactant from marine *Brevibacterium casei*. FEMS Immunol Med Microbiol 59:432–438

Knudsen GM, Ng Y, Gram L (2013) Survival of bactericidal antibiotic treatment by a persister subpopulation of *Listeria monocytogenes*. Appl Environ Microbiol 79:7390–7397

Kolter R, Greenberg EP (2006) Microbial sciences: the superficial life of microbes. Nature 441:300–302

Koutsoudis MD, Tsaltas D, Minogue TD, von Bodman SB (2006) Quorum-sensing regulation governs bacterial adhesion, biofilm development, and host colonization in *Pantoea stewartii* subspecies *stewartii*. Proc Natl Acad Sci USA 103:5983–5988

Kroupitski Y, Pinto R, Brandl MT, Belausov E, Sela S (2009) Interactions of *Salmonella enterica* with lettuce leaves. J Appl Microbiol 106:1876–1885

Kruijt M, Tran H, Raaijmakers JM (2009) Functional, genetic and chemical characterization of biosurfactants produced by plant growth-promoting *Pseudomonas putida* 267. J Appl Microbiol 107:546–556

Kubheka GC, Coutinho TA, Moleleki N, Moleleki LN (2013) Colonization patterns of an mCherry-tagged *Pectobacterium carotovorum* subsp. *brasiliense* strain in potato plants. Phytopathology 103:1268–1279

Kuiper I, Lagendijk EL, Pickford R, Derrick JP, Lamers GE, Thomas-Oates JE, Lugtenberg BJ, Bloemberg GV (2004) Characterization of two *Pseudomonas putida* lipopeptide biosurfactants, putisolvin I and II, which inhibit biofilm formation and break down existing biofilms. Mol Microbiol 51:97–113

Kumar A, Worobec EA (2002) Fluoroquinolone resistance of *Serratia marcescens*: involvement of a proton gradient-dependent efflux pump. J Antimicrob Chemother 50:593–596

Lasa I, Penadés JR (2006) Bap: a family of surface proteins involved in biofilm formation. Res Microbiol 157:99–107

Leclercq R, Courvalin P (1991) Bacterial resistance to macrolide, lincosamide and streptogramin antibiotics by target modification. Antimicrob Agents Chemother 35:1267–1272

Lee DH, Lim JA, Lee J, Roh E, Jung K, Choi M, Oh C, Ryu S, Yun J, Heu S (2013) Characterization of genes required for the pathogenicity of *Pectobacterium carotovorum* subsp. *carotovorum* Pcc21 in Chinese cabbage. Microbiology 159:1487–1496

Leid JG, Shirtliff ME, Costerton JW, Stoodley P (2002) Human leukocytes adhere to, penetrate, and respond to *Staphylococcus aureus* biofilms. Infect Immun 70:6339–6345

Leid JG, Willson CJ, Shirtliff ME, Hassett DJ, Parsek MR, Jeffers AK (2005) The exopolysaccharide alginate protects *Pseudomonas aeruginosa* biofilm bacteria from IFN-gamma-mediated macrophage killing. J Immunol 175:7512–7518

Licciardello G, Bertani I, Steindler L, Bella P, Venturi V, Catara V (2007) *Pseudomonas corrugata* contains a conserved N-acyl homoserine lactone quorum sensing system; its role in tomato pathogenicity and tobacco hypersensitivity response. FEMS Microbiol Ecol 61:222–234

Liu NT, Nou X, Bauchan GR, Murphy C, Lefcourt AM, Shelton DR, Lo YM (2015) Effects of environmental parameters on the dual-species biofilms formed by *Escherichia coli* O157:H7 and *Ralstonia insidiosa*, a strong biofilm producer isolated from a fresh-cut produce processing plant. J Food Prot 78:121–127

Lu XH, An SQ, Tang DJ, McCarthy Y, Tang JL, Dow JM, Ryan RP (2012) RsmA regulates biofilm formation in *Xanthomonas campestris* through a regulatory network involving cyclic di-GMP and the Clp transcription factor. PLoS ONE 7:e52646. doi:10.1371/journal.pone.0052646

Luna JM, Rufino RD, Sarubbo LA, Rodrigues LR, Teixeira JA, de Campos-Takaki GM (2011) Evaluation antimicrobial and antiadhesive properties of the biosurfactant Lunasan produced by *Candida sphaerica* UCP 0995. Curr Microbiol 62:1527–1534

Mack D, Rohde H, Dobinsky S, Riedewald J, Nedelmann M, Knobloch JK, Elsner HA, Feucht HH (2000) Identification of three essential regulatory gene loci governing expression of *Staphylococcus epidermidis* polysaccharide intercellular adhesin and biofilm formation. Infect Immun 68:3799–3807

Mah TF (2012) Biofilm-specific antibiotic resistance. Future Microbiol 7:1061–1072

Maheshwari DK (2013) Bacteria in agrobiology: disease management. Springer, Heidelberg. ISBN 978-3-642-33639-3

Malamud F, Torres PS, Roeschlin R, Rigano LA, Enrique R, Bonomi HR, Castagnaro AP, Marano MR, Vojnov AA (2010) The *Xanthomonas axonopodis* pv. *citri* flagellum is required for mature biofilm and canker development. Microbiology 157:819–829

Martinez-Gil M, Yousef-Coronado F, Espinosa-Urgel M (2010) *Lap*F, the second largest *Pseudomonas putida* protein, contributes to plant root colonization and determines biofilm architecture. Mol Microbiol 77:549–561

McCarthy Y, Ryan RP, O'Donovan K, He YQ, Jiang BL, Feng JX, Tang JL, Dow JM (2008) The role of *PilZ* domain proteins in the virulence of *Xanthomonas campestris* pv. campestris. Mol Plant Pathol 9:819–824

McNally RR, Toth IK, Cock PJ, Pritchard L, Hedley PE, Morris JA, Zhao Y, Sundin GW (2011) Genetic characterization of the *Hrp*L regulon of the fire blight pathogen *Erwinia amylovora* reveals novel virulence factors. Mol Plant Pathol 13:160–173

Meric G, Kemsley EK, Falush D, Saggers EJ, Lucchini S (2012) Phylogenetic distribution of traits associated with plant colonization in *Escherichia coli*. Environ Microbiol 15:487–501

Mireles JR 2nd, Toguchi A, Harshey RM (2001) *Salmonella enterica* serovar typhimurium swarming mutants with altered biofilm-forming abilities: surfactin inhibits biofilm formation. J Bacteriol 183:5848–5854

Monteiro AS, Miranda TT, Lula I, Denadai AM, Sinisterra RD, Santoro MM, Santos VL (2011) Inhibition of *Candida albicans* CC biofilms formation in polystyrene plate surfaces by biosurfactant produced by *Trichosporon montevideense* CLOA72. Colloids Surf B Biointerfaces 84:467–476

Morton ER, Fuqua C (2012) Phenotypic analyses of *Agrobacterium*. In: Current protocols in microbiology Chapter 3: unit 3D, vol 3. Wiley, New York

Mukhtar TA, Wright GD (2005) Streptogramins, oxazolidinones, and other inhibitors of bacterial protein synthesis. Chem Rev 105:529–542

Muranaka LS, Giorgiano TE, Takita MA, Forim MR, Silva LF, Coletta-Filho HD, Machado MA, de Souza AA (2013) N-acetylcysteine in agriculture, a novel use for an old molecule: focus on controlling the plant-pathogen *Xylella fastidiosa*. PLoS ONE 8:e72937. doi:10.1371/journal.pone.0072937

Nadell CD, Xavier JB, Levin SA, Foster KR (2008) The evolution of quorum sensing in bacterial biofilms. PLoS Biol 6:e14. doi:10.1371/journal.pbio.0060014

Niba ET, Naka Y, Nagase M, Mori H, Kitakawa M (2007) A genome-wide approach to identify the genes involved in biofilm formation in *E. coli*. DNA Res 14:237–246

Ongena M, Jourdan E, Adam A, Paquot M, Brans A, Joris B, Arpigny JL, Thonart P (2007) Surfactin and fengycin lipopeptides of *Bacillus subtilis* as elicitors of induced systemic resistance in plants. Environ Microbiol 9:1084–1090

O'Toole G, Kaplan HB, Kolter R (2000) Biofilm formation as microbial development. Annu Rev Microbiol 54:49–79

Ozanne B, Benveniste R, Tipper D, Davies J (1969) Aminoglycoside antibiotics: inactivation by phosphorylation in *Escherichia coli* carrying R factors. J Bacteriol 100:1144–1146

Padmavathi AR, Abinaya B, Pandian SK (2014) Phenol 2,4-bis(1,1-dimethylethyl) of marine bacterial origin inhibits quorum sensing mediated biofilm formation in the uropathogen *Serratia marcescens*. Biofouling 30:1111–1122

Padmavathi AR, Pandian SK (2014) Antibiofilm activity of biosurfactant producing coral associated bacteria isolated from Gulf of Mannar. Indian J Microbiol 54:376–382

Park HJ, Jung HW, Han SW (2014) Functional and proteomic analyses reveal that *wxc*B is involved in virulence, motility, detergent tolerance, and biofilm formation in *Xanthomonas campestris* pv. *vesicatoria*. Biochem Biophys Res Commun 452:389–394

Patel J, Singh M, Macarisin D, Sharma M, Shelton D (2013) Differences in biofilm formation of produce and poultry *Salmonella enterica* isolates and their persistence on spinach plants. Food Microbiol 36:388–394

Penaloza-Vazquez A, Sreedharan A, Bender CL (2010) Transcriptional studies of the *hrp*M/*opg*H gene in *Pseudomonas syringae* during biofilm formation and in response to different environmental challenges. Environ Microbiol 12:1452–1467

Pradhan AK, Pradhan N, Mall G, Panda HT, Sukla LB, Panda PK, Mishra BK (2013a) Application of lipopeptide biosurfactant isolated from a halophile: *Bacillus tequilensis* CH for inhibition of biofilm. Appl Biochem Biotechnol 171:1362–1375

Pradhan AK, Pradhan N, Sukla LB, Panda PK, Mishra BK (2013b) Inhibition of pathogenic bacterial biofilm by biosurfactant produced by *Lysinibacillus fusiformis* S9. Bioprocess Biosyst Eng 37:139–149

Purcino RP, Medina CL, Martins de Souza D, Winck FV, Machado EC, Novello JC, Machado MA, Mazzafera P (2007) *Xylella fastidiosa* disturbs nitrogen metabolism and causes a stress response in sweet orange *Citrus sinensis* cv. Pera. J Exp Bot 58:2733–2744

Qian G, Zhou Y, Zhao Y, Song Z, Wang S, Fan J, Hu B, Venturi V, Liu F (2013) Proteomic analysis reveals novel extracellular virulence-associated proteins and functions regulated by the diffusible signal factor (DSF) in *Xanthomonas oryzae* pv. *oryzicola*. J Proteome Res 12:3327–3341

Quinn GA, Maloy AP, Banat MM, Banat IM (2013) A comparison of effects of broad-spectrum antibiotics and biosurfactants on established bacterial biofilms. Curr Microbiol 67:614–623

Quinn GA, Maloy AP, McClean S, Carney B, Slater JW (2012) Lipopeptide biosurfactants from *Paenibacillus polymyxa* inhibit single and mixed species biofilms. Biofouling 28:1151–1166

Ramachandran R, Burke AK, Cormier G, Jensen RV, Stevens AM (2014) Transcriptome-based analysis of the *Pantoea stewartii* quorum-sensing regulon and identification of EsaR direct targets. Appl Environ Microbiol 80:5790–5800

Rautela R, Singh AK, Shukla A, Cameotra SS (2014) Lipopeptides from *Bacillus* strain AR2 inhibits biofilm formation by *Candida albicans*. Antonie Van Leeuwenhoek 105:809–821

Raya A, Sodagari M, Pinzon NM, He X, Zhang Newby BM, Ju LK (2010) Effects of rhamnolipids and shear on initial attachment of *Pseudomonas aeruginosa* PAO1 in glass flow chambers. Environ Sci Pollut Res Int 17:1529–1538

Rich JJ, Willis DK (1997) Multiple loci of *Pseudomonas syringae* pv. *syringae* are involved in pathogenicity on bean: restoration of one lesion-deficient mutant requires two tRNA genes. J Bacteriol 179:2247–2258

Rigano LA, Siciliano F, Enrique R, Sendin L, Filippone P, Torres PS, Questa J, Dow JM, Castagnaro AP, Vojnov AA, Marano MR (2007) Biofilm formation, epiphytic fitness, and canker development in *Xanthomonas axonopodis* pv. *citri*. Mol Plant Microbe Interact 20:1222–1230

Rivardo F, Martinotti MG, Turner RJ, Ceri H (2011) Synergistic effect of lipopeptide biosurfactant with antibiotics against *Escherichia coli* CFT073 biofilm. Int J Antimicrob Agents 37:324–331

Rivardo F, Turner RJ, Allegrone G, Ceri H, Martinotti MG (2009) Anti-adhesion activity of two biosurfactants produced by *Bacillus* spp. prevents biofilm formation of human bacterial pathogens. Appl Microbiol Biotechnol 83:541–553

Ronconi MC, Merino LA, Fernandez G (2002) Detection of *Enterococcus* with high-level aminoglycoside and glycopeptide resistance in *Lactuca sativa* (lettuce). Enferm Infecc Microbiol Clin 20:380–383

Roper MC (2011) *Pantoea stewartii* subsp. *stewartii*: lessons learned from a xylem-dwelling pathogen of sweet corn. Mol Plant Pathol 12:628–637

Rufino RD, Luna JM, Sarubbo LA, Rodrigues LR, Teixeira JA, Campos-Takaki GM (2011) Antimicrobial and anti-adhesive potential of a biosurfactant Rufisan produced by *Candida lipolytica* UCP 0988. Colloids Surf B Biointerfaces 84:1–5

Sachdev DP, Cameotra SS (2013) Biosurfactants in agriculture. Appl Microbiol Biotechnol 97:1005–1016

Salehi R, Savabi O, Kazemi M, Kamali S, Salehi AR, Eslami G, Tahmourespour A (2014) Effects of *Lactobacillus reuteri*-derived biosurfactant on the gene expression profile of essential adhesion genes (*gtf*B, *gtf*C and *ftf*) of *Streptococcus mutans*. Adv Biomed Res 3:169. doi:10.4103/2277-9175.139134

Sambanthamoorthy K, Feng X, Patel R, Patel S, Paranavitana C (2014) Antimicrobial and antibiofilm potential of biosurfactants isolated from *Lactobacilli* against multi-drug-resistant pathogens. BMC Microbiol 14:197. doi:10.1186/1471-2180-14-197

Sauer K, Camper AK, Ehrlich GD, Costerton JW, Davies DG (2002) *Pseudomonas aeruginosa* displays multiple phenotypes during development as a biofilm. J Bacteriol 184:1140–1154

Savabi O, Kazemi M, Kamali S, Salehi AR, Eslami G, Tahmourespour A, Salehi R (2014) Effects of biosurfactant produced by *Lactobacillus casei* on *gtf*B, *gtf*C, and *ftf* gene expression level in *S. mutans* by real-time RT-PCR. Adv. Biomed Res 3:231. doi:10.4103/2277-9175.145729

Schramm A, Larsen LH, Revsbech NP, Ramsing NB, Amann R, Schleifer KH (1996) Structure and function of a nitrifying biofilm as determined by in situ hybridization and the use of microelectrodes. Appl Environ Microbiol 62:4641–4647

Shemesh M, Tam A, Steinberg D (2007) Expression of biofilm-associated genes of *Streptococcus mutans* in response to glucose and sucrose. J Med Microbiol 56:1528–1535

Shrestha RK, Rosenberg T, Makarovsky D, Eckshtain-Levi N, Zelinger E, Kopelowitz J, Sikorski J, Burdman S (2013) Phenotypic variation in the plant pathogenic bacterium *Acidovorax citrulli*. PLoS ONE 8:e73189. doi:10.1371/journal.pone.0073189

Singh N, Pemmaraju SC, Pruthi PA, Cameotra SS, Pruthi V (2013) *Candida* biofilm disrupting ability of di-rhamnolipid (RL-2) produced from *Pseudomonas aeruginosa* DSVP20. Appl Biochem Biotechnol 169:2374–2391

Singh PK, Schaefer AL, Parsek MR, Moninger TO, Welsh MJ, Greenberg EP (2000) Quorum-sensing signals indicate that cystic fibrosis lungs are infected with bacterial biofilms. Nature 407:762–764

Sloan GP, Love CF, Sukumar N, Mishra M, Deora R (2007) The Bordetella Bps polysaccharide is critical for biofilm development in the mouse respiratory tract. J Bacteriol 189:8270–8276

Souza LC, Wulff NA, Gaurivaud P, Mariano AG, Virgilio AC, Azevedo JL, Monteiro PB (2006) Disruption of *Xylella fastidiosa* CVC *gum*B and *gum*F genes affects biofilm formation without a detectable influence on exopolysaccharide production. FEMS Microbiol Lett 257:236–242

Spoering AL, Gilmore MS (2006) Quorum sensing and DNA release in bacterial biofilms. Curr Opin Microbiol 9:133–137

Sriram MI, Kalishwaralal K, Deepak V, Gracerosepat R, Srisakthi K, Gurunathan S (2011) Biofilm inhibition and antimicrobial action of lipopeptide biosurfactant produced by heavy metal tolerant strain *Bacillus cereus* NK1. Colloids Surf B Biointerfaces 85:174–181

Srivastava D, Waters CM (2012) A tangled web: regulatory connections between quorum sensing and cyclic Di-GMP. J Bacteriol 194:4485–4493

Stewart PS, Costerton JW (2001) Antibiotic resistance of bacteria in biofilms. Lancet 358:135–138

Stickler DJ, Morris NS, McLean RJ, Fuqua C (1998) Biofilms on indwelling urethral catheters produce quorum-sensing signal molecules in situ and in vitro. Appl Environ Microbiol 64:3486–3490

Stoodley P, Cargo R, Rupp CJ, Wilson S, Klapper I (2002) Biofilm material properties as related to shear-induced deformation and detachment phenomena. J Ind Microbiol Biotechnol 29:361–367

Subramoni S, Nguyen DT, Sokol PA (2011) *Burkholderia cenocepacia* ShvR-regulated genes that influence colony morphology, biofilm formation, and virulence. Infect Immun 79:2984–2997

Sutherland IW (2001) The biofilm matrix-an immobilized but dynamic microbial environment. Trends Microbiol 9:222–227

Tahmourespour A, Salehi R, Kasra Kermanshahi R (2011a) *Lactobacillus acidophilus*-derived biosurfactant effect on GTFB and GTFC expression level in *Streptococcus mutans* biofilm cells. Braz J Microbiol 42:330–339

Tahmourespour A, Salehi R, Kermanshahi RK, Eslami G (2011b) The anti-biofouling effect of *Lactobacillus fermentum*-derived biosurfactant against *Streptococcus mutans*. Biofouling 27:385–392

Tao F, Swarup S, Zhang LH (2010) Quorum sensing modulation of a putative glycosyltransferase gene cluster essential for *Xanthomonas campestris* biofilm formation. Environ Microbiol 12:3159–3170

Tao J, He C (2010) Response regulator, VemR, positively regulates the virulence and adaptation of *Xanthomonas campestris* pv. *campestris*. FEMS Microbiol Lett 304:20–28

Tielen P, Strathmann M, Jaeger KE, Flemming HC, Wingender J (2005) Alginate acetylation influences initial surface colonization by mucoid *Pseudomonas aeruginosa*. Microbiol Res 160:165–176

Torres PS, Malamud F, Rigano LA, Russo DM, Marano MR, Castagnaro AP, Zorreguieta A, Bouarab K, Dow JM, Vojnov AA (2007) Controlled synthesis of the DSF cell-cell signal is required for biofilm formation and virulence in *Xanthomonas campestris*. Environ Microbiol 9:2101–2109

Toth IK, Bell KS, Holeva MC, Birch PR (2003) Soft rot Erwiniae: from genes to genomes. Mol Plant Pathol 4:17–30

Tran H, Ficke A, Asiimwe T, Hofte M, Raaijmakers JM (2007) Role of the cyclic lipopeptide massetolide A in biological control of *Phytophthora infestans* and in colonization of tomato plants by *Pseudomonas fluorescens*. New Phytol 175:731–742

Tsang JC, Sansing GA, Miller MA (1975) Relation of beta-lactamase activity to antimicrobial susceptibility in *Serratia marcescens*. Antimicrob Agents Chemother 8:277–281

Ude S, Arnold DL, Moon CD, Timms-Wilson T, Spiers AJ (2006) Biofilm formation and cellulose expression among diverse environmental *Pseudomonas* isolates. Environ Microbiol 8:1997–2011

Valle J, Latasa C, Gil C, Toledo-Arana A, Solano C, Penades JR, Lasa I (2012) Bap, a biofilm matrix protein of *Staphylococcus aureus* prevents cellular internalization through binding to GP96 host receptor. PLoS Pathog 8:e1002843. doi:10.1371/journal.ppat.1002843

Von Bodman SB, Bauer WD, Coplin DL (2003) Quorum sensing in plant-pathogenic bacteria. Annu Rev Phytopathol 41:455–482

Walker TS, Bais HP, Deziel E, Schweizer HP, Rahme LG, Fall R, Vivanco JM (2004) *Pseudomonas aeruginosa*-plant root interactions. Pathogenicity, biofilm formation, and root exudation. Plant Physiol 134:320–331

Watnick PI, Kolter R (1999) Steps in the development of a *Vibrio cholerae* El Tor biofilm. Mol Microbiol 34:586–595

Wei Q, Ma LZ (2013) Biofilm matrix and its regulation in *Pseudomonas aeruginosa*. Int J Mol Sci 14:20983–21005

Whitchurch CB, Tolker-Nielsen T, Ragas PC, Mattick JS (2002) Extracellular DNA required for bacterial biofilm formation. Science 295:1487

Wilking JN, Zaburdaev V, De Volder M, Losick R, Brenner MP, Weitz DA (2013) Liquid transport facilitated by channels in *Bacillus subtilis* biofilms. Proc Natl Acad Sci USA 110:848–852

Wozniak DJ, Wyckoff TJ, Starkey M, Keyser R, Azadi P, O'Toole GA, Parsek MR (2003) Alginate is not a significant component of the extracellular polysaccharide matrix of PA14 and PAO1 *Pseudomonas aeruginosa* biofilms. Proc Natl Acad Sci USA 100:7907–7912

Wu MC, Lin TL, Hsieh PF, Yang HC, Wang JT (2011) Isolation of genes involved in biofilm formation of a *Klebsiella pneumoniae* strain causing pyogenic liver abscess. PLoS ONE 6:e23500. doi:10.1371/journal.pone.0023500

Yang B, Bogdanove A (2012) Inoculation and virulence assay for bacterial blight and bacterial leaf streak of rice. Methods Mol Biol 956:249–255

Yao J, Allen C (2007) The plant pathogen *Ralstonia solanacearum* needs aerotaxis for normal biofilm formation and interactions with its tomato host. J Bacteriol 189:6415–6424

Yaron S, Romling U (2014) Biofilm formation by enteric pathogens and its role in plant colonization and persistence. Microb Biotechnol 7:496–516

Zaini PA, De La Fuente L, Hoch HC, Burr TJ (2009) Grapevine xylem sap enhances biofilm development by *Xylella fastidiosa*. FEMS Microbiol Lett 295:129–134

Zogaj X, Nimtz M, Rohde M, Bokranz W, Romling U (2001) The multicellular morphotypes of *Salmonella typhimurium* and *Escherichia coli* produce cellulose as the second component of the extracellular matrix. Mol Microbiol 39:1452–1463

Biofilm Formation and Biosurfactant Activity in Plant-Associated Bacteria

Emiliano D. Primo, Francisco Ruiz, Oscar Masciarelli and Walter Giordano

Abstract Biofilms are structurally complex communities of microbial cells that adhere to a surface and are surrounded by an extracellular polymeric matrix. Biofilm formation plays important roles in attachment and colonization of plant surfaces by both beneficial bacteria (e.g. plant growth-promoting rhizobacteria) and phytopathogenic bacteria. During the process of biofilm development and maturation, surface-attached cells undergo aggregation to form microcolonies. Biosurfactants are produced by many plant-associated bacterial species and play essential roles in bacterial motility, signaling, biofilm formation and control of plant-bacteria interactions. In this chapter, we review the biochemical and genetic mechanisms that underlie biofilm formation and biosurfactant activity in beneficial (symbiotic) bacteria and phytopathogenic bacteria, particularly *Pseudomonas* species.

Keywords Phytopathogens · Rhizobacteria · Microcolony · PGPR · *Pseudomonas* · Plant-microbe interaction

---

E.D. Primo · W. Giordano (✉)
Departamento de Biología Molecular, Universidad Nacional de Río Cuarto,
Ruta 36 Km 601, X5804BYA, Río Cuarto, Córdoba, Argentina
e-mail: wgiordano@exa.unrc.edu.ar

F. Ruiz
Departamento de Microbiología e Inmunología, Universidad Nacional de Río Cuarto,
Ruta 36 Km 601, X5804BYA, Río Cuarto, Córdoba, Argentina

O. Masciarelli
Departamento de Ciencias Naturales, Universidad Nacional de Río Cuarto,
Ruta 36 Km 601, X5804BYA, Río Cuarto, Córdoba, Argentina

D.K. Maheshwari (ed.), *Bacterial Metabolites in Sustainable Agroecosystem*,
Sustainable Development and Biodiversity 12, DOI 10.1007/978-3-319-24654-3_13

1 Introduction

Bacteria live and proliferate in natural environments either as individual cells (planktonic form) or as highly organized multicellular communities (termed "biofilms") that are enclosed in self-produced polymeric matrices and are closely associated with environmental surfaces or air–liquid interfaces (Burmølle et al. 2014).

In all species of soil bacteria (rhizobacteria) studied to date, biofilm formation depends on bacterial surface components (flagella, lipopolysaccharides and exopolysaccharides) in combination with bacterial quorum sensing (QS) signals (Rinaudi and Giordano 2010). This strategy allows the bacteria to colonize the surrounding habitat and survive common environmental stresses such as desiccation and nutrient limitation. *Pseudomonas* is a large and well-studied genus of bacteria that includes several phytopathogenic species. Biofilm formation enables these species to efficiently attach to and colonize plant surfaces and biosurfactant production is often involved in this process (D'aes et al. 2010).

In the course of biofilm development during plant-bacteria interactions both in the symbiotic models (rhizobia) and pathogenic models (e.g. *Pseudomonas*), an initial reversible attachment to the surface is followed by an irreversible attachment and multiplication of the bacteria to form microcolonies. The colonies then develop into mature communities with a three-dimensional (3-D) structure, in some cases permeated by channels that function as a circulatory system. These processes are all coordinated by bacterial QS systems (Stanley and Lazazzera 2004).

This chapter reviews studies demonstrating the importance of biofilm formation in plant colonization processes in various symbiotic and pathogenic bacterial models.

2 Four-Stage Process of Biofilm Formation in Rhizobacteria

Living in biofilms confers numerous advantages to rhizobacteria, including the ability to survive or grow under fluctuating or extreme environmental conditions (soil pH, temperature, water availability, redox potential and salt concentration), to accumulate nutrients and to dispose of metabolic wastes (Vanderlinde et al. 2009; Seneviratne et al. 2011; Bogino et al. 2013).

Four stages are generally recognized in the process of biofilm formation: (i) initial reversible attachment of an individual bacterium to a surface, leading to a stronger, irreversible attachment by bacteria, (ii) microcolony formation with early stage development of biofilm architecture, (iii) biofilm maturation and (iv) eventual dispersal of single cells from the biofilm (Schuster and Markx 2014). This series of biological stages involves coordination of many physical and molecular steps, as detailed in the following sections.

2.1 Stage 1: Initial Reversible Attachment and Transition to Stronger Irreversible Attachment

The initial reversible attachment step is mediated by flagellum, pili and several outer membrane proteins collectively termed as adhesins and by *Rhizobium*-adhering proteins (RAPs) such as $RapA_1$ (Fujishige et al. 2006; Mongiardini et al. 2008; Karatan and Watnick 2009). The irreversible attachment step primarily involves polysaccharides and other biopolymers on the bacterial surface. These cell surface components include lipopolysaccharides (LPSs), capsular polysaccharides, exopolysaccharides (EPSs), neutral polysaccharides, gel-forming polysaccharides and cyclic β-glycans (cellulose microfibrils). They mediate stronger adhesion of bacteria to root hairs and anchor bacteria to the root surface. The dynamic attachment also depends on proteins, glycoprotein and lectins on the root hair (De Holf et al. 2009; Robledo et al. 2011; Bogino et al. 2013).

Adhesins, flagellar and pilar proteins play important roles in motility because they help the bacteria to overcome repulsive forces, thus promoting initial interaction with the nearby surface and increasing the likelihood of close approach. In the rhizosphere, flagellar and/or pilar motility has been shown to accelerate surface adhesion in rhizobacteria such as *Agrobacterium tumefaciens* and *Pseudomonas aeruginosa*. These proteins are involved in biofilm formation as well as root colonization (Petrova and Sauer 2012).

The surface polysaccharides in rhizobacteria are highly diverse in terms of location and physico-chemical properties. In addition to the irreversible attachment step, they are involved in cell-cell interactions during the maturation process and interactions with matrix components in biofilms (Jefferson 2009). LPSs play important roles in symbiosis, as both structural components and signaling molecules (Fraysse et al. 2003).

Rhizobial EPSs are involved in adhesion to roots, root colonization and development of biofilms on roots (Rinaudi and Giordano 2010). Other reported functions of EPSs include nutrient gathering, protection against environmental stresses and antimicrobial activity. In nitrogen-fixing bacteria that establish symbioses with legume plants through formation of indeterminate-type nodules, EPS are essential for successful infection of host plant roots. Rhizobial EPSs are the major component of the biofilm matrix and are associated with successful biofilm formation on both abiotic surfaces and host plant roots (Robledo et al. 2012).

2.2 Stage 2: Microcolony Formation with Early Stage Development of Biofilm Architecture

During the attachment steps as mentioned above, rhizobacteria undergoes physiological changes that lead to EPS production and fix the cells to the root surface. The cells then divide and form microcolonies by clonal propagation. During this stage, bacteria often uses a QS system to accumulate from the surrounding environment various extracellular signaling molecules termed autoinducers, which make it possible to monitor population density and coordinate gene expression (Waters and Bassler 2005). The QS circuits are activated when the concentration of autoinducers produced by the bacteria reaches a critical level, allowing coordinated expression of genes and their consequent actions (Connell et al. 2010; Decho et al. 2010). QS systems modulate a wide variety of phenotypes, including biofilm formation, toxin production, EPS production, virulence, plasmid transfer and motility, which are essential for successful establishment of symbiotic or pathogenic relationships with eukaryotic hosts (Marketon et al. 2003; Rinaudi and Giordano 2010; Timmusk and Nevo 2011; Pérez Montaño et al. 2014).

Many different QS systems have been reported in Gram-negative plant-associated bacteria and other proteobacteria. The most commonly occurring diffusible signaling molecules are *N*-acyl homoserine lactones (AHLs). In *Sinorhizobium meliloti*, an ExpR/Sin QS system regulates many functions associated with root nodulation of *Medicago sativa* (alfalfa) by the bacteria. This system includes the response regulator ExpR and the autoinducer synthase SinI. Rhizobacteria may constitutively express multiple AHLs. For example, *S. meliloti* strain Rm8530 produces at least seven AHLs: C12-HSL, C14-HSL, 3-oxo-C14-HSL, C16-HSL, 3-oxo-C16:1-HSL, C16:1-HSL and C18-HSL. The concentration of the signal acts as an indicator of population density for the bacteria. Some strains synthesize ESP II, one of the two symbiotically active EPSs required for biofilm formation in roots (Gurich and González 2009; Sorroche et al. 2010; Gao et al. 2012). We recently characterized four novel AHLs (C6, 3OC10, 3OC12 and 3OC14) in peanut-nodulating *Bradyrhizobium* sp. strains. Comparisons of AHL-producing versus non-AHL-producing strains demonstrated a positive regulatory effect of these AHLs on motility and biofilm formation ability (Nievas et al. 2012). Another unique, conserved AHL system, (BraI/R) was found in all plant-associated *Burkholderia* species (Hirsch and Fujishige 2012).

2.3 Stage 3: Biofilm Maturation

Biofilm maturation requires two factors: QS signals and EPS accumulation through continued cell division. Differential gene expression between the two bacterial states (planktonic/sessile) is related to adhesive needs of populations during surface colonization. For example, production of surface appendages is inhibited in sessile forms because motility is no longer necessary. Expression of genes involved in production of cell surface proteins and excretion products increases concomitantly. Transport of extracellular products into the cell is facilitated by surface proteins (porins) such as *Opr*C and *Opr*E, whereas transport of excretion products out of the cell is facilitated by certain polysaccharides (Davies 2003; Garret et al. 2008). *Herbaspirillum seropedicae* is a nitrogen-fixing β-proteobacterium with plant growth-promoting rhizobacterial (PGPR) properties, associated with a variety of important agricultural crops (e.g. maize, rice, sorghum and sugarcane). Mutation of *eps* genes in this species prevented the development of a mature biofilm, demonstrating that EPSs are essential structural components (Balsanelli et al. 2014). In associative diazotrophs such as *Azospirillum brasilense* and *Gluconacetobacter diazotrophicus*, EPSs affected cell aggregation and biofilm formation on plant root surfaces (Burdman et al. 2000; Meneses et al. 2011). Knockout of rhamnose biosynthesis in *A. brasilense* caused reductions in EPS production and maize root colonization (Jofré et al. 2004). Continued EPS production contributes to later-stage 3-D architectural structure in biofilms. Within the 3-D architecture are various channels that allow circulation and exchange of water, nutrients, enzymes and signals and elimination of toxic metabolites (Sutherland 2001; Wagner et al. 2009). EPSs are essential for biofilm structure, adhesion and cohesion, and play key roles in mechanical stability and resistance to stressful environmental conditions.

2.4 Stage 4: Dispersal of Single Cells from the Biofilm

In the fourth and final stage of the biofilm life cycle, differentiated "dispersal cells" are produced and released. The formation of these cells and their eventual dispersal are controlled through a variety of sophisticated mechanisms, indicating strong evolutionary pressures for dispersal from mature and sessile biofilms. Dispersal of planktonic cells from mature biofilms allows rhizobacteria to reach and colonize new root substrates (McDougald et al. 2012; Bogino et al. 2013). The dispersal process is regulated by a variety of intracellular and extracellular factors. In the PGPR species *Gluconacetobacter xylinus*, changes from sessile to planktonic phenotype in response to intercellular signals or environmental conditions were associated with the intracellular concentration of cyclic di-guanosine monophosphate (c-di-GMP) (Römling et al. 2005). Reduction of the intracellular c-di-GMP level in mature biofilms triggered the dispersal process. *Pseudomonas aeruginosa* provides an advanced dispersal model. Dispersal from the biofilm in this species is activated by the lysis and death of a small cellular subpopulation found in the mature architecture (Webb et al. 2003). Dispersal bacteria are released through resulting, characteristic rupture points or hollow points in the biofilm. Sessile-phenotypes bacteria in the biofilm undergo upregulation of genes that encode flagellar and chemotactic proteins, and suppression of genes that encode EPSs and fimbria proteins (Rollet et al. 2009). Dispersal cells of *P. aeruginosa* are more similar to planktonic cells than to the mature biofilm cells, and revert to a planktonic growth mode (Webb et al. 2004). Many of the phenotypic and genotypic changes that characterize dispersal cells are clearly affected by QS signals (AHLs), physiological signals (D-amino acids, intracellular modulation of c-di-GMP level), nutritional factors (high nutrient levels, carbon or nitrogen limitation) and environmental factors (low iron concentration in soil). In addition to *P. aeruginosa*, dispersal of cells from biofilms into the surrounding rhizosphere has been reported for *Rhodobacter sphaeroides*, *Xanthomonas campestris*, *Bacillus subtilis*, *Pseudomonas fluorescens* and *P. putida* (Davies and Marques 2009; Gjermansen et al. 2010; Kolodkin-Gal et al. 2010).

Dispersal may occur as an "insurance policy" to seed new biofilms in the surrounding rhizosphere in response to resource limitation, or as a simple consequence of aging of the original biofilm. The dispersal event provides new areas of potential bacteria recruitment, and functions as a template for post-dispersal processes such as gene transfer, competition, invasion and development of new communities (McDougald et al. 2012).

3 Biosurfactants and Their Roles in Plant–Microbe Interactions

Among various microbial surfactants, glycolipids have been most extensively studied. These include rhamnolipids, trehalolipids, sophorolipids and mannosyl erythritol lipids (MELs) consisting, respectively, of mono- or di-saccharides of rhamnose, glucose, sophorose, or mannose combined with long-chain aliphatic acids or hydroxyaliphatic acids (Van Bogaert et al. 2011). These glycolipids and other known biosurfactants are summarized in Table 1.

Table 1 Microbial biosurfactants and their producers

Type of biosurfactant	Microbial producer	References
Glycolipids		
[a]Rhamnolipids	*Pseudomonas aeruginosa* *Burkholdheria* sp.	Abdel-Mawgoud et al. (2010), Dusane et al. (2012), Costa et al. (2011)
Trehalolipids	*Rhodococcus erythropolis* *Corynebacterium* sp. *Arthrobacter* sp.	Franzetti et al. (2010), Tokumoto et al. (2009)
Sophorolipids	*Candida bombicola*, other *Candida* sp.	Konishi et al. (2008), Kurztman et al. (2010)
Mannosyl erythritol lipids (MELs)	Various yeasts (*Pseudozyma*)	Konishi et al. (2007)
Lipopeptides and lipoproteins		
[a]Surfactin and subtilisin	*Bacillus subtilis* Bacillus amyloliquefaciens	Arguelles-Arias et al. (2009), Jacques (2011)
Iturin family	*B. subtilis*	Yuan et al. (2011)
Fengycin family	*B. subtilis*	Romero et al. (2007)
Bioemulsan RAG1	*Acinetobacter calcoaceticus* (ATCC31012)	Suthar et al. (2008)
Fatty acids (corynomicolic acids, spiculisporic acids)	*Corynebacterium* spp., *Nocardia erythropolis*, *Aspergillus* spp.	Wang et al. (2012), Mulligan and Gibbs (2004)
Phospholipids Pseudofactin Putisolvin	*Comamonas* spp. *Pseudomonas. fluorescens* *Pseudomonas putida*	Sun et al. (2013), Janek et al. (2010), Kuiper et al. (2004)

[a]Most important representative of its class

Soil microorganisms play essential roles in maintaining soil structure and fertility, and in remediating contaminated soils. In this regard, PGPR may exert a direct effect on plant growth dynamics, or an indirect effect through acidification, chelation, precipitation and immobilization of heavy metals, or mobilization of micronutrients or macronutrients in the rhizosphere. Plant-microbe interactions are important factors in soil health and sustainability, largely because of their promoting effects on plant growth and function (Tak et al. 2013).

Plant-rhizobacteria interactions are based on part of the coordinated bacterial mechanisms such as motility, the ability to form biofilms on root surfaces, and release of QS signals. The most abundant QS signaling molecules in the rhizosphere are the AHLs, which promote synthesis of antimicrobial substances and root surface colonization ability. In *Rhizobium etli*, the symbiotic partner of *Phaseolus vulgaris* (common bean plant), long-chain AHLs act as biosurfactants, regulating the swarming phenotype and surface movement, inducing liquid flow through a superficial tension gradient, and thereby facilitating attachment of the bacteria to surfaces (Daniels et al. 2006).

In various *Pseudomonas* spp., QS molecules regulate physiological processes involved in cell-social behavior and pathogenesis, as well as production of rhamnolipids as biosurfactants. The rhamnolipids are surface active agents that have antibacterial, antifungal and antiviral properties and are involved in motility, cell-cell interactions, differentiation and formation of water channels in *Pseudomonas* biofilms (Dusane et al. 2010).

Rhizobacterial motility is clearly important for colonization efficiency and fitness in plant symbioses, but motility mechanisms used by bacteria on and around plants remain poorly understood. In a recent study of the biosurfactant-producing strain *P. fluorescens* SBW25, Alsohim et al. (2014) demonstrated that depletion of the biosurfactant viscosin altered motility, root colonization ability and plant growth promotion by the bacteria. Positive associations were found among viscosin biosynthesis, root surface spreading efficiency and survival of germinating seedlings in soils infected with a phytopathogen (*Pythium* spp.).

Pseudomonas aeruginosa strain A11 is a rhamnolipid-producing bacterium with PGPR and multi-metal-resistant (MMR) properties, isolated from the rhizosphere of the allergy-producing invasive weed *Parthenium hysterophorus*. The biosurfactant-producing ability of *P. aeruginosa* A11 facilitated phytoremediation of heavy-metal-polluted soils, and plant growth promotion through effects on siderophores, hydrogen cyanide, catalase, ammonia production and phosphate solubility (Singh and Cameotra 2013).

4 Biofilm Formation in Phytopathogenic Bacteria

Many microbial species may exert harmful effects on plants by directly damaging plant tissues or competing for nutrients. Leaf surfaces typically support large populations of bacterial epiphytes, including plant pathogens that multiply on the surface before initiating a disease process.

Many plant-bacteria associations involve physical interaction of the bacteria with plant tissues. Microcolony formation, aggregates, and cell clusters of bacteria on plant surfaces can be directly observed (Morris and Monier 2003). Such multicellular structures may present the typical defining characteristics of biofilms, i.e. groups of cells located within an EPS matrix on a solid surface. In comparison to solitary cells, cells that belong to a large aggregate are more likely to survive water limitation (desiccation).

Ramey et al. (2004) reported that *Pseudomonas syringae* pv. *syringae* (Pss), the causative agent of brown spot disease on bean plants, may colonize the plant as small groups (<10 cells each) scattered on leaf surfaces, or as large populations (>1000 cells) located near trichomes or veins where nutrient availability is higher.

4.1 Biofilm Formation in Pseudomonas

Pseudomonads occupy many distinct niches in a wide variety of environments. Some *Pseudomonas* species are likely to cause disease, while others do not. Both pathogenic and nonpathogenic species are often attached to a surface and coated with a polymeric substance, characteristic of a biofilm (Mann and Wozniak 2012).

Pseudomonas species produce a variety of organic molecules, including polysaccharides, nucleic acids and proteins, that are used to form biofilm matrices. They may also synthesize accessory compounds involved in biofilm formation or adaptation to varying environmental conditions.

The opportunistic pathogen *P. aeruginosa* is the model most commonly utilized for studies of biofilm formation, composition and architecture. *P. aeruginosa* is also of interest because it causes various diseases in animals and humans (Donlan and Costerton 2002). Other *Pseudomonas* spp. has been studied in environments such as plant tissues, soil and freshwater streams.

Production of capsular polysaccharides is important for niche colonization by *Pseudomonas* spp. in many environments. Production of polysaccharides by biofilm-forming microbes in general promotes colonization of surfaces by facilitating aggregation, adhesion and surface tolerance (Keith et al. 2003; Laus et al. 2006).

4.2 Function, Structure and Functions of Biosurfactants in Plant-Pseudomonas Interactions

D'aes et al. (2010) proposed that biosurfactants have three distinct activities: modification of surface properties, alteration of compound bio-availability and interaction with membranes. The principal function of a particular biosurfactant depends on its structure and production characteristics.

The properties of biosurfactants arise from their basic structure, consisting of both hydrophobic and a hydrophilic motifs. These structural properties vary widely among different species and strains of biosurfactant-producing microbes. The involvement of biosurfactants in plant–bacteria interactions may be beneficial to both partners that have adverse effects on competing microorganisms (Banat et al. 2014), or harm the host plant, as in the case of pathogens such as *P. syringae*.

Pseudomonas spp. mainly produce cyclic lipopeptide (CLP) type biosurfactants, comprised of a cyclized oligopeptide lactone ring coupled to a fatty acid tail, and amino acids bound to the peptide chain. The biosurfactant activity depends on the type of amino acid and length of the fatty acid chain.

Biosynthesis of CLPs is controlled by nonribosomal peptide synthetases (NRPSs) (D'aes et al. 2010). Our knowledge of genetic regulation of CLP production in *Pseudomonas* spp. is quite limited. Studies of several strains have shown that a GacS/GacA two-component system is necessary for CLP production. Certain molecules with regulatory activity, including AHL QS signals, shock proteins, the global regulator GidA, and the protease ClpP are involved in genetic regulation of some (but not all) CLPs.

A variety of plant-associated bacteria, including PGPR and phytopathogens such as *Pseudomonas* spp., are able to produce biosurfactants. In early stages of plant-bacteria interactions, bacterial cells adhere to the phylloplane (for phytopathogens) or to the rhizoplane (for most beneficial species). Biofilm formation facilitates efficient colonization of plant surfaces by bacteria, and the process is sometimes accompanied by biosurfactant production. In *P. fluorescens* strains SS101 and SBW25, CLPs such as massetolide and viscosin are necessary for biofilm formation. In mutant strains of *Pseudomonas* sp. MIS38 and *P. putida* with defective synthesis of the CLPs arthrofactin and putisolvin, respectively, thick and unstable biofilms were formed. Lindow and Brandl (2003) proposed that biosurfactants enhance the epiphytic fitness of bacteria by promoting mobility and/or apoplast nutrient accumulation, because the waxy cuticle of plant leaves present a hydrophobic barrier that limits access of bacteria to nutrients in the absence of biosurfactants.

Burch et al. (2014) recently investigated the beneficial effects of the hygroscopic biosurfactant syringafactin synthesized by Pss B728a on leaves. Syringafactin on leaf surfaces had the ability to condense water vapor even during periods of low-atmospheric humidity. The increased humidity resulted in higher cuticle permeability and enhanced availability of apoplast nutrients.

5 Conclusions

The studies reviewed here demonstrate the importance of biofilm structures in initiating and maintaining contact of bacteria with the host plant, and the extent to which biofilm formation is an intrinsic component of plant–microbe interactions. Biosurfactant molecules are of particular interest because of their roles in plant disease, and their antimicrobial properties that can be utilized in biocontrol strategies. Biosurfactants may trigger membrane depolarization through formation of transmembrane pores, induce immune responses by the host plant, or alter the bio-availability of nutrients on plant tissue surfaces and thereby promote nutrient accumulation by bacteria. Through their modification of surface properties, biosurfactants affect important bacterial processes such as surface motility, biofilm formation, and colonization, which determine the efficiency and success of plant–bacteria interactions.

Acknowledgments This review and studies by the authors' group described herein were supported by grants from the Secretaría de Ciencia y Técnica (UNRC), Agencia Nacional de Promoción Científica y Tecnológica (ANPCyT) and Consejo Nacional de Investigaciones Científicas y Técnicas of the República Argentina (CONICET). EDP and FR have a CONICET fellowship. WG is a Career Member of CONICET. The authors are grateful to Dr. S. Anderson for English editing of the manuscript.

References

Abdel-Mawgoud AM, Lepine F, Deziel E (2010) Rhamnolipids: diversity of structures, microbial origins and roles. Appl Microbiol Biotechnol 86:1323–1336

Alsohim AS, Taylor TB, Barrett GA, Gallie J, Zhang XX, Altamirano-Junqueira AE, Johnson LJ, Rainey PB, Jackson RW (2014) The biosurfactant viscosin produced by *Pseudomonas fluorescens* SBW25 aids spreading motility and plant growth promotion. Environ Microbiol 16:2267–2281

Arguelles-Arias A, Ongena M, Halimi B, Yannick L, Brans A, Joris B, Fickers P (2009) *Bacillus amyloliquefaciens* GA1 as a source of potent antibiotics and other secondary metabolites for biocontrol of plant pathogens. Microb Cell Fact 8:63–74

Balsanelli E, Baura VA, Pedrosa FO, Souza EM, Monteiro RA (2014) Exopolysaccharide biosynthesis enables mature biofilm formation on abiotic surfaces by *Herbaspirillum seropedicae*. PLoS ONE 9:e110392. doi:10.1371/journal.pone.0110392

Banat IM, Díaz De Rienzo MA, Quinn GA (2014) Microbial biofilms: biosurfactants as antibiofilm agents. Appl Microbiol Biotechnol 98:9915–9929

Bogino P, Oliva MM, Sorroche FG, Giordano W (2013) The role of bacterial biofilms and surface components in plant-bacterial associations. Int J Mol Sci 14:15838–15859

Burch AY, Zeisler V, Yokota K, Schreiber L, Lindow SE (2014) The hygroscopic biosurfactant syringafactin produced by *Pseudomonas syringae* enhances fitness on leaf surfaces during fluctuating humidity. Environ Microbiol 16:2086–2098

Burdman S, Okon Y, Jurkevitch E (2000) Surface characteristics of *Azospirillum brasilense* in relation to cell aggregation and attachment to plant roots. Crit Rev Microbiol 26:91–110

Burmølle M, Ren D, Bjarnshol T, Sørensen SJ (2014) Interactions in multispecies biofilms: do they actually matter? Trends Microbiol 22:84–91

Connell JL, Wessel AK, Parsek MR, Ellington AD, Whiteley M, Shear JB (2010) Probing prokaryotic social behaviors with bacterial "lobster traps". Mol Biol 1:1–8

Costa SG, Deziel E, Lepine F (2011) Characterization of rhamnolipid production by *Burkholderia glumae*. Lett Appl Microbiol 53:620–627

D'aes J, De Maeyer K, Pauwelyn E, Höfte M (2010) Biosurfactants in plant-*Pseudomonas* interactions and their importance to biocontrol. Environ Microbiol Rep 2:359–372

Daniels R, Reynaert S, Hoekstra H, Verreth C, Janssens J, Braeken K, Fauvart M, Beullens S, Heusdens C, Lambrichts I, De Vos DE, Vanderleyden J, Vermant J, Michiels J (2006) Quorum signal molecules as biosurfactants affecting swarming in *Rhizobium etli*. Proc Natl Acad Sci USA 103:14965–14970

Davies D (2003) Understanding biofilm resistance to antibacterial agents. Nat Rev Drug Discov 2:114–122
Davies DG, Marques CNH (2009) A fatty acid messenger is responsible for inducing dispersion in microbial biofilms. J Bacteriol 191:1393–1403
De Hoff PL, Brill LM, Hirsch AM (2009) Plant lectins: the ties that bind in root symbiosis and plant defense. Mol Genet Genomics 282:1–15
Decho AW, Norman RS, Visscher PT (2010) Quorum sensing in natural environments: emerging views from microbial mats. Trends Microbiol 18:73–80
Donlan RM, Costerton JW (2002) Biofilms: survival mechanisms of clinically relevant microorganisms. Clin Microbiol Rev 15:167–193
Dusane DH, Nancharaiah YV, Zinjarde SS, Venugopalan VP (2010) Rhamnolipid mediated disruption of marine *Bacillus pumilus* biofilms. Colloids Surf B: Biointerfaces 81:242–248
Dusane DH, Dam S, Nancharaiah YV, Kumar AR, Venugopalan VP, Zinjarde SS (2012) Disruption of Yarrowia lipolytica biofilms by rhamnolipid biosurfactant. Aquat Biosyst 8:17
Franzetti A, Gandolfi I, Bestetti G, Smyth TJP, Banat IM (2010) Production and applications of trehalose lipid biosurfactants. Eur J Lipid Si Technol 112:617–627
Fraysse N, Couderc F, Poinsot V (2003) Surface polysaccharide involvement in establishing the rhizobium-legume symbiosis. Eur J Biochem 270:1365–1380
Fujishige NA, Kapadia NN, De Hoff PL, Hirsch AM (2006) Investigations of Rhizobium biofilm formation. FEMS Microbiol Ecol 56:195–206
Gao M, Coggin A, Yagnik K, Teplitski M (2012) Role of specific quorum-sensing signals in the regulation of exopolysaccharide II production within *Sinorhizobium meliloti* spreading colonies. PLoS ONE 7:e42611. doi:10.1371/journal.pone.0042611
Garrett TR, Bhakoo M, Zhang Z (2008) Bacterial adhesion and biofilms on surfaces. Prog Nat Sci 18:1049–1056
Gjermansen M, Nilsson M, Yang L, Tolker-Nielsen T (2010) Characterization of starvation-induced dispersion in *Pseudomonas putida* biofilms: genetic elements and molecular mechanisms. Mol Microbiol 75:815–826
Gurich N, González JE (2009) Role of quorum sensing in *Sinorhizobium meliloti*-Alfalfa symbiosis. J Bacteriol 191:4372–4382
Hirsch AM, Fujishige NA (2012) Molecular signals and receptors: communication between nitrogen-fixing bacteria and their plant hosts. In: Wizany G, Baluska F (eds) Biocommunication in plants, vol 14. Springer, Berlin, pp 255–280
Janek T, Lukaszewicz M, Rezanka T, Krasowska A (2010) Isolation and characterization of two new lipopeptide biosurfactants produced by *Pseudomonas fluorescens* BD5 isolated from water from the Arctic Archipelago of Svalbard. Bioresour Technol 101:6118–6123
Jacques P (2011) Surfactin and other lipopeptides from *Bacillus* spp. Microbiol Monogr 20:57–91
Jefferson KK (2009) In: Ullrich M (ed) Bacterial polysaccharides: current innovations and future trends. Caister Academic Press, Norfolk, pp 175–186
Jofré E, Lagares A, Mori G (2004) Disruption of dTDP-rhamnose biosynthesis modifies lipopolysaccharide core, exopolysaccharide production and root colonization in *Azospirillum brasilense*. FEMS Microbiol 231:267–275
Karatan E, Watnick P (2009) Signals, regulatory networks, and materials that build and break bacterial biofilms. Microbiol Mol Biol Rev 73:310–347
Keith RC, Keith LM, Hernandez-Guzman G, Uppalapati SR, Bender CL (2003) Alginate gene expression by *Pseudomonas syringae* pv. *tomato* DC3000 in host and non-host plants. Microbiology 149:1127–1138
Kolodkin-Gal I, Romero D, Cao S, Clardy J, Kolter R, Losick R (2010) D-amino acids trigger biofilm disassembly. Science 328:627–629
Konishi M, Fukuoka T, Morita T, Imura T, Kitamoto D (2008) Production of new types of sophorolipids by *Candida batistae*. J Oleo Sci 57:359–369
Konishi M, Morita T, Fukuoha T, Imura T, Kakugawa K, Kitamoto D (2007) Production of different types of mannosylerythritol lipids as biosurfactant by the newly isolated yeast strains belonging to the genus *Pseudozyma*. Appl Microbiol Biotechnol 75:521–531

Kuiper I, Lagendijk EL, Pickford R, Derrick JP, Lamers GEM, Thomas-Oates JE, Lugtenberg BJJ, Bloemberg GV (2004) Characterization of two *Pseudomonas putida* lipopeptide biosurfactants, putisolvin I and II, which inhibit biofilm formation and break down existing biofilms. Mol Microbiol 51:97–113

Kurtzman CP, Price NPJ, Ray KJ, Kuo TM (2010) Production of sophorolipid biosurfactants by multiple species of the *Starmerella (Candida) bombicola* yeast clade. FEMS Microbiol Lett 311:140–146

Laus MC, Logman TJ, Lamers GE, Van Brussel AA, Carlson RW, Kijne JW (2006) A novel polar surface polysaccharide from *Rhizobium leguminosarum* binds host plant lectin. Mol Microbiol 59:1704–1713

Lindow SE, Brandl MT (2003) Microbiology of the phyllosphere. Appl Environ Microbiol 69:1875–1883

Mann EE, Wozniak DJ (2012) *Pseudomonas* biofilm matrix composition and niche biology. FEMS Microbiol Rev 36:893–916

Marketon MM, Glenn SA, Eberhard A, Gonzalez JE (2003) Quorum sensing controls exopolysaccharide production in *Sinorhizobium meliloti*. J Bacteriol 185:325–331

McDougald D, Rice SA, Barraud N, Steinberg PD, Kjelleberg S (2012) Should we stay or should we go: mechanisms and ecological consequences for biofilm dispersal. Nat Rev Microbiol 10:39–50

Meneses CHSG, Rouws LFM, Simões-Araújo JL, Vidal MS, Baldani JI (2011) Exopolysaccharide production is required for biofilm formation and plant colonization by the nitrogen-fixing endophyte *Gluconacetobacter diazotrophicus*. Mol Plant Microbe Interac 24:1448–1458

Mongiardini EJ, Ausmees N, Pérez-Giménez J, Althabegoiti JM, Quelas JI, López-Garcia SL (2008) The rhizobial adhesion protein RapA1 is involved in adsorption of rhizobia to plant roots but not in nodulation. FEMS Microbiol Ecol 65:279–288

Morris CE, Monier JM (2003) The ecological significance of biofilm formation by plant-associated bacteria. Annu Rev Phytopathol 41:429–453

Mulligan CN, Gibbs BF (2004) Types, production and applications of biosurfactants. Proc Indian Natl Sci Acad 70:31–55

Nievas F, Bogino P, Sorroche F, Giordano W (2012) Detection, characterization, and biological effect of quorum-sensing signaling molecules in peanut-nodulating bradyrhizobia. Sensors 12:2851–2873

Pérez-Montaño F, Jiménez-Guerrero I, Del Cerro P, Baena-Ropero I, López-Baena FJ, Ollero FJ, Bellogín R, Lloret J, Espuny R (2014) The symbiotic biofilm of *Sinorhizobium fredii* SMH12, necessary for successful colonization and symbiosis of glycine max cv Osumi, is regulated by Quorum Sensing Systems and Inducing Flavonoids via NodD1. PLoS ONE 9:e105901. doi:10.1371/journal.pone.0105901

Petrova OE, Sauer K (2012) Sticky situations: key components that control bacterial surface attachment. J Bacteriol 194:2413–2425

Ramey BE, Koutsoudis M, von Bodman SB, Fuqua C (2004) Biofilm formation in plant-microbe associations. Curr Opin Microbiol 7:602–609

Rinaudi L, Giordano W (2010) An integrated view of biofilm formation in Rhizobia. FEMS Microbiol Letters 304:1–11

Robledo M, Jiménez-Zurdo JI, Soto MJ, Velázquez E, Dazzo F, Martínez-Molina E, Mateos PF (2011) Development of functional symbiotic White clover root hairs and nodules requires tightly regulated production of rhizobial cellulase CelC2. Mol Plant Microbe Interact 24:798–807

Robledo M, Rivera L, Jiménez-Zurdo JI, Rivas R, Dazzo F, Velázquez E, Martínez-Molina E, Hirsch AM, Mateos PF (2012) Role of *Rhizobium endoglucanase* CelC2 in cellulose biosynthesis and biofilm formation on plant roots and abiotic surfaces. Microb Cell Fact 11:125

Rollet C, Gal L, Guzzo J (2009) Biofilm-detached cells, a transition from a sessile to a planktonic phenotype: a comparative study of adhesion and physiological characteristics in *Pseudomonas aeruginosa*. FEMS Microbiol Lett 290:135–142

Romero D, de Vicente A, Rakotoaly RH, Dufour SE, Veening JW, Arrebola E, Cazorla FM, Kuipers OP, Paquot M, Perez-Garcia A (2007) The iturin and Fengycin families of lipopeptides are key factors in antagonism of *Bacillus subtilis* toward *Podosphaera fusca*. Mol Plant Microbe Interact 20:430–440

Römling U, Gomelsky M, Galperin MY (2005) C-di-GMP: the dawning of a novel bacterial signalling system. Mol Microbiol 57:629–639

Schuster JJ, Markx GH (2014) Biofilm architecture. Adv Biochem Eng Biotechnol 146:77–96

Seneviratne G, Weerasekara MLMAW, Seneviratne KACN, Zavahir JS, Kecskés ML, Kennedy IR (2011) Importance of biofilm formation in plant growth promoting rhizobacterial action. In: Maheshwari DK (ed) Plant growth and health promoting bacteria, vol 18. Springer, Berlin, pp 81–95

Singh AK, Cameotra SS (2013) Rhamnolipids production by multi-metal-resistant and plant-growth-promoting rhizobacteria. Appl Biochem Biotechnol 170:1038–1056

Sorroche FG, Rinaudi LV, Zorreguieta A, Giordano W (2010) EPS II dependent auto aggregation of *Sinorhizobium meliloti* planktonic cells. Curr Microbiol 61:465–470

Stanley NR, Lazazzera BA (2004) Environmental signals and regulatory pathways that influence biofilm formation. Mol Microbiol 52:917–924

Sun LN, Zhang J, Chen Q, He J, Li QF, Li SP (2013) *Comamonas jiangduensis* sp. nov., a biosurfactant-producing bacterium isolated from agricultural soil. Int J Syst Evol Microbiol 63:2168–2173

Suthar H, Hingurao K, Desai A, Nerurkar A (2008) Evaluation of bioemulsifier mediated microbial enhanced oil recovery using sand pack column. J Microbiol Methods 75:225–230

Sutherland IW (2001) The biofilm matrix—an immobilized but dynamic microbial environment. Trends Microbiol 9:222–227

Tak HI, Ahmad F, Babalola O (2013) Advances in the application of plant growth-promoting rhizobacteria in phytoremediation of heavy metals. Rev Environ Contam Toxicol 223:33–52

Timmusk S, Nevo E (2011) Plant root associated biofilms: perspectives for natural product mining. In: Maheshwari DK (ed) Bacteria in Agrobiology: plant nutrient management. Springer, Berlin, pp 285–300

Tokumoto Y, Nomura N, Uchiyama H, Imura T, Morita T, Fukuoka T, Kitamoto D (2009) Structural characterization and surface-active properties of a succinoyl trehalose lipid produced by *Rhodococcus* sp. SD-74. J Oleo Sci 58:97–102

Van Bogaert INA, Zhang JX, Soetaert W (2011) Microbial synthesis of sophorolipids. Process Biochem 46:821–833

Vanderlinde EM, Muszynski A, Harrison JJ, Koval SF, Foreman DL, Ceri H (2009) *Rhizobium leguminosarum* biovar viciae 3841, deficient in 27-hydroxyoctacosanoate-modified lipopolysaccharide, is impaired in desiccation tolerance, biofilm formation and motility. Microbiology 155:3055–3069

Wagner M, Ivleva NP, Haisch C, Niessner R, Horn H (2009) Combined use of confocal laser scanning microscopy (CLSM) and Raman microscopy (RM): investigations on EPS-matrix. Water Res 43:63–76

Wang R, Liu TM, Shen MH, Yang MQ, Feng QY, Tang XM, Li XM (2012) Spiculisporic acids B-D, three new γ-butenolide derivatives from a sea urchin-derived fungus Aspergillus sp. HDf2. Molecules 17:13175–13182

Waters CM, Bassler BL (2005) Quorum sensing: cell-to-cell communication in bacteria. Annu Rev Cell Dev Biol 21:319–346

Webb JS, Thompson LS, James S, Charlton T, Tolker-Nielsen T, Koch B, Givskov M, Kjelleberg S (2003) Cell death in *Pseudomonas aeruginosa* biofilm development. J Bacteriol 185:4585–4592

Webb JS, Lau M, Kjelleberg S (2004) Bacteriophage and phenotypic variation in *Pseudomonas aeruginosa* biofilm development. J Bacteriol 186:8066–8073

Yuan J, Raza W, Huang Q, Shen Q (2011) Quantification of the antifungal lipopeptide iturin A by high performance liquid chromatography coupled with aqueous two phase extraction. J Chromatogr B Analyt Technol Biomed Life Sci 879:2746–2750

Bioremediation Strategies Employed by *Pseudomonas* Species

Amina Meliani

Abstract With the need of green chemicals and sustainable agroecosystem, biosurfactants' study and application is becoming imperative. Nowadays, it is known that Rhamnolipids are potent biosurfactants with high potential for bioremediation applications. The elimination of a wide range of pollutants in different ecosystems is an absolute requirement to promote a sustainable development of our society with low environmental impact. Biological processes play a major role in the removal of contaminants and they take advantage of the astonishing catabolic versatility of microorganisms to degrade/convert such compounds. The need to remediate these pollutants from contaminated sites has led to the development of effective, economic and environmentally friendly technologies. One of the current strategies used to enhance this process is the application of *Pseudomonas* PGPR to remediate contaminated soils in association with plants. Of all the present contaminants, the profound impacts of organic and heavy metal pollutants have attracted worldwide attention. This review focuses on the progress of PGPR for remediation of soils contaminated with the description of certain mechanisms and strategies.

Keywords Rhizoremediation · PGPR · Biofilm · Biosurfactant · Rhamnolipids

---

A. Meliani (✉)
Department of Biology, University of Mustapha Stambouli, Mascara, Algeria
e-mail: ameliani2003@yahoo.fr

D.K. Maheshwari (ed.), *Bacterial Metabolites in Sustainable Agroecosystem*,
Sustainable Development and Biodiversity 12, DOI 10.1007/978-3-319-24654-3_14

1 Introduction

The genus *Pseudomonas* covers one of the most diverse and ecologically significant groups of bacteria. The taxonomy of the genus is complex, comprising at least 105 recognized species at the time this chapter was written. Members of this genus are found in large numbers in a wide range of environmental niches. This almost universal distribution of *Pseudomonas* suggests a remarkable degree of genomic diversity and genetic adaptability (Rehm Bernd 2008).

Pseudomonas is a large genus within the γ subclass of Proteobacteria known for its ubiquity in the environment, utilization of a striking variety of organic compounds as energy sources (Wu et al. 2010), and production of an array of secondary metabolites (Gross and Loper 2009). As such, *Pseudomonas* spp. function as key components of ecological processes that suppress plant diseases in agricultural and natural environments (Weller et al. 2002). Some of its members are well known for their beneficial role to plants; others are used for bioremediation and as biocontrol agents. This versatile group of Gram-negative bacteria is capable of synthesizing a variety of secondary metabolites and to degrade a wide range of macromolecules and various recalcitrant pollutant compounds. Furthermore, some strains of *Pseudomonas* spp. produce phytohormones (Loper and Schroth 1986), other strains induce resistance responses in plants against disease (Bakker et al. 2007).

Among the soil microbiota, fluorescent *Pseudomonas* species like the PGPR one have attracted a significant interest and have been considered for bioremediation. These Plant-surface associated microorganisms often form multicellular aggregates, generally described as biofilms and play important roles in bioremediation.

Consequently, the production of biosurfactants rhamnolipids depends essentially on understanding the full mechanism of their physiological and catabolic strategies towards the recalcitrant pollutants. An increased understanding of bioremediation strategies has driven a revolution that is now, more than ever, involvement of biosurfactants and biofilm in bioremediation processes. Biofilm-grown cells exhibit enhanced tolerance toward adverse environmental stress conditions, and thus there has been a growing interest in recent years to use biofilms for biotechnological applications (Halan et al. 2011). These multicellular aggregates embedded in the matrix are found to have applications in bioremediation of hazardous waste sites, waste water treatment and mining acids or metals (Singh et al. 2006). Several studies reported that biofilm formation, a surface life style for many bacteria (Watnick and Kolter 2000), was a promoting factor for biodegradation of Polycyclic Aromatic Hydrocarbons (PAHs) in that biofilm ensured higher bioavailability of PAHs (Johnsen and Karlson 2004).

As highlighted in this brief introduction, it would be difficult to present all domains and aspects of *Pseudomonas* PGPR role in bioremediation. Therefore, here we review some mechanisms and strategies employed by *Pseudomonas* PGPR related to bioremediation processes through a special focus on the Rhamnolipids

(RLs). The second focus discusses applications of biofilm-mediated bioremediation processes. This chapter aims also to assemble the known cellular processes that might contribute to chemotaxis and motility behavior toward pollutants.

2 Applications of *Pseudomonas* as PGPR

Beneficial rhizobacteria that stimulate plant growth are usually referred to as plant growth-promoting rhizobacteria or PGPR (Glick et al. 1995). PGPR are a heterogeneous group of bacteria that can be found in the rhizosphere, at root surfaces and in association with roots. Plant growth-promoting bacteria may facilitate plant growth and development either indirectly or directly (Glick 2012, 2014). Indirect plant growth promotion occurs when these bacteria decrease or prevent some of the deleterious effects of a plant pathogen (usually a fungus) by any one of several different mechanisms. The direct promotion of plant growth by plant

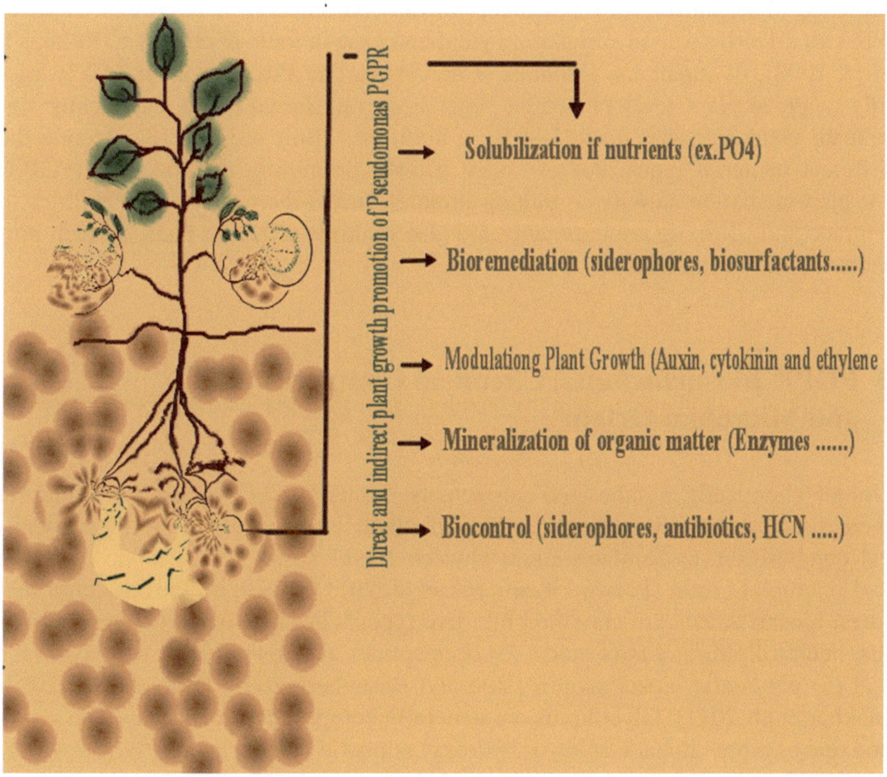

Fig. 1 The *pseudomonas* PGPR effect, the indirect plant growth promotion occurs with biocontrol action, whereas the direct promotion entails the releasing secondary metabolites in the rhizosphere or facilitating uptake of the released compounds from the environment (catabolic pathways)

growth-promoting bacteria generally entails facilitating the acquisition of nutrient resources from the environment (Fig. 1), such as nitrogen fixation (Compant et al. 2005), ammonia production (Marques et al. 2010), solubilization of mineral phosphate (Glick 2012), siderophore production (Lodewyckx et al. 2002), antibiotic production (Glick 2012) and production of plant hormones (Costacurta and Vanderleyden 1995) or in specifically modulating plant growth by altering plant hormone levels such as auxin, cytokinin and ethylene (Glick 2012).

Pseudomonas–plant interactions are found ubiquitously in nature and encompass a growing number of plant species, many of which are important to commercial horticulture (Rehm Bernd 2008). These interactions fall into two general groups those that are beneficial and those that are detrimental to the host plant's health (Rehm Bernd 2008). *Pseudomonas* sp. which successfully interacts with plant hosts ultimately gains an advantage over the competing microorganisms in the immediate environment like the PGPR agents. The most effective PGPR strains of *Pseudomonas* have been fluorescent *Pseudomonas* spp., and a considerable research is underway globally to exploit the potential of one group of bacteria that belongs to fluorescent Pseudomonads (Saharan and Nehra 2011). The presence of *P. fluorescence* inoculants in the combination of microbial fertilizer plays an effective role in stimulating yield and growth traits of chickpea (Rokhzadi et al. 2008), of sugarcane (Mehnaz et al. 2009). The Pseudomonads PGPR, rapidly colonize plant roots of potato, sugar beet, radish and cause statistically significant yield increases up to 144 % in field tests (Burr et al. 1978). Among the different bacterial types, use of plant growth promoting rhizobacteria (PGPR) for bioremediation activity is gaining imputes due to their differential abilities to degrade and detoxify contaminants and also multiple effect on plant growth promotion (Glick et al. 2010)

3 PGPR Biosurfactants, Chemical Composition and Microbial Origin

Unlike chemically synthesized surfactants, which are classified according to their dissociation pattern in water, biosurfactants are categorized by their chemical composition, molecular weight, physico-chemical properties, mode of action and microbial origin (Pacwa-Płociniczak et al. 2011). Based on the structural features, biosurfactants are classified into five types: (1) glycolipids, (2) phospholipids, neutral lipids and fatty acids, (3) lipopeptides and lipoproteins, (4) flavolipids and (5) polymeric biosurfactants (Ron and Rosenberg 2001; Bodour et al. 2003; Makkar et al. 2011). Glycolipids are generally composed of a carbohydrate group and one or more aliphatic acids or hydroxyl aliphatic acids. The mostly and widely studied glycolipids are sophorolipids and rhamnolipids. Furthermore, due to their molecular-mass, biosurfactants can be divided into two classes: low-molecular-mass molecules, which have the capability to lower surface and interfacial tension and high-molecular mass polymers, which are well known as emulsion-stabilizing agents (Abdel-Mawgoud et al. 2011). The classes of low-mass surfactants include glycolipids, lipopeptides and phospholipids whereas high-mass ones include polymeric and particulate surfactants (Abdel-Mawgoud et al. 2011).

As was pointed out by Rosenberg and Ron (1999) low-molecular-mass biosurfactants are efficient in lowering surface and interfacial tensions, whereas high-molecular-mass biosurfactants are more effective at stabilizing oil in water emulsions. It is obvious that biosurfactants are categorized mainly by their chemical composition and microbial origin (Maneerat 2005) as depicted in Table 1. The different types of biosurfactants include lipopeptides synthesized by many species of *Bacillus* and other species, glycolipids synthesized by *Pseudomonas* species and *Candida* species, phospholipids synthesized by *Thiobacillus thiooxidans*, and polysaccharide-lipid complexes synthesized by *Acinetobacter* species, or even the microbial cell surface (Mousa et al. 2006). To our knowledge, several biosurfactants have been reported to be produced by Pseudomonads, Rhamnolipids (*P. aeruginosa, P. putida, P. chlororaphis*), Viscosin (*P. fluorescens*), Acaterin (*Pseudomonas* sp. A92), Carbohydrate–protein–lipid (*P. fluorescens* 378) and Protein PA (*P. aeruginosa*) (Rehm Bernd 2008). According to Pamp and Nielsen (2007) *P. aeruginosa* produces a number of biosurfactants, of which the three most abundant are 3-(3-hydroxyalkanoyloxy) alcanoic acid (HAA), L-rhamnosyl-3-hydroxydecanoyl-3-hydroxydecanoate (mono-rhamnolipid) and L-rhamnosyl-L-rhamnosyl-3-hydroxyde-canoyl-3-hydroxydecanoate (di-rhamnolipid). HAA is synthesized via the *RhlA* enzyme and is converted to mono-rhamnolipid by the RhlB enzyme (Déziel et al. 1999). Mono-rhamnolipid is converted to di-rhamnolipid by the *RhlC* enzyme (Rahim et al. 2007) as described in Fig. 2. In addition, phosphate limitation and the presence of nitrate have been shown to promote the synthesis of rhamnolipids, while ammonium and high amounts of iron have been shown to repress the production of rhamnolipids (Pamp and Nielsen 2007).

Our focus in this review will be on rhamnolipid compounds, low-molecular-weight metabolites and the best studied and produced by other *Pseudomonas* species.

3.1 *Pseudomonas Biosurfactants*

Biosurfactants are surfactants that are produced extracellularly or as part of the cell membrane by bacteria, yeasts, and fungi (Mulligan 2005). They lower the surface tension of water; however, as commonly used, the term includes bioemulsifiers: substances which act as emulsifying agents but which do not necessarily have a significant effect on surface tension (Desai and Banat 1997). Most biosurfactants/bioemulsifiers are amphipathic cell components such as fatty acids, phospholipids, lipopolysaccharides, lipoteichoic acids, etc. Biosurfactants can potentially replace virtually any synthetic surfactant and, moreover, introduce some unique physico-chemical properties (Banat et al. 2010). Currently, the main

Table 1 Biosurfactants classification, microorganismes and environmental applications (Mulligan and Gibbs 2004; Pacwa-Płociniczak et al. 2011)

Biosurfactant group	Class	Microorganism	Environmental application
Glycolipids	Rhamnolipids	*Pseudomonas aeruginosa*, *Pseudomonas* sp.	Enhancement of the degradation and dispersion of different classes of hydrocarbons; emulsification of hydrocarbons and vegetable oils; removal of metals from soil
	Trehalolipids	*Mycobacterium tuberculosis, Rhodococcus erythropolis, Arthrobacter* sp., *Nocardia* sp., *Corynebacterium* sp.	Enhancement of the bioavailability of hydrocarbons
	Sophorolipids	*Torulopsis bombicola, (T. petrophilum, T. apicola)*	Recovery of hydrocarbons from dregs and muds; removal of heavy metals from sediments; enhancement of oil recovery
Fatty acids, phospholipids and neutral lipids	Corynomycolic acid	*Corynebacterium lepus*	Enhancement of bitumen recovery
	Spiculisporic acid	*Penicillium spiculisporum*	Removal of metal ions from aqueous solution; dispersion action for hydrophilic pigments; preparation of new emulsion-type organogels, superfine microcapsules (vesicles or liposomes), heavy metal sequestrants
	Phosphatidylethanolamine	*Acinetobacter* sp., *Rhodococcus erythropolis*	Increasing the tolerance of bacteria to heavy metals
Lipopeptides	Surfactin	*Bacillus subtilis*	Enhancement of the biodegradation of hydrocarbons and chlorinated pesticides; removal of heavy metals from a contaminated soil, sediment and water; increasing the effectiveness of phytoextraction
	Viscosin	*P. fluorescens*	Environmental and biomedical applications
		P. fluorescens SBW25	Spreading motility and plant growth promotion.
	Lichenysin	*Bacillus licheniformis*	Enhancement of oil recovery
Polymeric biosurfactants	Emulsan	*Acinetobacter calcoaceticus* AG-1	Stabilization of the hydrocarbon-inwater emulsions
	Alasa	*Acinetobacter radioresistens* KA-53	
	Biodispersan	*Acinetobacter calcoaceticus* A2	Dispersion of limestone in water
	Liposan	*Candida lipolytica*	Stabilization of hydrocarbon-in-water emulsions
	Mannoprotein	*Saccharomyces cerevisiae*	

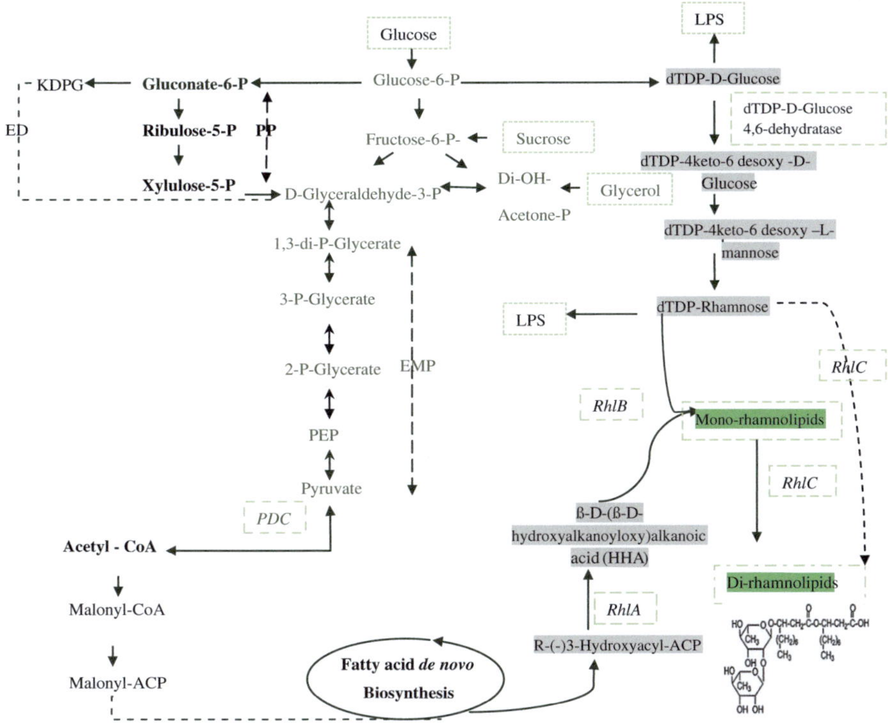

Fig. 2 Rhamnolipids biosynthesis pathway, highlighting the proposed contributions of *de novo* fatty acid biosynthesis and β-oxidation to the fatty acids incorporated into rhamnolipids. *EMP* Embden-Meyerhof-Parnas; *PP* Pentose phosphate; *ED* Entner-Doudoroff, *ACP* Acyl carrier protein; *CoA* Coenzyme A; *KDPG* 2-keto-3-deoxy-6-phosphogluconate, dTDP-L-rhamnose: deoxythymidine-diphospho-L-rhamnose, *HHA* β-hydroxyalkanoyl-β hydroxyalkanoic acid; *PDC* Pyruvate dehydrogenase complex

application is for the enhancement of oil recovery and hydrocarbon bioremediation due to their biodegradability and low critical micelle concentration (CMC) (Banat et al. 2010). Certain microorganisms produce extracellular biosurfactants or bioemulsifiers which may play a role e.g. in the adhesion of cells to and/or their detachment from surfaces, or in the utilization of hydrophobic substrates such as sulfur and hydrocarbons (Singleton and Sainsbury 2006).

Biosurfactants are a large heterogeneous group of microbial secondary metabolites and their most obvious property is their ability to effectively lower water surface tension (Fracchia et al. 2012). Biological surfactants possess several advantages over synthetic surfactants including high biodegradability, high emulsifying abilities, low toxicity and good general environmental compatibility (Pacwa-Plociniczak et al. 2011). Biosurfactants are therefore products with a broad potential of industrial (bioremediations, cosmetics, food and beverage manufacture) and pharmaceutical applications (Magalhaes and Nitschke 2013).

These secondary metabolites are a structurally diverse group of surface-active molecules and could have more or less specific roles in different ecological niches (Rosenberg and Ron 1999). Biosurfactants, produced by microorganisms (Nerurkar et al. 2009), are amphipathic surface-active molecules containing hydrophilic and hydrophobic moieties that act by emulsifying hydrocarbons, increasing their solubilization and subsequently rendering them available for microbial degradation (Darvishi et al. 2011). They consist of two parts: a polar (hydrophilic) moiety and non polar (hydrophobic) group. A hydrophilic group consists of mono-, oligo- or polysaccharides, peptides or proteins and a hydrophobic moiety usually contains saturated, unsaturated and hydroxylated fatty acids or fatty alcohols (Lang 2002). A characteristic feature of biosurfactants is a hydrophilic-lipophilic balance (HLB) which specifies the portion of hydrophilic and hydrophobic constituents in surface-active substances (Pacwa-Płociniczak et al. 2011).

Due to their amphiphilic structure, biosurfactants increase the surface area of hydrophobic water-insoluble substances, increase the water bioavailability of such substances and change the properties of the bacterial cell surface. Surface activity makes surfactants excellent emulsifiers, foaming and dispersing agents (Desai and Banat 1997).

3.1.1 Rhamnolipids (RLs) Biosurfactants

Among the biological surfactants, rhamnolipids reportedly have a good chance of being adopted by the industry as a new class of renewable resource-based surfactants (Müller et al. 2012). The rhamnolipid molecules reduce water surface tension and emulsify oil (Hang Pham et al. 2004), the glycolipid molecules comprising l-rhamnose and 3-hydroxylalkanoic acid were first identified in the mid 1900s in cultures of *P. aeruginosa* (Hauser and Karnovsky 1957). They are comprised of mono- and dirhamnose groups linked to 3-hydroxy fatty acids that vary in length, the most common being l-rhamnosyl-3-hydroxydecanoyl-3-hydroxydecanoate (monorhamnolipid) and l-rhamnosyl-l-rhamnosyl-3-hydroxydecanoyl-3-hydroxydecanoate (dirhamnolipid) (Maier and Soberon-Chavez 2000).

To date, the glycolipid biosurfactants produced mainly by *P. aeruginosa* are the most intensively studied biosurfactants. This arises from two contrasting facts (Abdel-Mawgoud et al. 2011). First, they display relatively high surface activities and are produced in relatively high yields after relatively short incubation periods by a well-understood, easy to cultivate microorganism. Second, they are one of the virulence factors contributing to the pathogenesis of *P. aeruginosa* infections, and consequently, many aspects of RL biosynthesis have been investigated, in part, to control their production and effects (Abdel-Mawgoud et al. 2011). However, many bacteria have been found to produce also rhamnolipids (Abdel-Mawgoud et al. 2010; Soberón-Chávez and Maier 2011). These include other *Pseudomonas* species or species that are taxonomically more distant from *Pseudomonas*, e.g. *Acinetobacter calcoaceticus*, *Pseudoxanthomonas* sp., *Enterobacter* sp., *Pantoea* sp., *Renibacterium salmoninarum*, *Nocardioides* sp., *Tetragenococcus koreensis* or *Burkholderia* sp. (Rooney et al. 2009).

***Pseudomonas* rhamnolipid producers** *Pseudomonas* species have long been the main sources of RLs, with *P. aeruginosa* being considered the primary producing species. However, many isolates from other bacterial species of a varying distance in their taxonomical classification are increasingly reported to be also RL producers (Abdel-Mawgoud et al. 2010). First, many *Pseudomonas* species other than *P. aeruginosa* have been reported to produce RLs (Onbasli and Aslim 2009).

Furthermore, Several attempts to produce *Pseudomonas* RLs in heterologous hosts have been reported. Yet, none produces RLs in comparable levels as the best *P. aeruginosa* strains. In view of a commercial production of RLs, there is still a huge potential for genetic optimization (Abdel-Mawgoud et al. 2011). Ochsner and Reiser (1995) cloned the *rhlAB* rhamnosyl transferase gene into various hosts, *P. fluorescens*, *P. oleovorans*, *P. putida* and *Escherichia coli*. The best RL production was 60 mg/L and was achieved with *P. putida*, whereas no production was obtained with *E. coli*.

The structure of RLs is highly diverse and those produced by *P. aeruginosa* have been extensively studied. These RLs are amphiphilic molecules typically composed of 3-hydroxyfatty acids linked through a beta-glycosidic bond to mono- or di-rhamnoses (Soberon-chavez et al. 2005) (Fig. 2). RLs have several potential functions in bacteria (Vasta et al. 2010). They are one of the virulence factors contributing to the pathogenesis of *P. aeruginosa* infections and are essential for surface motility and biofilm development (Abdel-Mawgoud et al. 2010). Inspite of that, rhamnolipids exhibit several useful industrial applications such as emulsification, detergency, wetting, foaming, dispersing, solubilization, antimicrobial and anti-adhesive activities in different areas from bioremediation to food additives. These topics have been extensively reviewed including some very recent articles (Banat et al. 2010; Pornsunthorntawee et al. 2010). Recent studies also reported a new role for RLs as potential players in the combat of plants and animals against microbes have recently emerged (Vatsa et al. 2010).

Non-*P. aeruginosa* rhamnolipid producers As described in the literature most RL-producing species belong to the closely related genera *Pseudomonas* and *Burkholderia* in the phylum proteobacteria (Walter et al. 2010). The genus *Burkholderia arose* from the genus *Pseudomonas* and was classified as a new genus in 1992 based on 16S rRNA sequence analysis (Yabuuchi et al. 1992). Consequently, bacteria of this genus have characteristics similar to *Pseudomonas*, and some species indeed produce RLs (Abdel-Mawgoud et al. 2011).

It is also interesting to note that another nonpathogenic species (Biosafety level 1) would represent a very interesting alternative—if sufficient RL yields can be obtained (Abdel-Mawgoud et al. 2011). The most prominent nonpathogenic RL producers from the genus *Pseudomonas* are *P. chlororaphis* (Gunther et al. 2005), *P. alcaligenes* (Oliveira et al. 2009), *P. putida* (Martinez-Toledo et al. 2006; Meliani and bensoltane 2014) and *P. fluorescens* (Meliani and bensoltane 2014).

It is interesting to point out that despite the apparent safety advantage of these RL producers, very little is yet known about the biotechnological potential of these species (Abdel-Mawgoud et al. 2011).

3.1.2 The Rhamnolipid Biosynthesis Pathway

Rhamnolipids are composed of one or two hydrophobic β-hydroxy fatty acids, which are linked through a β-glycosidic bond to one or two rhamnose molecules forming the hydrophilic moiety (Wang et al. 2007; Wittgens et al. 2011). According to the number of rhamnose moieties, mono- and di-rhamnolipids are differentiated (Wittgens et al. 2011). The fatty acids alkyl chain length in *P. aeruginosa* can vary from C8 to C14 (Abdel-Mawgoud et al. 2010); the most abundant species contains two β-hydroxy fatty acids with C10 chains. The alkyl chains can also contain up to two unsaturated C–C bonds (Wittgens et al. 2011).

Rhamnolipid production is transcriptionally regulated by quorum sensing (Ochsner and Reiser 1995; Pearson et al. 1997). The synthesis pathway of rhamnolipids consists of three dedicated enzymatic reactions (Wittgens et al. 2011). The biosynthesis of these tensio-active molecules proceeds by two sequential rhamnosyl transfer reactions, each catalysed by a specific rhamnosyltransferase (RhlB and RhlC, respectively), with deoxythymidine diphospho-l-rhamnose (dTDP-l-rhamnose) acting as rhamnosyl donor in both reactions and 3-(3-hydroxyalkanoyloxy) alkanoic acids (HAAs), the product of RhlA catalysis (Déziel et al. 2003; Cabrera-Valladares et al. 2006), or mono-rhamnolipid acting as the respective recipients (Aguirre-Ramı´rez et al. 2012). In the first step two activated β-hydroxy fatty acids are linked by *RhlA*, the 3-hydroxyacyl-ACP (acyl carrier protein):3-hydro-xyacyl-ACP O-3-hydroxyacyltransferase, to a dimer, called 3-(3-hydroxyalkanoyloxy) alkanoate (HAA) (Zhu and Rock 2008). Mono-rhamnolipids are produced by *RhlB*, the rhamno-syltransferase I, by the condensation of HAA and dTDP-L-rhamnose. The rhamnosyltransferase II (*RhlC*) adds another rhamnose moiety to the mono-rhamnolipid resulting in a di-rhamnolipid (Rahim et al. 2001) as outlined in Fig. 2 for *P. aeruginosa*. Notably, *RhlG*, ab-hydroxyacyl-ACP: CoA (coenzyme A) transacylase, previously associated with rhamnolipid synthesis, is not required for rhamnolipid synthesis in vitro. The genes *rhlA* and *rhlB* are organized in a single operon, while *rhlC* is localized in another region of the *P. aeruginosa* genome and forms an operon with a gene of unknown function (Wittgens et al. 2011).

The donor of the rhamnosyl moiety is dTDP-l-rhamnose, produced from d-glucose by a series of reactions catalyzed by the AlgC and Rml enzymes (Maier and Soberon-Chavez 2000). It has been suggested that the lipid moiety precursor is diverted from de novo fatty acid synthesis since the stereochemistry of the β-hydroxy fatty acid in the lipid moiety matches that of the intermediates in de novo fatty acid synthesis but not intermediates in the catabolic fatty acid β-oxidation pathway. RhlA has been identified as the enzyme that diverts the (*R*)-β-hydroxyacyl-acyl carrier protein (ACP) from de novo fatty acid synthesis to the rhamnolipid biosynthesis pathway by catalyzing the formation of HAA. *RhlB* and *RhlC* are two rhamnosyltransferases which are responsible for the consecutive additions of two rhamnosyl groups to HAA to form monorhamnolipid first and then dirhamnolipid (Zhang et al. 2012).

In *P. aeruginosa* the expression of all three genes involved in rhamnolipid synthesis is regulated by two quorum sensing systems and is active when *Paeruginosa* is cultivated under phosphate or nitrogen-limiting conditions (Benincasa et al. 2002). The σ^{54} factor is responsible for the expression of *rhlAB* under these conditions. LasI/R and RhlI/R are also regulating transcription of many virulence factors (Wittgens et al. 2011).

3.1.3 Focus on Rhamnolipids Market

Rhamnolipids are well-characterized and scientifically proven biosurfactants which are slowly and steadily becoming highly sought after biomolecules (Sekhon Randhawa and Rahman 2014). Among other biosurfactants rhamnolipids have the highest number of patents and research publications. However, cost-competitiveness is one of the major factors that is holding rhamnolipids back from becoming the champions of their field (Sekhon Randhawa and Rahman 2014).

It has been proven that rhamnolipids being produced using fermentation processes can be produced at a United States cost of $5–20/kg (Renfro 2013). Compared to a synthetic surfactant production at a United States cost of $1–3/kg, a decrease in Rhamnolipid manufacturing cost will increase production (Maier and Soberon-Chavez 2000; Soberon-Chavez et al. 2005). The current market price of rhamnolipid (R-95, 95 %) is $227/10 mg (Sigma-aldrich) and $200/10 mg (AGAE technologies, USA) calling for strenuous research. Rhamnolipids have favorable applications in various sectors and if made economically sustainable nothing can stop these biomolecules to rule the surface-active compounds market (Sekhon Randhawa and Rahman 2014).

3.2 Inimitable Applications of Rhamnolipids

In the past close to three decades, there has been a great body of research work carried out on rhamnolipids revealing many of their astonishing applications and making them reach the pinnacle of popularity among all the categories of biosurfactants in the global market (Sekhon Randhawa and Rahman 2014). The reason behind the current global interest in rhamnolipid production owes to their broad range of applications in various industries along with many spectacular "eco-friendly" properties (Sekhon Randhawa and Rahman 2014).

It is imperative to evaluate the major applications of rhamnolipids that make them noticeable among other biosurfactants (Sekhon Randhawa and Rahman 2014). Among the five major applications listed below, our attention had been focused on their application in Bioremediation and enhanced oil recovery (EOR):

(i) *Bioremediation and enhanced oil recovery* (*EOR*) Rhamnolipids show excellent emulsification properties, efficiently remove crude oil from contaminated soil and facilitate bioremediation of oil spills (Rahman et al. 2003).

(ii) *Pharmaceuticals and therapeutics* Clinical testing of RLs as pharmacoactive compounds has been performed. Some successful trials proved the potential applications of RLs for the treatment of ulcers (Piljac et al. 2008) and of full-thickness wounds (Stipcevic et al. 2006). Furthermore, rhamnolipids have been shown to display antibacterial activities against plant and human pathogenic bacteria (Vatsa et al. 2010)· They show low-toxicity, surface-active properties and antimicrobial activities against several microbes (*Bacillus cereus*, *Micrococcus luteus*, *Staphylococcus aureus*, *Listeria monocytogenes*), thereby showing promising applications in pharmaceuticals and therapeutics (Magalhaes and Nitschke 2013).

(iii) *Cosmetics* Rhamnolipid as an active ingredient is found to be effective for several skin treatments i.e. wound healing with reduced fibrosis, cure of burn shock and treatment of wrinkles; hence are in demand in the health and beauty industry (Sekhon Randhawa and Rahman 2014).

(iv) *Detergents and cleaners* Rhamnolipids are natural emulsifiers and surface-active agents leading to their wide spread usage in detergent compositions, laundry products, shampoos and soaps (Pacwa-Płociniczak et al. 2011; Parry et al. 2013).

(v) *Agriculture* Rhamnolipids are already used for soil remediation for improving soil quality and are now further getting explored for plant pathogen elimination, for aiding the absorption of fertilizers and nutrients through roots and as biopesticides (Sachdev and Cameotra 2013).

3.2.1 Rhamnolipids on Environmental Remediation

From a biotechnological point of view, RLs are powerful biosurfactants with applications related to environmental concerns, such as bioremediation of hydrocarbon, organic pollutants and heavy-metal-contaminated sites (Vatsa et al. 2010). Rhamnolipids have been shown to have potential use in several applications, but most of the research has focused on environmental remediation. Currently, bioremediation is thought to be a cost- and performance-effective technology to solve environmental pollution problems (Pornsunthorntawee et al. 2010). With the use of rhamnolipids, the biodegradation of these pollutants can be significantly enhanced (Pornsunthorntawee et al. 2010). Zhang et al. (1997) reported that rhamnolipids increased the solubility of phenanthrene (polycyclic aromatic hydrocarbons) in a test solution, resulting in the enhancement of the phenanthrene biodegradation rate.

In addition, rhamnolipids produced by *P. aeruginosa* UG2 was found to increase the solubilization of pesticides, resulting in the stimulation of biodegradation rate and extent (Mata-Sandoval et al. 2000). Furthermore, the enhancement of hexadecane biodegradation by rhamnolipids has also been reported by Abalos et al. (2001), however Rahman et al. (2002) showed that rhamnolipid-containing additives had positive effects on the bioremediation of gasoline-contaminated soil. Whang et al. (2008) signaled a potential use of rhamnolipids produced by *P. aeruginosa* J4 for the biodegradation of diesel-contaminated water and soil, whereas Clifford et al. (2007) found that rhamnolipids produced by *P. aeruginosa* ATCC 9027 significantly improved the solubilization of tetrachloroethylene (PCE), a common ground water pollutant, indicating the potential use of the tested biosurfactant in surfactant-enhanced aquifer remediation (SEAR) applications (Pornsunthorntawee et al. 2010).

Moreover, Cassidy et al. (2002) also suggested that rhamnolipids might be applied in intrinsic bioremediation using in situ rhamnolipid production at an abandoned petroleum refinery. Bai et al. (1997) reported that monorhamno-lipid produced by *P. aeruginosa* ATCC 9027 displayed efficiency in the removal of residual hexadecane from soil higher than three synthetic surfactants: SDS, polyoxyethylene and sorbitan monooleate. However, Noordman et al. showed that rhamnolipids produced by *P. aeruginosa* UG2 effectively removed phenanthrene from soil (Noordman et al. 1998).

For the removal of heavy metals such as copper, zinc and lead by rhamnolipids several studies have also been reported (Mulligan et al. 2001; Herman et al. 1995). Mulligan and Wang (2006) found that rhamnolipid foam effectively removed inorganic heavy metal, including cadmium and nickel, from a contaminated soil sample (Pornsunthorntawee et al. 2010).

4 Applications of Biofilms in Bioremediation

The pollutants can range from polycyclic aromatic hydrocarbons, refined petroleum products, acid mine drainage, pesticides, industrial waste and heavy metals to crude oil (Finnerty 1994). Interestingly, biofilms are found to have applications in bioremediation of hazardous waste sites, waste water treatment and mining acids or metals (Singh et al. 2006). Applications of biofilms in bioremediation are of recent interest because they exhibit better metabolic activity (Kirchman and Mitchell 1982), survival rate and rate of gene transfer (Molin and Tolker-Nielsen 2003).

Several studies emphasize the importance of the formation of biofilms in bioremediation of toxic pollutants and recalcitrant polymers. Furthermore, many reports are available on the bioremediation of heavy metals (Langley and Beveridge 1999) and hydrocarbons (Puhakka et al. 1995) by biofilms composed of *Pseudomonas* sp.

Similarly, applications of rhamnolipid in heavy metal removal and their role in enhancing the bioavailability of hydrocarbons have been reported (Mulligan 2004). Moreover, successful application of a bioremediation process relies upon an understanding of interactions among microorganisms, organic contaminants and soil or aquifer materials (Singh et al. 2006). Physiological properties of the microorganisms such as biosurfactant production and chemotaxis enhance bioavailability and, hence, degradation of hydrophobic compounds (Pandey and Jain 2002).

Microorganisms that secrete polymers and form biofilms on the surface of hydrocarbons are especially well suited for the treatment of recalcitrant or slow-degrading compounds because of their high microbial biomass and ability to immobilize compounds by biosorption (passive sequestration by interactions with biological matter), bioaccumulation (increased accumulation of microbes under influence) and biomineralization (formation of insoluble precipitates by interactions with microbial metabolic products) (Singh et al. 2006).

Furthermore, biofilms support a high biomass density that facilitates the mineralization processes by maintaining optimal conditions of pH, localized solute concentrations and redox potential in the vicinity of the cells. This is achieved by the unique architecture of the biofilm and controlled circulation of fluids within it (Flemming 1995). Biofilm-based reactors are commonly used for treating large volumes of dilute aqueous solutions such as industrial and municipal wastewaters (Singh et al. 2006).

In the case of *Pseudomonas* sp. (*P. putida, P. fluorescens* and *P. cepacia*) some bioreactors are currently being used for bioremediation of various hydrocarbon contaminants such as 2,4-dichloropheno, 2,4,6-trichlorophenol, 2,3,4,6-tetrachlorophenol, pentachlorophenol, pyrene, phenanthrene, o-cresol, naphthalene, phenol, 1,2,3-trimethylbenzene and carbon tetrachloride, with an overall efficiency of ~100 % (Kargi and Eker 2005; Puhakka et al. 1995; Singh et al. 2006).

4.1 Biofilm a Promoting Factor for Biodegradation

Biofilm formation is often thought to represent a protective mode of growth which may enhance bacterial survival under conditions of environmental stress (Webb et al. 2003). Recently it was found that rhamnolipids play a major role in the architecture of biofilms produced by *P. aeruginosa* and that the formation of water channels is strongly dependent on the presence of rhamnolipids (Davey et al. 2003). In this context, one obvious question is whether rhamnolipid regulates the biofilm formation and resistance toward pollutants. This will be addressed through a report of certain studies related to the promoting role of biofilm in biodegradation. Three gram-negative soil bacteria *P. fluorescens*, *P. putida* and *P. aeruginosa* are of intense interest for bioremediation, as they may degrade various organic pollutants and at the same time withstand the toxicity of many soluble metal ions. Growth in a biofilm modulates microbial metal resistance, often increasing the ability of microorganisms to withstand toxic metal ions by several orders of magnitude. Past studies have often focused on only a few metal ions at a time and hence little data are available for elucidating universal trends of metal toxicity in bacteria.

Several studies reported that biofilm formation, a surface life style for many bacteria (Watnick and Kolter 2000), was a promoting factor for biodegradation of Polycyclic Aromatic Hydrocarbons (PAHs) in that biofilm ensured higher bioavailability of PAHs (Johnsen and Karlson 2004). Meliani and Bensoltane (2014) study offers important insights into the biodegradation of monoaromatic, cyclic hydrocarbons compounds and indicates that the biofilm is able to grow more in the presence of xylene and thereby maintain itself over a long period of incubation with different aspects of thickness.

Biofilm-mediated bioremediation has been found to be a more effective and safer alternative to bioremediation with planktonic bacteria since cells growing within a biofilm have higher chances of adaptation to different environments and their subsequent survival (Singh et al. 2006). Biofilms have been found to protect the microbial community from environmental stresses (Flemming and Wingender 2002). This is why the formation of biofilms in natural and industrial environments allows bacteria to develop resistance to bacteriophage, amoebae, chemically diverse biocides, host immune responses and antibiotics (Costerton et al. 1999) and is increasingly recognized as an important virulence factor in a variety of chronic infections. Biofilm reactors are generally used to treat hydrocarbons, heavy metals and large volumes of dilute aqueous solutions such as industrial and municipal waste water (Singh et al. 2006).

Moreover, rhamnolipids, produced under control of the quorum-sensing regulator, *rhlR*, are known to contribute to the biofilm architecture and detachment, but their influence at early stages of biofilm formation is less certain.

4.2 Chemotaxis and Motility Behavior Toward Pollutants

Chemotaxis is the movement of organisms in response to a chemical nutrient or chemical gradient. It helps bacteria to find optimum conditions for growth and survival and is an integral feature of biodegradation (Pandey and Jain 2002). Cells displaying chemotaxis can sense chemicals such as those adsorbed to soil particles in a particular niche and swim toward them; hence, the mass-transfer limitations that impede the bioremediation process can be overcome (Al-Awadhi et al. 2003). Once the cells are brought into close contact with a surface, the mechanism of biofilm formation and surfactant production commences, which leads to enhanced bioavailability and biodegradation. When the target contaminants are dissolved in an aqueous medium, the rate of biodegradation is improved compared with those hydrophobic pollutants that remain adsorbed in the non-aqueous phase liquid (NAPL) associated with contaminated soils (Law and Aitken 2003; Singh et al. 2006). Bacteria access these target compounds either by dissolution of the target compounds in the aqueous phase or by direct adhesion to the NAPL–water inter-face, a process that is facilitated by biofilm formation (Singh et al. 2006).

Motile bacteria have the ability to sense changes in the concentration of chemicals in environments and respond to them by altering their pattern of motility. This behavioral response is called chemotaxis. The Pseudomonads also show chemotactic responses to various chemical compounds, including amino acids, organic acids, sugars, aromatic compounds and inorganic ions (Rehm Bernd 2008).

Since the Pseudomonads include important plant and animal pathogens and potential agents of geochemical cycles, biocontrol and bioremediation, and chemotaxis is thought to play an important role in microbe host and microbe substrate interactions, ecological aspects of chemotaxis have been intensively investigated in *Pseudomonas* species (Rehm Bernd 2008).

Soil-borne fluorescent pseudomonads, including *P. fluorescens* and *P. putida*, exhibit beneficial effects on plants such as the promotion of plant health and development, and biological control of soil-borne diseases (Bolwerk et al. 2003). These plant-deleterious and plant-beneficial Pseudomonads were found to exhibit chemotactic responses toward root and seed exudates, leaf surface water and organic compounds prevalent in them (Grewal and Rainey 1991). Furthermore, it was reported that ethylene, a plant hormone, also serves as a chemo attractant for *P. fluorescens*, *P. putida* and *P. syringae* (Kim et al. 2007).

Interestingly, chemotaxis of pollutant-degrading bacteria is a trait important for bioremediation, especially in situ bioremediation. The migration of pollutant degraders to pollutants is expected to speed the biodegradation process because it should bring the cells into contact with pollutants. Therefore, there has been much research on chemotaxis of Pseudomonads toward environmental pollutants. Benzoate-degrading *P. putida* PRS2000 is attracted by aromatic acids including benzoate, p-hydroxybenzoate, toluates, salicylate and chlorobenzoates (Rehm Bernd 2008). Parales et al. (2000) reported that the soil bacteria can sense and swim toward the toxic compounds toluene, benzene, trichloroethylene and related chemicals suggesting that the introduction of chemotactic bacteria into selected polluted sites may accelerate bioremediation processes.

Though chemotactic response of motile bacteria to a wide variety of attractants is well documented (Gupta et al. 1991), the behavioral aspect of different motility traits, such as speed, rate of change in direction and net to gross displacement ratio in response to recalcitrants compounds and pollutants is yet to be ascertained. Harrison et al. (2007) signaled that biofilm formation is a strategy that microorganisms might use to survive a toxic flux in the inorganic compounds. Evidence in the literature suggests that biofilm populations are protected from toxic compounds by the combined action of physiological phenomena that are, in some instances, linked to phenotypic variation among the constituent biofilm cells such as the swarming, swimming and twitching behavior. In this section, we have assembled the known cellular and physiological phenomena of *Pseudomonas* behavioral aspect related to Rhamnolipids synthesis.

Bacterial swarming motility has been shown to be important to biofilm formation (Shrout et al. 2006), where cells act not as individuals, but as coordinated groups to move across surfaces, often within a thin-liquid film (Kearns 2010). Many swarming bacteria are aided by the production of a surfactant that lowers surface tension of the liquid film to improve bacterial motility (Kearns 2010; Du et al. 2012). The swarming communities of *P. aeruginosa* represent a complex intersection of physical, biological and chemical phenomena (Du et al. 2012). The

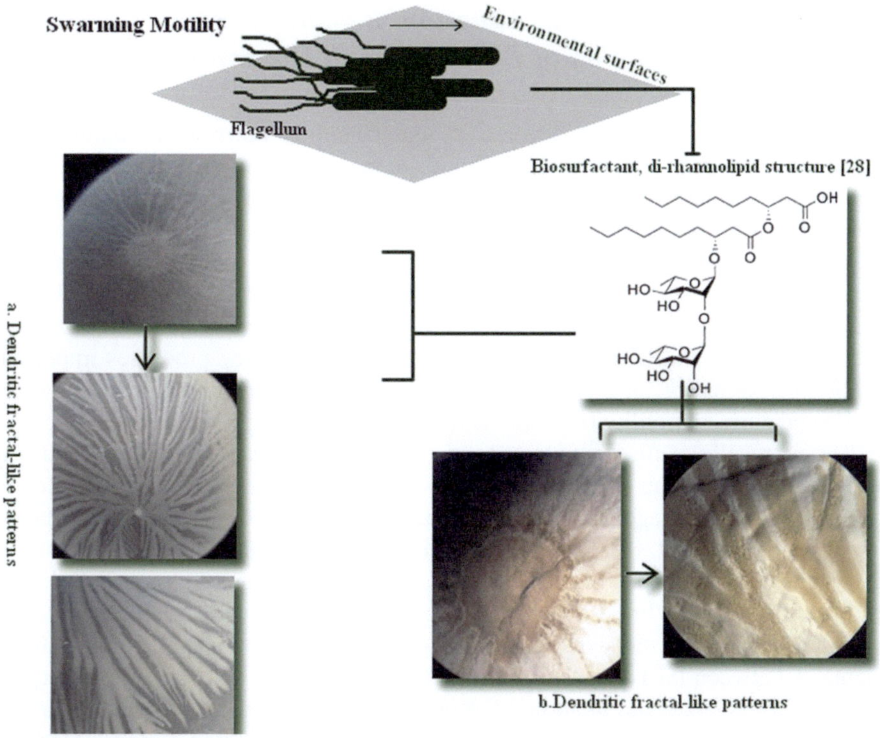

Fig. 3 Rhamnolipids are required for modulating swarming behavior in certain different species of *Pseudomonas*. Note the formation of dendritic fractal-like patterns formed by migrating swarms moving away from an initial location of these species. Swarming motility on TSA medium (1.5 % agar) for 72 h at 30 °C after a central spot of 5 ml of an overnight bacterial culture in TSB. Views of swarming motility are provided from my personal current study

branched tendril patterns that are often, but not always, observed in *P. aeruginosa* swarms require production of rhamnolipid (RL) (Xavier et al. 2011). As an extracellular lipid, RL acts as a surfactant to reduce surface tension in bacterial suspensions. In addition to RL, a functional bacterial flagellum is also required for swarms to form tendrils (Caiazza et al. 2005) (Fig. 3).

This complex type of motility is usually defined as a rapid and coordinated translocation of a bacterial population across a semi-solid surface (Tremblay and Déziel 2010). In addition to flagella, swarming of *P. aeruginosa* requires the release of two exoproducts, rhamnolipids (RLs) and 3-(3-hydroxyalka-noyloxy) alkanoic acids (HAAs), which act as wetting agents and chemotactic-like stimuli (Kohler et al. 2000) and are also implicated in many aspects of biofilm development (Tremblay and Déziel 2010).

4.3 Biofilm and Microbial Adhesion, Beneficial Tools for the Degradation of Environmental Hazardous

Another line of research is devoted to understanding the role of rhamnolipids in biofilm formation. Biofilms are abundant in nature and are of clinical, environmental and industrial importance (Glick et al. 2010). Biofilm development is known to follow a series of complex but discrete and tightly regulated steps (Harbron and Kent 1988; Kjelleberg and Givskov 2007), including (i) microbial attachment to the surface, (ii) growth and aggregation of cells into microcolonies, (iii) maturation and (iv) dissemination of progeny cells that can colonize new niches (Glick et al. 2010). Over the last decade, several key processes important for biofilm formation have been identified, including quorum sensing (Joseph et al. 2001) and surface motility (O'Toole and Kolter 1998; Rehm Bernd 2008).

The extracellular filamentous appendages produced by motile microorganisms are responsible for the attachment process and interact with surface in a different manner. Till date, Flagella and pili had been the subject of intense study mainly for two reasons. First, their responsibility in behavior motility. Second, because of their consideration as one of the three major matrix components.

Flagella are very fine threads of the protein flagellin with a helical structure extending out from the cytoplasm through the cell wall. Flagella may have a diameter between 0.01 and 0.02 mm, and a length of up to 10 mm. Many types of bacteria have flagella, including the genus *Pseudomonas*. It is possible that the flagellum itself may form an adhesive bond with the adhesion surface (Harbron and Kent 1988). The primary function of flagella in biofilm formation is assumed to be in transport and in initial cell–surface interactions (Sauer and Camper 2001). Flagella-mediated motility is believed to overcome repulsive forces at the surface of the substratum. Moreover, pili or fimbriae are found on many Gram-negative bacteria including *Pseudomonas* species. They are fine, filamentous appendages, also of protein, 4–35-nm wide and up to several micrometers long (Harbron and Kent 1988). These structures are usually straight, and are not involved in motility. Their only known general function is to make cells more adhesive, since bacteria with pili can adhere strongly to other bacterial cells and inorganic particles (Rogers 1979).

Biofilm development involves several stages which must be understood in order to achieve biofilm control and which begin by the attachment of pioneer bacteria to the surface. Biofilm formation and swarming motility are inversely regulated (Kuchma et al. 2007). This phenotype requires coordination of several pathways, including flagellar motility, cell–cell signaling/quorum sensing and biosurfactant secretion (Déziel et al. 2003). Swarming motility is operationally defined as a rapid multicellular movement of bacteria across a surface, powered by rotating flagella (Kearns 2010). Swimming motility is a mode of bacterial movement that is also powered by rotating flagella but, unlike swarming motility, swimming takes place as individual cells moving in liquid environments. Twitching motility is a surface motility powered by the extension and retraction of type IV pili, which confers slow cell movement, often with a jerky or 'twitchy' appearance (Mattick 2002). Gliding motility is a catch-all definition for active surface movement that occurs along the long axis of the cell without the aid of flagella or pili (Kearns 2010). Gliding seems to have evolved independently in multiple lineages but generally involves the cell body moving through the use of focal-adhesion complexes that bind to a surface substrate. Sliding motility is a passive form of surface spreading that does not require an active motor but instead relies on surfactants to reduce surface tension, enabling the colony to spread away from the origin, driven by the outward pressure of cell growth (Kearns 2010).

Biofilm-forming bacteria are generally known to employ both extracellular and intracellular biofilm factors including membrane appendages and extracellular matrices (Déziel et al. 2003). As is the case with most bacteria, environmental isolates of *Pseudomonas* are capable of forming different types of biofilm, including pellicles (floating biofilms at the air liquid interface) or wrinkly spreaders (WSs, or solid surface-associated submerged biofilms) (Kearns 2010). Recent studies involving *P. fluorescens* WSs have shown that certain factors including cellulose matrix, fimbriae and lipopolysaccharides (LPS) might be extremely relevant to the strength and integrity of WS (Mattick 2002).

Biofilms can be detrimental to both human life and industrial processes, causing infection associated with medical implants (Dankert et al. 1986), pathogen interaction with host cells (Soto and Hultgren 1999), periodontitis, contamination of food from processing equipment, enhancement of metal corrosion (Hori and Matsumoto 2010), and so on. However, microbial adhesion can be also beneficial, for example, in the degradation of environmental hazardous chemicals in soil (Bouwer and Zehnder 1993) or in bioreactors for waste water treatment or off-gas treatment (Hori and Matsumoto 2010), in agricultural uses of root nodule bacteria in the rhizosphere (Espinosa-Urgel et al. 2000) and in bioflocculants used for the separation of coal particles (Hori and Matsumoto 2010).

4.3.1 Role of Biofilms in Remediation of Heavy Metals

Metal contamination has been linked to birth defects, cancer, skin lesions, mental and physical retardation, learning disabilities, liver and kidney damage and a host of other diseases (Singh and Cameotra 2004). Heavy metals are the primary inorganic contaminants, which include cadmium, chromium, copper, lead, mercury, nickel and zinc. Heavy metal bioremediation can be achieved by immobilization, concentration and partitioning to an environmental compartment, thereby minimizing the anticipated hazards (Barkay and Schaefer 2001; Lloyd 2003).

Another promising application of biofilms is in heavy metal and radionuclide remediation (Barkay and Schaefer 2001). The distribution and diversity of microbes that inhabit contaminated sites and of the genes that encode for phenotypes responsible for metal–microbe interactions are crucial elements in metal and radionuclide bioremediation (Singh et al. 2006). Déziel et al. (1996) demonstrated the potential of rhamnolipids in bioremediation of sites contaminated with toxic heavy metals such as uranium, cadmium and lead.

Furthermore and because of their anionic nature, rhamnolipids can be used to remove heavy metalions, i.e. cadmium, lead and zinc (Miller 1995). Along with their potential in removing heavy metals, their addition to the hydrophobic substrates helps microorganisms with uptake and assimilation of insoluble hydrocarbons such as linear alkanes, which are very insoluble in water but are good nutrient sources for *P. aeruginosa* (Hommel 1994).

Bioremediation and enhanced oil recovery (EOR): Rhamnolipids show excellent emulsification properties, efficiently remove crude oil from contaminated soil and facilitate bioremediation of oil spills (Rahman et al. 2003; Costa et al. 2010).

Furthermore, a simultaneous increase in the EPS content of the biofilm was also observed, which suggested the role of EPS and biofilms in the entrapment of metal precipitates (Singh et al. 2006). Macaskie et al. (2000) observed that polycrystalline $NaUO_2PO_4$ accumulated in and around the cell wall of Citrobacter sp. N14 by adsorption to lipopolysaccharide and, hence, aided in its bioprecipitation of the uranium salt. In another study of metal precipitation, Labrenz et al. (2000) observed the formation of sphalerite (ZnS) by members of the aerotolerant *Desulfo bacteriaceae* in a natural biofilm. In this process, Zn was concentrated and metal sulfides were then precipitated by sulfate-reducing bacteria in the second phase of a combined sulfur oxidation–reduction biotreatment technique (Singh et al. 2006).

In White and Gadd (2000) study by a simultaneous increase in the EPS content of the biofilm was also observed, which suggested the role of EPS and biofilms in the entrapment of metal precipitates. Biofilms can also affect the fate of other compounds in their vicinity as a consequence of their physiological response during the absorption of water and inorganic or organic solutes (Flemming 1995).

According to the United States Agency for Toxic Substances and Disease Registry, the Comprehensive Environmental Response, Compensation and Liability Act 2005 Priority List names As, Pb, Hg and Cd as 4 of the top 10 most prevalent environmental toxins that are hazardous to public health (Harrison et al. 2007). Microbial biofilms, natural or engineered, could be used to remediate heavy metal pollution by biochemical modification and/or the accumulation of toxic metal ions (Chang et al. 2006), which notably include the radioactive actinides as well as other radionuclides (Anderson et al. 2006).

An understanding of metal toxicity in biofilms is crucial to the successful design of bioreactors that are used for biomining (Harrison et al. 2007), as well as those reactors that are used for biodegrading organic contaminants that are frequently intermingled with metals (Singh et al. 2006). Moreover, Biofilm-mediated bioremediation presents a proficient and safer alternative to bioremediation with planktonic microorganisms because cells in a biofilm have a better chance of adaptation and survival (especially during periods of stress) as they are protected within the matrix (Decho 2000).

Owing to the close, mutually beneficial physical and physiological interactions among organisms in biofilms, the usage of xenobiotics is accelerated and, consequently, biofilms are used in industrial plants to help in immobilization and degradation of pollutants (Das et al. 2012). However, it is only during the past few decades that biofilm reactors have become a focus of interest for researchers in the field of bioremediation (Singh et al. 2006).

4.3.2 Role of Biofilms in Remediation of Hydrocarbons

Polycyclic aromatic hydrocarbons (PAHs) are ubiquitous contaminants in the environment with potential mutagenicity and carcinogenicity. They are generated from natural combustion processes as well as from human activities (Luan et al. 2006). Most of the pollutant hydrocarbons in the environment are often composed of mixtures of numerous homologous compounds and bound to particulates in soil and sediments, restricting their availability for biodegradation (Mao et al. 2012).

To our knowledge microbial bioremediation technology depends mainly on aerobic microorganisms; rhamnolipids can be used in the microbial remediation of hydrocarbon- and crude-oil-contaminated soils (Itoh and Suzuki 1974). Biodegradation of hydrocarbons by native microbial populations is the primary mechanism by which hydrocarbon contamination can be removed from the environment (Banat 1993).

In another experiment, Shabtai and Gutnick demonstrated a 25–70 % and 40–80 % increase in the recovery of hydrocarbons from contaminated sandy-loam and silt-loam soil, respectively (Shabtai and Gutnick 1985). Furthermore, and in another report, 56 and 73 % of the aliphatic and aromatic hydrocarbons, respectively, were recovered from contaminated sandy-loam soil when treated with rhamnolipids (Scheibenbogen et al. 1994). On the other hand, the degradation of hexadecane and octadecane by different *Pseudomonas* strains has also been studied (Miller 1995).

The ability of rhamnolipid biosurfactants to emulsify hydrocarbon-water mixtures, degrade hydrocarbons in oil spill management, and remediate metal-contaminated soil has been well documented (Long et al. 2013).

The recent report by Cameotra and Singh (2009) throws more light on the uptake mechanism of *n*-alkane by *P. aeruginosa* and the role of rhamnolipids in the process. The authors reported a new and exciting research for hydrocarbon uptake involving internalization of hydrocarbon inside the cell for subsequent degradation (Pacwa-Plociniczak et al. 2011). Biosurfactant action dispersed hexadecane into micro droplets, increasing the availability of the hydrocarbon to the bacterial cells. The electron microscopic studies indicated that uptake of the biosurfactant-coated hydrocarbon droplets occurred. Interestingly, "internalization" of "biosurfactant layered hydrocarbon droplets" was taking place by a mechanism similar in appearance to active pinocytosis. This mechanism was not earlier visually reported in bacterial modes for hydrocarbon uptake (Pacwa-Płociniczak et al. 2011).

5 Remediation of Organic Contaminants by PGPR/Rhizoremediation

Recently, the combination of microbial remediation and phytoremediation has become a general practice in the field treatment of petroleum-contaminated soils. This technique can be defined as rhizoremediation, which is a specific type of phytoremediation that involves both plants and their associated rhizosphere microbes. It is interesting to point out that different approaches such as rhizoremediation, combination of PGPR and specific contaminant degrading bacteria, genetically engineered microbes, transgenic plants and enzyme technology can be used to improve the efficiency of bioremediation (Divya and Deepak Kumar 2011).

Phytoremediation is an emerging technology that uses plants and associated bacteria for the treatment of soil and groundwater contaminated by toxic pollutants (Salt et al. 1997). Moreover, this strategy uses plants to degrade, stabilize and/or remove soil contaminants (Gurska et al. 2009). Depending upon the type of contaminant and underlying process, phytoremediation is broadly categorized into several areas such as: phytoextraction, phytoaccumulation, phytostabilization, phytostimulation/rhizostimulation, phytovolatilization and rhizofiltration (Akpor and Muchie 2010). On the basis of understanding these mechanisms, researchers have focused on the relationships between plants and their microbial rhizospheric symbionts. They have speculated that application of rhizoremediation process is a simple and rational choice.

As a specific form of phytoremediation, rhizoremediation can either occur naturally or can be facilitated by inoculating soil with microorganisms capable of degrading environmental contaminants. The presence of *Pseudomonas* PGPR in rhizosphere with particular traits of the uptake of certain nutrients is considered a promising method to improve bioremediation effectiveness of hydrocarbon-contaminated environments (Fig. 4). Due to their excellent root-colonizing ability and ecological competence, pseudomonads are ideal candidates for rhizoremediation. Several species of *Pseudomonas* have been identified as carrying the abilities to degrade a number of environmental pollutants. Genetic engineering technology has the potential to increase the bioremediation ability of microorganisms in the degradation of three important organic pollutants, i.e. polycyclic aromatic hydrocarbons (PAHs), polychlorinated biphenyls (PCBs) and pesticides (Rehm Bernd 2008).

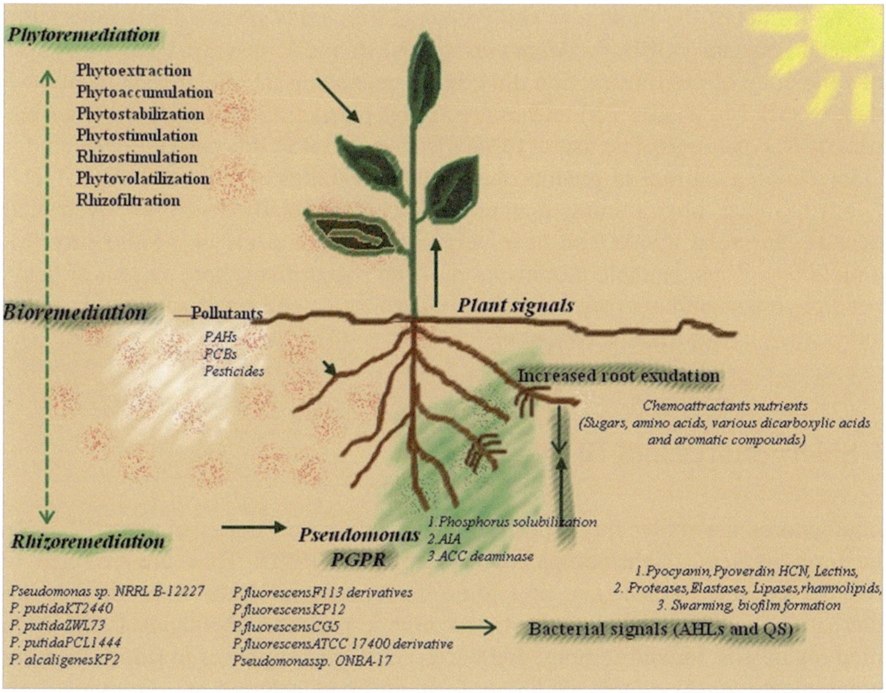

Fig. 4 The feasibility of using PGPR *Pseudomonas*, inoculation of these species increases the efficiency of phytoextraction directly by enhancing the metal accumulation in plant tissues, *note that* phytoremediation is broadly categorized into several areas such as: phytoextraction, phytoaccumulation, phytostabilization, phytostimulation/rhizostimulation, phytovolatilization and rhizofiltration. Furthermore, chemotaxis and biofilm formation governed by the QS system are selected as an advantageous behavior in bacteria, along with xenobiotic degradation capabilities after exposure to pollutants compounds

5.1 Heavy Metal Solubilizing Pseudomonas

Heavy metals pose a critical concern to human health and environmental issues due to their high occurrence as a contaminant, low solubility in biota and the classification of several heavy metals as carcinogenic and mutagenic (Diels et al. 2002). Therefore, the application of heavy metal solubilizing microorganisms is a promising approach for increasing heavy metal bioavailability in heavy metal-amended soils. In addition, bacteria producing indole acetic acid, siderophores and 1-aminocyclopropane-1-carbox-ylate deaminase and phosphate-solubilizing bacteria are capable of stimulating plant growth (Glick et al. 1995). Generally, PGPR function in three different ways (Glick 2001), synthesizing particular compounds for the plants, lessening or preventing the plants from diseases Raj et al. (2003), and facilitating the uptake of certain nutrients from the environment (Çakmakçi et al.2006). The feasibility of using a PGPR *Pseudomonas*, for the removal of heavy metals from a contaminated soil and sediments was evaluated by several scientific reports.

The results of the Rajkumar and Freitas (2008) study revealed that inoculation of metal resistant PGPB *Pseudomonas* sp. PsM6 and *P. jessenii* PjM15 increases the efficiency of phytoextraction directly by enhancing the metal accumulation in plant tissues (especially Zn) and indirectly by promoting the shoot and root biomass of *R. communis*. The use of these metal resistant PGPB can be considered as a biotechnological tool of great economical and ecological relevance. As the technology of metal 'phytomining' matures and is commercially developed, even small increases in metal uptake can have very significant impacts on profitability (Ma et al. 2009). Thus, suitable modification of the roots/rhizosphere system of heavy metal phytoaccumulators with beneficial microflora could promote metal bioavailability and phytoextraction (Ma et al. 2009).

6 Siderophores and Bioremediation

Siderophores are molecules especially designed to trap traces of iron(III) under the form of very stable complexes (Winkelmann 1991). They are excreted by iron-starved microorganisms, and after the complexes have formed, they are internalized into the cells by specific membrane receptors (Neilands 1982). Most microorganisms, including fungi and bacteria, use siderophores to fulfill their iron requirements and a couple of hundred of different siderophore structures have been described (Winkelmann 1991).

Pyoverdine, the well-known yellow-green fluorescent pigment characteristic of the fluorescent *Pseudomonas* species (Elliott 1958), is the major siderophore of these bacteria (Meyer and Stintzi 1998). In the structure of pyoverdine, there is a quinoleinic chromophore which imparts the color and fluorescence to the molecule, associated with a peptide chain of L-,D-, and uncommon amino acids, such as δ-*N*-hydroxyornithine and β-hydroxyaspartic acid (9). Both parts of the molecule participate in the complexation of the iron(III) ion (Meyer et al. 2002).

During the last years, there had been an increasing interest in investigating the potential of using siderophores in metal bioremediation (Neubauer et al. 2000). Siderophores become a useful tool in bioremediation, which is a cost-effective and environmentally friendly technique (Rajkumar et al. 2010). Schalk et al. (2011) reported that the siderophores are extremely effective in solubilizing and increasing the mobility of a wide range of metals such as Cd, Cu, Ni, Pb, Zn and the actinides Th(IV), U(IV) and Pu(IV) Because of their ligand functionalities, siderophores may have a strong affinity or selectivity for a particular metal other than Fe with regard to the stability constants of this metal siderophore complex.

Not surprisingly, pseudomonads have a wide range of iron acquisition mechanisms (Rehm Bernd 2008). Furthermore, siderophore pyochelin produced by *P. aeruginosa* is capable to chelate a wide range of metals, i.e. Ag^{+}, Al^{3+}, Cd^{2+}, Co^{2+}, Cr^{2+}, Cu^{2+}, Eu^{3+}, Ga^{3+}, Hg^{2+}, Mn^{2+}, Ni^{2+}, Pb^{2+}, Sn^{2+}, Tb^{3+}, Tl^{+} and $Zn^{2+;}$ however, the uptake process did not appear to assimilate any metal other than Fe^{3+} (Braud et al. 2009).

As was pointed out by Edberg et al. (2010), the siderophores also played a significant role in mobilizing metals from mine waste material or metal-contaminated soils. Several metals (i.e. Fe, Ni and Co) were mobilized from waste material (acid-leachedore) of a former uranium mine in the presence of siderophores produced by *P. fluorescens*. Moreover, It has also been shown that pyoverdines mobilized U(VI), Np(V) and other metals from uranium mine waste (Behrends et al. 2012). It is also interesting to note that microbial siderophores participate in the biodegradation of petroleum hydrocarbons through an indirect mechanism, by facilitating the Fe acquisition for the degraded microorganisms under Fe-limiting conditions (Barbeau et al. 2002). Both pyoverdine and pyochelin can decompose organotin pesticides (Inoue et al. 2003) and monothiocarboxylic acid (PDTC) was shown to dechlorinate CCl4 in the presence of reducing agents (Lee et al. 1999). The possibility of applying siderophores from *Pseudomonas* in xenobiotic degradation deserves further investigation (Rehm Bernd 2008).

7 Conclusion

During the last few decades, a great body of research work was carried out on PGPR rhamnolipids revealing many of their astonishing applications and making them reach the pinnacle of popularity among all the categories of biosurfactants. However, the importance of *Pseudomonas* rhamnolipids (RLs) is obvious, and they play a significant role in the bioremediation applications, even if there are many questions remaining to be answered. What is the specific role of Biofilm formation, chemotaxis and motility behavior toward the different nature of pollutants? It is important to note that future endeavors are needed to answer this question, to elucidate mechanisms that govern this link and might also explain and improve strategies employed by *Pseudomonas* PGPR toward multiple pollutants.

References

Abalos A, Pinazo A, Infante MR et al (2001) Physicochemical and antimicrobial properties of new rhamno-lipids produced by Pseudomonas aeruginosa AT10 from soybean oil refinery wastes. Langmuir 17:1367–1371

Abdel-Mawgoud AM, Lepine F, Deziel E (2010) Rhamnolipids: diversity of structures, microbial origins and roles. Appl Microbiol Biotechnol 86:1323–1336

Abdel-Mawgoud AM, Hausmann R, Lépine F, Müller MM, Déziel E (2011) Rhamnolipids: detection, analysis, biosynthesis, genetic regulation, and bioengineering of production. In: Biosurfactants, Microbiology Monographs 20. Springer-Verlag, Berlin. doi:10.1007/978-3-642-14490-5_2

Aguirre-Ramı´rez M, Medina G, Gonza´lez-Valdez A, Grosso-Becerra V, Sobero´n-Cha´vez G (2012) The *Pseudomonas aeruginosarml BDA* Coperon, encoding dTDP-L-rhamnose biosynthetic enzymes, is regulated by the quorum-sensing transcriptional regulator RhlR and the alternative sigma factor σ^S. Microbiology 158:908–916

Akpor OB, Muchie M (2010) Remediation of heavy metals in drinking water and wastewater treatment systems: processes and application. Int J Phys Sci 5(12):1807–1817

Al-Awadhi H, Al-Hasan RH, Sorkhoh NA, Salamah S, Radwan SS (2003) Establishing oil-degrading biofilms on gravel particles and glass plates. Int Biodeterior Biodegradation 51(3):181–185

Anderson C, Pederson K, Jakobsson AM (2006) Autoradiographic comparisons of radionuclide adsorption between subsurface anaerobic biofilms and granitic host rocks. Geomicrobiol J 23:15–29

Bai G, Brusseau ML, Miller RM (1997) Biosurfactant-enhanced removal of residual hydrocarbon from soil. J Contain Hydrol 25:157–170

Bakker PAHM, Pieterse CMJ, van Loon LC (2007) Induced systemic resistance by fluorescent Pseudomonas spp. Phytopathology 97:239–243

Banat IM (1993) The isolation of a thermophilic biosurfactant producing Bacillus sp. Biotechnol Lett 15(6):591–594

Banat IM, Franzetti A, Gandolfi I, Bestetti G, Martinotti MG, Fracchia L, Smyth TJ, Marchant R (2010) Microbial biosurfactants production, applications and future potential. Appl Microbiol Biotechnol 87:427–444

Barbeau K, Zhang GP, Live DH, Butler A (2002) Petrobactin, a photoreactive siderophore produced by the oil-degrading marine bacterium *Marinobacter hydrocarbonoclasticus*. J Am Chem Soc 124:378–379

Barkay T, Schaefer J (2001) Metal and radionuclide bioremediation: issues, considerations and potentials. Curr Opin Microbiol 4:318–323

Behrends T, Krawczyk-Bärsch E, Arnold T (2012) Implementation of microbial processes in the performance assessment of spent nuclear fuel repositories. Appl Geochem 27:453–462

Benincasa M, Contiero J, Manresa MA, Moraes IO (2002) Rhamnolipid production by *Pseudomonas aeruginosa* LBI growing on soap stock as the sole carbon source. J Food Eng 54:283–288

Bodour AA, Drees KP, Maier RM (2003) Distribution of biosurfactant-producing bacteria in undisturbed and contaminated arid southwestern soils. Appl Environ Microbiol 69:3280–3287

Bolwerk A, Lagopodi AL, Wijfjes AHM, Lamers GEM, Chin-A-Woeng TFC, Lugtenberg BJJ, Bloemberg GV (2003) Interactions in the tomato rhizosphere of two *Pseudomonas* biocontrol strains with the phytopathogenic fungus *Fusarium oxysporum* f. sp. *radicis-lycopersici*. Mol Plant Microbe Interact 11:983–993

Bouwer EJ, Zehnder AJB (1993) Bioremediation of organic compounds—putting microbial metabolism to work. Trends in Biotech 11:360–367

Budzikiewicz H (2004) Siderophores of the Pseudomonadaceae sensu stricto (fluorescent and non-fluorescent *Pseudomonas* spp.). Fortschr Chem Org Naturst 87:81–237

Burr TJ, Schroth MN, Suslow T (1978) Increased potato yields by treatment of seedpieces with specific strains of *Pseudomonas fluorescens* and *P. purida*. Phytopathology 68:1377–1383

Braud A, Jézéquel K, Bazot S, Lebeau T (2009) Enhanced phytoextraction of an agricultural Cr- and Pb-contaminated soil by bioaugmentation with siderophore-producing bacteria. Chemosphere 74:280–286

Cabrera-Valladares N, Richardson AP, Olvera C, Trevino LG, De´ziel E, Le´pine F, Sobero´n-Cha´vez G (2006) Monorhamnolipids and 3-(3-hydroxyalkanoyloxy)alkanoic acids (HAAs) production using *Escherichia coli* as a heterologous host. Appl Microbiol Biotechnol 73:187–194

Caiazza NC, Shanks RM, O'Toole GA (2005) Rhamnolipids modulate swarming motility patterns of *Pseudomonas aeruginosa*. J Bacteriol 187:7351–7361

Çakmakçı R, Dönmez F, Aydın A, fiahin F (2006) Growth promotion of plants by plant growth-promoting rhizobacteria under greenhouse and two different field soil conditions. Soil Biol Biochem 38:1482–1487

Cameotra SS, Singh P (2009) Synthesis of rhamnolipid biosurfactant and mode of hexadecane uptake by *Pseudomonas* species. Microbial Cell Fact 8:1–7

Cassidy DP, Hudak AJ, Werkema DD et al (2002) In situ rhamnolipid production at an abandoned petroleum refinery. Soil Sediment Contam 11:769–787

Chang WC, Hsu GS, Chiang SM, Su MC (2006) Heavy metal removal from aqueous solution by wasted biomass from a combined AS-biofilm process. Bioresour Technol 97:1503–1508

Clifford JS, Ioannidis MA, Legge RL (2007) Enhanced aqueous solubilization of tetrachloroethylene by a rhamnolipid biosurfactant. J Colloid Interf Sci 305:361–365

Compant S, Reiter B, Sessitsch A, Nowak J, Clement C, Ait Barka E (2005) Endophytic colonization of Vitis vinifera L. by a plant growth-promoting bacterium, Burkholderia sp. strain PsJN. Appl Environ Microbiol 71:1685–1693

Cornelis P, Matthijs M (2002) Diversity of siderophore-mediated iron uptake systems in fluorescent pseudomonads: not only pyoverdines. Environ Microbiol 4:777–923. doi:10.1046/j.1462-2920.2002.00369

Costa S, VAO G, Nitschke M, Lepine F, Deziel E, Contiero J (2010) Structure, properties and applications of rhamnolipids produced by *Pseudomonas aeruginosa* L2–1 from cassava wastewater. Proc Biochem 45:1511–1516. doi:10.1016/j.procbio.2010.05.033

Costacurta A, Vanderleyden J (1995) Synthesis of phytohormones by plant-associated bacteria. Crit Rev Microbiol 21:1–18

Costerton JW, Stewart PS, Greenberg EP (1999) Bacterial biofilms: a common cause of persistent infections. Science 284:1318–1322

Dankert J, Hogt AH, Feijen J (1986) Biomedical polymers: bacterial adhesion, colonization, and infection. CRC Critical Rev Biocompat 2:219–301

Darvishi P et al (2011) Biosurfactant production under extreme environmental conditions by an efficient microbial consortium, ERCPPI-2. Colloids Surf B: Biointerf. doi:10.1016/j.colsurfb.01.011

Das N, Geetanjali Basak LV, SalamJ Abdul, Abigail MEA (2012) Application of biofilms on remediation of pollutants–an overview. J Microbiol Biotech Res 2(5):783–790

Davey ME, Caiazza NC, O'Toole GA (2003) Rhamnolipid surfactant production affects biofilm architecture in *Pseudomonas aeruginosa* PAO1. J Bacteriol 185:1027–1036

Decho AW (2000) Microbial biofilms in intertidal systems: an overview. Cont Shelf Res 20:1257–1273

Desai JD, Banat IM (1997) Microbial production of surfactants and their commercial potential. Microbiol Mol Biol Rev 61:47–64

Deziel E, Paquette G, Villemur R, et al (1996) Biosurfactant production by a soil *Pseudomonas* strain growing on polycyclic aromatic hydrocarbons. Appl Environ Microbiol 62(6):1908–1912

Déziel E, Lepine F, Milot S, Villemur R (2003) rhlA is required for the production of a novel biosurfactant promoting swarming motility in *Pseudomonas aeruginosa*: 3-(3-hydroxyalkanoyloxy)alk-anoic acids (HAAs), the precursors of rhamnolipids. Microbiology 149:2005–2013

Deziel E, Lepine F, Dennie D, Boismenu D, Mamer OA, Villemur R (1999) Liquid chromatography/mass spectrometry analysis of mixtures of rhamnolipids produced by *Pseudomonas aeruginosa* strain 57RP grown on mannitol or naphthalene. Biochim Biophys Acta 1440:244–252

Diels L, van der Lelie N, Bastiaens L (2002) New development in treatment of heavy metal contaminated soils. Rev Environ Sci Biotechnol 75–82

Divya B, Deepak Kumar M (2011) Plant-microbe interaction with enhanced bioremediation. Res J BioTechnol 6(4)

Du H, Zhiliang Xu, Morgen Anyan, Oleg Kim, Matthew Leevy W, Shrout JD, Mark Alber (2012) High density waves of the bacterium *Pseudomonas aeruginosa* in propagating swarms result in efficient colonization of surfaces. Biophys J 103(3):601–609

Edberg F, Kalinowski BE, Holmström SJM, Holm K (2010) Mobilization of metals from uranium mine waste: the role of pyoverdines produced by *Pseudomonas fluorescens*. Geobiology 8:278–292

Elliott RP (1958) Some properties of pyoverdine, the water soluble pigment of the *Pseudomonas*. Appl Microbiol 6:241–246

Espinosa-Urgel M, Salido A, Ramos JL (2000) Genetic analysis of functions involved in adhesion of *Pseudomonas putida* to seeds. J Bacteriol 9:2363–2369

Finnerty WR (1994) Biosurfactants in environmental biotechnology. Curr Opin Biotechnol 5:291–295

Flemming HC (1995) Sorption sites in biofilms. Water Sci Technol 32:27–33

Flemming HC, Wingender J (2002) Extracellular polymeric substances: structure, ecological functions, technical relevance. In: Bitton G (ed) Encyclopedia of environmental microbiology. Wiley, New York, p 3

Fracchia L, Cavallo M, Giovanna Martinotti M, Banat Ibrahim M (2012) Biosurfactants and bioemulsifiers biomedical and related applications. In: Prof. Dhanjoo N Ghista (ed) Present status and future potentials, Biomedical Science, Engineering and Technology. ISBN: 978-953-307-471- 9

Glick BR (2001) Phytoremediation: synergistic use of plants and bacteria to clean up the environment. Biotechnol Adv 21:383–393

Glick BR (2012) Plant growth-promoting bacteria: mechanisms and applications. Scientifica ID963401, 15

Glick BR (2014) Bacteria with ACC deaminase can promote plant growth and help to feed the world. Microbiol Res 169:30–39

Glick BR, Karaturovic DM, Newell PC (1995) A novel procedure for rapid isolation of plant growth promoting pseudomonads. Can J Microbiol 41:533–536

Glick R, Gilmour C, Tremblay J, Satanower S, Avidan O, Déziel E, Peter Greenberg E, Poole K, Banin E (2010) Increase in rhamnolipid synthesis under iron-limiting conditions influences surface motility and biofilm formation in *Pseudomonas aeruginosa*. J Bacteriol 192(12):2973–2980

Grewal SI, Rainey PB (1991) Phenotypic variation of *Pseudomonas putida* and *P. tolaasii* affects the chemotactic response to *Agaricus bisporus* mycelial exudate. J Gen Microbiol 137(12):2761–2768

Gross H, Loper JE (2009) Genomics of secondary metabolite production by Pseudomonas spp. Nat Prod Rep 26:1408–1446

Gunther NW, Nunez A, Fett W, Solaiman DKY (2005) Production of rhamnolipids by *Pseudomonas chlororaphis*, a nonpathogenic bacterium. Appl Environ Microbiol 71:2288–2293

Gupta Gupta S, Arora DK, Rai B et al (1991) Ecological aspects of microbial chemotaxis. J Sci Res Banaras Hindu University 41:21–39

Gurska J, Wang WX, Gerhardt KE, Khalid AM, Isher-wood DM, Huang XD, Glick BR, Greenberg BM (2009) Three year field test of a plant growth promoting rhizobacteria enhanced phytoremediation system at a land farm for treatment of hydrocarbon waste. Environ Sci Technol 43(12):4472–4479

Halan B, Schmid A, Buehler K (2011) Real-time solvent tolerance analysis of *Pseudomonas* sp. strain VLB120ΔC catalytic biofilms. Appl Environ Microbiol 77(5):1563–1571

Hang Pham T, Webb Jeremy S, Bernd Rehm HA (2004) The role of polyhydroxyalkanoate biosynthesis by *Pseudomonas aeruginosa* in rhamnolipid and alginate production as well as stress tolerance and biofilm formation. Microbiology 150:3405–3413

Harbron RS, Kent CA (1988) Aspects of cell adhesion. In Melo LF, Bott TR, Bernardo CA (Eds) NATO ASI series 145:125–140

Harrison JJ, Turner RJ, Ceri H (2007) Multimetal resistance and tolerance in microbial biofilms. Nat Rev Microbiol 5:928–938

Herman DC, Artiola JF, Miller RM (1995) Removal of cadmium, lead and zinc from soil by a rhamnolipid biosurfactant. Environ Sci Technol 29:2280–2285

Hommel RK (1994) Formation and function of biosurfactants for degradation of water-insoluble substrates. In: Ratledge C, (ed) Biochemistry of microbial degradation, pp 63–87. Springer, New York

Hori K, Matsumoto S (2010) Bacterial adhesion: from mechanism to control. Biochem Eng J 48:424–434

Inoue H et al (2003) Tin–carbon cleavage of organotin compounds by pyoverdine from *Pseudomonas chlororaphis*. Appl Environ Microbiol 69:878–883

Itoh S, Suzuki T (1974) Fructose-lipids of arthrobacter, *Corynebacteria*. Nocardia and mycobacteria grown on fructose, Agric Biol Chem 38

Johnsen AR, Karlson U (2004) Evaluation of bacterial strategies to promote the bioavailability of polycyclic aromatic hydrocarbons. Appl Microbiol Biotechnol 63(4):452–459

Joseph B et al (2001) Biofilm formation by *Salmonella* spp. on food contact surfaces and their sensitivity to sanitizers. Intl J Food Microbiol 64:367–372

Kargi F, Eker S (2005) Removal of 2,4-dichlorophenol and toxicity from synthetic wastewater in a rotating perforated tube biofilm reactor. Process Biochem 40:2105–2111

Kearns DB (2010) A field guide to bacterial swarming motility. Nat Rev Microbiol 8:634–644

Kim J, Kang Y, Choi O, Jeong Y, Jeong JE, Lim JY, Kim M, Moon JS, Suga H, Hwang I (2007) Regulation of polar flagellum genes is mediated by quorum sensing and *FlhDC* in *Burkholderia glumae*. Mol Microbiol 64(1):165–179

Kirchman D, Mitchell R (1982) Contribution of particle-bound bacteria to total micro-heterotrophic activity in five ponds and two marshes. Appl Environ Microbiol 43:200–209

Kjelleberg S, Givskov M (2007) The biofilm mode of life: mechanisms and adaptations. In: Kjelleberg S, Givskov M (eds) Horizon bioscience, pp 5–21. Wymondham

Kohler T, Curty LK, Barja F, van Delden C, Pechere JC (2000) Swarming of*Pseudomonas aeruginosa* is dependent on cell-to-cell signaling and requires flagella and pili. J Bacteriol 182(21):5990–5996

Kuchma SL, Brothers KM, Merritt JH, Liberati NT, Ausubel FM, O'Toole GA (2007) BifA, a cyclic-Di-GMP phosphodiesterase, inversely regulates biofilm formation and swarming motility by *Pseudomonas aeruginosa* PA14. J Bacteriol 189(22):8165–8178

Labrenz M et al (2000) Formation of sphalerite (ZnS) deposits in natural biofilms of sulfate-reducing bacteria. Science 290:1744–1747

Lang S (2002) Biological amphiphiles (microbial biosurfactants). Curr Opin Colloid Inter Sci 7:12–20

Langley S, Beveridge TJ (1999) Metal binding by *Pseudomonas aeruginosa* PAO1 is influenced by growth of the cells as a biofilm. Can J Microbiol 45:616–622

Law AMJ, Aitken MD (2003) Bacterial chemotaxis to naphthalene desorbing from a nonaqueous liquid. Appl Environ Microbiol 69:5968–5973

Lee CH, Lewis TA, Paszczynski A, Crawford RL (1999) Identification of an extracellular agent [correction of catalyst] of carbon tetrachloride dehalogenation from *Pseudomonas stutzeri* strain KC as pyridine-2,6-bis (thiocarboxylate). Biochem Biophys Res Commun 261:562–566

Lloyd JR (2003) Microbial reduction of metals and radionuclides. FEMS Microbiol Rev 27:411–425

Lodewyckx C, Mergeay M, Vangronsveld J, Clijsters H, van der Lelie D (2002) Isolation, characterization, and identification of bacteria associated with the zinc hyperaccumulator *Thlaspi caerulescens* subsp. *calaminaria*. Int J Phytorem 4:101–115

Long X, Zhang G, Shen C et al (2013) Application of rhamnolipid as a novel biodemulsifier for destabilizing waste crude oil. Bioresour Technol 131:1–5

Loper JE, Schroth MN (1986) Influence of bacterial sources of indole-3-acetic acid on root elongation of sugar beet. Phytopathology 76:386–389

Luan TG, Yu KS, Zhong Y, Zhou HW, Lan CY, Tam NF (2006) Study of metabolites from the degradation of polycyclic aromatic hydrocarbons (PAHs) by bacterial consortium enriched from mangrove sediments. Chemosphere 65:2289–2296

Ma Y et al (2009) Improvement of plant growth and nickel uptake by nickel resistant-plant-growth promoting bacteria. J Hazard Mater. doi:10.1016/j.jhazmat.2008.12.018
Magalhaes L, Nitschke M (2013) Antimicrobial activity of rhamnolipids against *Listeria monocytogenes* and their synergistic interaction with nisin. Food Control 29:138–142. doi:10.1016/j.foodcont.2012.06.009
Maier RM, Soberon-Chavez G (2000) *Pseudomonas aeruginosa* rhamnolipids: biosynthesis and potential applications. Appl Microbiol Biotechnol 54:625–633
Makkar RS, Cameotra SS, Banat IM (2011) Advances in utilization of renewable substrates for biosurfactant production. AMB Express 1:5
Macaskie LE et al (2000) Enzymatically mediated bioprecipitation of uranium by *Citrobacter* sp.: a concerted role for exocellular lipopolysaccharide and associated phosphatase in biomineral formation. Microbiology 146:1855–1867
Maneerat S (2005) Biosurfactants from marine microorganisms. Songklanakarin J Sci Technol 27:1263–1272
Mao J, Luo Y, Teng Y, Li Z (2012) Bioremediation of polycyclic aromatic hydrocarbon-contaminated soil by a bacterial consortium and associated microbial community changes. Int Biodeterior Biodegrad 70:141–147
Marques APC, Pires C, Moreira H, Rangel AO, Castro PML (2010) Assessment of the plant growth promotion abilities of six bacterial isolates using *Zea mays* as indicator plant. Soil Biol Biochem 42:1229–1235
Martinez-Toledo A, Rios-Leal E, Vazquez-Duhalt R, Gonzalez-Chavez Mdel C, Esparza-Garcia JF, Rodriguez-Vazquez R (2006) Role of phenanthrene in rhamnolipid production by *P. putida* in different media. Environ Technol 27:137–142
Mata-Sandoval JC, Karns J, Torrents A (2000) Effect of rhamnolipids produced by *Pseudomonas aeruginosa* UG2 on the solubilization of pesticides. Environ Sci Technol 34:4923–4930
Mattick JS (2002) Type IV pili and twitching motility. Annu Rev Biochem 56:289–314
Mehnaz S, Baig DN, Jamil F, Weselowski B, Lazarovits G (2009) Characterization of a phenazine and hexanoyl homoserine lactone producing *Pseudomonas aurantiaca* strain PB-St2, isolated from sugarcane stem. J Microbiol Biotechnol 19:1688–1694. doi:10.4014/jmb.0904.04022
Meliani A, Bensoltane A (2014) Enhancement of hydrocarbons degradation by use of pseudomonas biosurfactants and biofilms. J Pet Environ Biotechnol 5:68. doi:10.4172/2157-7463.1000168
Meyer JM, and Stintzi A (1998) Iron metabolism and siderophores in Pseudomonas and related species, pp 201–243. In Montie TC (ed) Biotechnology handbooks, vol. 10: *Pseudomonas*. Plenum Publishing Co., New York
Meyer JM, Geoffroy VA, Baida N, Gardan L, Izard D, Lemanceau P, Achouak W, Palleroni NJ (2002) Siderophore typing, a powerful tool for the identification of fluorescent and non fluorescent Pseudomonads. Appl Environ Microbiol 68(6):2745–2753
Miller RM (1995) Biosurfactant-facilitated remediation of metal-contaminated soils. Environ Health Perspect 103(Suppl 1):59
Molin S, Tolker-Nielsen T (2003) Gene transfer occurs with enhanced efficiency in biofilms and induces enhanced stabilization of the biofilms structure. Curr Opin Biotechnol 14:255–261
Moussa TAA, Ahmed AM, Abdel-hamid SMS (2006) Opt imization of cultural conditions for biosurfactant production from *Nocardia amarae*. J Appl Sci Res 2:844–850
Müller MM, Kügler JH, Henkel M, Gerlitzki M, Hörmann B, Pöhnlein M et al (2012) Rhamnolipids-Next generation surfactants? J Biotechnol
Mulligan CN (2004) Environmental applications for biosurfactants. Environ Pollut 133:183–198
Mulligan CN, Wang S (2006) Remediation of a heavy metal-contaminated soil by a rhamnolipid foam. Eng Geol 85:75–81
Mulligan CN, Yong RN, Gibbs BF (2001) Heavy metal removal from sediments by biosurfactants. J Hazard 68 Mater 85:111–125
Neilands JB (1982) Microbial envelope proteins related to iron. Annu Rev Microbiol 36:285–309

Nerurkar AS, Hingurao KS, Suthar HG (2009) Bioemulsifiers from marine microorganisms. J Sci Ind Res 68:273–277

Neubauer U, Nowak B, Furrer G, Schulin R (2000) Heavy metal sorption on clay minerals affected by the siderophore desferroixamine B. Environ Sci Technol 34:2749–2755

Noordman WH, Ji W, Brusseau ML et al (1998) Effects of rhamnolipid biosurfactants on removal of phenanthrene from soil. Environ Sci Technol 32:1806–1812

Ochsner UA, Reiser J (1995) Autoinducer-mediated regulation of rhamnolipid biosurfactant synthesis in Pseudomonas aeruginosa. Proc Natl Acad Sci USA 92:6424–6428

Oliveira FJS, Vazquez L, de Campos NP, de Franca FP (2009) Production of rhamnolipids by a *Pseudomonas alcaligenes* strain. Process Biochem 44:383–389

Onbasli D, Aslim B (2009) Biosurfactant production in sugar beet molasses by some *Pseudomonas* spp. J Environ Biol 30:161–163

O'Toole GA, Kolter R (1998) Flagellar and twitching motility are necessary for Pseudomonas aeruginosa biofilm development. Mol Microbiol 30:295–304

Pacwa-Plociniczak PGA, Piotrowska-Seget Z, Cameotra SS (2011) Environmental applications of biosurfactants: recent advances. Int J Mol Sci 12:633–654

Pamp SJ, Tolker-Nielsen T (2007) Multiple roles of biosurfactants in structural biofilm development by *Pseudomonas aeruginosa*. J Bacteriol 189:2531–2539

Pandey G, Jain RK (2002) Bacterial chemotaxis toward environmental pollutants: role in bioremediation. Appl Environ Microbiol 68:5789–5795

Parales RE, Ditty JL, Harwood CS (2000) Toluene-degrading bacteria are chemotactic towards the environmental pollutants benzene, toluene, and trichloroethylene. Appl Environ Microbiol 66:4098–4104

Parry AJ, Parry NJ, Peilow C, Stevenson PS (2013) Combinations of rhamnolipids and enzymes for improved cleaning. Patentno EP 2596087:A1

Pearson JP, Pesci EC, Iglewski BH (1997) Roles of *Pseudomonas aeruginosa las* and *rhl* quorum-sensing systems in control of elastase and rhamnolipid biosynthesis genes. J Bacteriol 179:5756–5767

Piljac A, Stipcevic T, Piljac-Zegarac J, Piljac G (2008) Successful treatment of chronic decubitus ulcer with 0.1 % dirhamnolipid ointment. J Cutan Med Surg 12:142–146

Pornsunthorntawee O, Wongpanit P, Rujiravanit Ratana (2010) Rhamnolipid biosurfactants: production and their potential in environmental biotechnology biosurfactants. In: Ramkrishna S (ed) Landes Bioscience and Springer Science, pp 217–219

Puhakka JA, Melin ES, Jarvinen KT, Koro PM, Rintala JA, Hartikainen P, Shieh WK, Ferguson JF (1995) Fluidized bed biofilms for chlorophenol mineralization. Water Sci Technol 31:227–235

Rahim R, Ochsner UA, Olvera C, Graninger M, Messner P, Lam JS, Soberon-Chavez G (2001) Cloning and functional characterization of the *Pseudomonas aeruginosa rhlC* gene that encodes rhamnosyl transferase 2, an enzyme responsible for di-rhamnolipid biosynthesis. Mol Microbiol 40:708–718

Rahman KSM, Banat IM, Thahira J et al (2002) Bioremediation of gasoline contaminated soil by a bacterial consortium amended with poultry litter, coir pith and rhamnolipid biosurfactant. Bioresour Technol 81:25–32

Rahman KSM, Rahman TJ, Kourkoutas Y, Petsas I, Marchant R, Banat IM (2003) Enhanced bioremediation of n-alkane in petroleum sludge using bacterial consortium amended with rhamnolipid and micronutrients. Bioresour Technol 90:159–168. doi:10.1016/S0960-8524(03)00114-7

Raj SN, Deepak SA, Basavaraju P, Shetty HS, Reddy MS, Kloepper JW (2003) Comparative performance of formulations of plant growth promoting rhizobacteria in growth promotion and suppression of downy mildew in pearl millet. Crop Prot 22:579–588

Rajkumar M, Freitas H (2008) Influence of metal resistant-plant growth-promoting bacteria on the growth of *Ricinus communis* in soil contaminated with heavy metals. Chemosphere 71:834–842

Rajkumar MAeN, Prasad MNV, Freitas H (2010) Potential of siderophore-producing bacteria for improving heavy metal phytoextraction. Trends Biotechnol 28:142–149

Rehm Bernd HA (2008) Pseudomonas. In: Model organism, pathogen, cell factory. WILEY-VCH Verlag, GmbH & Co. KGaA, pp 1–153

Renfro TD (2013) Rhamnolipid biosurfactant transport. In: Agricultural Soils. Electronic Theses, Treatises and Dissertations. Paper 7571

Rogers HJ (1979) Adhesion of microorganisms to surfaces. Academic Press

Rokhzadi A, Asgharzadeh A, Darvish F, Nour-Mohammadi G, Majidi E (2008) Influence of plant growth promoting Rhizobacteria on dry matter accumulation of Chickpea (*Cicer arietinum* L) under field conditions. J Agricul Environ Sci 3(Suppl 2):253–257

Ron EZ, Rosenberg E (2001) Natural roles of biosurfactants. Environ Microbiol 3:229–236

Rooney AP, Price NP, Ray KJ, Kuo TM (2009) Isolation and characterization of rhamnolipid-producing bacterial strains from a biodiesel facility. FEMS Microbiol Lett 295:82–87

Rosenberg E, Ron EZ (1999) High- and low-molecular-mass microbial surfactants. Appl Microbiol Biotechnol 52(2):154–162

Sachdev DP, Cameotra SS (2013) Biosurfactants in agriculture. Appl Microbiol Biotechnol 97:1005–1016. doi:10.1007/s00253-012-4641-8

Saharan BS, Nehra V (2011) Plant growth promoting rhizobacteria: a critical review. Life Sci Med Res 21:1–31

Salt DE, Pickering IJ, Prince RC, Gleba D, Dushenkov S, Smith RD, Raskin I (1997) Metal accumulation by aquacultured seedlings of Indian Mustard. Environ Sci Technol 31(6):1636–1644

Sauer K, Camper AK (2001) Characterization of phenotypic changes in *Pseudomonas putida* in response to surface-associated growth. J Bacteriol 183:6579–6589

Schalk IJ, Hannauer M, Braud A (2011) Minireview new roles for bacterial siderophores in metal transport and tolerance. Environ Microbiol 13:2844–2854

Scheibenbogen K, Zytner R, Lee H, Trevors J (1994) Enhanced removal of selected hydrocarbons from soil by *Pseudomonas aeruginosa* UG2 biosurfactants and some chemical surfactants. J Chem Technol Biotechnol 59(1):53–59

Sekhon Randhawa KK, Rahman Pattanathu KSM (2014) Rhamnolipid biosurfactants—past, present, and future scenario of global market. Front Microbiol. doi:10.3389/fmicb.2014.00454

Shabtai Y, Gutnick DL (1985) Tolerance of *Acinetobacter calcoaceticus* RAG-1 to the cationic surfactant cetyltrimethylammonium bromide: Role of the bioemulsifier emulsan. Appl Environ Microbiol 49(1):192–197

Shrout JD, Chopp DL, Just CL, Hentzer M, Givskov M, Parsek MR (2006) The impact of quorum sensing and swarming motility on *Pseudomonas aeruginosa* biofilm formation is nutritionally conditional. Mol Microbiol 62:1264–1277

Singh P, Cameotra SS (2004) Enhancement of metal bioremediation by use of microbial surfactants. Biochem Biophys Res Commun 319:291–297

Singh R, Paul D, Jain R (2006) Biofilms: implications in bioremediation. Trends Microbiol 14:389–397

Singleton P, Sainsbury D (2006) Dictionary of microbiology and molecular biology, 3rd edn. Wiley, Paul Singleton and Diana Sainsbury. ISBN pp 637-638

Soberón-Chávez G, Maier RM (2011) Biosurfactants: a general overview. In: Soberón-Chávez G (ed) Biosurfactants. Springer-Verlag, Berlin, Germany

Soberón-Chávez G, Lépine F, Déziel E (2005) Production of rhamnolipids by *Pseudomonas aeruginosa*. Appl Microbiol Biotechnol 68(6):718–725

Soto GE, Hultgren SJ (1999) Bacterial adhesins: common themes and variations in architecture and assembly. J Bacteriol 181:1059–1071

Stipcevic T, Pijac A, Pijac G (2006) Enhanced healing of full-thickness burn wounds using di-rhamnolipid. Burns 32:24–34

Tremblay J, Déziel E (2010) Gene expression in *Pseudomonas aeruginosa* swarming motility. BMC Genom 11:587. doi:10.1186/1471-2164-11-587

Vatsa P, Sanchez L, Clement C, Baillieul F, Dorey S (2010) Rhamnolipid biosurfactants as new players in animal and plant defense against microbes. Int J Mol Sci 11:5095–5108. doi:10.3390/ijms11125095

Walter V, Syldatk C, Hausmann R (2010) Microbial production of rhamnolipid biosurfactants. In: FlickingerMC (ed) Encyclopedia of industrial biotechnology. Wiley-VCH, Weinheim, Germany

Wang Q, Fang X, Bai B, Liang X, Shuler PJ, Goddard WA, Tang Y (2007) Engineering bacteria for production of rhamnolipid as an agent for enhanced oil recovery. Biotechnol Bioeng 98(4):842–853

Watnick P, Kolter R (2000) Biofilm, city of microbes. J Bacteriol 182(10):2675–2679

Webb JS, Givskov M, Kjelleberg S (2003) Bacterial biofilms: prokaryotic adventures in multicellularity. Curr Opin Microbiol 6:578–585

Weller DM, Raaijmakers JM, Gardener BBM, Thomashow LS (2002) Microbial populations responsible for specific soil suppressiveness to plant pathogens. Annu Rev Phytopathol 40:309–348

Whang LM, Liu PWG, Ma CC et al (2008) Application of biosurfactants, rhamnolipid and surfactin, for enhanced biodegradation of dieselcontaminated water and soil. J Hazard Mater 151:155–163

White C, Gadd GM (2000) Copper accumulation by sulfate-reducing bacterial biofilms. FEMS Microbiol Lett 183(2):313–318

Winkelmann G (1991) Handbook of microbial iron chelates. CRC Press, Boca Raton

Wittgens A, Tiso T, Arndt TT, Wenk P, Hemmerich J, Muller C, Wichmann R, Küpper B, ZwickM, Wilhelm S, HausmannR, SyldatkC, Rosenau FandBlankL M (2011) Growth independent rhamnolipid production from glucose using the non-pathogenic *Pseudomonas putida* KT2440. Microb Cell Fact 10:80–97. doi:10.1186/1475-2859-10-80

Wu X, Monchy S, Taghavi S, Zhu W, Ramos J et al (2010) Comparative genomics and functional analysis of niche-specific adaptation in *Pseudomonas putida*. FEMS Microbiol Rev 35:299–323

Xavier JB, Kim W, Foster KR (2011) A molecular mechanism that stabilizes cooperative secretions in *Pseudomonas aeruginosa*. Mol Microbiol 79:166–179

Yabuuchi E, Kosako Y, Oyaizu H, Yano I, Hotta H, Hashimoto Y, Ezaki T, Arakawa M (1992) Proposal of *Burkholderia* Gen-Nov and transfer of 7 species of the genus *Pseudomonas* homology group-II to the new genus, with the type species *Burkholderia cepacia* (Palleroni and Holmes 1981) Comb-Nov. Microbiol Immunol 36:1251–1275

Zhu K, Rock CO (2008) RhlA convertsb-hydroxyacyl-acyl carrier protein intermediates in fatty acid synthesis to the b-hydroxydecanoyl-b-hydroxydecanoate component of rhamnolipids in *Pseudomonas aeruginosa*. J Bacteriol 190:3147–3154

Zhang Y, Maier WJ, Miller RM (1997) Effect of rhamnolipids on the dissolution, bioavailability and biodegradation of phenanthrene. Environ Sci Technol 31:2211–2217

Zhang L, Veres-Schalnat Tracey A, Somogyi A, Pemberton JE, Maier RM (2012) Fatty acid cosubstrates provide β-oxidation precursors for rhamnolipid biosynthesis in *Pseudomonas aeruginosa*, as evidenced by isotope tracing and gene expression assays. Appl Environ Microbiol 78(24):8611. doi:10.1128/AEM.02111-12

Index

D.K. Maheshwari (ed.), *Bacterial Metabolites in Sustainable Agroecosystem*, Sustainable Development and Biodiversity 12, DOI 10.1007/978-3-319-24654-3

MIX
Papier aus verantwortungsvollen Quellen
Paper from responsible sources
FSC® C105338

If you have any concerns about our products,
you can contact us on
ProductSafety@springernature.com

In case Publisher is established outside the EU,
the EU authorized representative is:
Springer Nature Customer Service Center GmbH
Europaplatz 3, 69115 Heidelberg, Germany

Printed by Libri Plureos GmbH
in Hamburg, Germany